THE IDENTIFICATION OF
MOLECULAR SPECTRA

THE IDENTIFICATION OF MOLECULAR SPECTRA

R.W.B. PEARSE
D.Sc., F.R.A.S.

Formerly Reader in Spectroscopy,
Imperial College, London

and

A.G. GAYDON
D.Sc., F.R.S.

Emeritus Professor of Molecular Spectroscopy,
Imperial College, London

Formerly Warren Research Fellow
of the Royal Society, Imperial College, London

FOURTH EDITION

LONDON
CHAPMAN AND HALL

A HALSTED PRESS BOOK
John Wiley & Sons, Inc., New York

First published 1941
Second edition 1950
Third edition 1963
Reprinted with supplement 1965
Fourth edition 1976
by Chapman and Hall Ltd,
11 New Fetter Lane, London EC4P 4EE

Set by Hope Services, Wantage
Printed in Great Britain
at the University Printing House
Cambridge

ISBN 0 412 14350 X

Distributed in the U.S.A.
by Halsted Press, a Division
of John Wiley & Sons, Inc. New York

Library of Congress Cataloging in Publication Data
Pearse, Reginald William Blake.
    The identification of molecular spectra.

    Includes indexes.
    1. Molecular spectra—Tables. I. Gaydon,
Alfred Gordon, joint author. II. Title.
QC454.M6P4   1976   535'.84   76-18734
ISBN 0-470-15164-1

Contents

List of Plates

(Between pages 392–393)

Preface to First Edition

These tables have been constructed with the aim of facilitating the identification of molecular spectra. Several excellent books have been written dealing with the theory of molecular spectra and some have included collections of molecular constants derived from the analysis of such spectra, yet it has hitherto remained necessary to search through original papers or to calculate the positions of bands from the tables of derived constants in order to identify a given system of bands. This task is usually tedious and sometimes impossible to one without considerable experience.

Originally we prepared for use in the laboratory a list of the wavelengths of the heads of a limited number of band systems which we frequently encountered as impurities in the course of spectroscopic research. This has proved so useful that it seems worth while to extend the list to cover, as far as possible, all known band systems. Since it appears, moreover, that such a list can be of service, not only to pure spectroscopists, but also to those who use spectroscopy as a tool for research in other fields such as astrophysics, chemistry and chemical technology, we have ventured to gather together in book form such information about known band spectra as may assist in their identification.

In the first list the bands were given in order of wavelength; all bands of the systems considered being included. This arrangement was soon found to possess practical disadvantages. A more useful arrangement was obtained by dividing the data into two sections. The advantages of the division are discussed in the introduction preceding the tables.

As a first stage in the compilation of the available data we have been obliged to limit the scope of the tables in several directions. Thus there are limits to the range of spectrum considered and to the complexity of the molecules whose spectra are included. The wavelength region considered is from 10 000 Å to 2000 Å, that is roughly from the photographic infra-red to the ultra-violet limit of quartz spectrographs, except that in a few cases, where the origin of a system lies near the border line, one or two bands have been included which are just outside the range. As to complexity we have endeavoured to include all recorded systems of diatomic molecules, but only those of triatomic and more complex molecules which show well-defined banded structure and are of frequent occurrence in spectroscopic investigations. The absorption spectra of complex organic molecules and of solutions have been omitted.

In addition to the wavelengths of the band heads, the tables include information about the appearance and occurrence of each band spectrum. Though the information thus given is often useful for reference for other purposes, the object of identification has been kept foremost throughout in making decisions relating to the selection and arrangement of material.

For some systems we have found that the existing data are very incomplete. Where these systems are of frequent occurrence we have made new wavelength measurements. In a large number of cases where no estimates of intensities are given in the original paper, but a photograph is included, we have included estimates of intensities made from the photograph. In other cases where the analysis alone is given without mention of the positions and intensities of the most prominent heads, we have located the positions of the heads from the analysis where possible, and if necessary converted the corresponding wave-numbers to wavelengths. In this connection we should like to point out that it would be of great assistance for purposes of identification if authors of papers reporting new band systems would always in future include a brief description of the appearance of the system with wavelengths and intensities of the strongest heads, a few notes on the sources with which it is obtained, and, if possible, publish a photograph with a wavelength scale or a comparison spectrum.

In addition to photographs which we have taken ourselves, we have been very fortunate in having access to numerous spectrograms taken by Professor A. Fowler and his colleagues and students in the Astrophysics Department of the Royal College of Science. Several of the reproductions of common band spectra have been taken from these plates.

Finally, it is with pleasure that we acknowledge our indebtedness to the late Professor A. Fowler for a thorough introduction to the study of spectroscopy and for turning our attention to many of the spectra dealt with herein; to Professor H. Dingle for interest and encouragement in the preparation of these tables; to Dr. W. Jevons, Dr. R. W. Lunt, Dr. E. C. W. Smith, Dr. R. F. Barrow and Dr. R. C. Pankhurst for the use of spectrograms and unpublished data as well as for useful criticism during trial of the tables, and to Mr. E. S. Parke for very valuable assistance in the preparation of the plates. One of us (A. G. G.) is also indebted to the Trustees of the Beit Fellowships for Scientific Research for a special grant, during the tenure of which a large part of the manuscript was compiled.

R. W. B. P.
A. G. G.

LONDON.
SEPTEMBER 1940.

Preface to Fourth Edition

The first edition of this book appeared in 1941, the second in 1950, the third in 1963, and, as this sold out quickly, we had it reprinted with a 20-page supplement in 1965. This edition has been very fully revised. The first edition contained information about spectra of around 250 diatomic molecules and 35 polyatomic; this one covers about 490 diatomic and 127 polyatomic. Developments in techniques, such as flash photolysis, and interest in new subjects such as upper-atmosphere chemistry have contributed to the expansion of the amount of data available.

We have done our best to comb the literature up to about the end of 1974. Data listed are based on original papers or, in some cases, on our own measurements, but the compilations edited by B. Rosen et al. (*Données Spectroscopiques Relatives aux Molécules Diatomiques, 1970*) and its supplement edited by R. F. Barrow et al. (*Diatomic Molecules; A Critical Bibliography of Spectroscopic Data*, 1973), the book by G. Herzberg (*Electronic Spectra of Polyatomic Molecules*, Van Nostrand, 1966) and the Berkeley Newsletters on *Analysis of Molecular Spectra*, prepared regularly by J. G. Phillips and S. P. Davis have been of great value in tracing references to the literature.

Although other works, such as that edited by Rosen, give constants and some data for diatomic molecules, and the book by Herzberg gives molecular constants for polyatomic molecules, this is still the only work aimed primarily at Identification, giving both diatomic and polyatomic molecules and including a considerable collection of reproductions of the actual spectra.

A. G. G.
R. W. B. P.

LONDON.
AUGUST 1975.

Introduction

Experience in using the first list which we drew up of band heads arranged in order of wavelength showed that in extending this list to include a large number of molecules it was desirable to modify the system. The Tables for the Identification of Molecular Spectra are therefore divided into two sections.

The first section consists of a list of the most prominent heads of the more persistent and better-known band systems of each molecule. These heads are listed in order of wavelength, with abbreviated information about the direction of degradation of the bands and their appearance, and, of course, the molecule responsible. In earlier editions we made an attempt to include estimates of relative intensity of the heads within the system, listing these intensities under the sources in which the head was likely to be observed; however, since only the most prominent heads of each system were included, most of the intensities were '10' and in many cases information was not really adequate. This information about intensities has therefore been omitted from the first section in this Fourth Edition. This has enabled the table to be set more compactly.

The second section consists of individual lists of band heads for each system of each molecule, accompanied by notes about the occurrence and appearance of the system, the nature of the electronic transition involved, the vibrational assignment of the bands of the system, intensity estimates on a visual scale and references to the source of the data. The lists are arranged in alphabetical order of the chemical symbols of the molecules.

The general considerations leading to this division are briefly as follows. For practical reasons it is preferable to identify the molecular contribution to a given spectrum system by system, rather than band by band. It is the practice to identify the atomic contribution line by line, with the aid of tables of atomic lines in order of wavelength and there is a natural tendency to proceed to identify bands in a similar way. Such a procedure, however, often leads to incorrect identification. In an atom, each change of electronic state gives rise to a line, whereas in a molecule each change in electronic state gives rise to a whole band system. The various bands within the system arise from changes of the vibrational energy of the molecule and in general involve much smaller energy intervals than the electronic changes. Thus in respect of variation of intensity from source to source the bands of one system behave somewhat like the components of a fairly close multiplet, appearing and disappearing together. But whereas the multiplet contains relatively few lines of the whole spectrum, a single band system often contains many heads, perhaps several hundreds, and may comprise all the radiation that is readily excited for that particular molecule. Inclusion of all such bands in a single list leads to a large number of chance coincidences in wavelength. Such coincidences are more troublesome in the case of bands than in the case of lines, since the wavelength recorded for a band head is seldom as precise as that for an atomic line, depending, because of the structure of the head itself, very considerably on the judgement of the observer and on the dispersion used. This makes it much less safe to identify a single band by wavelength alone than it does to identify a single atomic line in this way. Supporting evidence should always be sought. Such evidence can be obtained by considering the system as a whole. The list of Section I has therefore been restricted to a few of the most prominent heads of each system so that it is somewhat analogous to the list of persistent lines of the elements. The purpose of the list is to provide a clue to the identity of an unrecognised system. The most prominent head of the unknown system is selected. This will usually be the front head of the strongest band, but not necessarily the strongest individual head for a multi-headed band. This head is then compared with the Persistent Heads list and a close agreement of wavelength and direction of degradation may suggest that it is a member of a certain system of a given molecule. Reference is then made to the individual list for that system and the presence or absence of other members is checked. The process is then continued with outstanding heads of the strongest of the remaining unidentified bands, and so on. It is also advisable to look for other systems of the molecules for which systems are found and also for systems of other molecules containing the same elements; thus if a system of C_2 is found and also one of N_2, it is desirable to look for systems of CN as well as others systems of C_2 and N_2. Or, again, if a trace of oxygen is suspected, systems of NO and CO may be checked. This procedure often leads to the discovery of weak bands, masked by stronger ones, which would otherwise have passed unnoticed. In following up other systems in this manner, and indeed in all cases where interest lies in the spectrum of a given molecule, the arrangement of Section II is especially convenient.

The actual number and choice of bands which should be included in the two lists is mainly a matter for experience to decide. We have tried to place most emphasis on those molecules which are cosmically abundant or are known to occur readily as impurities. Since the first edition of the book a very large number of new molecular spectra have been reported. In some cases this may almost degenerate to 'stamp collecting'; studies of rare molecules involving unusual combinations of elements form convenient Ph.D. projects for students. However, it is often very difficult to predict what will and what will not be useful. For example the prominence of ZrO bands in some stellar spectra could hardly have been foreseen. Unusual molecules may arise from use of rare elements in transistor materials; development of special laser system may require unusual species; pollution by toxic heavy metals, like cadmium, may involve unusual spectra in their estimation. For common molecules we have tended to give the prominent bands of all known systems; for the rarer species we have usually included only the strongest band or two of the strongest system in the first list of Persistent Heads, and to give all systems but only a few bands of each in the second list, Individual Systems. Some band systems are very extensive, involving many bands of comparable intensity. These present special difficulty and it is not possible readily to identify such systems from the main Persistent Heads table.

It is well to emphasise that in making identifications, the evidence of the presence of atomic lines may be very helpful. To facilitate the checking of the presence of atoms a table of persistent lines of the elements has been included in the Appendix. If it is desired to check the line spectra more fully, recourse should be made to the various tables of atomic lines that are available.

Finally, in as much as direct comparison of photographs is the quickest and most certain way of identification, a number of plates are included showing many of the more frequently encountered band systems. Also those references to papers which contain a useful reproduction of the spectrum are indicated by a dagger, †, following the date.

Table of Persistent Band Heads

The object of this table is to provide a clue to the nature of the unknown band system as quickly as possible, so that it may be compared directly with the appropriate detailed list or with one of the plates. For this purpose the list contains for all suitable band systems which are of frequent occurrence a few (sometimes only one) of the outstanding band heads, arranged in decreasing order of wavelength.

In this fourth edition the structure of this table has been substantially modified because of the large increase in the number of known band systems. For the simplest systems with strong $(0, 0)$ sequence we have included just this one head. In a greater number of cases we have given the heads of the $(0, 1)$, $(0, 0)$ and $(1, 0)$ sequences. For systems with more open Franck-Condon intensity parabolae we have had to include more heads, but have kept the number down to the minimum. To include many heads not only makes the table long and expensive to set but increases the risk of chance meaningless coincidences.

Following the wavelength we have given a brief indication of the appearance of the head. This always includes the direction of degradation

R degraded to longer wavelengths (red)

V degraded to shorter wavelengths (violet)

or that the band measurement is for the maximum, **M**, of a headless band, or is the origin, **O**, of a band with a clear region near the origin and branches spreading in each direction from this origin. **L** indicates a very narrow, line-like, feature.

Other remarks in this appearance column are:

CD.	Close double head (separation usually less than 5 Å).
CT.	Close triple head.
D.	Double head.
DCD.	Double head, each component a close double.
F.	Group of four or more heads.
L.	Narrow band resembling an atomic line.
T.	Triple head.
wr.	Accompanied by weaker head to the red.
wv.	Accompanied by weaker head to shorter wavelengths (violet).

We should stress that we have usually given the *first* or most prominent head of the sequence, not necessarily the strongest. Some molecules have very extensive band systems consisting of many bands of comparable intensity and it has not been possible to include these. It often happens that homonuclear molecules have spectra of this type and examples are As_2, Br_2, Cl_2, I_2, K_2, Li_2, Na_2, P_2, Se_2, Te_2. Some mixed molecules of the same group of the periodic table also tend to give extensive systems, e.g. IBr, ICl, NaK. We have also excluded from this table the very rare molecules, the collector's pieces.

In previous editions we listed intensities, often hypothetical, for the band in each of six sources. However, since we are always selecting just the strongest bands, these intensities were mostly 10 or 9 and not of much help, and since this arrangement took up a lot of space and required a lot of rather arbitrary decisions on our part, we have now dropped the inclusion of intensities from this list of persistent heads.

We have, however, retained the asterisk, *, to denote those bands which in our experience occur rather frequently as impurities in other spectra.

For those molecules of importance possessing a number of strong systems, such as CO, N_2, CN, we have given in very abbreviated form an indication of the system following the symbols for the molecule, e.g.

      N_2 1st Pos.   Nitrogen First Positive System

      C_2 Desl.-d'A.   Deslandres-d'Azambuja's System

     In making an identification using this table, the procedure should be to select *several* of the most prominent heads of the unknown system and to compare these with the table. If the wavelength of one of them coincides with that in the table, and the direction of degradation is right, then further comparison should be made with the individual lists. This is just a table of persistent heads, not all heads.

λ	App.	Mol.	System	λ	App.	Mol.	System
9874·4	R, T	BaH		8675	R	N_2	I-R Aft.
9868·6	R	TaO		*8652·2	R	CaO	
9849·3	R	TaO					
*9835·0	R	CaO		8626·0	R, T	TiS	
9589·1	R	RhC		8624·0	R	VO	
				8619·8	V	NO	I-R Quartet
9776·1	R	SrO		8613·3	R	SO	
9725	R, D	TiO	δ	8611·1	R	CrH	
*9669	R	H_2O					
9658·0	R	BaCl		*8597·8	R	O_2	Atmosph.
9647·5	R, D	BeO		8585·5	V	C_2O	
				8571·5	R	BaF	
9513·5	R	BaH		8557·8	V	C_2O	
9494	R	SeO		*8541·8	V, T	N_2	1st Pos.
9400	R	OH	Meinel				
9357·2	R, T	BaH		8488·9	R	PrO	
9299·6	R	ZrO		8451·2	R	BaH	
				8448·1	V	PH_2	
*9277	R	H_2O		8435·9	O	SiF	
9195·8	R	SrO		8421·6	R, D	BaH	
9152	R	HCl					
9145·3	R, T	N_2^+	Meinel Aur.	8420·8	R	BaCl	
*9140·6	R, T, wv	CN	Red	8394·6	R	NiCl	
				8283·6	R	TiS	
9123·1	R	He_2		8282·7	R	BiO	
9101·7	R	BaCl		8230	R	FeO	
9098	R	BaCl					
9062·4	R	TiS		8226	M	P_2	
9026·1	R	BaH		8217·1	R	IrC	
				8192	R	ZrO	
8981·7	R, D	BaH		8182	R	AgBi	
*8980·5	R	C_2	Phillips	8164	R	BeS	
8859·6	R, D	TiO	δ				
8829·4	R, D	OH	Meinel	*8153·0	R	CaO	
8742·4	V, wr	NF		8137·0	R, D	BaF	
				8112	R	FeO	
*8722·3	V, T	N_2	1st Pos.	8108·2	R	C_2	Phillips
8700·0	R	SrO		*8097	R	H_2O	
8685·4	R	NiCl					

λ	App.	Mol.	System	λ	App.	Mol.	System
8066·0	R	ScCl		7550·3	R, D	ArH	
8057·6	V	N_2	Herman I-R	7506·8	V	SiH	
8053·6	R, T	N_2^+	Meinel Aur.	*7503·9	V, T	N_2	1st Pos.
8020·9	V, DCD	NO	Ogawa	7468	M	H_2O^+	
7991·4	V	PO		7463	R	HCl	
7973·8	R, DCD	BS		7430	V	S_2	
7953·3	R, wv	BeO		7420·2	V, DCD	NO	Ogawa
7936·7	R	ScCl		7393·2	R	VO	
7919	M	NH_3		*7386·6	V, T	N_2	1st Pos.
7918·5	R	OH	Meinel	7379·8	R, D	LaO	
7915·2	R	SO		7348·0	V	SrH	
*7907·7	R	C_2	Phillips	7346·7	V	SrH	
7901	M	C_2H_2		7334·7	V	O_2^+	1st Neg.
7900	R	NiO		7330	R	NiO	
7899·3	R	PtC		*7318·5	R	CaO	
7896·0	R, wv	VO		7311·7	R	BS	
7877·2	R, D	LaO		7303·1	V	C_2O	
7874	M	C_2H_2		7299·5	R	PbF	
*7872·7	R, T, wv	CN	Red	7297·2	R	CeO	
7850·9	R, F	VO		7284·2	R	OH	Meinel
*7850·2	R, T	CN	Red	7282·1	V	C_2O	
7845	R	BiO		7277·3	R	BiO	
7831·8	R	CeO		7260·0	R	LaS	
7825·7	R, T	N_2^+	Meinel Aur.	7245·9	V, D	AlF	
7820·1	R	CO_2	Venus	7235·8	R	CeO	
7780	R	N_2	I-R Aft.	7210·4	R, F	CO	Asundi
7776·3	V	SbF		7200	R	TmO	
7763·1	V	PH_2		7173·1	V	PH_2	
7755·8	R	OH	Meinel	*7164·5	R	H_2O	
*7753·2	V, T	N_2	1st Pos.	7125·6	R	TiO	γ
*7714·6	R	C_2	Phillips	7116·0	R, D	BaF	
7674·4	R, wv	ArH		7083·2	V	C_2	High P.
7672·1	R	TiO	γ	7081·8	R	PbS	
7662·8	R	PrO		7069·6	V	S_2	
7633·7	R	ScCl		7051·0	R	LaS	
*7626·2	V, T	N_2	1st Pos.	7036·8	R, T	N_2^+	Meinel Aur.
*7593·7	R	O_2	Atmosph.	7029·6	R	PbCl	
7591	R	N_2	I-R Aft.	7028·4	V	NCl	
7588·6	R	PbF		7018·1	V	SrH	
7570·1	R	PtC		7011	R	NdO	

λ	App.	Mol.	System	λ	App.	Mol.	System
6987	M	H_2O^+		6676·3	R	PbTe	
6983·8	V	S_2		6675	M	SrOH	
6984·7	V	SrH		6666·7	V	SrBr	
6960·4	R	IrC		6659·6	V	BiS	
6942·6	V, wr	CaH		6655·6	V, D	SrF	
6936·5	R	OsO		6652	M	NH_2	Ammonia α
6936·5	R	BiO		6646·7	V, D	NCl	
6931·6	V	SrI		*6631·3	R, T, wv	CN	Red
6926·1	R	NiCl		6625	M	NO_3	
6921·9	R	PbCl		*6623·6	V, T	N_2	1st Pos.
6918·3	R	BaS		6613·7	V, wr	SrCl	
6898·4	V, DCD	NO	Ogawa	6612·1	R	IrCl	
6896	R, T	N_2	I-R Aft.	6607·6	R	RaCl	
6894·3	R, wv	SnH		6594	M	H_2O^+	
6891·6	R	MnF		6590·2	V	YbH	
6879	R	OH	Meinel	6590	M	SrOH	
6876·4	O	PbS		6585·2		NdO	
6872·9	R	Rb_2		6578·3	R	CaS	
6870·4	R	PbH		6576·3	R	YCl	
*6867·2	R	O_2	Atmosph.	6569	R, F	TiO	γ'
6866·4	R	HfI		6561·5	R	Na_2	
*6856·3	V	O_2^+	1st Neg.	6549·3	R	IF	
6850·2	V, F, wr	BaH		*6544·8	V, T	N_2	1st Pos.
6820	M	SrOH		6542	M	H_2O^+	
6818·7	R	PbCl		6533·5		SmO	
6804·0	R, F	CO	Asundi	6522·5	R	AsS	
6800·2	V	SrBr		6521·7	R	PbTe	
6797·8	R	Rb_2		6517·2	R	F_2	
6782·8	R	CaS		6513·5	R, F	CO	Asundi
6778·8	V	SrI		6513·2	R	Na_2	
6777·3	R	BaS		6513·0	V	SrBr	
6775·7	R	Rb_2		6510·9		SmO	
6763·3	R	RaCl		6509·4	R	MoO	
6747·2	V	SrI		6498·1	R	RaCl	
6744·9	V	SrCl		6494·8	R	NbO	
6718·8	R	VCl		6493·1	R	BaO	
6708·8	R	BiO		6485·8	R	Pb_2	
*6704·8	V, T	N_2	1st Pos.	6481·2	R	IF	
6689·5	V, D	BaH		*6478·5	R, T, wv	CN	Red
6686	M	H_2O^+		6474·1	V	FeCl	

λ	App.	Mol.	System	λ	App.	Mol.	System
6473·7	R	ZrO	γ	6257·8	V, D	MnBr ?	
*6468·5	V, T	N_2	1st Pos.	6252·9	V, wr	CaBr ?	
6460	M	SrOH		*6252·8	V, T	N_2	1st Pos.
6457·5	R	H_2O		6249·3	R	IF	
6445·4	R	MoO		6246·0	R	NiH	
6425·1	R	NiH		6244·7	R	MoN	
*6418·7	V	O_2^+	1st Neg.	6233	M	NO_3	
6418·4	R	Na_2		6230	M	CaOH	
6417·5	R	Pb_2		6229·4	R, T	ZrO	γ
6412·9	V	CaI		6228·5	R	MoN	
6406·3	R	RuO		*6224·9	V, DCD	CaCl	
6406·1	R	CaS		6219·8	R	FeO	
6405	R	BrF		6216·0	R	NaRb	
*6401·0	R, T	CO	Triplet	*6211·6	V, wr	CaCl	
6398·7	R	He_2		6211·5	R	ScF	
*6394·7	V, T	N_2	1st Pos.	6210·6	R	PtO	
6394·3	R	CrO		6210	M	H_2O^+	
6389·3	V	CaH		6205·5	R	HfBr	
6388·8	V	CaI		6199·5	R	TiN	
6383·8	R	AgSb		6199	M	H_2O^+	
6377·8	V, DCD	NO	Ogawa	6196·0	V	$ZnTl_2$	
6370·4	R	Pb_2		*6189·4	R, D	CO^+	Comet-t.
6368	R	O_2	Liquid	*6184·9	V, wr	CaCl	
6362·4	V	SrCl		6180·5	R	FeO	
6349·5		SmO		6174·4	R, D	NbO^+	
6342·2	R	NiO		6165·7	R	H_2O	
*6322·9	V, T	N_2	1st Pos.	6161·6	R	CuO	
6315·4	V	CaI		6159·1	R, CT	TiO	α
6302	M	NH_2	Ammonia α	6158	M	H_2O^+	
6301·5	R	AsS		6149·9	R	SnBr	
6293	R	BrF		6148·5	O	InH	
6283·6	R	Tl_2		6144·7	V	AlAu	
6277·7	V, wr	CaBr ?		6138·6	R, D	TiN	
6277·2	V, D	MnBr ?		6138·0	R, D	CHO	
*6276·6	R	O_2	Atmosph.	6122·9	R	AgSb	
6272·0	V	AlAu		6117·5	V	BiH	
6265·9	V	NBr		6116·3	R	CuSe	
6263·3	R	CuTe		6110·7	R	MoN	
6261·7	V, D	NCl		6109·9	M	FeO	
6258·5	V	—	Ca oxide	6105·9	R	PbI	

λ	App.	Mol.	System	λ	App.	Mol.	System
6103·7	M	CrCl		*5982	R, T	CO	Triplet
6102·8	R	F$_2$		5979·9	R	TbO	
6098	R	AgBi		5977·7	V	FeBr	
6097·3	M	FeO		5969	R	—	Sr oxide
6097	V	—	Ca oxide	5962·4	V	NBr	
6096·8	R, D	YO		*5959·0	V, T	N$_2$	1st Pos.
6095·0	R	SnH		5957·1	V	FeBr	
6092·0	R	AsS		5950	M	—	Sr oxide
*6086·9	V, DCD	CaF		5949·4	R	SbO	
6086·4	R, D	VO		5939·1	R, D	YO	
6084·7	V, CD	SrOH		5938	R	—	Sr oxide
*6079·9	V	CO	Angstrom	*5934·0	R	CaCl	
6076·6	R	TbO		5933·8	V	NBr	
6076·6	V	SrOH		5923·9	V	BrF	
6073	V	La$_2$		5920·8	R	TbO	
*6064·4	V, wr	CaF		5916·0	R, D	NbO$^+$	
6063·9	V	FeCl		5914	M	H$_2$O$^+$	
6060·3	V	MgO		5913·8	L	InH	
6059·3	R	CuO		5910·7	R	PbO	
6051·6	R	CrO		*5906·0	V, T	N$_2$	1st Pos.
6050	M	SrOH		5905·0	V	NBr	
6047	V	La$_2$		5902·3	R	PtO	
6045·0	R	CuO		5899·3	V	C$_2$	High P.
6043·2		NbO$^+$		5893	M	NO$_3$	
6039·6	R	BaO		5880·6	R	Br$_2$$^+$	
6036·1	R, D, wv	ScO		5869·2	R	BiO	
6031·1	R	IF		5868·1	R	FeO	
6028·6	R, D	NbO$^+$		5864·5	R	BaO	
*6026·4	V	O$_2$$^+$	1st Neg.	*5858·2	R, T, wv	CN	Red
6021·9	R	SnH		*5853·1	R	MnO	
6017·5	V	AlAu		5852·6	R	F$_2$	
*6006	V	—	Ca oxide	5847	R, F	TiO	γ'
6004·6	R	Br$_2$$^+$		5845·7	R	PbI	
*6003	R	—	Ca oxide	5835	R	ZrN	
6001	O	NO		5834·8	R	DyO	
5997·6	V, T	YbH		*5830	R	CaF	
5991	M	PHO		5829·7	R	FeBr	
5990·7	V	NBr		5827·5	R	SnF	
5989·5	V	FeCl		5827	M	H$_2$O$^+$	
*5983	R	—	Ca oxide	5815·1	V, wr	N$_2$	Gaydon G.

λ	App.	Mol.	System	λ	App.	Mol.	System
5808	V	P_2		*5635·5	V	C_2	Swan
*5804·3	V, T	N_2	1st Pos.	*5631·9	V	O_2^+	1st Neg.
5804·1	R	BrF		5629·3	R, D	TiO	β
5802·2	R	BiI		*5610·2	V	CO	Angstrom
5800·9	R	Al_2		5609·5	V	BaI	
5799·8	R	RuO		5599·9	R, CD	LaO	
5799	M	H_2O^+		5597·6	R, D	TiO	β
5789·8	R	FeO		5597	M	PHO	
5789·6	R	SnO		5596	M	ErO	
5772·0	R, CD	SrF		5595·0	V, F	N_2	Gaydon G.
5763·5	O	GaH		5593·5	R, DCD	BS	
5752·4	V	FeCl		5589	R	TaO^+	
5736·7	R, D	VO		5586·2	R	PF	
5731·7	R	PbTe		5584·7	R	PO	
5731·4	R	F_2		5582·2	R	MnO	
5728·2	R	NaK		5574·8	V, wr	N_2	Gaydon G.
5718·1	R	ZrO	β	5574·6	R	CuS	
5713	M	NH_2	Ammonia α	5574·1	R	CuSe	
5712·1	R	CuS		5572·3	R	SnBr	
5710·1	R	SnTe		5570	M	ScH	
5706·2	O	TlH		5561·1	R	SnCl	
5699·9	R	Br_2^+		5561·0	O	PbH	
5697·7	V, D	GaH		*5547·4	M	CaOH	
5696·2	O	RbH		*5542·6	M	CaOH	
5696	V	HoO		5540·5	R	AsSe	
5694·3	V	CuF		*5538·6	M	CaOH	
5691·1	R	PrO		5529·5	R	NiO	
5688·6	O	InH		5525·8	R	RuO	
5687·8	R	ScCl		5506·8	R, D	SnF	
5678·9	R	ZnO		*5498	R	—	Ca oxide
5677·5	V, F	MnH		5497·3	R	SnBr	
5668·3	R	SnCl		5492·7	R	BaO	
5667·8	R	CoCl		5489	M	H_2O^+	
5663·0	R	PtO		5473	R	—	Ca oxide
5661·5	R, D	TiO	β	5470·3	R	NbO	
5660	M	HoO		5469·3	R, D	VO	
5648	R	ZrN		*5461·4	R, wv	CO^+	Comet-t.
5644·1	R	BaO		5459·4	R	PbO	
5640·2	V, T	YbH		*5458·7	M	CuOH	
5635·9	R, D	SnF		5457·1	R	SnCl	

λ	App.	Mol.	System	λ	App.	Mol.	System
5450	M	BO_2		*5262·3	R	CuCl	
5448·3	R, CT	TiO	α	5258·2	R	Bi_2	
5442·7	R	SiBr		5248·7	M	PHO	
5440·9	R	MnS		5240·8	R	Au_2	
*5418·6	M	CuOH		5240·5	R	BaCl	
5415·9	V	C_2^-		5240·0	R	AuCl	
5399·5	R	PtC		5236·0	R	SbCl	
5382	V	TiH		5226·4	V	HgIn	
5381·7	V	BaI		5224·1	R	Bi_2	
5376·0	R	XeO		5213·9	R	PS	
*5372·8	V, T	N_2	1st Pos.	5211·0	V, D	MgH	
5372·6	R	PbSe		5208·2	R	BaBr	
5370	V	TiH		5205·5	R	AuCl	
5368	R	NaRb		5200·9	R	NbO	
5364	R	O_2	Liquid	5199·8	R	MnS	
5361	V	TiH		*5198·2	V	CO	Angstrom
5360·1	R	BaBr		5193·8	R	IrC	
5359·6	R	MnO		5189·5	R	Au_2	
*5356·5	M	CuOH		5185·0	R	ZrO	
5352	R	TmO		5180·1	R	SnSe	
5347	R	TmO		5180	M	BO_2	
5346·6	R	AuCl		5174·5	R	NiO	
5341·2	R	AsSe		5172·7	V	HgIn	
5334·3	V, DCD	NO	Ogawa	*5172·6	R, D	SiN	
5333·8	R	SbCl		5171·4	R	LuO	
5331·0	R	SnSe		5166·9	R, CT	TiO	α
5325·1	R	PbSe		*5165·2	V	C_2	Swan
5307·3	R	IO		5162·2	R	SiCl	
5300·8	R	MnS		5158·6	R	PbSe	
*5295·7	V	O_2^+	1st Neg.	5148·0	R	SnO	
5293·4	R	Bi_2		*5145·4	R, CD	CaF	
*5291·0	R	CaF		5141·1	R	CuI	
5287·8	V, L	NF		5139·2	V	BaCl	
5283·2	R, DCD	BS		5134·0	R	SbS	
5278·7	R	PbSe		5132·6	R, D	SnF	
5277·7	R	SbO		5123·5	R	OH	Schuler-W
5273·4	R	DyO		5120	M	BaOH	
5268·9	O	RbH		5102·6	R	NSe	
5267·8	R, DCD	AlCl		5088·6	R	PtH	
5263·4	R	DyO		*5079·4	R	AlO	

λ	App.	Mol.	System	λ	App.	Mol.	System
5076·8	V	AlN		4926·2	R	Pb_2	
5074·7	R	HfO		4925	V	XeN	
5072·8	R	CuI		4924·2	R	Na_2	
5069	R	SbBr		4919·9	R	PbS	
5067·9	R	$C_4H_2^+$		4917·2	R	ScF	
5067	R	ErO		4916·5	R	PbCl	
5065·0	R	SbO		4915·1	R	NbO	
5061·1	V	CuF		4902·0	V	C_2^-	
5054·4	R	BeO		4901·5	R	Cu_2	
5037·8	R	SiBr		4894·5	R	Na_2	
*5033	R, T	CO	Triplet	4892·1	O	RbH	
5033·0	R	SnTe		4890·2	R	Pb_2	
5031·5	R	MnF		4883·3	R	SnF	
5031·4	R	CoCl		*4881·5	R	CuCl	
5031·1	R	SiCl		4879·3	R	CuBr	
5025·8	R	HgI		4878·5	R, D	NSe	
5021·4	R	SbF		4874·6	R	HfI	
5019·7	R	CuI		4870·2	R	KH	
*5007·3	V	MgO	Green	4870	M	BaOH	
5000·6	R	BaF		4863·2	R	CeO	
4995·6	V	BS		4862·5	R, F	FeCl	
4993·7	R	MnBr		4858·5	R	PbS	
4992·1	R	BaF		4857·8	R	ScO	
4990·8	V, L	BeH		4855·0	R	Pb_2	
4988·2	R	PbCl		4847·4	R	NaK	
4985·6	R, D	SnF		*4846·9	R	CuCl	
4984·9	R	NSe		4846·0	R	PbCl	
4983·8	R	PbO		4844·5	R	IO	
4982·7	R	PbS		*4842·2	R	AlO	
4977·4	R	SiC_2		*4842·1	R	S_2	
4969·7	R	F_2^+		4842·0	O	BiH	
4968·5	R	SnTe		4840	R	YbO	
4966·5	R	MnCl		4837·1	R	N_2	Vegard-K
4962·8	R	Na_2		*4835·3	V	CO	Angstrom
4950·8	R	BaF		4818·5	R	PbBr	
4940·4	V, wr	BCl		4817·4	R, CD	YO	
4937·6	R	SiI		4806·3	R	WO	
4936·3	R	MnF		4805·4	R	Rb_2	
4932·0	V	CuF		4802·8	R	KH	
4930·8	R	PbTe		4801·2	R	IrC	

λ	App.	Mol.	System	λ	App.	Mol.	System
4798·5	R	SnTe		4648·5	R	He_2	
4797·1	R	Rb_2		4648·4	R	CuH	
4791·7	R	CeO		*4648·1	R	AlO	
*4787·3	R, T	CO	Triplet	4647·6	R, D	PbF	
4778	R	YbO		4640·6	R, wv	ZrO	α
4777	V	$C_2H_2O_2$	Glyoxal	4639·8	R	SiC_2	
4775·8	R	Rb_2		4635·0	R	CuO	
4769·9	R	CuO		4631·5	R	SbS	
4768·3	R, D	NSe		4630·3	R	PF	
4763·9	R	SnO		*4628·9	R	NH^+	
4761·3	R	ClF		4627·8	V	AsH_2	
4740·4	V	PH_2		4625·6	R, D	He_2	
*4737·1	V	C_2	Swan	4619·8	R, T	ZrO	α
4736·7	R	KH		4615·6	R	GdO	
4736·4	R	PS		4614·4	R	BiCl	
4731·8	V	AlH		4612·7	R, CD	BO	α
4729·8	V, DCD	MgCl		4605·8	V	PO	
4729·7	R	OH	Schuler-W	4597·0	R	NiCl	
4726·0	R	F_2^+		4596·9	R	Cu_2	
4710·8	R	CuO		4596·0	R	NiBr	
4709·7	R	WO		4590·2	R	WO	
4709	M	VH		4588·0	R	NiBr	
4708·6	R	BeO		4586·2	R	IO	
4708·0	V	C_2N		4585·7	R, DCD	BO	α
4701·5	R	CoCl		4583·8	R	CuO	
4699·0	V	C_2N		4582·3	O	Si_2	
4698	O, wr	BiH		4576·6	R	MnS	
4695·5	R	RhC		4575·1	R	CuI	
4692·4	R	H_2S^+		*4571	R, T	CO	Triplet
4689	M	VH		4566·9	R	PbBr	
4689·0	R	NbO		4565·0	V	AlI	
4685	M	VH		4555	V	$C_2H_2O_2$	Glyoxal
4680·2	V	C_2	High P.	4553·7	R	PbO	
4673·0	R, wv	NiF		4547·2	R	PtH	
4663·7	V	ZnTl		4544·1	R	SrO	
4661·7	R	LuO		4543·2	V	PH_2	
4660·3	R	PbCl		4542·5	R, D	PbF	
4658·7	R	RhC		*4539·4	R, D	CO^+	Comet-t
4658·0	R	PbO		4537·3	R	InH	
4653·4	R	ScCl		4535·8	R	SnTe	

λ	App.	Mol.	System	λ	App.	Mol.	System
4535·5	R	N_2	Vegard-K.	4410·4	R	PbO	
4533	R	BrO		4410·3	R, CD	SrI	
4524·6	R, DCD, wv	CP		4407·9	R	AlSe	
4524·0	V	AlI		*4406·9	R	SiN	
4523·2	R	NH		4403·6	V	NCO	
4520·0	O	ZnH		4403	R	B_2O_2	
4517·8		SiI		4399·6	R	SrO	
4517·5	V, DCD	CP		4398	R	BrO	
4515	M	POBr		4396·6	R	ZrI	
4514·6	R, DCD	NiF		4396·4	R	SnSe	
4514·0	R	CdCl		*4393·1	V	CO	Angstrom
*4510·9	V	CO	Angstrom	4390	R	NO_2	
4510·7	R	NbO		4388·4	R	MnS	
4507·2	R	H_2S^+		4387·3	R	Ag_2	
4505·3	R	SnS		*4385	V	CH	
4504·6	R	SnSe		4384·8	V	NCO	
*4502·0	R	NH		4380·3	V	CO	Herzberg
4491·3	V, wr	CdH		4373·0	R	PF	
4487·4	R	TeO		4371·9	R	LaO	
4487·3	R	SbS		4368·2	R	SiF	
4487·2	R	IO		4368·1	R	PbS	
4484·3	R	CdCl		4366·7	R	BiF	
4480	M	NO_2		4366·7	R	CaO	
4479	R	MnS		4364·8	R, D	PbF	
4477·9	R, D	CoH		4364·3	V	PH_2	
4476·7	R	PS		4363·4	R, CD	BO	α
4465·5	R	BiF		4363	M	POCl	
4461·7	R	CdBr		4359·9	R	CuI	
4459·5	R	WO		4354·1	R	NiBr	
4453·5	R	PF		*4353·2	R	CuCl	
4449·1	R	AlAu		4351·2	R	CaO	
4448	R	NO_2		4348·5	R	NH^+	
4447·3	R	SnH		4347·0	R	Ag_2	
4444·4	R	SbF		4344·7	R	SnSe	
4441·2	R, D	PbF		4344	R	B_2O_2	
*4434	R	CH_2O	Cool fl.	*4343·6	V, CT	N_2	2nd Pos.
*4433·8	R, D	CuCl		4342·8	R	TeO	
4427·3	R	BeO		4341·1	R	CuBr	
4418·1	R	LaO		4338·6	R	MgS	
4416·8	R	NiBr		4336·6	R	OH	Schuler-W.

λ	App.	Mol.	System
*4333·2	R	CuCl	
4331·6	R	BH	
4326·8	R	Rb$_2$	
*4324	L	CH	
4322·1	R	Ag$_2$	
4316·5	R	PbS	
*4314·2	V	CH	
4309·6	R	CuI	
4308·1	R	F$_2^+$	
4304·9	R	NiCl	
4304·6	O	P$_2^+$	
4301	V, wr	ZnH	
4297·6	V, wr	CdH	
4295	M	NCS	
4293·6	V, DCD	AlBr	
4292·4	V	CN	Le Blanc
4288·6	R	CuBr	
4288·2	R	Rb$_2$	
4287·9	R, CD	CaI	
4283·1	O	Si$_2$	
4281·0	R	SrO	
4279·6	R	CuH	
*4278·1	V	N$_2^+$	1st Neg.
4273·6	R	CrS	
4270	R	BrO	
4267	M	POCl	
4266·0	V	TiBr	
4262·8	R	CuBr	
4261·7	R	CuI	
4260	R, D	ZnH	
*4259·5	R, wv	AlH	
4258	V	NCO	
4252·1	R	HfO	
*4248·9	R, DCD	CO$^+$	Comet-t.
4248·6	R	SnS	
4246·8	R, wv	BeI	
4245·9	R, D	BH	
*4245	M	CH$_2$O	Cool fl.
4244·6	R	SO$_2$	
4243·2	R	TlH	
*4241·0	R, D	AlH	
4240	V, wr	ZnH	
*4239·1	R	SiN	
4239·0	V	CN	Le Blanc
4238·7	R	HeNe	
4236·2	V, DCD	CO$^+$	Baldet-J.
4230·9	O	P$_2^+$	
*4230	M	CH$_2$O	Cool fl.
4229·6	R, F	CrF	
4227·5	R, DCD	BO	α
4225·7	R	TlH	
4225·7	V, DCD	AlBr	
4225·3	R	CH$^+$	
*4216·0	V	CN	Violet
4214·6	R	CuI	
4211·0	R, DCD	CaI	
4210·2	R	CuBr	
4208·5	R	Po$_2$	
4207·5	R	AgO	
4205·1	R	CaO	
4205·1	R	TeO	
*4204·1	R	SiN	
4202·7	V, CD	AlBr	
4200·6	R	NiH	
4194·7	R	PtH	
4192·7	V, F, wv	TiCl	
4186	R	AgI	
4184·5	R	AlS	
4183·8	R	SnS	
4182	R	CuO	
4180·5	R, D, wv	BeI	
4179·9	R	AlSe	
4177·5	R, D	AgO	
4175·6	R	CdBr	
4167·2	R	SrO	
4166·9	V	NCO	
4156·0	R	Po$_2$	
4154·5	R	TaO	
4154·1	R, DCD	AgBr	
4142·3	R	SiH	

λ	App.	Mol.	System	λ	App.	Mol.	System
4138·7	R	ZrCl		*4045·6	M	TlCl	
4135·8	R	TeO		*4045·6	R	S_2	
4135·0	R	ZrBr		*4039·1	M	C_3	Comet-h.
4131·5	V	AlBr		4027·8	R, D	NO	β
4130·2	R	SbO		4025·3	R	CH	
4129·0	R, D	NS		*4021·8	R	CuCl	
4128·0	R, D	SiH		4017	V, D	HgH	
4126·9	R	AgO		4015·0	R, DCD	BO	α
*4126·6	R	SiN		4008·5	M	SiF_2	
4124·8	V	CO	Herzberg	4005·4	R	CuH	
*4121	R	CH_2O	Cool fl.	*3997·3	R, DCD	CO^+	Comet-t.
*4116·3	R	SiN		3994·7	R, D	SiN	
4113·7	R	TiO		3992·9	R	SiH^+	
4112·1	R	SnBr		3991·8	R	SbS	
4110·9	R	NiBr		3990·8	R	C_3	Comet-h.
4110·3	V	MgI		3984·8	R, D	COS^+	
*4108·3	R	TlCl		3983	R	OH^+	
4108	R	SrBr		3977·7	V, DCD	CO^+	Baldet-J.
4102·3	V	C_2	Desl.-d'A.	3972·8	R	ClO_2	
4102·3	R	HfO		3972·8	R	AuH	
4098·1	R	ZrBr		*3968·5	M	Ca^+	Scattered sunlight
4094·5	R, D	AgO		3965	M	NCS	
*4087·4	R	SiN		3960·9	R	CO_2^+	Fox-D.-B.
*4085·9	R	TlCl		3960·2	R	TlBr	
4084·3	R	CaO		3960·2	R	Po_2	
*4082·4	R, D	O_2^+	2nd Neg.	3955·9	R	CuCl	F
4079·9	R	SnO		3955·0	R	PbO	C
4076·1	V, F, wr	TiF		3954·6	M	SiF_2	
4070·7	R	SnBr		3954·4	R	CH^+	
4066·5	R	AlAu		3953·6	V, wr	CO^+	Baldet-J.
4062·6	V, CD	AlCl		*3952	R	CH_2O	Cool fl.
4061·4	R	NiCl		3950·4	R	CaBr	Violet
*4060	M	TlCl		3948·5	R	SnO	
*4059·4	V, CT	N_2	2nd Pos.	3945·2	R	TlBr	
4058·8	R	SnTe		3944·4	R	Po_2	
4053·9	R	SnH					
				3943·1	R	P_2	Violet
4053	R	SrBr		3942·4	R, D	NS	
*4050·7	R	SiN		3942·1	O	Si_2	Visible
*4049·8	R	C_3	Comet-h.	3938·9	R, wv	S_2	
4048·3	R	SO_2		3934·9	O	NaH	
4046·7	R	ZrCl					

λ	App.	Mol.	System	λ	App.	Mol.	System
*3933·7	M	Ca^+	Scattered sunlight	3850·7	R, T	BeS	
3933·2	R	ZrI	B	3846·2	R, D	COS^+	
3930·7	R, D	SeO	Main	3841·7	R	CrF	
3923·0	V	BaCl	U-V	3841·0	V, CT	BiF	B
3915·9	R	CaBr	Violet	3837·2	V	TiBr	
*3914·4	V	N_2^+	1st Neg.	3837·1	R, wv	S_2	
3914·3	R	BiCl	U-V	3834·8	M	MgOH	
3911·4	V	GaI		3832·9	R, CD	SiO^+	
*3909·5	R	CN	Violet	3830·5	R	O_2^+	2nd Neg.
3907·9	R	NaK	U-V	3830·2	V	BaI	D
3902·1	R, T	SiO		3828·1	R	P_2	Violet
3901·5	M	SiF_2		3828·0	R, DCD	BO	α
3896·7	R	PH^+		3827·1	R	NSF	
3893·2	V	BaBr		3825·9	R	ZnBr	Visible
3893·1	V	CO	Herzberg	3819·8	R	NCS	
3892·9	R	Au_2	B	3815·7	R	MgO	
3889·3	R	GaO		3812·6	L	AlH	
3889·2	R	N_2	Vegard-K.	3810·2	V	MgOH	
*3883·4	V	CN	Violet	3809·9	V	BaF	
3883·0	R	SO_2		3808·3	V	SrH	
3881·4	R	PbS	C	3807·3	R	ZnBr	Visible
3881·1	R	SrS		3807·3	R, D	SeO	Main
3880·5	V, D	MgBr		3805·3	R	MgO	
3879·9	O	NaH		*3804·9	V, CT	N_2	2nd Pos.
3877·8	R	PbO	C	3804·7	V	BaI	D
*3871·3	R, D	CH		3803·2	R	BN	Triplet
3871·1	R	AgCu		3798·6	R	CuH	
3870·5	R	CO_2^+	Fox-D.-B.	3798·2	R, D	MgO	
3864·6	O	BeS	Violet	3796·9	V, T	VCl	A
3864·1	V	MgBr		3793·9	R, wv	BeBr	
3863·5	R	SnO		*3788·5	R, D	NO	β
3863·2	R	BeS	Violet	3785·8	R	SnBr	Violet
3863·1	R	SiH		3778·9	V, D	MgCl	
3862·2	V	GaI		3778·5	V	GaO	
3857	V	TiCl		*3777·8	R, DCD	CO^+	Comet-t.
3857	M	NCS		3776·5	R	VCl	B
3854·0	R	SnTe	D	3775·2	R	ZrBr	B
3853·7	R, F, wv	PH^+		3774·0	R	SbO	B
3852·2	V	C_2	Desl.-d'A.	3770·0	R, D	NS	
3851·8	R	ZrI		3768·1	R, D	BH^+	

λ	App.	Mol.	System	λ	App.	Mol.	System
3766·1	R	MgO		3678	V	CrH	U-V
3765·8	R	BeBr		3675	V	FeOH	
3763·5	R, DCD	CaCl		3671·0	R	BeBr	
3760·3	R, CD	HBr⁺		3671	R	O₂	Schumann-R.
3758·5	R	SnCl		3668·7	R	SO₂	
3757·5	R	PbS	D	3662·4	R, D	BH	
3756·1	R, CD	AsSb		3661·6	R	CO₂⁺	Fox-D.-B.
3753·0	R	NSF		3660·6	R, DCD	BO	α
3751·7	R	PtH		3659·8	R, D	COS⁺	
3751·1	V, CT	BiF	B	3656·4	R	SF	
3750·8	R	SnBr	Violet	3651·5	R	AuH	
3750·6	V, T	VCl	D	3647·2	V, D	PbF	B
3747·2	R	TaO		3644·3	R	AgAl	
3744·4	R	InO⁺		3644·3	R	SbSe	I
3742·8	O	BeS	Violet	3641·0	R, DCD	SrF	C
3741·6	R	PbTe	D	3632·3	V, DCD	AlH⁺	
3739·8	R, wv	S₂		3630·0	R	SbSe	I
3735·1	V	BaI	E	3627·2	R, D	CH	
3734	R	O₂	Herzberg I	3625·9	R	Si₂	U-V
3729·7	V, DCD	CO⁺	Baldet-J.	3623·6	R	SbP	
3728·5	R	ScF		3621·2	R, CD	CS	
3728·3	R	SrS		3617·3	R	SiTe	
3720·7	V	MgO		3612·9	R	PbO	
3720·0	R	P₂	Violet	3612·4	R	CP	A
3719·6	V	MgOH		3611·9	R	NCS	
3719·3	V	BaBr		3607·3	V	C₂	Desl.-d'A.
3718·9	V, D	PbF	B	3603·7	R	O₂⁺	2nd Neg.
3716·8	V, F	MnCl		3599·2	R	BN	Triplet
3713·8	R	ZrCl	B	3597·0	R	SF	
3712·7	R	NCS		3594·2	V, CD	MgF	
3708·7	R, CD	AsSb		3593·2	R	PbSe	D
3708	R	ZrBr		*3590·4	V	CN	Violet
3707·0	V, D	N₂O⁺		3588·3	R	SbSe	I
3706·2	R, T	N₂	Goldstein-K.	3582·8	V, F	FeCl	II
3702·3	R	BaS	B	*3582·1	V	N₂⁺	1st Neg.
3693·3	R	PbTe	D	3580·7	R	SbO	C
3691·7	V	BaCl		*3576·9	V, CT	N₂	2nd Pos.
3690·2	R, D	SeO	Main	3576·7	R	BiO	
3684·0	R	CuH		*3572·4	R, D	NO	β
3682·1	R	ZrO	A	3572·2	R	CuH	

λ	App.	Mol.	System	λ	App.	Mol.	System
3571·0	R	SF		3487·8	R	SnCl	
3568·7	V, D	PbF	B	3485·7	R	PbO	D
3567·9	R, DCD, wv	BeCl		3484·5	R	SnO	
3567·4	O, wr	BO$^+$		3481·5	R, CD	PF	
3565	R	OH$^+$		3475·0	R	CaO	U-V
3564·0	R	CaO	U-V	3473·2	V	SrI	
3559·6	R	SbP		3465·9	V	SrCl	U-V
3559·2	R	NCS		3459·2	R	CP	A
3558·4	V, D	N$_2$O$^+$		3457·6	R	AgSe	
3557·8	R	PbSe	D	3457·6	R	SbSe	I
3556·5	R	Si$_2$	U-V	3453·2	R	SiBr$^+$	
3556·1	R	SiTe		3453·1	O	P$_2^+$	
3554·8	R	PbO		3452·4	V	SrI	
3553·1	V	AgO	U-V	3452·3	R	TlBr	
3549·4	M	Br$_2$		3450·4	V, CD	SrBr	U-V
3549·3	R	GaBr	A	3447·6	V, F	FeCl	I
3549·0	V	BaF		3442·1	R	AgSe	
3545·9	R	CO$_2^+$	Fox-D.-B.	3436	V	I$_2$	
3542·5	M	HNO$_2$		3435·8	R	BN	Triplet
3541·0	R	CdI		3435·8	R	ZnO	
*3535·0	R, D	SiN	Weak	3434·4	R	BiO	
3533·6	V	CaH	C	3432·0	R	Se$_2$	Main
3517·8	V, F	MnF	Near U-V	3431·6	R	AgTe	A
3517·7	R	SiBr$^+$		3431·6	R	SbSe	I
3517	R	O$_2$	Schumann-R.	3429·6	R	TlBr	
3516	R	AgH		3428·8	V	SrI	
3514·3	R	SiTe		3428·3	R	PbS	E
3514	R	CH$_3$O		3425	M	I$_2$	
3510·4	R	XeF$_2$		3419·6	R	PH	
3510·4	R	SbP		3418·8	V	AgF	A
3506·1	R, T	N$_2$	Goldstein-K.	3417·6	R, CD	HBr$^+$	
3505·1	O	P$_2^+$		3415·1	V	BH	
3503·8	R	SrO		3414·7	R	AgSe	
3502·3	V	XeF$_2$		3405·7	V	PO	β
3501·5	R	GeCl		3404·4	R	SiBr$^+$	
3501·3	R	AgTe	A	3403·8	R	Na$_2$	U-V
*3501·0	R	CHO		3403·3	O	P$_2^+$	
3499·0	V	InCl	B	3401·9	R	PbO	D
3497·4	R	CuH		3397·8	R	O$_2^+$	2nd Neg.
3489·1	R	Si$_2$	Near U-V	3395·3	V	BaF	

λ	App.	Mol.	System	λ	App.	Mol.	System
3393·9	R	PbS	E	3305·5	V	FeF	
3392·7	R	GeCl		*3298·2	R	CHO	
3392·0	R	SeCl		3298·1	V	CdBr	C
3388·3	R	SnO		3295	M	AlI	U-V
3386·6	V	BaCl		3294·5	R	CN_2	
3384·8	V	CdI		3291·1	V	ZnI	
3384·4	V	GaCl	A	3290·4	V	SnBr	
3381·0	R	BiO		3290·1	R	CN_2	
3379·8	V	PO	β	3286·3	V	CN_2	
3379·4	R, CD	AlH		3286·2	R	TiO	U-V
*3376·4	R, D	NO	β	3279·8	M	O_3	
*3376·3	R	CHO		3277·2	R, CD	AsO	
3374·8	R	SbO	C	3272·8	R	B_2	
*3371·3	V, CT	N_2	2nd Pos.	*3271·0	R	SO	
3370·0	R	CO_2^+	Fox-D.-B.	3270·5	V	PO	β
3370	R	O_2	Herzberg I	3262·4	R	SiSe	
3370	R	O_2	Schumann-R.	*3253·4	R	NH	
3369·2	R	Na_2	U-V	3252·1	R, F	AsH	
3363·5	R	CP	A	3251·7	R	AgI	
3363	M	SiF		*3250·8	R	TlCl	
*3360·1	M, L	NH		3249·7	M	O_3	
3346·8	V	GaCl	B	3248·9	R	Si_2	
3346	M	SiF		3247·7	R	BrO	
3343·4	R	SO_2		3246·9	R	CO_2^+	Fox-D.-B.
3342·6	V	ZnI		3246·2	V	PO	β
3336·6	M	Br_2		3245·9	R, DCD	HCl^+	
3332	R	OH^+		*3242·1	V, F	CO	5B
3330	R	AgH		*3240·1	R	NH	
3327·9	R	SeCl		3236·6	R	SH	
3325·2	R, DCD	AlS		3234·4	R	S_2O	
3325·2	R, T	N_2	Goldstein-K.	3232	R	O_2	Schumann-R.
3323·4	R	SnO		3230·3	R	AgI	
3320·0	R	BaO		3221·5	M	O_3	
3320	M	C_3H_3		*3220·9	R	TlCl	
3318·0	V	ZnI		3219·9	R	SeH	
3314·3	O	ZnH		3219·3	O	ZnH	
3314·0	R	Na_2	U-V	3217	M	C_3H_3	
3313·2	R	PbS	E	3215·0	R	AgF	B
3310·0	R	CaF		3211	R	O_2	Herzberg I
*3305·7	V, F	CO	3rd Pos.	3210·8	R	O_2^+	2nd Neg.

λ	App.	Mol.	System	λ	App.	Mol.	System
3209·8	V	SbF		3126·7	R	AgF	B
3208·9	R	AgI		3124·6	V	SnI	
3208·4	R	AsP		3122·1	V, DCD, wr	BF	
3207·9	R	BrO		3122·0	R	SO_2	
3203·0	V	CaBr	U-V	3120·5	R	Na_2	
3203·0	R	SiSe		3119	M	C_3H_3	
3199·5	R	AgBr		3117·5	V, CD	SrBr	
3199·3	R	FeCl		3117·1	V	SiI	
*3198·0	R, D	NO	β	*3114·8	R	CHO	
3197·8	V	SnCl		3113·7	R, DCD	TiO	
3189·5	M	CS_2		3110·7	M	Br_2	
*3186·0	R	CHO		3106·4	R	SiSe	
3181	V	CdCl		3105·2	V	SnI	
3180·2	R, D	C_2H_4O		3102·4	V	SrCl	U-V
3177·0	M	O_3		3100·1	V	BH	
3176·6	V	CdBr	D	3089·8	R	AsP	
3176·5	V	BaF		3082·3	V	SnI	
3172	V	CdCl		3082·2	V	SiI	
3170·6	R, CD	AsO		3079·8	R	AgIn	
3166·2	R	AgCl		3077·0	R	SbO	D
*3164·8	R	SO		3072	V	CdCl	
3164	R	CH_2O		3067	V	BaCl	
3163·5	R	AgF	B	*3064·1	R	SO	
3163·4	V	BaF		*3063·6	R, DCD, wv	OH	
3163·4	V	BaH		3063·3	R	CN^+	
3161·9	R, T	N_2	Goldstein-K.	3060·8	R	S_2	U-V
*3159·3	V, CT	N_2	2nd Pos.	3059·4	R	PtC	
3154·5	V	SnCl		3054·2	R	BaO	
3151·0	R	AgAl		3052·3	V, D	CBr	
3148·2	R	AsP		3047·3	V	SiBr	
3145·3	R	SiSe		3043·6	V	SnBr	
*3144	M	CH		3043·6	R	O_2^+	2nd Neg.
*3143·4	R	CH		3043	R	TiN	U-V
3141·2	V	SnF	A	3039·0	R	BeO	
3139	R	BS		3038·3	V	SiI	
3137·4	M	O_3		*3034·9	R, D	NO	β
3136·4	R	SbBr		3022·4	V	GeCl	
*3134·4	V, F	CO	3rd Pos.	3021·6	R, D	AlO	
3134·0	R	BeO		3017·5	R, D	PS	
3131·2	R	Na_2		3014·5	M	CBr	

λ	App.	Mol.	System	λ	App.	Mol.	System
3014	R	TiN	U-V	2927·9	V	SnF	A
3009·6	R, DCD	BeF		2926·1	R	SiS	
3008·8	V, wr	SiBr		2923·5	R	CdF	
*3008·8	V, DCD	NO	γ	2920·3	R	Pb_2	U-V
3007·5	R	SbBr		2915·6	V	SrCl	U-V
3004·8	R, T	N_2	Goldstein-K.	2915·0	V	PbI	U-V
3003·9	R	SbO	D	2912·1	R, CD	NSe	
3003·3	V	GeI	U-V	2911·0	V	PbBr	U-V
*3001·0	M	SO_2	Main	2910·0	R	ZrCl	A
2995·6	V	SiI		2904·3	R, CD	SbN	
2989·8	V, CD	GaF		2903·1	R	NSe	
2989·6	R, CD	NSe		2903·0	R	P_2	
2988·8	V, DCD	SiCl		*2900·4	M	Br_2	
2987	R	C_2	Fox-H.	2898·7	V	PbI	U-V
2985·7	R, CD	SbN		2894·4	V, D	SiF	β
*2977·4	V, F	CO	3rd Pos.	2893·8	V	PbBr	U-V
*2976·8	V, CT	N_2	2nd Pos.	*2887·7	M	SO_2	
2976·5	V	GeI	U-V	*2885·2	R, D	NO	β
2976·2	R	ZnCl		2885	R	NH^+	
2973·4	V	SnCl		2882·9	V, DCD	SiCl	
2969·8	R	P_2		2882·4	V	CaH	
*2961·2	M	SO_2	Main	2881·7	R, CD	GeO	
2957·5	R, D	AlO		2872·6	V	PbBr	U-V
2954	M	BBr		2871·4	R	ZrCl	A
2953·8	R	SiO		2869·5	V, D	AlO	
2953·6	R	P_2	Main	2867	V	Xe_2	
2952·6	R, D	PS		2866·2	R	PbO	E
2952·5	R	CN^+		2863·7	R	SiS	
2949·5	R	HgH		2861·5	R	ScF	
2948·1	V	BBr		2860·4	R	BCl	
2945·5	R	Na_2		2860·0	R, wv	S_2	
2943·9	V	CaBr	U-V	2858·0	V	PbCl	U-V
2942·6	R	ZnCl		2855	R	C_2	Fox-H.
2942·2	V, DCD	SiCl		2854·7	R	$CS_2{}^+$	
2936·0	V	GeCl		2853·2	R	P_2	
2935·8	V	SnCl		*2852·0	M	SO_2	
2935·7	R	N_2	Vegard-K.	2848·0	V	TlF	
2931·8	V	PbI	U-V	2847·5	R	S_2	U-V
2931	R	CH_2O		2846·6	R, D	PS	
2929·3	R	Pb_2	U-V	2844·5	R, CD	CSe	

λ	App.	Mol.	System	λ	App.	Mol.	System
2843·5	V	TlF		2757·1	R	P_2	Main
2842·9	R, T	CN		2754·5	V, D	PbF	B
2841·2	V, D	PbF	B	*2747·6	R, D	NO	β
2839	R	CH_2O		*2739·1	R	C_6H_6	
*2833·1	V, F	CO	3rd Pos.	2739·0	V	SbSe	II
2830·5	R, D	PS		*2731·6	R	Bi_2	U-V
2829·1	R, wv	S_2		2730·0	R	AsS^+	
2827·9	R	AsS^+		2727·4	R, F	BCl	
2827·1	V	N_2	Gaydon-H.	2724·0	V, F	BCl	
2824·8	R, D	PF^+		2723·0	V	N_3	
2823·7	R	O_2^+	2nd Neg.	*2722·2	V, DCD	NO	γ
2823·5	V, DCD	SiCl		2721	V	Xe_2	
2819·3	R, D	NS		*2720·7	R	Bi_2	U-V
2819·2	R	CS_2^+		2720·2	V	CaH	F
2818·3	R, CD	AlBr		2720·0	V, F	BCl	
2813·2	R	S_2	U-V	2719·5	V	N_3	
2813·2	R, D	PO		2713·8	R	BO	β
*2811·3	R, DCD, wv	OH		2713·7	R, CD	TlF	
2807·2	V, DCD	SiCl		2711·3	R	CF_2	
2807·0	R, CD	HgH		2708·2	V	SbSe	II
2805·2	V	BiSe		2703·2	R	ZnF	
2804·9	R, D	PS		2685·0	V	PbF	
2804·6	V, D	AlO		2682·9	R	OH	U-V
2804·2	R, CD	GeO		*2678·6	R	C_6H_6	
2801·2	V, D	PbF	B	2676·3	R	ZnF	
2799·0	R	AlS		2671·7	V	N_2	Gaydon-H.
2792·4	R	BCl		*2667·4	R	C_6H_6	
2792·0	R	Ag_2		2665·9	V	SbO	E
2788·8	R, CD	AlBr		*2665·3	V, F	CO	5B
2788·3	V	CCl		*2662·6	R, CD	CS	
2784·2	R, CD	AsN		2660·5	V, F	N_2	4th Pos.
2783·2	R	SiS		2660·1	R	CN^+	
2782·3	V	CCl		2656·8	R	Ag_2	
2778·2	R	AsS^+		2648·5	R, DCD	BS	γ
2777·6	V	CCl		2634·1	R, D	NS	
2773·3	R	AuH		2632·3	V	SbF	
2770·3	V	SbSe	II	2631·4	V, F	BF	
2767·2	R, CD	AlBr		2628·5	R	CF_2	
2761·8	R, D	NS		2619·8	R	S_2	U-V
2760·6	R	N_2	Vegard-K.	2612·8	R	AuH	

λ	App.	Mol.	System	λ	App.	Mol.	System
2612·2	V	PF		2520·6	R	PS^+	
*2610·2	R, CD	AlCl		2519·3	O	O_2	Herzberg
2605·0	R, CD	PN		2518·6	R	CF_2	
2603·9	R	MgBr	U-V	2518·2	R, CD	PN	
2603·8	R	N_2	Vegard-K.	2516·6	R	P_2	Main
*2602·6	R	C_6H_6		2515·3	V, D	MgH	
2599·6	R	OH	U-V	2513·2	V	SbO	E
*2596·9	V, F	CO	3A	2510·9	R, F	CO	Cameron
*2595·7	V, CD	NO	γ	*2507·3	R, CD	CS	
2595·5	V, CD	SnF	B	2506·7	R	NS	β
2595·0	R	CF_2		*2504·6	R	CO^+	1st Neg.
2590·7	V	BiO		2500·1	R, D	AlO	
*2589·0	R	C_6H_6		*2492·9	R	CO	4th Pos.
2585·4	R, T	SO	U-V	2491·1	R	NO_2	U-V
2584·8	R	NS	β	2490·7	R	Ag_2	
2581·0	R, D	O_2^+	2nd Neg.	2490·6	R	GaCl	C
2580·5	R	AgTe	B	2490·5	R	C_2^+	
*2577·7	R	CO^+	1st Neg.	*2489·9	V, F	CO	3A
2575·8	R	ScF		2488·9	O	O_2	Herzberg
*2575·6	R, CD	CS		2487·8	R	CF_2	
2575·3	R, F	CO	Cameron	2487·3	R	AlO	
2574·6	R	PS^+		*2486·8	R	SiO	
2570·9	V, DCD	AsO		2479·4	V, DCD	CF	
2569·1	R, DCD	BS		*2478·7	V, CD	NO	γ
2562·6	R	Ag_2		2477·9	V, DCD	PO	γ
2559·4	R	BeH^+		*2474·2	R	CO^+	1st Neg.
2558·5	R, T	SO	U-V	2464·2	V, DCD	PO	γ
2558·2	V, DCD	CF		*2463·2	R	CO	4th Pos.
*2557·3	V, D	NH		2463·2	O	O_2	Herzberg
2556·9	R, D	NS		2461·6	R	N_2	Vegard-K.
2555·0	V, DCD	PO	γ	2460·3	R, D	CO^+	Marchand-d'I.-J.
2551·4	R	BO	β	2456·9	R	P_2	Main
2550·7	V, F	N_2	4th Pos.	2456·4	V	TeSe	
2544·7	R	OH	U-V	2456·0	V	SiCl	
2543·4	R, CD	N_2^+	Janin-d'I.	2453·0	V, DCD	NS	γ
2540·4	V, DCD	PO	γ	2451·1	R, CD	PN	
2539·2	V, D	SiF	γ	2448·7	R	NS	β
2532·5	R, T	SO	U-V	2448·0	V, F	N_2	4th Pos.
*2530·2	V, L, wr	NH		2446·3	V	AsCl	
*2528·6	R	C_6H_6		2446·1	V	SbI	

λ	App.	Mol.	System	λ	App.	Mol.	System
*2445·8	R	CO^+	1st Neg.	2380·8	R, D	NS^+	
2440·1	V	BiO		2380·8	R	PbTe	F
2438·3	R	AlO		2380·4	R, DCD	AlO	
2437·9	R	CdH^+		2379·8	V	SnI	
2437·1	R	BO	β	2373·5	V	GeBr	
2436·4	V	SbSe	III	2371·7	R	SiS	
2435·3	R	AgTe	C	2371·6	V	SiBr	
*2433·9	R	CO	4th Pos.	*2370·2	V, DCD	NO	γ
2433·0	M	C_4H_2		2369·9	V	TeSe	
2432·5	R	PS^+		2367·8	R	CCl	
2428·0	O	AgH		2367·5	V	TeS	
2426·5	V	ZnH		2367·3	R	HgH^+	
2425·4	R	PbTe	F	2365·3	V	SbBr	
2425·2	R, D	PO		2361·8	R	NSF	
2424·7	R	AgSe		2361·7	R	PbTe	F
2422·6	V	SnCl		2355·6	R	AsO^+	
2421	R	NO_2	U-V	2351·4	V, F	N_2	4th Pos.
2418·6	V	PCl		2351	R	NO_2	U-V
2418·4	V	TeI		2349·6	R	SiS	
2417·4	V	SbSe	III	2349·2	R	AgH	
2413·9	V	TeBr		2345·4	V	TeCl	
*2413·8	R	SiO		*2344·3	R	SiO	
2410·9	R	AgSe		2340·9	R	CdH^+	
2409·2	R, F	CO	Cameron	2337·3	V, D	InF	
2405·6	V	BiTe		2331·3	R	BO	β
2404·1	V, DCD	CF		2326·5	R, D	NS^+	
2401·3	V	SiBr		2326·4	R, DCD	NS	
2400·3	R	AgSe		2326·2	V, CD	HgF	
2398·2	V	SnI		2324·8	V	SbTe	
2397·1	R, L	N_2	Gaydon-H.	2323·9	R	SO_2	
2396·3	V, DCD	PO	γ	2319·3	V	SbCl	
2396·0	V	SbSe	III	2317·2	V, CD	NS	γ
2394·5	R	SiS		*2316·4	V	AlF	
2393·8	V	BiTe		2313·8	R	AsO^+	
2392·8	V	GeBr		*2312·7	O	C_2	Mulliken
*2389·7	V, F	CO	3A	2311·6	V	BiTe	
2388·8	R, F	CO	Cameron	*2311·5	R	CO	4th Pos.
2383·6	V, DCD	NS	γ	2309·6	V	SbTe	
2383·5	R, D	CN		2308·2	V	TeS	
2383·5	V, DCD	PO	γ	2303·9	R	AgH	

λ	App.	Mol.	System	λ	App.	Mol.	System
2299·6	R	CO^+	1st Neg.	*2221·5	R	CO	4th Pos.
2298·9	R	SiO		2206·0	R, CD	C_2H_2	
2295·3	V	SeO		2205·0	R	CdCl	
2295·1	V	SbF		2202·9	R, CD	N_2^+	Janin-d'I.
2294·6	V	SbTe		2199·9	V	PbF	E
2283·7	V	SbSe	IV	2199·7	V, D	SiCl	
2281·5	R, L	N_2	Gaydon-H.	2198·9	V	N_2	5th Pos.
2278·2	V	SbTe		2198·0	R	TlF	
2276·8	R	AgH		*2196·8	R	CO	4th Pos.
2276·4	R, CD	N_2^+	Janin-d'I.	2192·9	R	PO^+	
2275·9	V	TeI		2190·0	R	PbSe	F
*2274·7	V, L	AlF		2189·8	R	CO^+	1st Neg.
2271·7	R	N_2	Lyman-B.-H.	2188·9	R	PbTe	
*2269·4	V, DCD	NO	γ	2188·9	R, CD	C_2H_2	
2268·6	V	SnCl		2186·6	R	SO_2	
2267·5	V	SiBr		2185·6	R, CD	C_2H_2	
2266·3	V, CD	SnF	D	*2181·8	V, DCD	NO	ε
2264·8	V	SbSe	IV	2180·6	O	CN^+	
2263·9	R	HgH^+		2179·5	R, CD	HgF	
*2261·7	R	CO	4th Pos.	2173·6	V	PbF	E
2258·5	R	Sb_2		2169·9	V	O_2^+	Hopfield
*2255·9	R	SiO		2164·1	V	SeBr	
2249·0	V	BiSe		2163·6	M	CH_3	
2246·4	V	SbSe	IV	2157·9	R, CT	CN	Carroll
2245·3	V	PF		2157·6	M	CH_3	
2242	M	C_2H_5		2157·2	R, CD	HgF	
2240·2	R	ZnH^+		2157·1	V, CD	SnF	D
2239·3	R	PbSe	F	*2154·9	V, DCD	NO	γ
2239·2	R	CuH		2151·9	R	ZnH^+	
2238·6	R	SO_2		2147·0	V, DCD	SiF	
2238·4	R, CD	N_2^+	Janin-d'I.	2142·9	V	C_2	Freymark
2236·7	R	AgI		*2141·3	V, DCD	NO	δ
2231·6	V, D	SiCl		2140·5	R, CD	C_2H_2	
2231·6	V	SbF		2138·4	R	Si_2	
*2228·6	R	AlH		2135·1	R, CD	HgF	
2228	M	C_2H_5		2129·0	R	PO^+	
*2226·8	V, DCD	NO	δ	2125·9	R	N_2	Lyman-B.-H.
2225·8	V	N_2	Kaplan I	2122·7	V	O_2^+	Hopfield
2224·2	R	AgCl		2115·0	V	SeBr	
2222·8	R	Sb_2		2115·0	R	Sb_2	

λ	App.	Mol.	System	λ	App.	Mol.	System
2110·2	R, D	GaF		*2060·9	V, DCD	NO	δ
2108·7	V, DCD	SiF		2059·0	R	N_2^+	2nd Neg.
2108·1	R	P_2	Main	2053·1	R, D	PO	
2105·9	R, D	PO		*2046·3	R	CO	4th Pos.
*2099·8	V, DCD	NO	ε	2045·7	R, D	NH_3	
2091·7	R	ZnH^+		2033·6	V	N_2	5th Pos.
*2089·9	R	CO	4th Pos.	2033	V	AlH	
2089·1	V	SeO		2028·2	O	AlH	
2075·9	V	O_2^+	Hopfield	*2022·3	V, DCD	NO	ε
2075·7	R, DCD	CF		2016·7	R	BO	γ
2075·6	R	ZnCl		2016·6	V	BeF	
*2067·6	R	CO	4th Pos.	2011·0	R, CD	BF	
2065·5	V, DCD	SiF		2010·9	R, D	NH_3	
2063·0	R, D	PO					
2061·9	R, F	CO	Cameron				

*This band occurs readily as an impurity. For other abbreviations, see page 3.

Individual Band Systems

In this section detailed lists of bands are given for each system separately and are arranged in alphabetical order of the chemical symbols of the molecules concerned. The elements in the molecule are taken in the order which appears most frequently in the chemical literature, thus usually with the more electropositive (i.e. more metallic) element first as in CaCl, AgO, MgH, while OH is used in preference to HO and HCl rather than ClH, but H_2O not OH_2. Most organic molecules and radicals are grouped under C with other elements in the order of H, N and O atoms.

For each molecule a few general introductory remarks are made at the beginning if the spectrum possesses several systems or is particularly important, and then the various band systems are dealt with separately, the more important systems usually being given first, or when systems are of comparable ease of being observed the longer wavelength ones come first. The treatment for the systems varies somewhat according to the particular case, but the electronic transition and region covered are often given in the system heading and then the conditions of occurrence and the appearance are described before the tables of wavelengths, intensities and vibrational quantum numbers.

Transitions. The type of electronic transition, where fairly certainly known, is stated for all diatomic molecules, as this gives additional information about the band structure and thus the appearance, especially when the system is observed under large dispersion. The ground state of the molecule is normally denoted X, thus $X^2\Sigma$; systems involving the ground state are likely to occur in absorption (except in the case of ionized molecules) whereas transitions between two excited states are much less readily observed in this way.

A detailed account of the rotational structure and appearance of the various electronic transitions is beyond the scope of this work, but the following brief indications may be helpful; the appearance depends greatly on the relative values of the molecular constants, especially of B' and B'', in the two electronic states and upon the dispersion with which the band is studied, and so these brief generalisations serve only as a guide and some systems may depart from the typical forms.

$^1\Sigma-^1\Sigma$ and $^1\Pi-^1\Pi$. Single P and R branches. Single headed.

$^1\Sigma-^1\Pi$ and $^1\Pi-^1\Sigma$. Single P, Q and R branches. Usually double-headed.

$^2\Sigma-^2\Sigma$. Double P and R branches with extremely weak satellite branches. Usually close double-headed.

$^2\Pi-^2\Pi$, $^2\Delta-^2\Delta$. Two P and two R branches and short weak Q branches and satellite branches. Molecules with small spin (multiplet) splitting, in Hund's case b coupling, show double-headed bands. Molecules with large spin splitting, in Hund's case a, show two separate single-headed bands.

$^2\Sigma-^2\Pi$, $^2\Pi-^2\Sigma$, $^2\Pi-^2\Delta$ and $^2\Delta-^2\Pi$. Double P, Q and R branches and weak satellite branches. Often double double-headed. In Hund's case b the doubling is small and the satellite branches of O or S form are quite weak. In Hund's case a the main doubling is large so that the appearance is of two separate double-headed bands with the ^OP and ^SR branches nearly as strong as the P, Q, and R branches, one of these (i.e. ^OP or ^SR) forming a definite head in front of the main heads.

$^3\Sigma-^3\Sigma$. Triple P and R branches. Weak Q and satellite branches. Often close triple-headed.

$^3\Sigma-^3\Pi$, $^3\Pi-^3\Delta$, etc. Triple P, Q and R branches and often fairly strong satellite branches so that bands have complex structure with multiple heads.

Systems of higher multiplicity (quartet, quintet, etc.) and higher Λ (Φ, Γ, etc.) are less common, but tend to show very complex structure. Some molecules of manganese show septet multiplicity and may show seven separate sub-bands for each main vibrational transition.

In addition to the Greek symbols, indicating the value of Λ, and the superscript values of the multiplicity, Π and Δ states have an added subscript, when in case a coupling, indicating the value of the quantum number Ω. Transitions with larger Ω have a greater number of 'missing lines' near the origin.

Electronic energy levels are named by an ordinary letter, placed before the symbols indicating the type of electronic level. As already noted the ground state is denoted by a capital X and higher electronic states of the same multiplicity are, ideally, denoted by the capital letters A, B, C, etc. in order. In practice higher states are sometimes discovered and named A, B, etc. before lower energy states are discovered; these lower states may then be denoted A′, B′, etc., or in some cases the alphabetical sequence may be broken; sometimes when this happens later authors will try to straighten out the nomenclature, but usually this just makes the confusion worse. When molecules show levels of two multiplicities then capital letters are again used for terms with the same multiplicity as the ground state and small italics letters a, b, c, etc. are used for the other multi-plicity; again departures from this rule are not uncommon (e.g. N_2).

Occurrence. Mention is made of various sources in which the system has been obtained. This serves to give a general idea of the conditions which are favourable to the production of the system, but does not of course imply that the bands cannot be obtained, perhaps weakly, in other sources. Arc sources usually mean low-voltage arcs (100-250 V) running in air at atmospheric pressure. Where a band system is of frequent occurrence as an impurity in other spectra this is indicated.

Appearance. The direction in which the bands are shaded is recorded, e.g. degraded to shorter wavelengths indicates that the bands have sharp heads on the red side and are shaded to the violet. A brief description of the system is given, calling attention to any characteristics which may help to identify it, such as obvious clear sequences, double-headed bands of line-like heads which may resemble atomic lines.

Wavelengths. These are given in angstroms (1 Å = 10^{-10} metre). The S.I. unit is the nanometre (1 nm = 10 Å) but the use of angstroms is widespread and they are a conveniently sized unit for spectra in the visible and ultra-violet regions. All values are rounded to 0·1 Å, as location of band heads, unlike atomic lines, rarely justifies use of more significant figures.

For analyses of band spectra into energy levels the wavenumber (number of waves in a centimetre, cm^{-1}) **in vacuum** is used, and some papers quote only these values. In such cases we have converted to the usual wavelengths **in air**; for values of the refractive index of air and method of conversion see the Appendix. There is of course a case for listing both wave-lengths and wave-numbers, but this would make the book much larger and more expensive to set.

Many elements possess isotopes of comparable abundances and when this is so molecular compounds of these elements show spectra which are split by this isotope effect. Bands near the system origin (e.g. the (0, 0) band) show only a very small isotope splitting, but the splitting increases, roughly linearly, with distance from the system origin so that band heads show an increasing doubling (or more complex splitting if there are more than two isotopes). Often, under small dispersion the effect of isotope splitting is to make band heads appear diffuse. The effects are usually not very important and where measurements of heads of separate isotopic components are available we have, without comment, listed the head of the most abundant isotope although sometimes the head of a less abundant isotope may be more prominent and has been listed. In a few cases, for compounds of boron and chlorine, we have sometimes listed individual isotope heads. For the hydrogen isotope, deuterium, a separate table is given.

Intensities. These are usually listed on a 'visual' (i.e. non-quantitative) scale of 0 to 10, the strongest band of the system being denoted 10 and the weakest just visible band as 0. This system is really an order of precedence of strength of the bands and is somewhat similar to the use of stellar magnitudes, which, however, run the other way from 1 for the brightest to 6 for the weakest visible star. Although not quantitative and often distorted by variation in the sensitivity of the photographic plate at different wavelengths, these intensity estimates are most valuable, indeed almost essential for identification. In a few cases we have retained author's use of other scales, e.g. 0 to 5 or 0 to 6. For some molecules several systems may be placed on one scale, adopting 10 for the strongest band of the group of systems.

In many cases the intensity estimates have been made either from our own spectrograms or from spectrograms published in the literature.

References. These are given to those papers which are most useful for the purpose of identification. Those which contain useful photographs of the spectrum are indicated by a dagger, †, following the date. Often an early paper gives a far better general description of a system than later papers.

Abbreviations are given in commonly accepted forms except for the following specially shortened ones:

P.R.   *Physical Review.*
P.R.S.   *Proceedings of the Royal Society* (*London*), *Series A.*
Z.P.   *Zeitschrift für Physik.*

Ag_2

Occurrence. Emission in a discharge tube, and emission and absorption in a King furnace.

Appearance. Five systems are known, all degraded to the red and consisting of numerous close bands. The sequences are usually obvious.

References. B. Kleman and S. Lindkvist, *Ark. Fys.*, **9**, 385 (1955)†.
J. Ruamps, *C. R. Acad. Sci.* (*Paris*), **238**, 1489 (1952); and *Ann. Phys.* (*Paris*), **4**, 1111 (1959)†.
Choong Shin-Piaw, Wang Loong-Seng and Lin Yoke-Seng, *Nature*, **209**, 1300 (1966).

VISIBLE SYSTEM, A

The following are among the strongest bands, with our estimates of intensity from published photographs:

λ	I	v', v''	λ	I	v', v''	λ	I	v', v''
4461·9	7	0, 3	4387·3	9	0, 1	4347·0	9	0, 0
4424·2	8	0, 2	4365·3	8	2, 2	4322·1	10	1, 0
4401·2	7	2, 3	4357·9	8	1, 1	4293·8	9	2, 0
						4265·9	8	3, 0

ULTRA-VIOLET SYSTEMS

In all cases, only the outstanding sequence heads are given.

System B.   λλ 2821·9 (0, 2), 2806·9 (0, 1), 2792·0 (0, 0), 2780·4 (1, 0), 2768·2 (2, 0).
System C.   λλ 2656·8 (0, 0), 2645·1 (1, 0).
System D.   λ 2562·6 (0, 0).
System E.   λλ 2502·8 (0, 1), 2490·7 (0, 0), 2482·0 (1, 0).

Choong Shin-Piaw *et al.* have given a different analysis for system B and have also found bands associated with the Ag lines at 3280·7 and 3382·9 Å.

AgAl

Occurrence. In absorption.
Reference. R. M. Clements and R. F. Barrow, *Trans. Faraday Soc.*, **64**, 2893 (1968).

3750-3590 Å SYSTEM, $B^1\Sigma - X^1\Sigma$

Degraded to the red. 3678·4 (0, 1), 3644·3 (0, 0), 3617·9 (1, 0).

3150 Å SYSTEM, $C^1\Pi - X^1\Sigma$

Degraded to the red. 3151·0 (0, 0).

AgBi

Occurrence. Emission in a King furnace.
Appearance. Three systems, all degraded to the red. Long sequences.
Reference. J. Lochet, *J. Phys.*, B, **7**, 505 (1974)†.

INDIGO SYSTEM, 4950—4700 Å

4853 (0, 1), 4818 (0, 0), 4785 (1, 0).

RED SYSTEM, 6400—6000 Å

6147 (0, 1) 6098 (0, 0).

INFRA-RED SYSTEM, 8300—8000 Å

The (0, 0) sequence has heads, R 8182, Q* 8189 and Q 8210 Å.

AgBr

Occurrence. In absorption and fluorescence.
Appearance. Single-headed bands degraded to the red.
Transition. Probably $^1\Sigma - ^1\Sigma$, ground state.
References. J. Franck and H. Kuhn, *Z.P.*, **44**, 607 (1927).
B. A. Brice, *P.R.*, **38**, 658 (1931)†.
R. F. Barrow and M. F. R. Mulcahy, *Nature*, **162**, 336 (1948).

The bands observed in absorption by Brice are numerous and extend from 3500 Å to 3165 Å. Those observed by Franck and Kuhn are in the region λλ3393—3182. No intensities are given, but the following bands are probably among those most easily observed:

λ	v', v''	λ	v', v''	λ	v', v''
3310·8	1, 5	3258·3	1, 3	3225·2	0, 1
3302·8	0, 4	3250·8	0, 2	3199·5	0, 0
3284·2	1, 4	3232·7	1, 2	3182·1	1, 0
3276·7	0, 3				

Barrow and Mulcahy have reported a system of bands, degraded to the red, λλ2475—2150, in absorption.

AgCl

STRONG SYSTEM, λλ3379–3114

Occurrence. In discharge tubes (including high-frequency discharge) containing silver chloride, in fluorescence and in absorption.

Appearance. Degraded to the red. Marked sequences.

Reference. B. A. Brice, *P.R.*, **35**, 960 (1930)†.

Strong bands as given by Brice:

λ	I	v', v''	λ	I	v', v''	λ	I	v', v''
3251·4	4	2, 4	3200·8	9	0, 1	3157·5	7	3, 2
3243·2	5	1, 3	3181·9	4	2, 2	3147·9	8	2, 1
3236·0	3	0, 2	3173·3	5	1, 1	3139·7	9	1, 0
3216·3	8	2, 3	3166·2	10	0, 0	3124·2	3	2, 0
3208·0	9	1, 2						

WEAKER SYSTEMS

Occurrence. In absorption.

Appearance. Degraded to red.

Reference. F. A. Jenkins and G. D. Rochester, *P.R.*, **52**, 1141 (1937).

λλ2400–2200. No intensities are given; the following may be the strong bands:

λ	v', v''	λ	v', v''	λ	v', v''
2390·3	0, 5	2300·6	1, 1	2238·4	4, 0
2365·9	0, 4	2285·3	2, 1	2224·2	5, 0
2352·9	0, 3	2270·4	3, 1	2210·2	6, 0
2318·7	1, 2	2252·9	3, 0	2196·5	7, 0
2303·2	2, 2				

λλ2200–2100. No intensities given. The following may be the strong bands:

λ	v', v''
2150·3	0, 8
2135·4	0, 7
2120·7	0, 6

AgCu

Reference. K. C. Joshi and K. Majumdar, *Proc. Phys. Soc.*, **78**, 197 (1961).

Occurrence. Emission in King furnace.

Forty red-degraded bands, λλ4094–3792. Strongest sequence heads 3844·6 (1, 0), 3871·1 (0, 0), 3906·2 (0, 1).

AgF

Occurrence. Absorption in King furnace.

Reference. R. F. Barrow and R. M. Clements, *P.R.S.*, **322**, 243 (1971)†.

SYSTEM A 0^+–$X^1\Sigma$

Well developed sequences of bands which are barely degraded either way. Strong features:

AgF (*contd.*)

λ	v', v''	Character	λ	v', v''	Character
3540·9	0, 2	P head, deg. v.	3418·8	0, 0	P head, deg. v.
3479·0	0, 1	P head, deg. v.	3365·4	1, 0	origin
3423·9	1, 1	origin	3361·2	1, 0	R head, deg. r.

SYSTEM B $0^+-X^1\Sigma$

Degraded to the red with marked sequences. 3267·5 (0, 2), 3215·0 (0, 1), 3163·5 (0, 0), 3126·7 (1, 0).

There is also a strong continuum centred at 3030 Å and a weak system below 2600 Å.

AgGa

Reference.   M. Biron, *C. R. Acad. Sci. (Paris)*, B, **264**, 1097 (1967)†.

Absorption in a King furnace. Bands are degraded to the red with marked sequences. 3042·2 (0, 1), 3025·2 (0, 0), 3011·5 (1, 0).

AgH

References.   E. Bengtsson and E. Olsson, *Z.P.*, **72**, 163 (1931).
E. Bengtsson, *Nova. Acta Reg. Soc. Sci. Uppsala* (IV), **8**, No. 4 (1932).

MAIN SYSTEM, 3330 Å, $A^1\Sigma^+-X^1\Sigma^+$

Occurs in discharges where silver vapour is mixed with hydrogen. The bands given below were obtained by Bengtsson and Olsson in the spectrum of an arc in hydrogen between electrodes of silver aluminium alloy.

Bands with single R and P branches degraded to the red.

λ	v', v''	λ	v', v''	λ	v', v''
3179	1, 0	3583	2, 3	4108	4, 7
3220	2, 1	3637	3, 4	4190	5, 8
3275	3, 2	3710	4, 5	4273	6, 9
3330	0, 0	3740	1, 3	4328	4, 8
3357	1, 1	3781	2, 4	4397	5, 9
3396	2, 2	3833	3, 5	4472	6, 10
3451	3, 3	3905	4, 6	4536	7, 11
3516	0, 1	3990	5, 7	4669	6, 11
3546	1, 2	4039	3, 6		

FAR ULTRA-VIOLET SYSTEMS

Occurrence.   Absorption in King furnace. Not found in emission.
Reference.   U. Ringström and N. Åslund, *Ark. Fys.*, **32**, 19 (1966)†.

Five overlapping systems, with heads not usually prominent. Intensities on a scale of 0 to 5.

AgH (*contd.*)

System	v', v''	λ_{origin}	λ_{head}	*deg.*	*I*
$C^1\Pi - X^1\Sigma$	2, 2	2448·4			2
	1, 1	2438·2	Q 2438·3	R	3
	0, 0	2428·0	{Q 2428·0	V	4
			Q 2425·6	R	
	1, 0	2341·6	Q 2341·8	R	0
$B^1\Sigma^+ - X^1\Sigma^+$	0, 2	2443·7			2
	1, 2	2379·8	2378·9	R	2
	0, 1	2350·4	2349·2	R	5
	0, 0	2260·5	2259·5	R	4
$b^3\Pi_1 - X^1\Sigma^+$	0, 1	2333·7	2332·6	R	2
	0, 0	2245·0	2244·0	R	3
$D^1\Pi - X^1\Sigma^+$	1, 3	2361·4	2360·3	R	4
	0, 2	2322·4	2320·7	R	3
	1, 2	2277·7	2276·8	R	5
	0, 1	2238·0	2236·6	R	4
$c_1{}^3\Pi_1 - X^1\Sigma^+$	0, 3	2390·4			2
	0, 2	2304·7	2303·9	R	5
	0, 1	2221·5			3
$c_0{}^3\Pi_0 - X^1\Sigma^+$	0, 3	2374·0	2372·3	R	1
	0, 2	2289·4	2288·0	R	3

AgI

References. J. Franck and H. Kuhn, *Z.P.*, **43**, 164 (1927).
B. A. Brice, *P.R.*, **38**, 658 (1931)†.
C. R. Sastry and K. R. Rao, *Indian J. Phys.*, **19**, 136 (1945).
R. F. Barrow and M. F. R. Mulcahy, *Proc. Phys. Soc.*, **61**, 99 (1948).
N. Metropolis, *P.R.*, **55**, 636 (1939).
R. F. Barrow and M. F. R. Mulcahy, *Nature*, **162**, 336 (1948).
N. Metropolis and H. Beutler, *P.R.*, **55**, 1113 (1939).

VISIBLE SYSTEM, 4600–4170 Å

Metropolis and Beutler mention 45 bands in absorption; their formula indicates (0, 0) at 4186 Å and that they are degraded to the red.

NEAR ULTRA-VIOLET SYSTEM, 3556–3160 Å

Occurrence. In absorption, in emission from high-frequency discharge and in fluorescence.
Appearance. Degraded to the red. Long sequences. Main sequence heads 3294·5 (0, 4), 3273·4 (0, 3), 3251·7 (0, 2), 3230·3 (0, 1), 3208·9 (0, 0), 3196·8 (1, 0), 3185·4 (2, 0).

FAR ULTRA-VIOLET SYSTEM, 2350–2175 Å

Occurrence. In absorption 700–900°C. At higher temperature the system is overlaid by continuum.

AgI (*contd.*)

Appearance. Degraded to the red, with well formed sequences and a long $v' = 0$ progression. 39 sharp heads have been recorded. Sequence heads 2309·6 (0, 7), 2298·9 (0, 6), 2288·4 (0, 5), 2277·9 (0, 4), 2267·5 (0, 3), 2257·2 (0, 2), 2246·9 (0, 1), 2236·7 (0, 0), 2229·2 (1, 0), 2221·9 (2, 0).

Metropolis also records bands further to the ultra-violet, $\lambda\lambda$2211·2, 2209·8, 2203·7, 2201·9, 2200·1, 2194·2, 2192·4, 2190·3.

AgIn

Reference.   M. Biron, *C. R. Acad. Sci. (Paris)*, B, **265**, 1026 (1967).
Two systems are reported in absorption.
System A, 3170–3060, degraded to the red, has sequence heads 3094·6 (0, 1), 3079·8 (0, 0) and 3067·2 (1, 0).
System B, 3025–2960; no details given.

AgO

Occurrence. Arc between silver poles in an atmosphere of oxygen.
Reference.   U. Uhler, *Ark. Fys.*, 7, 125 (1953)†.
There are strong systems in the blue and near ultra-violet. There is also mention of a very weak system in the red.

BLUE SYSTEM, $A^2\Pi - X^2\Pi$

Appearance. Degraded to the red.
Strongest heads, with own estimates of intensity.

λ	I	v', v''		λ	I	v', v''		λ	I	v', v''	
4484·6	3	1, 5	R_2	4262·8	4	0, 2	R_2	4177·5	8	0, 1	R_2
4337·9	3	1, 3	R_1	4250·7	3	1, 2	R_1	4126·9	10	0, 0	R_1
4305·4	3	1, 3	R_2	4219·4	4	1, 2	R_2	4094·5	10	0, 0	R_2
4294·0	3	0, 2	R_1	4207·5	10	0, 1	R_1	4083·5	6	1, 0	R_1

ULTRA-VIOLET SYSTEM, $B^2\Pi - X^2\Pi$

Appearance. Degraded to shorter wavelengths. Marked sequences.
The following are the strongest heads; own estimates of intensity.

λ	I	v', v''		λ	I	v', v''		λ	I	v', v''	
3620·8	5	0, 1	P_1	3558·4	10	0, 0	P_1	3490·1	5	2, 1	P_1
3615·3	5	0, 1	P_2	3553·1	10	0, 0	P_2	3487·7	6	1, 0	P_2
3609·1	6	1, 2	P_2	3493·4	4	1, 0	P_1	3484·2	6	2, 1	P_2

AgSb

Reference.   Y. Lefebvre and J. Lochet, *C. R. Acad. Sci. (Paris)*, B, **275**, 85 (1972)†.
Two red-degraded systems are observed in emission in a King furnace. System A, 6350–6100 Å, has sequence heads 6200·8 (0, 1), and 6122·0 (0, 0). System B, 6450–6350, has a (0, 0) sequence at 6383·8, but the (0, 0) band also has a violet-degraded head of the piled up Q branch at 6386·3.

AgSe

Reference. R. C. Maheshwari, *Proc. Phys. Soc.*, **81**, 514 (1963).

Two systems of red-degraded bands in absorption. The first has nine bands 3520–3400 Å, strongest 3457·6 (0, 2), 3442·1 (1, 2), 3414·7 (1, 1). The second has bands 2424·7 (0, 1), 2410·9 (0, 0), 2400·3 (1, 0).

AgTe

Reference. R. C. Maheshwari, *Proc. Phys. Soc.*, **81**, 514 (1963).

Three red-degraded systems in absorption at $1900°$ C. System A, 3650–3400, has 54 bands, sequence heads 3501·3 (0, 3), 3478·1 (0, 2), 3454·7 (0, 1), 3431·6 (0, 0). System B, 2640–2530, has 31 bands with sequence heads 2593·5 (0, 1), 2580·5 (0, 0), 2570·8 (1, 0). System C, 2475–2360 has 22 bands, sequences 2458·5 (0, 2), 2446·8 (0, 1), 2435·3 (0, 0), 2427·4 (1, 0).

Al_2

References. D. S. Ginter, M. L. Ginter and K. K. Innes, *Astrophys J.*, **139**, 365 (1964)†.
P. B. Zeeman, *Canad. J. Phys.*, **32**, 9 (1954)†.

Occurrence. Emission in King furnace.

An extensive system of red-degraded bands 6600–5650 due to a $^3\Sigma - ^3\Sigma$ transition. The following may be outstanding heads: 6042·0 (0, 2), 5942·2 (1, 2), 5919·8 (0, 1), 5800·9 (0, 0), 5732·2 (2, 1), 5709·1 (1, 0).

AlBr

VIOLET SYSTEM, $a^3\Pi - X^1\Sigma$

Occurrence. Discharge tube containing AlBr vapour and He.

Appearance. Degraded to violet. Some sequences of single-headed bands ($^3\Pi_0 - ^1\Sigma$), some of double-headed bands ($^3\Pi_1 - ^1\Sigma$).

Reference. D. Sharma, *Astrophys., J.*, **113**, 219 (1951)†.

Outstanding heads. Own estimates of intensity from published photograph.

λ	I	v', v''		λ	I	v', v''		λ	I	v', v''	
4363·4	3	P_1	0, 2	4263·3	7	P_2	1, 2	4154·1	8	P_1	1, 0
4338·4	3	P_2	0, 2	4225·7	10	P_1	0, 0	4149·1	8	P_1	2, 1
4331·6	4	P_2	1, 3	4219·8	9	P_1	1, 1	4144·3	7	P_1	3, 2
4293·6	6	P_1	0, 1	4202·3	10	P_2	0, 0	4131·5	5	P_2	1, 0
4287·1	5	P_1	1, 2	4201·4	10	Q_2	0, 0	4130·7	5	Q_2	1, 0
4269·4	8	P_2	0, 1	4196·7	9	P_2	1, 1	4085·5	2	P_1	2, 0
4268·6	8	Q_2	0, 1	4195·8	9	Q_2	1, 1				

ULTRA-VIOLET SYSTEM, $A^1\Pi - X^1\Sigma$

Occurrence. High-frequency discharge through aluminium tribromide vapour.

Appearance. Degraded to the red. Close double-headed bands, separation between the R and Q heads being about 0·2 Å.

References. H. G. Howell, *P.R.S.*, **148**, 696 (1935)†.
C. G. Jennergren, *Nature*, **161**, 315 (1948).

AlBr (*contd.*)

The following are the R heads of the strong bands:

λ	I	v', v''	λ	I	v', v''
2855·3	7	1, 3	2804·9	8	2, 2
2848·0	6	0, 2	2796·3	9	1, 1
2834·1	8	2, 3	2788·8	10	0, 0
2825·6	8	1, 2	2775·8	8	2, 1
2818·3	9	0, 1	2767·2	9	1, 0

AlC

Bands initially attributed to AlC are now assigned to Al_2.

AlCl

ULTRA-VIOLET SYSTEM, $A^1\Pi - X^1\Sigma^+$

Occurrence. Uncondensed discharge or microwave discharge through $AlCl_3$ vapour. This system is frequently observed as an impurity in discharge tubes with aluminium electrodes.

Appearance. As may be seen from Plate 1, this system is of rather complex structure. The (2, 0), (1, 0) and (0, 0) sequences are degraded to the red, and the bands are close double-headed, with additional weak heads due to the less abundant isotope of chlorine. Some of the bands of the (0, 2) and (0, 1) sequences are degraded to shorter wavelengths. The heads of the (0, 0) band at 2610 and 2614 Å are usually outstanding if the system is only weakly present, *e.g.*, as an impurity.

Transition. $^1\Pi - {}^1\Sigma$, probably ground state.

References.   B. N. Bhaduri and A. Fowler, *P.R.S.*, **145**, 321 (1934)†.
W. Holst, *Z.P.*, **93**, 55 (1934–35).

The following measurements of the outstanding heads are from Bhaduri and Fowler. Intensities are on a scale of 8. Weaker isotope heads are omitted, and for the close double-headed bands degraded to the red, only the R heads are given, the Q heads being usually less than 1 Å to the red. The letters R or V after the wavelength indicate the direction of degradation of the band, while the nature of the head (R or Q) is indicated before the vibrational quantum numbers.

λ	I	v', v''	λ	I	v', v''	λ	I	v', v''
2708·9 R	2	R 7, 9	2638·1 R	3	R 6, 6	2614·4 R	8	Q 0, 0
2702·3 R	3	R 6, 8	2632·8 R	3	Q 5, 5	2610·2 R	6	R 0, 0
2696·4 R	3	R 5, 7	2632·2 R	3	R 5, 5	2606·7 R	2	R 6, 5
2692·8 R	5	Q 4, 6	2627·8 R	4	Q 4, 4	2600·7 R	3	R 5, 4
2685·7 R	6	Q 2, 4	2627·0 R	3	R 4, 4	2595·4 R	2	R 4, 3
2683·1 V	6	Q 1, 3	2623·5 R	5	Q 3, 3	2590·8 R	2	R 3, 2
2681·1 V	4	Q 0, 2	2622·4 R	4	R 3, 3	2586·7 R	2	R 2, 1
2649·7 V	4	Q 1, 2	2620·0 R	4	Q 2, 2	2564·3 R	1	R 4, 2
2647·5 V	6	Q 0, 1	2618·2 R	3	R 2, 2	2559·6 R	1	R 3, 1
2644·9 R	2	R 7, 7	2617·0 R	4	Q 1, 1	2555·5 R	1	R 2, 0

VIOLET SYSTEM, $a^3\Pi - X^1\Sigma^+$

Occurrence. Discharge through $AlCl_3$ and He.

Appearance. Degraded to shorter wavelengths. Sequences of single-headed bands ($^3\Pi_0 - {}^1\Sigma$), and stronger sequences of double-headed bands ($^3\Pi_1 - {}^1\Sigma$).

AlCl (*contd.*)

Reference. D. Sharma, *Astrophys. J.*, **113**, 210 (1951)†.

Outstanding heads.

λ	I	v', v''	λ	I	v', v''	λ	I	v', v''
4154·0	1	P_1 0, 1	4066·5	5	P_1 1, 1	3989·4	2	P_1 1, 0
4142·8	4	P_2 0, 1	4062·6	10	P_2 0, 0	3978·7	3	P_2 1, 0
4141·8	2	Q_2 0, 1	4061·4	8	Q_2 0, 0	3977·4	3	$\left\{\begin{array}{l} Q_2 \ \ 1, 0 \\ P_1 \ \ 3, 2 \end{array}\right.$
4134·8	3	P_2 1, 2	4055·6	7	P_2 1, 1			
4073·6	8	P_1 0, 0	4054·3	5	Q_2 1, 1	3972·8	3	P_2 2, 1

GREEN SYSTEM, $b^3\Sigma - a^3\Pi$

Occurrence. Discharge through $AlCl_3$ and He.

Appearance. Degraded to the red. A single progression of bands, each consisting of three widely separated sub-bands, each of these having apparently double heads, but in reality a more complex structure.

Reference. D. Sharma, *Astrophys., J.*, **113**, 210 (1951)†.

S, R and Q heads with isotope splitting give a complex structure to the bands; the following are a few of the outstanding features. In denoting the heads, subscripts $_{1, 2, 3}$, indicate transitions to $^3\Pi_0$, $^3\Pi_1$ and $^3\Pi_2$ sub-states respectively.

\multicolumn{3}{c}{*0, 2 band*}						

λ	I	head	λ	I	head	λ	I	head
5612·5	2	Q_3	5454·9	3	Q_3	5304·3	5	Q_3
5611·8	5	R_3	5454·1	7	R_3	5303·7	8	R_3
5591·8	5	R_2	5451·9	6	S_3	5301·8	4	S_3
5589·2	4	S_2	5435·2	7	R_2	5285·8	9	R_2
5572·8	2	Q_1	5432·9	6	S_2	5283·7	6	S_2
5571·8	4	R_1	5416·3	7	R_1	5268·6	5	Q_1
5569·0	2	S_1	5413·9	3	S_1	5267·8	10	R_1
						5265·8	4	S_1

0, 2 band — *0, 1 band* — *0, 0 band*

OTHER SYSTEMS

Sharma (see above) also reports bands λλ3510–2950, which may be AlCl or possibly some other emitter, such as Al_2.

AlF

The main resonance system $A^1\Pi - X^1\Sigma^+$ lies in the far ultra-violet near 2278 Å and there are several other strong systems in the vacuum ultra-violet (e.g. $B^1\Sigma - X^1\Sigma$ 1925–1780 Å). Six much weaker singlet systems are known in the visible or near infra-red and eight triplet systems.

MAIN SYSTEM, 2360–2190 Å, $A^1\Pi - X^1\Sigma^+$

Occurrence. In emission in discharge tubes, and in absorption.

Appearance. Strong sequences. Barely degraded either way.

References. G. D. Rochester, *P.R.*, **56**, 305 (1939)†.

S. M. Naudé and T. J. Hugo, *Canad. J. Phys.*, **31**, 1106 (1953).

H. C. Rowlinson and R. F. Barrow, *Proc. Phys. Soc.* A, **66**, 437 (1953)†.

There is a strong (0, 0) sequence with line-like Q branches between 2274 and 2278 Å, a (0, 1) sequence around 2317 Å and a (1, 0) sequence near 2235 Å. The following outstanding

AlF (*contd.*)

heads, with direction of degradation, are from Rowlinson and Barrow's long list.

λ	I	v′, v″	λ	I	v′, v″	λ	I	v′, v″
2360·5	2V	P 0, 2	2316·0	10V	Q 1, 2	2274·9	8V	Q 2, 2
2359·2	3V	Q 0, 2	2278·8	4R	Q 7, 7	2274·7	10V	Q 1, 1
2358·2	4V	Q 1, 3	2277·6	7R	Q 6, 6			Q 0, 0
2318·3	8V	P 0, 1	2276·6	8L*	Q 5, 5	2235·9	2R	Q 3, 2
2317·9	8V	P 1, 2	2275·8	8V	Q 4, 4	2235·1	2R	Q 2, 1
2316·4	10V	Q 0, 1	2275·2	8V	Q 3, 3	2234·4	4R	Q 1, 0

* Line-like.

WEAK SINGLET SYSTEMS

Occurrence. Hollow cathode discharge.

Reference. S. M. Naudé and T. J. Hugo, *Canad. J. Phys.*, **31**, 1106 (1953); **32**, 246 (1954)†.

All bands are degraded to the violet. The following are the regions covered by each system, the transition, and the (0, 0) band head, which is usually the strongest;

8200–6790 Å	$C^1\Sigma^+ - A^1\Pi$	7245·9
6050–5490	$D^1\Delta - A^1\Pi$	5773·7
5055–5020	$E^1\Pi - A^1\Pi$	5051·8
4565–4510	$F^1\Pi - A^1\Pi$	4562·3
4455–4425	$G^1\Sigma - A^1\Pi$	4453·2
4265–4235	$H^1\Sigma - A^1\Pi$	4265·1

WEAK TRIPLET SYSTEMS

Occurrence. Hollow-cathode discharge.

References. P. G. Dodsworth and R. F. Barrow, *Proc. Phys. Soc.*, A, **68**, 824 (1955).

H. C. Rowlinson and R. F. Barrow, *Proc. Phys. Soc.*, A, **66**, 437 (1953)†.

I. Kopp and R. F. Barrow, *J. Phys.*, B, **3**, L 118 (1970).

R. F. Barrow, I. Kopp and R. Scullman, *Proc. Phys. Soc.*, **82**, 635 (1963).

All systems are weakly degraded to shorter wavelengths, and most show well developed (1, 0), (0, 0) and (0, 1) sequences. The following are the probably transitions, regions covered and first (but not necessarily the strongest) heads of selected sequences:

$x^3\Sigma - v^3\Sigma.$ 9975–9835 Å; 9973·3 (0, 0).

$v^3\Sigma - w^3\Sigma.$ 10600–8600; 9794·8 (0, 0).

$w^3\Sigma - a^3\Pi.$ 6052–5462; heads 5774·0, 5770·0, 5705·3, 5696·2, 5687·5, 5680·2 and also (degraded to the red) 5711·0.

$x^3\Sigma - w^3\Sigma.$ 4935–4905; 4931·0 (0, 0).

$y^3\Sigma - w^3\Sigma.$ 4930–4530; 4746·7 (0, 0).

$v^3\Sigma - a^3\Pi.$ 3717·9 (0, 1), 3608·2 (0, 0), 3492·2 (1, 0).

$c \quad - a^3\Pi.$ 2845·0 (0, 1), 2780·3 (0, 0), 2710·6 (1, 0).

$y^3\Pi - a^3\Pi.$ 2648·6 (0, 1), 2592·2 (0, 0), 2531·0 (1, 0).

VACUUM ULTRA-VIOLET SYSTEM

Reference. S. M. Naudé and T. J. Hugo, *Canad. J. Phys.*, **35**, 64 (1957).

AlH

Some seven band systems are attributed to AlH. Under conditions of mild excitation the λ4241 system is the strongest. The prominent heads at 4241 Å and 4259 Å are frequently observed in discharge tubes with aluminium electrodes.

References.   E. Bengtsson and R. Rydberg, *Z.P.*, **59**, 540 (1930).
J. W. I. Holst, *Dissertation, Stockholm* (1935).
W. Holst and E. Hulthen, *Z.P.*, **90**, 712 (1934)†.
E. Bengtsson, *Z.P.*, **51**, 889 (1928).
B. E. Nilsson, *Ark. Mat. Astr. Fys.*, A, **35**, (Paper 19), 10 (1948).
M. A. Khan, *Proc. Phys. Soc.*, **71**, 65 (1958)†.
P. B. Zeeman and G. J. Ritter, *Canad. J. Phys.*, **32**, 555 (1954).

4241 Å SYSTEM, $A^1\Pi - X^1\Sigma$

Double-headed bands of open structure degraded to the red. Occurs readily in emission in discharges where aluminium vapour and hydrogen are present together. May also be obtained in absorption. Found in late-type stars.

v', v''	R heads	Q heads		v', v''	R heads	Q heads
1, 4	5418·6	5438·3		0, 1	4546·5	4576·4
0, 3		5314·6		1, 1	4353·1	4360·5
1, 3	5024·3	5037·9		0, 0	4241·0	4259·5
0, 2	4886	4929·1		1, 0	4066·3	4072·6
1, 2	4670·9	4680·7				

2229 Å SYSTEM, $C^1\Sigma - X^1\Sigma$

Weak bands with P and R branches slightly degraded towards the red.

v', v''	Origins	R heads
1, 2	2326·3	
1, 1	2255·8	
0, 0	2241·6	2228·6
1, 0	2176·9	2173·7

2033 Å SYSTEM, $D^1\Sigma - X^1\Sigma$

Weak bands with P and R branches. The P branch of the (0, 0) stretches to about 2033 Å.

v', v''	Origins	P heads
0, 1	2097	2101·1
0, 0	2028·2	2033

4752 Å SYSTEM, $C^1\Sigma - A^1\Pi$

Bands show P, Q and R branches degraded to the violet. (0, 0) Q head, 4731·8.

3380 Å SYSTEM, $E^1\Pi - A^1\Pi$

Bands degraded to the red, each with two P and two R branches.

v', v''	Origin	Heads	
0, 0	3390	3379·4	3384·8

AlH (*contd.*)

3800 Å SYSTEM, $b^3\Sigma - a^3\Pi$

Reference.   W. Holst, *Z.P.*, **86**, 338 (1933).

Triplet bands with three P and three R branches symmetrically arranged about piled up Q branches; similar to the 3360 Å band of NH. The (1, 1) band shows a weaker Q maximum on the long wave side of the (0, 0).

v', v''	*Origin*
0, 0	3812·6

2700 Å SYSTEM. Brief mention of a band about 2700 Å.

Reference.   W. Holst, *Z.P.*, **89**, 47 (1934).

AlH⁺

References.   W. Holst, *Z.P.*, **89**, 40 (1934)[†].
G. M. Almy and M. C. Watson, *P.R.*, **45**, 871 (1934)[†].

3602 Å SYSTEM, $^2\Pi - {}^2\Sigma$

Bands degraded to the violet. Observed in arc between Al electrodes in hydrogen at reduced pressures, in discharge through a mixture of $AlCl_3$ vapour and hydrogen, and in hollow aluminium cathode containing hydrogen and helium.

(0, 0) Band	*(1, 1) Band*
3632·3 P_1	3600·5 P_2
3630·2 P_2	3593·0 Q_1
3611·9 Q_1	3583·0 Q_2
3602·4 Q_2	

AlI

Occurrence.   In discharge tubes of various types containing aluminium iodide, and in absorption.

Appearance.   Two systems in the blue, degraded to the violet, and a single progression of diffuse bands in the ultra-violet.

References.   E. Miescher, *Helv. Phys. Acta*, **8**, 279 (1935)[†].
E. Miescher, *Helv. Phys. Acta*, **9**, 693 (1936)[†].

SYSTEM A, A $^3\Pi - X^1\Sigma$

Strongest bands:

λ	I	v', v''
4631·4	7	0, 1
4565·0	10	0, 0
4561·1	8	1, 1
4496·6	8	1, 0
4431·0	3	2, 0

AlI (*contd.*)

SYSTEM B, B $^3\Pi - X^1\Sigma$

Strongest bands:

λ	I	v', v''
4589·2	2	0, 1
4524·0	7	0, 0
4520·9	4	1, 1

ULTRA-VIOLET SYSTEM, $^1\Pi - X^1\Sigma$

λ	v', v''	λ	v', v''
3426	0, 5	3295	0, 2
3382	0, 4	3244	0, 1
3339	0, 3	3175	0, 0

OTHER BANDS

Miescher also records numerous bands, degraded to the red, in the blue and violet.

AlN

Occurrence. In microwave discharge through $AlCl_3 + N_2$.

Reference. J. D. Simmons and J. K. McDonald, *J. Mol. Spec.*, **41**, 584 (1972)†.

Two violet-degraded bands of a $^3\Pi - {}^3\Pi$ transition have been found; the stronger (0, 0) band has a head at 5076·8, and the weaker (1, 0) band is near 5276 Å.

AlO

The main system of AlO in the green occurs readily. Recently a number of other systems have been found in the ultra-violet. The naming of the electronic states has been changed by Singh, causing some confusion, but his notation is used here.

GREEN SYSTEM, $B^2\Sigma^+ - X^2\Sigma^+$

Occurrence. Readily in a variety of sources including arcs, flames, exploding wires, shock tubes, photo-flash bulbs and M-type stars.

Appearance. Marked sequences of red-degraded single-headed bands. See Plate 1.

References. A. Lagerqvist, N. E. L. Nilsson and R. F. Barrow, *Ark. Fys.*, **12**, 543 (1957)†.
D. C. Tyte and R. W. Nicholls, *Identification Atlas of Molecular Spectra*, 1, Univ. Western Ontario (1964)†.

The following measurements of the strongest heads, selected from the photographs, are from Tyte and Nicholls with our estimates of intensity.

λ	I	v', v''	λ	I	v', v''	λ	I	v', v''
5409·7	4	4, 6	5102·0	6	1, 2	4537·6	4	5, 3
5392·3	4	3, 5	5079·4	5	0, 1	4516·2	5	4, 2
5376·9	4	2, 4	4866·2	8	1, 1	4493·8	5	3, 1
5357·8	3	1, 3	4842·2	10	0, 0	4470·4	4	2, 0
5337·0	3	0, 2	4694·4	7	3, 2	4373·8	3	6, 3
5142·6	5	3, 4	4671·9	8	2, 1	4352·4	3	5, 2
5123·1	6	2, 3	4648·1	9	1, 0	4330·4	2	4, 1

AlO (*contd.*)

ULTRA-VIOLET SYSTEMS

References. S. L. N. G. Krishnamachari, N. A. Narasimham and M. Singh, *Canad. J. Phys.*, **44**, 2513 (1966)†.
J. K. McDonald and K. K. Innes, *J. Mol. Spec.*, **32**, 501 (1969)†.
M. Singh, *Proc. Indian Acad. Sci.*, A, **71**, 82 (1970).
M. Singh, *J. Phys.*, B, **6**, 521 (1973)†.
M. Singh and M. K. Saksena, *Proc. Indian Acad. Sci.*, A, **77**, 139 (1973).
V. W. Goodlett and K. K. Innes, *Nature*, **183**, 243 (1959)†.

These systems occur in emission from arcs, hollow-cathode discharges and microwave excitation.

$C^2\Pi - X^2\Sigma$, 3320–2870 Å

Degraded to the red. Double headed. Main bands 3112·6 (0, 1), 3021·6 (0, 0) and 2957·5 (1, 0).

$D^2\Sigma - A^2\Pi$, 3000–2800 Å

Degraded to shorter wavelengths. Double headed. First heads of strong bands 2930·1 (0, 1), 2869·5 (0, 0), 2862·4 (1, 1), 2804·6 (1, 0).

$E^2\Delta - A^2\Pi$, 2500 Å

Degraded to the red. Double headed. 2592·9 (0, 2), 2545·9 (0, 1), 2500·1 (0, 0).

$D^2\Sigma - X^2\Sigma$, 2800–2300 Å

18 single-headed bands, degraded to the red. 2677·4 (0, 3), 2611·8 (0, 2), 2548·5 (0, 1), 2487·3 (0, 0), 2438·3 (1, 0) and 2391·9 (2, 0).

$F^2\Pi - A^2\Pi$, 2380 Å

A double double-headed band, degraded to the red. (0, 0) heads 2389·0, 2387·9, 2381·6 and 2380·4.

There are several mentions of bands in the red or near infra-red, but data are not available.

AlS

Three systems of red-degraded bands are known. All occur in absorption and the first is also found in emission in a King furnace.

References. C. M. MacKinney and K. K. Innes, *J. Mol. Spec.*, **3**, 235 (1959).
A. A. Mal'tsev, V. F. Shevelkov and E. D. Krupnikov, *Opt. Spectrosc. Mol. Spectrosc.*, Suppl. 2, 4 (1966).
M. Kronekvist and A. Lagerqvist, *Ark. Fys.*, **39**, 133 (1969).

4800–3700 Å, $A^2\Sigma - X^2\Sigma$

Strong $v' = 0$ and $v'' = 0$ progressions; 22 bands are known. Strongest, 4634·2 (0, 3), 4540·7 (0, 2), 4390·0 (0, 1), 4275·4 (0, 0), 4184·5 (1, 0), 4097·8 (2, 0) and 4015·1 (3, 0).

3400–3100 Å, $B^2\Pi - X^2\Sigma$

Double double-headed. First heads, 3394·6 (0, 1), 3325·2 (0, 0), 3267·6 (1, 0), 3218·1 (2, 0).

3000–2700 Å, $C^2\Sigma - X^2\Sigma$

A long $v' = 0$ progression. Strongest, 2897·5 (0, 2), 2847·6 (0, 1), 2813·1 (1, 1), 2799·0 (0, 0), 2765·7 (1, 0).

AlSe

References.   J. Singh, D. P. Tewari and H. Mohan, *J. Phys.*, B, **2**, 627 (1969); *Indian J. Pure Appl. Phys.*, **10**, 386 (1972).

27 bands, degraded to the red, have been observed in thermal emission. Strongest, 4593·0 (0, 3), 4498·8 (0, 2), 4435·5 (2, 3), 4407·9 (0, 1), 4179·9 (2, 0), 4114·0 (3, 0), 4050·9 (4, 0).

ArH

7674 Å, $B^2\Pi - A^2\Sigma^+$

Occurrence.   Electrical discharge through mixture of argon and hydrogen.

Reference.   J. W. C. Johns, *J. Mol. Spec.*, **36**, 488 (1970)†.

Degraded to the red. The (0, 0) band shows a weak R head 7550·3 and a strong Q head at 7674·4.

ArO

Occurrence.   Afterglow of argon containing a trace of oxygen.

Appearance.   Quasi-continuum near auroral green line at 5577 Å.

References.   R. Herman, C. Weniger and L. Herman, *P.R.* **82**, 751 (1951).

L. Vegard and G. Kvifte, *Skr. Norske Vidensk Akad. Oslo*, **1**, No. 2 (1955).

Herman *et al.* note heads λλ5555, 5529, 5487. Analysis indicates following heads near auroral line; λλ5580·4 (0, 1), 5574·8 (1, 1), 5572·9 (0, 0), 5570·1 (2, 1), 5567·6 (1, 0).

ArXe

Occurrence.   Negative glow of discharge through mixed argon and xenon.

Appearance.   Two bands 5507 to 5120 Å and 5081 to 4960 are reported. The published photograph shows diffuse features, shaded to the violet, superposed on strong Ar and Xe lines.

Reference.   H. M. Jongerius, H. M. Van Koeveringe and H. J. Oskam, *Physica*, **25**, 406 (1959)†.

As$_2$

There are two strong extensive overlapping singlet systems A — X and B — X and a number of weaker triplet or triplet-singlet systems. The notation of Perdigan and d'Incan is used here; some states approach case *c* coupling.

MAIN SYSTEMS

Occurrence.   In discharges through heated arsenic and a carrier gas (H$_2$, He, Ne). Also absorption and fluorescence.

Appearance.   Degraded to the red. Two very extensive systems of overlapping bands. Over 100 bands of A — X and about 50 of B — X are known.

References.   G. E. Gibson and A. MacFarlane, *P.R.*, **46**, 1059 (1934)†.

G. M. Almy and G. D. Kinzer, *P.R.*, **47**, 721 (1935).

G. D. Kinzer and G. M. Almy, *P.R.*, **52**, 814 (1937)†.

The following are the strongest bands, listed intensity 5 or 6, from Almy and Kinzer.

As$_2$ (*contd.*)

SYSTEM $A^1\Sigma_u^+ - X^1\Sigma_g^+$, 5555 to 2240 Å

λ	v', v''	λ	v', v''	λ	v', v''	λ	v', v''
3318·0	8, 31	2922·9	8, 20	2570·4	6, 7	2464·7	6, 3
3104·5	6, 24	2922·3	5, 16	2551·6	4, 5	2449·9	7, 3
3093·5	5, 23	2889·1	2, 15	2506·9	5, 4	2439·4	6, 2
3069·0	9, 25	2856·4	2, 14	2480·7	5, 3	2424·9	7, 2

The following are strong in absorption: 2357·1 (10, 1), 2333·8 (10, 0), 2319·7 (11, 0), 2306·0 (12, 0), 2292·7 (13, 0), 2279·6 (14, 0), 2266·7 (15, 0), 2254·0 (16, 0).

SYSTEM $B^1\Sigma_u^+ - X^1\Sigma_g^+$, 5530–2350 Å

3150·6 (2, 24), 3141·8 (1, 23), 3113·8 (2, 23), 3105·1 (1, 22), 2998·3 (1, 19), 2638·6 (0, 7), 2554·4 (0, 4).

TRIPLET SYSTEMS

Occurrence. Relatively weak systems observed in discharge tubes, especially with neon as carrier.

References. P. Perdigan and J. d'Incan, *Canad. J. Phys.*, **48**, 1140 (1970). S. Mrozowski and C. Santaram, *J. Opt. Soc. Amer.*, **57**, 522 (1967)†.

$a\ 0_u^+ - c^3\Sigma_u^+$, 10000–7400 Å

Numerous weak bands, degraded to the violet.

$d\ 1_g - c^3\Sigma_u^+$, 7000–5600 Å

Degraded to the violet. The *c* state is split into two components, forming two sub-systems, i and ii. Strongest heads 6297·7 (0, 1 i), 6234·6 (0, 1 ii), 6052·7 (1, 0 i), 5994·3 (1, 0 ii), 5933·5 (2, 0 i), 5927·2 (3, 1 ii), 5877·5 (2, 0 ii), 5870·8 (3, 1 ii).

$c\ ^3\Sigma_u^+ - X^1\Sigma_g^+$, 8400–5380 Å

Degraded to the red. 32 bands.

$e - X^1\Sigma_g^+$, 6240–4440 Å

Degraded to the red. 21 bands.

$a\ 0_u^+ - X^1\Sigma_g^+$, 5580–3700 Å

Degraded to the red. 40 bands. 4317·9 (2, 5), 3870·5 (5, 1).

$b\ ^3\Pi - X^1\Sigma_g^+$, 3840–2980 Å

Degraded to the red. 3341·4 (0, 2), 3294·7 (0, 1), 3258·7 (1, 1), 3248·9 (0, 0), 3214·1 (1, 0), 3180·1 (2, 0).

$d\ 1_g - X^1\Sigma_g^+$, 3390–2980 Å

Degraded to the red. Strongest heads from Mrozowski and Santaram, 3389·1 (0, 3), 3341·4 (0, 2), 3294·7 (0, 1), 3258·7 (1, 1), 3248·9 (0, 0), 3214·1 (1, 0), 3180·1 (2, 0).

As_2^+

Bands provisionally attributed to As_2^+ by Kinzer and Almy appear to be the $d\,1_g - c\,^3\Sigma_u^+$ system of the neutral molecule.

AsCl

Reference. N. Basco and K. K. Yee, *Chem. Comm.*, 1255 (1967).

Violet-degraded bands 2500–2400 Å are observed in absorption following flash photolysis of $AsCl_3$. Heads 2499·8 (0, 2), 2472·9 (0, 1), 2446·3 (0, 0), 2415·6 (1, 0).

AsF

TRIPLET SYSTEMS

Occurrence. Discharge through AsF_3.

Reference. G. Pannetier, P. Deschamps and J. Guillaume, *C. R. Acad. Sci. (Paris)*, **261**, 3396 (1965)†.

4750–3150 Å, $A^3\Pi - X^3\Sigma$

About 300 red-degraded bands grouped in three's. No data available apart from small photo.

2300–2050 Å, $B^3\Pi - X^3\Sigma$

30 bands, degraded to shorter wavelengths. No data. The spectrogram shows widely spaced triplets.

WEAK SINGLET AND SINGLET-TRIPLET SYSTEMS

In high-frequency discharge, and in some cases in absorption with a synchronized flash background.

References. K. K. Yee, D. S. Liu and W. E. Jones, *J. Mol. Spec.*, **35**, 153 (1970).

A. Chatalic, N. Danon and G. Pannetier, *C. R. Acad. Sci. (Paris)*, C, **273**, 874 (1971).

D. S. Liu and W. E. Jones, *Canad. J. Phys.*, **50**, 1230 (1972)†.

$b^1\Sigma - X^3\Sigma$, 7400–7000 Å

Degraded to the violet. 7396·7 (0, 0), 7390·8 (1, 1), 7384·2 (2, 2).

$c'^1\Pi - a^1\Delta$, 4500–3500 Å

28 bands. 4181·7 (0, 2), 3834·8 (2, 0).

$c^1\Pi - b^1\Sigma$, 2910–2780 Å

Degraded red. 2907·2 (0, 1), 2849·9 (0, 0), 2785·3 (1, 0).

$d^1\Pi - b^1\Sigma$, 2755–2350 Å

2750·9 (0, 1), 2699·5 (0, 0).

$c^1\Pi - a^1\Delta$, 2530–2350 Å

Degraded violet. A relatively strong system. 2480·3 (0, 2), 2439·1 (0, 1), 2431·5 (1, 2), 2398·8 (0, 0), 2353·2 (1, 0).

$d^1\Pi - a^1\Delta$, 2405–2290 Å

Very weak. 2328·4 (0, 1), 2291·6 (0, 0).

AsF (*contd.*)

$c\,^1\Pi - X\,^3\Sigma$, 2150–1950 Å

Degraded to the violet. 2080·1 (0, 1), 2051·1 (0, 0), 2017·7 (1, 0), 2012·9 (2, 1).

AsH

Occurrence. These bands are very difficult to obtain, but have been observed in absorption following flash photolysis of AsH_3.

Appearance. A multi-headed (0, 0) band with open rotational structure, degraded to the red, and a weaker (1, 0) band.

Reference. R. N. Dixon and H. M. Lamberton, *J. Mol. Spec.*, **25**, 12 (1968)†.

Heads of (0, 0) band, λλ3252·1, 3270·4, 3272·8, 3336·8, 3402·0, 3410·3, 3413·8.

AsH₂

Reference. R. N. Dixon, G. Duxbury and H. M. Lamberton, *P.R.S.*, **305**, 271 (1968)†.

A single progression, $(0, v_2', 0$–$0, 0, 0)$, of violet-degraded bands observed in flash photolysis of AsH_3. Heads, 5022·3, 4816·3, 4627·8, 4460·1, 4295·4, 4148·2, 4011·5.

AsN

3400–2400 Å, $A\,^1\Pi - X\,^1\Sigma$

Occurrence. In heavy-current discharge tubes containing arsenic and nitrogen.

Appearance. Degraded to the red. The bands presumably have close double heads.

Reference. J. W. T. Spinks, *Z.P.*, **88**, 511 (1934)†.

The following are the bands as listed by Spinks:

λ	I	v', v''	λ	I	v', v''
3051·0	2	0, 3	2784·2	10	0, 0
3007·8	1	3, 5	2719·5	7	1, 0
2884·7	3	1, 2	2675·6	4	3, 1
2868·7	6	0, 1	2656·5	5	2, 0
2833·5	1	3, 3	2602·0	2	3, 0

3431 Å, $B\,^1\Sigma - X\,^1\Sigma$

Occurrence. High-frequency discharge.

Reference. J. d'Incan and B. Fémelot, *C. R. Acad. Sci.* (*Paris*), B, **267**, 796 (1968)†.

A single red-degraded band with head at 3431·8 Å.

AsO

A large number of systems have been observed. The A and B are most readily obtained; 7 weaker ones are grouped together. The notation is that used by Anderson and Callomon.

SYSTEM A, 3450–2950 Å, $A\,^2\Sigma^+ - X\,^2\Pi$

Occurrence. Arsenic in carbon arc, high tension arc between As electrodes, high frequency discharge through As_2O_3 in He, and in $AsCl_3/O_2$ flame.

Appearance. Degraded to the red. Evenly spaced bands with close double heads.

AsO (*contd.*)

References. F. C. Connelly, *Proc. Phys. Soc.*, **46**, 790 (1934)†.
F. A. Jenkins and L. A. Strait, *P. R.*, **47**, 136 (1935)†.

Strong heads from Connelly, with intensities on scale of 9. Sub-bands indicated i and ii.

λ	I	v', v''		λ	I	v', v''		λ	I	v', v''	
3310·7	4	1, 1	ii Q	3207·8	4	1, 0	ii R	3137·2	4	2, 1	i Q
3279·1	9	0, 0	ii Q	3202·0	4	1, 1	i Q	3135·6	4	2, 1	i R
3277·2	8	0, 0	ii R	3172·4	8	0, 0	i Q	3106·8	6	1, 0	i Q
3241·4	4	2, 1	ii Q	3170·6	5	0, 0	i R	3105·6	4	1, 0	i R
3209·1	6	1, 0	ii Q	3144·1	4	2, 0	ii Q				

SYSTEM B, 3450–2950 Å, $B^2\Sigma^+ - X^2\Pi$

Occurrence and References. As system A.

Appearance. Degraded to shorter wavelengths. Marked sequences of close double-headed bands.

λ	I	v', v''		λ	I	v', v''		λ	I	v', v''	
2635·5	4	0, 1	ii P	2570·9	6	0, 0	ii P	2504·7	6	0, 0	i P
2634·4	5	0, 1	ii Q	2569·7	8	0, 0	ii Q	2503·6	7	0, 0	i Q
2624·7	4	1, 2	ii Q	2565·2	4	0, 1	i Q	2438·5	3	1, 0	i P

WEAKER SYSTEMS

Occurrence. These have mostly been obtained in high-frequency discharges. Systems A″ and G occur in the $AsCl_3/O_2$ flame.

Appearance. All these systems are degraded to the red. There is considerable overlapping. Only a few main heads are listed for each system.

References. S. V. J. Lakshman and P. T. Rao, *Indian J. Phys.*, **34**, 278 (1960).
M. Venkataramanaiah and S. V. J. Lakshman, *Indian J. Phys.*, **38**, 209 (1964).
S. Mrozowski and C. Santaram, *J. Opt. Soc. Amer.*, **56**, 1174 (1966)†.
J. d'Incan and J. P. Goure, *C. R. Acad. Sci.* (*Paris*), B, **268**, 1647 (1969).
J. P. Goure, J. Figuet, J. N. Massot and J. d'Incan, *Canad. J. Phys.*, **50**, 1926 (1972).
V. S. Kushawaha, B. P. Asthana and C. M. Pathak, *J. Mol. Spec.*, **41**, 577 (1972).
V. M. Anderson and J. H. Callomon, *J. Phys.*, B, **6**, 1664 (1973).

SYSTEM A″, 6550–5800 Å, A″ $(?^2\Sigma) - X^2\Pi$

$A'' - X^2\Pi_{\frac{3}{2}}$. 6533 (0, 0), 6217 (2, 1), 6178 (1, 0).
$A'' - X^2\Pi_{\frac{1}{2}}$. 6493 (0, 1), 6151 (1, 1), 6111 (0, 0), 5876 (3, 2), 5846 (2, 1), 5808 (1, 0).

SYSTEM G (formerly A′), 4700–3900 Å, $G^2\Pi - X^2\Pi$

$G - X^2\Pi_{\frac{3}{2}}$. 5404·8 (0, 7), 5155·4 (0, 6), 4925·4 (0, 5), 4712·4 (0, 4).
$G - X^2\Pi_{\frac{1}{2}}$. 5042·1 (0, 7), 4824·4 (0, 6), 4621·5 (0, 5), 4433·7 (0, 4).

SYSTEM E, 4100–3400 Å, E $(?^2\Pi) - X^2\Pi$

$E - X^2\Pi_{\frac{3}{2}}$. 3842·1 (0, 1), 3706·3 (0, 0), 3611·4 (1, 0).
$E - X^2\Pi_{\frac{1}{2}}$. 3747·1 (0, 0), 3650·5 (1, 0).

SYSTEM F, 4100–3400 Å, $F - X^2\Pi$

3628·8, 3596·9, 3570·9, 3510·2. Goure and d'Incan doubt the reality of this system.

AsO (*contd.*)

SYSTEM H, 4000–3300 Å, $H^2\Pi - X^2\Pi$

$H^2\Pi_{\frac{3}{2}} - X^2\Pi_{\frac{3}{2}}$. 3775·9 (1, 1), 3621·8 (0, 9), 3510·4 (0, 8), 3404·4 (0, 7), 3303·5 (0, 6).

SYSTEM C, 3100–2400 Å, $C^2\Delta - X^2\Pi$

$C^2\Delta_{\frac{5}{2}} - X^2\Pi_{\frac{3}{2}}$. 2962·9 (0, 4), 2882·5 (0, 3), 2807·7 (0, 2), 2735·0 (0, 1), 2687·4 (1, 1), 2620·1 (1, 0), 2557·8 (2, 0), 2536·0 (3, 0), 2496·7 (4, 0).
$C^2\Delta_{\frac{3}{2}} - X^2\Pi_{\frac{1}{2}}$. 2879·8 (0, 4), 2804·8 (0, 3), 2732·9 (0, 2).

SYSTEM D, 3100–2400 Å, $D^2\Sigma^- - X^2\Pi$

$D - X^2\Pi_{\frac{3}{2}}$. 2893·4 (0, 3), 2817·0 (0, 2), 2768·8 (1, 2), 2743·4 (0, 1), 2611·6 (3, 1), 2587·9 (2, 0), 2547·9 (3, 0), 2509·7 (4, 0).
$D - X^2\Pi_{\frac{1}{2}}$. 2927·7 (1, 3), 2900·5 (0, 2).

AsO⁺

2500–2200 Å, $A^1\Pi - X^1\Sigma$

Occurrence. Various types of discharge.
References. S. V. J. Lakshman, *Proc. Phys. Soc.*, **89**, 774 (1966).
D. V. K. Rao and P. T. Rao, *J. Phys.*, B, **3**, 430 (1970).
Degraded to the red. The following are the strongest R heads: 2482·7 (0, 2), 2417·8 (0, 1), 2373·8 (1, 1), 2355·6 (0, 0), 2313·8 (1, 0), 2274·2 (2, 0), 2236·8 (3, 0).

AsP

3220–2960 Å, $A^1\Pi - X^1\Sigma$

Reference. L. M. Harding, W. E. Jones and K. K. Yee, *Canad. J. Phys.*, **48**, 2842 (1970).
12 red-degraded bands have been obtained in a high-frequency discharge through $PCl_3 + AsCl_3$. Strongest R heads: 3208·4 (0, 2), 3148·2 (0, 1), 3089·8 (0, 0), 3045·5 (1, 0), 3002·8 (2, 0).

AsS

7000–4300 Å, $A'\,^2\Pi - X^2\Pi$

References. M. Shimauchi, *Science of Light*, **18**, 90 (1969)†.
M. Shimauchi and S. Karasawa, *Science of Light*, **22**, 127 (1973)†.
Over 60 red-degraded bands were obtained with As_2S_2 in various types of discharge. Strongest heads:
$^2\Pi_{\frac{3}{2}} - {}^2\Pi_{\frac{3}{2}}$. 6522·5 (0, 7), 6301·5 (0, 6), 6092·0 (0, 5), 5946·0 (1, 5), 5758·0 (1, 4).
$^2\Pi_{\frac{1}{2}} - {}^2\Pi_{\frac{1}{2}}$. 2858·5 (0, 6), 5676·8 (0, 5), 5505·5 (0, 4).

AsS⁺

3050–2900 Å, $A^1\Pi - X^1\Sigma$

Reference. M. Shimauchi, *Science of Light*, **18**, 90 (1969)†.
With As_2S_2 in a hollow cathode discharge about 35 bands occur. Degraded to the red. Strongest, 2931·7 (0, 5), 2879·0 (0, 4), 2827·9 (0, 3), 2778·2 (0, 2), 2730·0 (0, 1), 2698·0 (1, 1).

AsSb

Reference. K. K. Yee and W. E. Jones, *Chem. Comm.*, 752 (1969).

Close double-headed bands, degraded to the red, observed in a microwave discharge through $AsCl_3 + SbCl_3$. R heads of strongest, 3804·5 (0, 3), 3756·1 (0, 2), 3708·7 (0, 1), 3662·3 (0, 0).

AsSe

Reference. R. Vasudev and W. E. Jones, *J. Mol. Spec.*, **54**, 144 (1975).

Six red-degraded bands 5700–5300 Å obtained in a microwave discharge. Heads:

$^2\Pi_{\frac{3}{2}} - {}^2\Pi_{\frac{3}{2}}$. 5717·6 (0, 2), 5627·8 (0, 1), 5540·5 (0, 0).

$^2\Pi_{\frac{1}{2}} - {}^2\Pi_{\frac{1}{2}}$. 5505·7 (0, 2), 5422·4 (0, 1), 5341·2 (0, 0).

Au_2

Occurrence. Emission from King furnace containing gold.

Appearance. Two extensive systems of about 100 red-degraded bands.

References. B. Kleman, S. Lindkvist and L. Esselin, *Ark. Fys.*, **8**, 505 (1954)†.
J. Ruamps, *C. R. Acad. Sci. (Paris)*, **238**, 1489 (1954), and *Ann. Phys. (Paris)*, Ser. **13**, **4**, 1111 (1959)†.

SYSTEM A, A $0^+ - X^1\Sigma^+$

The following are among the strongest heads, with intensities on a scale of 5. S indicates head of sequence.

λ	I	v', v''	λ	I	v', v''	λ	I	v', v''
5371·2	2	2, 7	5240·8	5S	0, 3	5089·3	2S	0, 0
5318·2	4	2, 6	5189·5	4S	0, 2	5052·8	3S	1, 0
5305·4	4	1, 5	5138·9	3S	0, 1	5017·2	3S	2, 0
5292·9	2S	0, 4	5111·2	2	1, 1	4982·3	2S	3, 0
5253·5	4	1, 4						

SYSTEM B, B $0^+ - X^1\Sigma^+$

The following are the main sequence heads. $\lambda\lambda$3980·9 (0, 3), 3951·2 (0, 2), 3922·0 (0, 1), 3892·9 (0, 0), 3866·0 (1, 0), 3839·7 (2, 0).

AuAl etc. see AuM (gold/metal compounds)

AuCl

Occurrence. Gold chloride in active nitrogen.

Appearance. Two overlapping systems in the green, both degraded to the red.

Reference. W. F. C. Ferguson, *P.R.*, **31**, 969 (1928)†.

Strong bands only. Intensities on a scale of 5.

AuCl (*contd.*)

SYSTEM A, A−X

λ	I	v′, v″	λ	I	v′, v″
5590·2	2	1, 4	5346·6	5	0, 1
5570·4	2	0, 3	5240·0	5	0, 0
5476·0	2	1, 3	5155·9	4	1, 0
5456·8	4	0, 2	5075·0	1	2, 0

SYSTEM B, B−X

λ	I	v′, v″	λ	I	v′, v″
5531·7	1	0, 3	5205·5	5	0, 0
5437·6	2	1, 3	5121·9	4	1, 0
5419·4	3	0, 2	5041·7	1	2, 0

AuH

3656 Å SYSTEM, $A^1\Sigma^+ - X^1\Sigma^+$

Occurrence. Gold arc in H_2 and in absorption in a furnace.
Appearance. Degraded to the red. Single-headed bands with open structure.
References. E. Bengtsson, *Ark. Mat. Astr. Fys.*, **18**, 27 (1925).
T. Heimer, *Z.P.*, **104**, 303 (1937).
    R heads of stronger bands,

λ	I	v′, v″	λ	I	v′, v″
4436·6	2	1, 3	3972·8	10	0, 1
4339·4	3	0, 2	3651·5	9	0, 0
4068·2	3	1, 2	3457·4	2	1, 0

2615 Å SYSTEM, $B\,0^+ - X^1\Sigma^+$

Occurrence and Appearance. Similar to 3656 System.
Reference. U. Ringström, *Ark. Fys.*, **27**, 227 (1964)†.
    R heads of stronger bands,

λ	I	v′, v″	λ	I	v′, v″
2819·4	6	1, 2	2564·7	4	2, 1
2773·3	7	0, 1	2511·7	5	1, 0
2612·8	6	0, 0	2426·8	4	2, 0

WEAKER SYSTEMS

Obtained by Ringström (see above) in absorption. All degraded to the red.
$C^1\Sigma - X^1\Sigma$. R head 2317·6 (0, 0).
$c\,0^- - X^1\Sigma$. Origins, 2486·1 (0, 1), 2412·6 (1, 1), 2356·3 (0, 0), 2290·0 (1, 0).
$b\,1 - X^1\Sigma$. R heads, 2491·1 (0, 1), 2360·7 (0, 0).
$a\,2 - X^1\Sigma$. Origins, 2508·1 (0, 1), 2375·9 (0, 0), 2322·9 (4, 1), 2292·4 (5, 0).

AuM

Gold forms diatomic compounds with a large number of other metals, and thermal emission or absorption are usually obtained in a King furnace. These spectra are relatively unimportant for identification, so only the barest details are given; for some further summarized details see also the compilation by Rosen *et al.* Usually there are several systems of bands with compact sequences, the (0, 0) being strongest. Here the first heads of the (0, 0) bands of each system only are given, with direction of degradation (R = to the red, V = to the violet or shorter wavelength).

References. J. Ruamps, *C. R. Acad. Sci. (Paris)*, **239**, 1200 (1954), AuCu.

J. Ruamps, *Ann. Phys. (Paris)*, Ser 13, **4**, 1111 (1959)†, AuAl.

J. Schiltz, *Ann. Phys. (Paris)*, **8**, 67 (1963)†, AuBa, AuCa, AuSr.

M. Collette and J. Schiltz, *C. R. Acad. Sci. (Paris)*, **257**, 2902 (1963)†, AuSn.

R. F. Barrow, W. J. M. Gissane and D. N. Travis, *Nature*, **201**, 603 (1964), AuBe, Ga, Ge, In, Mg, Si.

R. F. Barrow, W. J. M. Gissane and D. N. Travis, *P.R.S.*, **287**, 240 (1965)†, AuBe, AuMg.

R. Houdart and P. Carette, *C. R. Acad. Sci. (Paris)*, **260**, 5746 (1965)†, AuPb.

R. Houdart and J. L. Bocquet, *C. R. Acad. Sci. (Paris)*, B, **263**, 151 (1966)†, AuBi; B, **264**, 860 (1967), AuSb; B, **264**, 1717 (1967), AuAs.

J. L. Bocquet and R. Houdart, *C. R. Acad. Sci. (Paris)*, B, **265**, 979 (1967), AuGa.

AuAl 6145·0 deg. V; 6449·1 R; 6066·5 R.
AuAs 6070·9, 6031·4, 5734·3, 5417·4 all deg. R.
AuBa 7902·1 deg. R; 4584·2 V.
AuBe 5819·1, 5281·1, 5274·5 all deg. V.
AuBi 5621·7 deg. R.
AuCa 6890·5, 6655·6, 5787·1 all deg. V.
AuCu 4845·6, 4838·7, 4508·7, 4223·2 all deg. R.
AuGa 5536·2, head deg. V, but sequence deg. R.
AuGe 8189·5, 7273·4 deg. R.
AuIn 5827·5 deg. R.
AuMg 5431·8 deg. V; 5126·1, deg. V; 3222 deg. R.
AuPb 6112·7 deg. R.
AuSb 6647·6, 6584·0, 6207·1, 5967·7 all deg. R.
AuSi 7961·7 deg. V. 7331·9 deg. R.
AuSn 8810, 7144·8 both deg. R.
AuSr 7229·0, 7058·6, both deg. R.

AuSe

Reference. B. Rosen *et al.*, *Spectroscopic data relative to diatomic molecules*, (1970).

R. F. Barrow has obtained a red-degraded system in thermal emission. 6326 Å (0, 0).

AuTe

Reference. R. C. Maheshwari and D. Sharma, *Proc. Phys. Soc.*, **81**, 898 (1963)†.

27 red-degraded bands observed in a King furnace. Sequence heads 6651·3 (0, 2), 6559·3 (0, 1), 6469·4 (0, 0), 6404·6 (1, 0).

B_2

Occurrence. Boron trichloride in active nitrogen. Discharges in helium with a trace of boron trichloride.

Appearance. Bands with single P and R branches degraded to the red.

Transition. $^3\Sigma_u^- - {}^3\Sigma_g^-$.

References. A. E. Douglas and G. Herzberg, *P.R.*, **57**, 752 (1940).

A. E. Douglas and G. Herzberg, *Canad. J. Res.*, A, **18**, 165 (1940)†.

Bands were observed for the molecules $B^{11}B^{11}$ and $B^{10}B^{11}$. The heads for the more abundant molecule $B^{11}B^{11}$ are given below.

λ	I	v', v''	λ	I	v', v''
3300·4	1	3, 3	3204·2	0	4, 3
3292·7	5	2, 2	3196·4	2	3, 2
3283·4	8	1, 1	3187·0	2	2, 1
3272·8	10	0, 0	3176·5	3	1, 0

BBr

MAIN SYSTEM, $A^1\Pi - X^1\Sigma^+$

Occurrence. In electrodeless high-frequency discharge through BBr_3 and in absorption.

Appearance. Bands degraded in each direction.

Reference. E. Miescher, *Helv. Phys. Acta*, **8**, 279 (1935)†.

E. Miescher and E. Rosenthaler, *Nature*, **145**, 642 (1940).

Prominent heads as listed by Miescher. The letters R or V following the wavelength indicate that the head is degraded to longer or shorter wavelengths respectively.

λ	I	v', v''	λ	I	v', v''
3094·5 R	5	3, 5 Q	2954·4 V	10	1, 1 Q
3082·7 R	5	2, 4 Q'	2954·0 R	10	1, 1 Q'
3010·3 V	5	0, 1 P	2951 –	8	0, 0 P
2973·0 R	7	3, 3 R	2948·1 V	10	0, 0 Q
2963·1 R	8	2, 2 Q	2944·2 R	9	0, 0 Q
2959·9 R	7	2, 2 R	2935·7 R	7	0, 0 R

WEAKER SYSTEM, 5600–5170 Å, $a^3\Pi - X^1\Sigma$

Occurrence. Discharge through BBr_3.

Appearance. Two sub-systems. $^3\Pi_1 - {}^1\Sigma$ gives double-headed bands, but $^3\Pi_0 - {}^1\Sigma$ gives single-headed ones. Degraded to the violet.

Reference. J. Lebreton, L. Marsigny and G. Bosser, *C. R. Acad. Sci. (Paris)*, C, **271**, 1113 (1970)†.

Q_2 heads of $^3\Pi_1 - {}^1\Sigma$			P_1 heads of $^3\Pi_0 - {}^1\Sigma$		
λ	I	v', v''	λ	I	v', v''
5489·8	3	(0, 1)	5347·2	6	(0, 0)
5466·8	2	(1, 2)	5326·3	4	(1, 1)
5293·1	10	(0, 0)	5306·2	2	(2, 2)
5273·5	7	(1, 1)			
5254·7	5	(2, 2)			

BCl

MAIN SYSTEM, $A^1\Pi - X^1\Sigma^+$

Occurrence. In high-frequency electrodeless discharge through BCl_3. Also in uncondensed discharge through helium with small amount of BCl_3 present.

Appearance. Well-developed sequences degraded to the violet with some bands degraded to the red.

References. E. Miescher, *Helv. Phys. Acta*, **8**, 279 (1935)†.

G. Herzberg and W. Hushley, *Canad. J. Res.*, A, **19**, 127 (1941)†.

G. Pannetier *et al.*, *C. R. Acad. Sci. (Paris)*, **258**, 1201 (1964).

Prominent heads as listed by Miescher. The letters R and V following the wavelength indicate that the head is degraded to longer or shorter wavelengths respectively.

λ	I	v', v''	λ	I	v', v''	λ	I	v', v''
2914·7 V		0, 3 Q	2786·7 V	8		2723·6 R	8	2, 2 ?
2880·6 R	6	7, 9 Q	2786·4 R	8		2722·2 V	10	0, 0 P
2867·1 R	7	5, 7 R	2784·4 V	8		2721·7 V	10	1, 1 Q
2860·4 R	10	5, 7 Q	2784·1 V	8		2720·0 V	10	0, 0 Q
2859·2 R	8	5, 7 Q*	2783·7 V	9		2714·2 R	8	
2857·3 R	7		2733·3 R	9	4, 4 Q	2669·7 R	8	3, 2 Q
2847·5 V	7	0, 2 Q	2727·8 V	9	3, 3 Q	2665·3 V	9	2, 1 Q
2796·1 R	8	4, 5 Q**	2727·4 R	10	3, 3 Q′	2664·9 R	9	2, 1 Q′
2792·7 R	8	4, 5 Q	2727·2 R	8	3, 3 Q′*	2660·2 V	8	1, 0 Q**
2792·4 R	9	4, 5 Q′	2724·0 V	10	2, 2 Q	2659·8 R	8	1, 0 Q′**

* Head due to isotope $B^{11}Cl^{35}$.

** Head due to isotope $B^{10}Cl^{37}$.

Other bands due to $B^{11}Cl^{37}$.

Bands whose analysis is not given are due to the piled up (0, 2), (0, 1) or (0, 0) sequences, the bands of which are very close at the head of the sequence.

Q heads of $B^{11}Cl^{35}$ and $B^{10}Cl^{35}$(‡) as given by Herzberg and Hushley.

λ	v', v''	λ	v', v''	λ	v', v''
2790·7	4, 5‡	2728·6	4, 4‡	2675·1	5, 4‡
2787·8	4, 5	2727·8	4, 4	2669·7	4, 3
2787·0	3, 4‡	2724·4	3, 3‡	2668·5	4, 3‡
2285·0	0, 1‡	2724·0	3, 3	2665·3	3, 2
2784·4	3, 4	2721·5	2, 2	2663·7	3, 2‡
2782·7	0, 1; 2, 3	2720·3	1, 1‡	2662·1	2, 1
2782·2	1, 2	2720·2	1, 1	2660·2	1, 0; 2, 1‡
2734·7	5, 5‡	2720·0	0, 0; 0, 0‡	2658·1	1, 0‡
2733·4	5, 5	2675·7	5, 4	2604·7	2, 0
				2660·7	2, 0‡

WEAKER SYSTEM, $a^3\Pi_1 - X^1\Sigma^+$

Reference. J. Lebreton, L. Marsigny and J. Ferran, *C. R. Acad. Sci. (Paris)*, C, **272**, 1094 (1971)†.

A sequence of double-headed violet-degraded bands obtained in a discharge through BCl_3 + He. Q heads, 4940·4 (0, 0), 4923·1 (1, 1), 4906·4 (2, 2).

BF

The main resonance system lies at the limit of the quartz ultra-violet. Two weaker triplet systems lie in the near ultra-violet. There are many weak singlet and triplet systems involving highly excited states in the visible and near infra-red. There are also many further systems involving Rydberg series in the extreme ultra-violet. This summary and the notation used are based on the detailed paper by Caton and Douglas.

Occurrence. In discharges of various types through BF_3 or $BF_3 + He$. Also some systems have been observed in flash absorption.

References. R. B. Dull, *P.R.*, **47**, 458 (1935)†.
H. M. Strong and H. P. Knauss, *P.R.*, **49**, 740 (1936)†.
M. Chrétien, *Helv. Phys. Acta*, **23**, 259 (1950)†.
R. F. Barrow *et al.*, *Proc. Phys. Soc.*, **71**, 61 (1958).
D. W. Robinson, *J. Mol. Spec.*, **11**, 275 (1963)†.
R. B. Caton and A. E. Douglas, *Canad. J. Phys.*, **48**, 432 (1970)†.

MAIN SYSTEM, $A^1\Pi - X^1\Sigma$

Appearance. Degraded to red. Marked sequences of close double-headed bands, somewhat complicated by the isotope splitting.

The following are the Q heads of the strongest bands of $B^{11}F$. R heads lie about 1 Å to longer wavelengths. Own estimates of intensity.

λ_{air}	I	v', v''	$\lambda_{vac.}$	I	v', v''
2071·5	3	1, 3	1962·7	9	1, 1
2067·4	4	0, 2	1957·4	10	0, 0
2015·8	7	1, 2	1911·0	10	1, 0
2011·0	8	0, 1	1867·8	8	2, 0

$b\,^3\Sigma - a\,^3\Pi$ SYSTEM

Appearance. Degraded to shorter wavelengths. Bands with five heads, similar in appearance to the CO third positive bands.

The following are the five heads of the (0, 0) band and P_3 heads of other strong bands from Strong and Knauss.

λ	I	v', v''	λ	I	v', v''	λ	I	v', v''
3549·8	4	0, 3	3222·9	1	1, 2	3120·3	9	0, 0 P_1
3396·9	3	0, 2	3124·1	2	0, 0 O	3118·4	9	0, 0 Q_2
3359·7	2	1, 3	3122·1	7	0, 0 P_3	2974·8	4	1, 0
3254·8	5	0, 1	3121·2	9	0, 0 P_2	2844·5	2	2, 0

$c\,^3\Sigma - a\,^3\Pi$ SYSTEM

Appearance. Similar to $b^3\Sigma - a^3\Pi$.

The following are the P_3 heads from Strong and Knauss.

λ	I	v', v''
2824·0	3	0, 2
2724·9	4	0, 1
2631·4	7	0, 0

BF (*contd.*)

WEAKER SINGLET SYSTEMS

Sequence heads from Caton and Douglas, with their notation for the transitions. No intensities are available. The direction of degradation is indicated as R or V.

$(4s\sigma)^1\Sigma^+ - (3s\sigma)^1\Sigma^+$ deg. R. 8620·2 (0, 0).

$(3s\sigma)^1\Sigma^+ - A^1\Pi$ deg. V. 8364·8 (0, 2), 7595·3 (0, 1), 6942·1 (0, 0), 6221·0 (1, 0), 5644·5 (2, 0), 5172·2 (3, 0).

$(3p\sigma)^1\Sigma^+ - A^1\Pi$ deg. V. 6414·2 (0, 2), 5951·6 (0, 1), 5543·0 (0, 0), 5098·6 (1, 0).

$(3d\delta)^1\Delta - A^1\Pi$ deg. V. 4377·9 (0, 2), 4157·7 (0, 1), 3953·9 (0, 0).

$(3d\pi)^1\Pi - A^1\Pi$ deg. V. 4166·3 (0, 2), 3966·2 (0, 1), 3780·2 (0, 0).

WEAK TRIPLET SYSTEMS

Heads of (0, 0) sequences from Caton and Douglas.

$d^3\Pi - b^3\Sigma^+$ deg. V. 10293·9.

$f^3\Pi - c^3\Sigma^+$ deg. V. 9586·7.

$e^3\Sigma^+ - b^3\Sigma^+$ deg. V. 6709·8.

$f^3\Pi - b^3\Sigma^+$ deg. V. 6095·8.

$g^3\Sigma^+ - b^3\Sigma^+$ deg. V. 5905·8.

$h^3\Pi - b^3\Sigma^+$ deg. V. 5200·0.

BH

References. S. F. Thunberg, *Z.P.*, **100**, 471 (1936).

G. M. Almy and R. B. Horsfall, *P.R.*, **51**, 491 (1937).

A. E. Douglas and G. Herzberg, *P.R.*, **57**, 752 (1940).

A. E. Douglas, *Canad. J. Res.*, A, **19**, 27 (1941).

J. W. C. Johns, F. A. Grimm and R. F. Porter, *J. Mol. Spec.*, **22**, 435 (1967).

4332 Å SYSTEM, $A^1\Pi - X^1\Sigma$

Bands degraded to the red with single P, Q and R branches. The Q heads are very intense through superposition of several lines. Observed in hollow cathode with boron and hydrogen, and in a discharge through $H_2 + BCl_3$. Also in absorption following flash photolysis of BH_3.

v', v''	Q heads	R heads
0, 0	4331·6	4245·9
1, 1	4367·3	4319·2
2, 2	4433·7	

3662 Å SYSTEM, $b^3\Sigma - a^3\Pi$

Emission under similar conditions to above system. Degraded to the red. The (0, 0) band has a Q head at 3693·8 and R head 3662·4.

3415 Å SYSTEM, $B^1\Sigma^+ - A^1\Pi$

Obtained in uncondensed discharge through He containing small amounts of H_2 and BCl_3. Single P, Q and R branches, degraded to shorter wavelengths. The structure is very open and the first lines of the Q branches are given: 3415·1 (0, 0), 3396·4 (1, 1).

BH (*contd.*)

3099 Å SYSTEM, $C^1\Sigma^+ - A^1\Pi$

Weakly degraded to shorter wavelengths. Origins 3100·1 (0, 0), 3073·4 (1, 1), 2972·4 (2, 2).

4340 Å SYSTEM, $C'^1\Delta - A^1\Pi$

This weak system is overlaid by the much stronger 4332 band. It was found by Johns *et al.* in absorption following flash photolysis. Weakly degraded to shorter wavelengths. Q heads 4340·4 (0, 0), 4260·2 (1, 1), 4154·5 (2, 2).

BH⁺

> *Reference.* G. M. Almy and R. B. Horsfall, *P.R.*, **51**, 491 (1937).

3768 Å SYSTEM, $^2\Pi - {}^2\Sigma$

Bands showing P, Q and R branches degraded to the red, each consisting of narrow doublets. Obtained with a hollow cathode of the Schüler type containing boron, hydrogen and helium.

v', v''	R heads	Q head
0, 0	3768·1	3792
1, 1	3803	

BH₂

> *Occurrence.* Absorption following flash photolysis of borine carbonyl H_3BCO.
> *Appearance.* A progression of headless bands with widely spaced fine structure 8700–6400 Å. Bands alternate in structure and there are not easily recognisable features. Absorption is strong around 8520, 7360 and 6870 Å.
> *Reference.* G. Herzberg and J. W. C. Johns, *P.R.S.*, **298**, 142 (1967)†.

BI

> *Reference.* A. G. Briggs and R. Piercy, *Spectrochim. Acta*, A, **29**, 851 (1973).
> In absorption following flash photolysis of BI_3, 21 maxima were recorded in three sequences, not clearly degraded either way. (0, 1) sequence 2786·4 to 2779·1, (0, 0) 2741·1 to 2713·2 and (1, 0) 2675·5 to 2659·9.

BN

> *Occurrence.* In a discharge through helium containing traces of BCl_3 and nitrogen.
> *Reference.* A. E. Douglas and G. Herzberg, *Canad. J. Res.*, A, **18**, 179 (1940)†.

TRIPLET SYSTEM, 3400–4000 Å, $A^3\Pi - X^3\Pi$

> *Appearance.* Bands with R heads degraded to longer wavelengths.
> R heads $B^{11}N$ and $B^{10}N$ (*),

BN (*contd.*)

λ	v', v''	λ	v', v''	λ	v', v''
3496·3	3, 2	3681·8	3, 3	3856·1	2, 3
3494·1	3, 2*	3653·2	2, 2	3836·2	1, 2*
3467·4	2, 1	3625·6	1, 1	3829·3	1, 2
3464·4	2, 1*	3599·2	0, 0	3809·3	0, 1*
3439·7	1, 0			3803·2	0, 1
3435·8	1, 0*				

SINGLET SYSTEM, 2900–3250 Å

Under the same conditions as for the triplet system three further bands were observed with their main heads at 3228·7, 3046·3 and 2897·8 and weaker heads at 3226·4, 3044·3 and 2895·9 Å. Possibly the (0, 1), (0, 0) and (1,0) bands of the same system, but the first is degraded violet, the others degraded red.

BO

There are three strong resonance systems, usually known as the α, β and γ systems, and a fourth weaker intercombination system.

α SYSTEM, λλ8519–3136, $A^2\Pi - X^2\Sigma$

Occurrence. In arcs containing B_2O_3 and in boron arc in air. The bands are also very well developed when a volatile boron compound such as BCl_3 is introduced into active nitrogen containing a trace of oxygen.

Appearance. Degraded to the red. Double double-headed bands (see Plate 10).

References. R. S. Mulliken, *P.R.*, **25**, 259 (1925)†.
F. A. Jenkins and A. McKellar, *P.R.*, **42**, 464 (1932)†.
A. Lagerqvist, N. E. L. Nilsson and K. Wigartz, *Ark. Fys.*, **13**, 379 (1958).
T. M. Dunn and L. Hanson, *Canad. J. Phys.*, **47**, 1657 (1969).
A. A. Mal'tsev, D. I. Kataev and V. M. Tatevskii, *Opt. Spectrosc.*, **9**, 376 (1960)†.

The strong heads are listed below. Intensities, by Mulliken, are for BCl_3 in active nitrogen. Only data for $B^{11}O$ are given.

λ	I	v', v''	λ	I	v', v''	λ	I	v', v''
6165·4	5	0, 4 Q_1	4585·7	7	0, 1 R_{21}	3950·5	4	3, 1 Q_1
6159·7	5	0, 4 R_1	4365·9	8	1, 1 Q_1	3848·7	10	2, 0 Q_1
5551·5	8	0, 3 Q_1	4363·4	10	1, 1 R_1	3847·0	9	2, 0 R_1
5547·5	7	0, 3 R_1	4341·9	8	1, 1 R_2	3829·9	8	2, 0 R_2
5513·0	5	0, 3 R_2	4339·4	8	1, 1 R_{21}	3828·0	6	2, 0 R_{21}
5043·5	6	0, 2 Q_1	4250·4	5	0, 0 Q_1	3679·1	10	3, 0 Q_1
5040·1	9 ?	0, 2 R_1	4247·9	4	0, 0 R_1	3677·8	8	3, 0 R_1
5011·6	4	0, 2 R_2	4227·5	4	0, 0 R_2	3662·3	6	3, 0 R_2
4746·9	8	1, 2 Q_1	4145·5	7	2, 1 Q_1	3660·6	5	3, 0 R_{21}
4744·0	8	1, 2 R_1	4143·4	6	2, 1 R_1	3526·8	7	4, 0 Q_1
4718·7	5	1, 2 R_2	4124·1	4	2, 1 R_2	3525·5	7	4, 0 R_1
4715·5	5	1, 2 R_{21}	4037·4	8	1, 0 Q_1	3511·3	6	4, 0 R_2
4615·4	10	0, 1 Q_1	4035·5	7	1, 0 R_1	3510·0	5	4, 0 R_{21}
4612·7	10	0, 1 R_1	4017·1	6	1, 0 R_2	3389·1	5	5, 0 Q_1
4588·8	8	0, 1 R_2	4015·0	5	1, 0 R_{21}	3387·6	7	5, 0 R_1
						3374·7	5	5, 0 R_2

BO (*contd.*)

β SYSTEM, $\lambda\lambda3645-2120$, $B^2\Sigma - X^2\Sigma$

Occurrence. As for α system.

Appearance. Degraded to the red. Single-headed bands.

Reference. R. S. Mulliken, *P.R.*, **25**, 259 (1925)†.

The strong bands as obtained by Mulliken in active nitrogen are listed below. The more abundant isotope $B^{11}O$ is given.

λ	I	v', v''	λ	I	v', v''	λ	I	v', v''
3493·1	4	4, 11	2934·9	9	3, 7	2551·4	9	0, 2
3441·6	6	3, 10	2892·2	10	2, 6	2544·3	7	3, 4
3391·2	5	2, 9	2850·6	8	1, 5	2507·7	6	2, 3
3354·6	5	5, 11	2809·9	8	0, 4	2472·0	5	1, 2
3305·4	8	4, 10	2793·9	7	3, 6	2437·1	10	0, 1
3256·9	9	3, 9	2753·4	9	2, 5	2433·3	6	3, 3
3209·3	7	2, 8	2713·8	10	1, 4	2398·5	10	2, 2
3134·6	6	4, 9	2703·4	4	4, 6	2364·5	8	1, 1
3088·6	9	3, 8	2675·3	8	0, 3	2331·3	7	0, 0
3043·6	9	2, 7	2664·1	4	3, 5	2330·4	6	3, 2
2999·7	7	1, 6	2625·6	6	2, 4	2264·8	6	1, 0
2978·5	6	4, 8	2588·0	8	1, 3	2234·6	4	3, 1
2956·6	6	0, 5	2581·6	4	4, 5	2203·0	4	2, 0

γ SYSTEM, 2090–1360 Å, $C^2\Pi - X^2\Sigma$

Occurrence. In discharges and probably arcs.

Appearance. Degraded to the red. Several heads to each band. Fairly obvious sequences.

R_2 heads from Mal'tsev *et al.*,

λ_{air}	v', v''	λ_{vac}	v', v''	λ_{vac}	v', v''
2037·4	1, 4	1996·0	1, 3	1878·9	0, 1
2016·7	0, 3	1946·1	0, 2	1815·4	0, 0

COMBINATION SYSTEM, $B^2\Sigma - A^2\Pi$

Occurrence. BCl_3 in active nitrogen containing oxygen.

Appearance. This is a weak system and is usually masked by the overlapping α system. The bands are not clearly degraded, but some are shaded slightly to the violet (see Plate 10).

Reference. R. S. Mulliken, *P.R.*, **25**, 259 (1925)†.

The following are the strongest bands as listed by Mulliken:

λ	I	v', v''	λ	I	v', v''
5916·2	1*	0, 2 B_1	5155·5	1	2, 2 B_1
5895·3	$\frac{1}{2}$	1, 3 B_1	4881·2	4*	1, 0 2,1 3,2 A_1
5493·7	$\frac{1}{2}$	2, 3 B_1	4850·9	5	1, 0 2,1 3,2 B_1
5201·1	1	0, 0 A_1	4580·8	1	3, 1 B_1
5189·3	1	2, 2 A_1	4576·5	$\frac{1}{2}$	2, 0 A_1

* Masked by α band.

BO⁺

Reference.   D. I. Kataev and A. A. Mal'tsev, *Vestnik. Moskov. Univ. Ser II, Khim.*, **22**, 23 (1967).

Violet-degraded bands observed in emission from a discharge. The origin of the (0, 0) band is at 3567·4. The (0, 1) and (1, 0) sequences lie near 3811 and 3335 Å.

B_2O_2 ?

PORTER AND DOWS BANDS

Reference.   R. F. Porter and D. A. Dows, *J. Chem. Phys.*, **24**, 1270 (1956).

In the violet glow of finely divided boron oxidising in air in a low-temperature furnace, emission was observed between 4250 and 4500 Å and between 4750 and 5100 Å. The tracing of the first group shows strong red-degraded heads near 4403 and 4344.

BO_2

BORIC ACID FLUCTUATION BANDS, BO_2

References.   W. Jevons, *P.R.S.*, **91**, 120 (1915)†.
N. L. Singh, *Curr. Sci.*, **11**, 276 (1942), *Proc. Indian Acad. Sci.*, A, **29**, 424 (1949)†.
W. E. Kaskan and R. C. Milliken, *J. Chem. Phys.*, **32**, 1273 (1960).
J. W. C. Johns, *Canad. J. Phys.*, **39**, 1738 (1961)†.

Waves of narrow bands are observed when boric acid is introduced into an arc or flame, or when finely divided boron is burnt. Work by Kaskan and Milliken indicates that the emitter is BO_2, not B_2O_3 as formerly supposed. Maxima of waves λλ6390, 6200, 6030, 5800, 5450, 5180, 4930, 4710, 4520. (See Plate 10).

The waves show some structure and BO bands are usually present as well. Singh has measured some features (strongest heads 5789·5, 5781·0, 5478·5, 5361·2) and suggested that they form a new system of BO with a transition ending on $B^2\Sigma$. The intensity distribution is unsatisfactory, and we do not think that excitation to a high electronic level of BO is likely in a flame source.

Johns has obtained the bands in absorption after flash photolysis of $BCl_3 + O_2$ and observed clear heads including ones at λλ5470·9, 5456·8, 5207·2, 5196·1, 5180·7, 5168·8 and a new system with the strongest heads at 4090·7 and 4065·5.

BS

Occurrence.   Discharge through B_2S_3 vapour in Ar or He.
References.   P. B. Zeeman, *Canad. J. Phys.*, **29**, 336 (1951)†.
J. K. McDonald and K. K. Innes, *J. Mol. Spec.*, **29**, 251 (1969)†.

α SYSTEM, $A^2\Pi - X^2\Sigma$

Appearance.   Extensive system of double double-headed bands, degraded to red. The structure is rather complex, partly because of overlapping of bands of B^{10} and B^{11} isotopes.

The following are probably among the outstanding heads for the stronger isotope.

BS (*contd.*)

λ	v', v''	λ	v', v''	λ	v', v''
8195·6	0, 3 Q_1	6258·1	2, 1 Q_1	5379·4	4, 0 Q_1
8191·6	0, 3 R_1	6130·1	2, 1 R_2	5378·0	4, 0 R_1
7978·4	0, 3 R_2	5986·1	3, 1 Q_1	5284·6	4, 0 R_2
7973·8	0, 3 $^{s}R_{21}$	5869·9	3, 1 R_2	5283·2	4, 0 $^{s}R_{21}$
7493·8	0, 2 Q_1	5594·9	3, 0 Q_1	5182·2	5, 0 Q_1
7311·7	0, 2 R_2	5593·5	3, 0 R_2	5093·5	5, 0 R_2

γ SYSTEM, $C^2\Pi - X^2\Sigma$

Appearance. Double double-headed bands, degraded to the red.

No intensities are available; the following are the $^{s}R_{21}$ heads of bands which are likely to be prominent, with all heads of the (0, 0) and (0, 1) bands.

λ	v', v''	λ	v', v''	λ	v', v''
2912·7	0, 4	2648·5	0, 1 $^{s}R_{21}$	2531·4	2, 1
2820·0	0, 3	2577·5	0, 0 Q_1	2512·4	1, 0
2732·1	0, 2	2576·8	0, 0 R_1	2478·2	3, 1
2657·6	0, 1 Q_1	2570·0	0, 0 R_2	2458·8	2, 0
2656·7	0, 1 R_1	2569·1	0, 0 $^{s}R_{21}$	2408·5	3, 0
2649·5	0, 1 R_2				

5000 Å SYSTEM, $B^2\Sigma - A^2\Pi$

Double double-headed bands forming sequences. Degraded to the violet. McDonald and Innes show good photographs but wavelengths are difficult to extract. Main heads of (0, 0) band 4995·6 and 4913·5. The (0, 1) has first head at 5164·2 Å.

5100 Å SYSTEM, $^2\Sigma - ^2\Pi$

Double double-headed band, degraded to the red. Heads of (0, 0) band 5130·5, 5125·7, 5046·4 and 5042·5.

3140 Å SYSTEM, $^2\Delta - A^2\Pi$

Degraded to the red. Heads of (0, 0) band near 3143, 3141 and 3139 Å.

BaBr

GREEN SYSTEM, $C^2\Pi - X^2\Sigma$

Occurrence. When barium bromide is introduced into a flame or arc, and in absorption.

Appearance. Marked close sequences; the rotational structure appears to be degraded to the red, while the vibrational structure is degraded to the violet.

References. K. Hedfeld, *Z.P.*, **68**, 610 (1931).
O. H. Walters and S. Barratt, *P.R.S.*, **118**, 120 (1928).

The following measurements of the Q heads of the sequences are by Hedfeld. Intensities I_a and I_e are for absorption and emission in an arc respectively, the former being by Walters and Barratt.

BaBr (*contd.*)

λ	I_a	I_e	Sequence
5415·9	7	4	0, 1 i
5360·1	10	10	0, 0 i
5305·5	6	2	1, 0 i
5260·6	4	2	0, 1 ii
5208·2	10	10	0, 0 ii
5156·4	5	1	1, 0 ii

ULTRA-VIOLET SYSTEMS, $D^2\Sigma - X^2\Sigma$ AND $E^2\Sigma - X^2\Sigma$

R. E. Harrington (see B. Rosen *et al.*, Ed., *Spectroscopic Data Relative to Diatomic Molecules*, (1970)) has reported single-headed violet-degraded bands in absorption. Sequence heads:

$D^2\Sigma - X^2\Sigma$		$E^2\Sigma - X^2\Sigma$	
3922·6	(0, 1)	3773·3	(0, 2)
3893·2	(0, 0)	3746·2	(0, 1)
3861·9	(1, 0)	3719·3	(0, 0)
3831·3	(2, 0)	3689·2	(1, 0)

BaCl

GREEN SYSTEM, $C^2\Pi - X^2\Sigma$

Occurrence. Barium chloride in carbon arc or flame, and in absorption.

Appearance. The system is only slightly degraded, and hence the heads are rather indefinite, the rotational and vibrational structure being degraded in opposite directions for some sequences.

References. O. H. Walters and S. Barratt, *P.R.S.*, **118**, 120 (1928)†.

K. Hedfeld, *Z.P.*, **68**, 610 (1931).

A. E. Parker, *P.R.*, **46**, 301 (1934)†.

The analyses proposed by Hedfeld and by Parker differ in some details. The following are probably the most obvious points for measurement under low dispersion. The letters R and V indicate that the sequences are degraded to longer or shorter wavelengths.

λ	I	Sequence
5320·8 V	3	0, 1 i
5240·5 R	10	0, 0 i
5213 V	1	0, 1 ii
5167 R	2	1, 0 i
5139·2 V ⎱ 5136 R ⎰	10	0, 0 ii
5066 V	1	1, 0 ii

ULTRA-VIOLET SYSTEMS

References. A. E. Parker, *P.R.*, **46**, 301 (1934)†.

R. E. Harrington (see B. Rosen *et al.*, Ed., *Spectroscopic Data Relative to Diatomic Molecules* (1970)).

Four $^2\Sigma - {}^2\Sigma$ systems of single-headed bands, all degraded to shorter wavelengths have been reported in absorption, and the first two also occur in emission in a carbon arc.

P heads of main sequences,

BaCl (*contd.*)

$D^2\Sigma - X^2\Sigma$		$E^2\Sigma - X^2\Sigma$		$F^2\Sigma - X^2\Sigma$		$G^2\Sigma - X^2\Sigma$	
3923·0	(0, 0)	3691·7	(0, 0)	3418·8	(0, 1)	3098·1	(0, 1)
3876·8	(1, 0)	3649·9	(1, 0)	3386·6	(0, 0)	3067	(0, 0)
3832·0	(2, 0)			3349·3	(1, 0)	3041·8	(1, 0)
				3313·0	(2, 0)	3011·8	(2, 0)

INFRA-RED SYSTEMS

References. R. F. Barrow and D. V. Crawford, *Nature*, **157**, 339 (1946).
B. Rosen *et al.*, Ed., *Spectroscopic Data Relative to Diatomic Molecules*, (1970).
Occurrence. In flames of pyrotechnic compositions.
Appearance. A number of rather compact sequences of bands degraded to longer wavelengths. Some sequences are of close double-headed bands. The violet edges of the two strongest sequences are at 8421 and 9098 Å. At least 80 heads have been found.
Strongest sequence heads:

$A^2\Pi_{\frac{1}{2}} - X^2\Sigma$		$A^2\Pi_{\frac{3}{2}} - X^2\Sigma$		$B^2\Sigma - X^2\Sigma$		$B^2\Sigma - X^2\Sigma$	
9658·0	(0, 0)	9334·4	(0, 1)	8832·8	(0, 2)	8244·7	(1, 0)
		9101·7	(0, 0)	8622·4	(0, 1)	8076·9	(2, 0)
		8895·3	(1, 0)	8420·8	(0, 0)		

BaF

Occurrence. All systems have been reported in absorption and thermal emission from a furnace. The green, extreme red and infra-red systems are also emitted by a carbon arc or flame containing BaF_2.
References. S. Datta, *P.R.S.*, **99**, 436 (1921)†.
T. E. Nevin, *Proc. Phys. Soc.*, **43**, 554 (1931)†.
F. A. Jenkins and A. Harvey, *P.R.*, **39**, 922 (1932)†.
C. A. Fowler, *P.R.*, **59**, 645 (1941)†.
J. Singh and H. Mohan, *J. Phys.*, B, **4**, 1395 (1971).

GREEN SYSTEM, 5139–4842 Å, $C^2\Pi - X^2\Sigma$

Degraded to the red. Closely grouped bands forming sequences with 'tails'. Appearance best seen from Datta's photographs. The strong (0, 0) sequence appears split into three sequences. The strongest heads, with our estimates of intensity are:

λ	I	Sequence
5000·6	8	0, 0 Q_1
4992·1	5	0, 0 R_1
4950·8	10	0, 0 R_2

EXTREME RED SYSTEM, λλ7734–6716, $B^2\Sigma - X^2\Sigma$

Appearance. Degraded to longer wavelengths. Double-headed bands.
R_2 and R_1 heads of strong bands. Intensities on scale of 8.

BaF (*contd.*)

λ	I	v', v''		λ	I	v', v''
7430·8	6	3, 4		6958·7	5	3, 2
7426·9				6955·9		
7142·0	8	1, 1		6935·1	5	2, 1
7138·8				6932·5		
7119·2	8	0, 0				
7116·0						

INFRA-RED SYSTEM, $\lambda\lambda 8738-7862$

Appearance. Degraded to longer wave-lengths. Marked sequences.

Transition.   A $^2\Pi - {}^2\Sigma$, ground state.

Strong bands only. Intensities on scale of 8.

λ	I	v', v''		λ	I	v', v''
8618·8	7	2, 2 i Q		8172·3	8	1, 1 ii Q
8595·3	8	1, 1 i Q		8158·7	7	1, 1 ii R
8571·5	7	0, 0 i Q		8151·0	8	0, 0 ii Q
8193·6	8	2, 2 ii Q		8137·0	7	0, 0 ii R

ULTRA-VIOLET SYSTEMS

Fowler obtained five of these systems in absorption. Singh and Mohan have obtained them in thermal emission, confirmed the splitting of D into D and D′, and found an extra system, $I - X^2\Sigma$. All systems are degraded to shorter wavelengths and probably have compact well-marked sequences.

$D^2\Sigma - X^2\Sigma$. 46 bands, 4290–3900 Å. Strongest 4135·8 (0, 0), 4051·3 (1, 0).

$D'^2\Sigma - X^2\Sigma$. 3949–3500 Å. 38 bands including:

λ	I	v', v''
3878·6	3	0, 1
3809·9	10	0, 0
3804·5	5	1, 1
3738·3	6	2, 2

$E^2\Sigma - X^2\Sigma$. 10 bands, 3608–3475 Å. Sequence heads:

λ	I	v', v''
3608·6	3	0, 1
3549·0	10	0, 0
3482·9	5	1, 0

$F^2\Sigma - X^2\Sigma$. 34 bands, 3451–3278 Å. Strongest:

λ	I	v', v''		λ	I	v', v''
3450·2	4	0, 1		3336·0	7	1, 0
3395·3	10	0, 0		3329·8	8	2, 1
3388·8	8	1, 1		3278·9	2	2, 0

BaF (*contd.*)

$G^2\Sigma - X^2\Sigma$. 19 bands, 3210–3069 Å. Strongest:

λ	I	v', v''	λ	I	v', v''
3224·2		0, 1	3125·9	6	1, 0
3176·5	10	0, 0	3121·9	7	2, 1
3172·1	6	1, 1	3077·1	2	2, 0

$H^2\Sigma - X^2\Sigma$. 10 bands. Strongest 3210·7 (0, 1), 3163·4 (0, 0), 3113·6 (1, 0).

$I - X^2\Sigma$. 3057–2965 Å. Sequence heads 3007 and 2965 Å.

BaH

10000 Å SYSTEM, $A^2\Pi - X^2\Sigma$.

Reference.　I. Kopp, M. Kronekvist and A. Guntsch, *Ark. Fys.*, **32**, 371 (1966)†.

Overlapping triple-headed bands, degraded to longer wavelengths obtained in absorption and in an arc (Ba or Cu poles) in H_2. Heads:

	$A^2\Pi_{\frac{3}{2}} - X^2\Sigma$			$A^2\Pi_{\frac{1}{2}} - X^2\Sigma$		
v', v''	R_{21}	R_2	Q_{21}	R_1	R_{12}	Q_1
1, 0				9357·2	9513·5	9513·7
2, 1				9446·0	9596·1	9594·7
0, 0	9874·4	10052·6	10056·0	10332·1	10602·9	10603·3
1, 1	9950·0	10117·3	10119·9	10411·9	10669·9	10670·5

8924 Å SYSTEM, $B^2\Sigma - X^2\Sigma$

Reference.　I. Kopp and R. Wirked, *Ark. Fys.*, **32**, 307 (1966)†.

Double-headed bands, degraded to longer wavelengths, in absorption and in an arc.

v', v''	R_2 head	R_1 head	v', v''	R_2 head	R_1 head
1, 2	–	9779·9	0, 0	8981·7	9026·1
0, 1	9666·8	9737·9	3, 2	8530·3	8558·2
2, 2	9081·5	9123·1	2, 1	8475·3	8504·1
1, 1	9030·9	9074·4	1, 0	8421·6	8451·2

6700 Å SYSTEM, $E^2\Pi - X^2\Sigma$

References.　G. W. Funke, *Z.P.*, **84**, 610 (1933).
A. Schaafsma, *Z.P.*, **74**, 254 (1932).
W. R. Fredrickson and W. W. Watson, *P.R.*, **39**, 753 (1932).

System of complex bands degraded to the violet obtained in barium arc in hydrogen at reduced pressure.

			Heads			
v', v''	$^OP_{12}$	P_1	$^PQ_{12}$	P_2	$^QP_{12}$	Q_2
0, 1	7481	7423	7422	7222	–	7174
0, 0	6923·5	6850·2	6848·6	6689·5	6635·1	6634·3
1, 1	–	6827·4	6825·8	6665·0	6610·8	6610·0
1, 0	–	–	–	–	6152·4	6151·2

BaH (*contd.*)

4500 Å SYSTEM, $D^2\Sigma - X^2\Sigma$

Reference. I. Kopp, N. Åslund, G. Edvinsson and B. Lindgren, *Ark. Fys.*, **30**, 321 (1965).

A weak progression, degraded to the red, observed in absorption in a King furnace and in emission from an arc at 100 torr pressure. Some bands are strongly perturbed, making location of heads difficult. 4552·6 (1, 0), 4473·0 (2, 0), 4394·4 (3, 0), 4330 (4, 0), 4250 (5, 0), 4178·3 (6, 0), 4120·4 (7, 0), 4060 (8, 0), 4010·0 (9, 0).

4228 Å SYSTEM, $C^2\Sigma - X^2\Sigma$

References. I. Kopp *et al.*, *Ark. Fys.*, **30**, 321 (1965)†.
G. W. Funke and B. Grundström, *Z.P.*, **100**, 293 (1936).

Weak bands, not clearly degraded either way, but strongly perturbed. They occur in absorption and in an arc at 100 torr pressure. Heads are difficult to locate because of the perturbations. Structure is strong from 4240 to 4140 and 4020 to 3940. The following heads are visible: 4208·9 (0, 0) deg. red, 4202 (1, 1), deg. V, 4013 (1, 0) deg. V and 3999·6 (2, 1) deg. V.

3380 Å SYSTEM, $F^2\Sigma - X^2\Sigma$

Reference. M. A. Khan, *J. Phys.*, B, **1**, 985 (1968).

Absorption in a King furnace. The (0, 1) band is degraded to shorter wavelengths and has its head at 3380·3.

3163 Å SYSTEM, $G^2\Sigma - X^2\Sigma$

Reference. M. A. Khan, as above.

Absorption in a King furnace. The (0, 0) band is degraded to shorter wavelengths and has a head at 3163·4.

BaI

Four systems are known; the green one is strongest.

GREEN SYSTEM, 5705–5280 Å, C (? $^2\Pi$) $- X^2\Sigma$

Occurrence. In absorption and in flames.
Appearance. Degraded to shorter wavelengths.
References. O. H. Walters and S. Barratt, *P.R.S.*, **118**, 120 (1928).
C. M. Olmsted, *Z. wiss. Photogr.*, **4**, 255 (1906).

Measurements by Walters and Barratt. Intensities I_a and I_f are for absorption and emission in a flame respectively, the latter being by Olmsted.

λ	I_a	I_f	v', v''
5609·5	10	10	0, 0 i
5381·7	10	7	0, 0 ii
5260	0		
5160	0		

SYSTEM D, 3970–3750 Å, $D^2\Sigma - X^2\Sigma$

Reference. B. R. K. Reddy and P. T. Rao, *J. Phys.*, B, **3**, 1008 (1970).
Occurrence. High frequency discharge, and in absorption.
Appearance. Degraded to shorter wavelengths. Single headed.

BaI (*contd.*)

Strongest heads:

λ	I	v', v''	λ	I	v', v''	λ	I	v', v''
3878·2	6	0, 2	3830·2	10	0, 0	3779·7	6	2, 0
3854·1	8	0, 1	3828·4	10	1, 1	3778·1	7	3, 1
3852·2	8	1, 2	3804·7	10	1, 0	3755·1	5	3, 0

SYSTEM E, 3800–3565 Å, $E^2\Sigma - X^2\Sigma$

Reference, occurrence and appearance as previous system.

λ	I	v', v''
3735·1	3	0, 0
3710·8	2	1, 0
3709·0	2	2, 1
3686·9	1	2, 0

4500–4300 Å SYSTEM, $?^2\Pi - E^2\Sigma$

Reference.   S. G. Shah, M. M. Patel and A. B. Darji, *J. Phys.*, B, **5**, L 191 (1972)†.

Two relatively strong (0, 0) sequences at 4411·5 and 4304·8, and weaker (0, 1) and (1, 0). Degraded to the red. Obtained in a high frequency discharge.

BaO

MAIN SYSTEM, 7100–4600 Å, $A^1\Sigma - X^1\Sigma$

Occurrence.   When barium salts are introduced into carbon arc or flame. Also in emission and absorption in shock tubes.

Appearance.   Degraded to longer wavelengths.

References.   P. C. Mahanti, *Proc. Phys. Soc.*, **46**, 51 (1934)†.

A. Lagerqvist, E. Lind and R. F. Barrow, *Proc. Phys. Soc.*, A, **63**, 1132 (1950).

The system extends from λ7905 to λ4269. Only the strong bands are listed below.

λ	I	v', v''	λ	I	v', v''	λ	I	v', v''
7097·4	5	0, 4	6039·6	9	1, 1	5349·7	8	4, 0
6782·8	8	0, 3	5976·3	3	0, 0	5214·7	7	5, 0
6493·1	9	0, 2	5864·5	10	2, 1	5086·7	6	6, 0
6291·0	8	1, 2	5805·1	6	1, 0	4965·4	3	7, 0
6224·7	6	0, 1	5701·0	8	3, 1	4850·6	6	8, 0
6165·1	6	3, 3	5644·1	9	2, 0	4680·3	5	11, 1
6102·3	5	2, 2	5492·7	10	3, 0			

PARKINSON'S SYSTEM, 3900–2900 Å, $B(?\ ^1\Pi) - X^1\Sigma$

Occurrence.   Shock heating of barium peroxide.

Appearance.   Degraded to the red. 27 bands obtained.

Reference.   W. H. Parkinson, *Proc. Phys. Soc.*, **78**, 705 (1961)†.

BaO (*contd.*)

Strongest heads:

λ	I	v', v''	λ	I	v', v''	λ	I	v', v''
3630·4	4	0, 8	3320·0	4	0, 4	3116·5	8	3, 3
3548·1	5	0, 7	3273·2	5	1, 4	3054·2	10	3, 2
3468·9	7	0, 6	3226·8	7	2, 4	2994·0	5	3, 1
3393·0	8	0, 5	3204·0	6	1, 3	2957·0	9	4, 1
3344·5	7	1, 5	3159·6	8	2, 3	2922·6	7	5, 1

VIOLET SYSTEM, $A'^1\Pi - X^1\Sigma$

Reference. C. J. Hsu *et al.*, *J. Mol. Spec.*, **53**, 273 (1974)†.

A long (v', 0) progression of 24 red-degraded bands observed in emission from barium metal vapour burning in N_2O, NO_2 or O_2. The following are strong on the photograph: λλ4265·9 (14, 0), 4195·7 (15, 0), 4128·0 (16, 0), 4063·4 (17, 0), 4001·1 (18, 0).

Barium Oxide

In addition to the above system of BaO, arc spectra also show more complex bands; the emitter is uncertain, but may be polyatomic and Gaydon favours Ba_2O_2.

There are band groups near 4800, 5020, 5500 and in the infra-red. Heads, degraded to the red, have been observed at:

λλ4729·9, 4739·6, 4789·1, 4830·5, 4844·5, 4846·4, 4855·1
λλ5007·0, 5009·6, 5012·6, 5015·1, 5017·6
λλ8176, 8652.

The 5500 group shows two main bands 5445–5452 and 5482–5488.

Reference. M. Charton and A. G. Gaydon, *Proc. Phys. Soc.*, A, **69**, 520 (1956)†.

BaOH

The strong green colouration of flames containing barium salts has recently been shown to be due to the triatomic hydroxide, not the oxide.

Occurrence. Strongly in hydrogen or hydrocarbon flames containing barium salts, and also in an arc in an atmosphere of water vapour.

Appearance. Diffuse headless bands in the green and infra-red.

References. C. G. James and T. M. Sugden, *Nature*, **175**, 333 (1955).
E. M. Bulewicz, *Nature*, **177**, 670, (1956).
M. Charton and A. G. Gaydon, *Proc. Phys. Soc.*, A, **69**, 520 (1956)†.

Green bands, maxima λλ4870 and 5120.
Infra-red bands, 7120–7580 and 7830–8390.

BaS

Occurrence. Absorption in furnace.

Appearance. Two systems, both degraded to the red.

Reference. R. F. Barrow, N. G. Burton and P. A. Jones, *Trans. Faraday Soc.*, **67**, 902 (1971).

BaS (*contd.*)

SYSTEM A, $A^1\Sigma - X^1\Sigma$

R heads:

λ	v', v''	λ	v', v''
7503·8	0, 3	6918·3	0, 0
7298·9	0, 2	6777·3	1, 0
7103·9	0, 1	6652·4	2, 0

SYSTEM B, $B^1\Sigma - X^1\Sigma$

28 bands. The following R heads probably serve for identification:

λ	v', v''	λ	v', v''	λ	v', v''	λ	v', v''
4037·6	0, 6	3826·5	1, 3	3702·3	0, 0	3601·4	3, 0
3978·2	0, 5	3808·8	0, 2	3651·7	3, 1	3585·7	5, 1
3920·1	0, 4	3772·7	1, 2	3635·1	2, 0	3569·1	4, 0
3864·0	0, 3	3719·7	1, 1	3618·3	4, 1	3537·1	5, 0

BeBr

VIOLET SYSTEM, 3920–3660 Å, $A^2\Pi - X^2\Sigma$

Occurrence. High-frequency discharge.

Appearance. Sequences of double double-headed bands, slightly degraded to the red.

References. Y. P. Reddy and P. T. Rao, *J. Phys.*, B, **1**, 482 (1968)†.

B. R. Reddy, Y. P. Reddy and P. T. Rao, *J. Phys.*, B, **3**, L 1 (1970)†.

Prominent heads:

λ	I	v', v''	λ	I	v', v''	λ	I	v', v''
3898·4	5	0, 1 Q_1	3793·9	10	0, 0 Q_1	3758·9	33	0, 0 ^SR
3868·6	4	0, 1 R_2	3787·8	5	0, 0 R_1	3696·7	33	1, 0 Q_1
3798·4	5	2, 2 Q_1	3769·1	5	1, 1 R_2	3675·4	33	2, 1 R_2
3796·0	8	1, 1 Q_1	3765·8	8	0, 0 R_2	3671·0	33	1, 0 R_2

BeCl

MAIN SYSTEM, 3686–3468 Å, $A^2\Pi - X^2\Sigma$

Occurrence. Be arc in Cl_2, microwave discharge, and absorption.

Appearance. Double double-headed bands in well-marked sequences. Degraded to the red.

References. W. R. Fredrickson and M. E. Hogan, *P.R.*, **46**, 454 (1934)†.

M. M. Novikov and L. N. Tunitskii, *Opt. Spectrosc.*, **8**, 396 (1960)†.

R. Colin, M. Carleer and F. Prevot, *Canad. J. Phys.*, **50**, 171 (1972)†.

Main heads of $BeCl^{35}$ with own estimates of intensity:

λ	I	v', v''	λ	I	v', v''
3676·8	6	0, 1 R_2	3570·9	8	0, 0 R_1
3578·4	7	1, 1 Q_1	3567·9	10	0, 0 R_2
3575·7	9	0, 0 Q_1	3559·2	1	0, 0 $^SR_{21}$
3571·5	8	1, 1 R_2	3468·3	4	1, 1 R_2

BeCl (*contd.*)

2610 Å SYSTEM

Novikov and Tunitskii (as above) have found a red-degraded sequence with first head at 2610·2 Å. 22 heads are listed 2610–2638.

BeF

MAIN SYSTEM, 3393–2816 Å, $A^2\Pi - X^2\Sigma$

Occurrence. Beryllium fluoride in carbon arc and in absorption; also in a discharge tube.
Appearance. Degraded to red. Marked sequences.
References. W. Jevons, *P.R.S.*, **122**, 211 (1929)†.
F. A. Jenkins, *P.R.*, **35**, 315 (1930)†.
C. A. Fowler, *P.R.*, **59**, 645 (1941)†.
V. M. Tatevskii, L. K. Tunitskii and M. M. Novikov, *Optika i Spektrosk*, **5**, 520 (1958).

Bands in ultra-violet, λλ3393–2816. Strongest bands only listed. The R_2 heads only are given except for (0, 0) band. Own intensities from published photographs.

λ	I	v', v''		λ	I	v', v''
3126·1	8	0, 1		3009·6	9	0, 0 R_2
3018·0	9	1, 1		2909·0	7	1, 0
3013·0	6	0, 0 Q_1		2816·0	3	2, 0
3009·9	10	0, 0 R_1				

FAR ULTRA-VIOLET SYSTEMS, $B^2\Sigma - X^2\Sigma$ AND $C^2\Sigma - X^2\Sigma$

Reference. M. M. Novikov and L. V. Gurvich, *Opt. Spectrosc.*, **23**, 173 (1967)†.

Two overlapping systems, obtained in a Schuler-type discharge. Both are degraded to shorter wavelengths with marked sequences and strong (0, 0).

$B^2\Sigma - X^2\Sigma$		$C^2\Sigma - X^2\Sigma$	
λ	v', v''	λ	v', v''
2068·1	0, 1	2032·7	0, 1
2016·6	0, 0	1983·4 (λ vac)	0, 0

WEAK VISIBLE SYSTEM, $C^2\Sigma - A^2\Pi$

Reference. K. M. Rao and P. T. Rao, *Indian J. Pure Appl. Phys.*, **3**, 177 (1965).
Emission in a furnace. Sequence heads 6214 (0, 1), 5799 (0, 0), 5264 (1, 0), 4969 (2, 0).

BeH

References. W. W. Watson, *P.R.*, **32**, 600 (1928).
E. Olsson, *Z.P.*, **73**, 732 (1932).
W. W. Watson and R. F. Humphreys, *P.R.*, **52**, 318 (1937).

4988 Å SYSTEM, $A^2\Pi - X^2\Sigma$

Bands show P, Q and R branches degraded at first to the violet but turning to the red at high values of the rotational quantum number.

Obtained in arc between beryllium poles in hydrogen at a few cm. pressure.

BeH (*contd.*)

v', v''	Origins	Q Heads
0, 1	5537·2	
1, 2	5507·9	
0, 0	4988·3	4990·8
1, 1	4983·3	4985·7
2, 2	4980·5	4982·8

1960 Å SYSTEM, $B^2\Pi - X^2\Sigma$

Principal feature of this system is a single strong Q branch degraded to short wavelengths from a head at 1960 Å.

Obtained in a hollow cathode of molybdenum and in a beryllium arc in hydrogen.

v', v''	Q Heads
0, 0	1960
1, 1	1956
1, 0	1882

BeH⁺

References. W. W. Watson and R. F. Humphreys, *P.R.*, **52**, 318 (1937).
W. W. Watson, *P.R.*, **32**, 600 (1928).

2559 Å SYSTEM, $A^1\Sigma - X^1\Sigma$

An extensive system of singlet bands degraded to the red. Obtained from an arc between beryllium electrodes in hydrogen at low pressures.

v', v''	R Heads	v', v''	R Heads
2, 0	2384·6	1, 3	2910·9
1, 0	2468·1	0, 3	3039·6
0, 0	2559·4	1, 4	3081·8
0, 1	2707·4		

BeI

4500–4100 Å, $A^2\Pi - X^2\Sigma$

Occurrence. High-frequency discharge through I_2 over heated metal.
References. P. S. Murty and P. T. Rao, *Curr. Sci. (India)*, **38**, 187 (1969).
P. S. Murty and P. T. Rao, *Proc. Roy. Irish Acad.*, **72A**, 71 (1972)†.

The strongest feature is the almost headless (0, 0) band of the $^2\Pi_{\frac{1}{2}} - {}^2\Sigma$ sub-system with origin at 4247·4 Å. Other red-degraded heads are:

λ	I	v', v''	λ	I	v', v''
4359·5	4	0, 1 Q_1	4170·2	2	0, 0 $^SR_{21}$
4288·9	2	0, 1 R_2	4141·2	2	1, 0 Q_1
4246·8	10	0, 0 Q_1	4135·5	1	1, 0 R_1
4180·5	8	0, 0 R_2			

BeO

There is a strong system in tbe blue-green, a moderately intense system in the far red, a weaker intercombination system in the same region and two or more systems in the ultra-violet.

Occurrence. Be salts in carbon arc and in arc between Be electrodes in air.

References. E. Bengtsson, *Ark. Mat. Astr. Fys.*, **20A**, No. 28 (1928).

L. Herzberg, *Z.P.*, **84**, 571 (1933)†.

A. Harvey and H. Bell, *Proc. Phys. Soc.*, **47**, 415 (1935).

A. Lagerqvist and R. Westöö, *Ark. Mat. Astr. Fys.*, **32A**, No. 10 (1945)†.

A. Lagerqvist, *Ark. Mat. Astr. Fys.*, **34B**, No. 23 (1947).

A. Lagerqvist, *Dissertation, Uppsala* (1948).

BLUE-GREEN SYSTEM, 5495–4180 Å, $B^1\Sigma - X^1\Sigma$

Appearance. Degraded to the red. Marked sequences of single-headed bands.

The following wavelengths and intensities are for the system as observed with an arc in air between beryllium electrodes:

λ	I	v', v''	λ	I	v', v''	λ	I	v', v''
5475·7	5	2, 4	4794·9	3	4, 4	4451·7	8	2, 1
5461·7	5	1, 3	4775·7	5	3, 3	4427·3	6	1, 0
5444·9	4	0, 2	4755·0	7	2, 2	4290·9	4	7, 5
5127·7	5	4, 5	4732·6	9	1, 1	4271·0	4	6, 4
5112·4	7	3, 4	4708·6	10	0, 0	4250·1	5	5, 3
5095·1	8	2, 3	4516·5	5	5, 4	4227·9	5	4, 2
5075·7	8	1, 2	4496·3	7	4, 3	4204·6	4	3, 1
5054·4	7	0, 1	4474·7	8	3, 2	4180·1	3	2, 0

FAR RED SYSTEM, 11600–5600 Å, $A^1\Pi - X^1\Sigma$

References. L. Herzberg, *Z.P.*, **84**, 571 (1933)†.

A. Lagerqvist and R. Westöö, *Ark. Mat. Astr. Fys.*, **31A**, No. 21 (1945)†.

A. Lagerqvist, *Ark. Mat. Astr. Fys.*, **34B**, No. 23 (1947).

Appearance. Degraded to longer wavelengths. Double-headed bands, but with R heads rather faint and diffuse.

The origins of the strong bands are listed below. The wavelengths shorter than 7954 Å are due to Herzberg and the others to Lagerqvist and Westöö. The v' values are those given by Lagerqvist (1947).

λ_0	v', v''	λ_0	v', v''	λ	I	v', v''
11234·6	1, 1	8468·5	5, 2	7953·3	6	3, 0
10826·6	0, 0	8206·0	4, 1	7324·8	3	4, 0
10344·7	3, 2	7954·5	3, 0	6523·5	4	7, 1
9647·5	1, 0			6344·4	3	6, 0
9002·7	3, 1			6286·9	3	9, 2
8713·4	2, 0			6117·8	3	8, 1

WEAK FAR RED SYSTEM, $B^1\Sigma - A^1\Pi$

Lagerqvist records 16 weak violet-degraded bands. Origins at 8358 (0, 0), 7507 (1, 0) and 6821 (2, 0) may assist identification.

BeO (*contd.*)

ULTRA-VIOLET SYSTEMS

Lagerqvist observed many bands 3600–2600 Å, mostly degraded to the red but some possibly shaded to shorter wavelengths. They fall into five groups similar to sequences, but the structure is complex and they may be due to two or more singlet systems and an overlapping triplet system. The more probable singlet systems are:

Bengtsson's Ultra-violet System, $D(? \, ^1\Pi) - A^1\Pi$

Degraded to the red. Strong heads:

λ	I	v', v''	λ	I	v', v''
3367·6	2	0, 2	3145·7	3	1, 1
3258·1	3	1, 2	3134·0	3	0, 0
3247·7	3	0, 1	3039·0	2	1, 0

Harvey and Bell's Ultra-violet System, $C(?^1\Pi) - A^1\Pi$

Degraded to the red. No intensities are given but the (0, 0) is probably the strongest. 3496·0 (0, 1), 3371·1 (1, 1), 3363·7 (0, 0), 3247·6 (1, 0).

BeS

Reference.   C. J. Cheetham, W. J. M. Gissane and R. F. Barrow, *Trans. Faraday Soc.*, **61**, 1308 (1965)†.

Singlet systems in the violet and near infra-red occur in absorption and thermal emission in a furnace. A weak triplet band at 3851 Å occurs in absorption.

VIOLET SYSTEM, $B^1\Sigma - X^1\Sigma$

Degraded to the red. The strongest band is the (0, 0) with head at 3863·2 Å. Band origins:

λ_0	v', v''	λ_0	v', v''
4039·1	1, 2	3786·0	3, 2
4017·7	0, 1	3764·5	2, 1
3864·6	0, 0	3742·8	1, 0

INFRA-RED SYSTEM, $A^1\Pi - X^1\Sigma$

Strongly degraded to longer wavelengths, so that heads are close to the origins. Origins of the principal bands:

λ_0	v', v''	λ_0	v', v''	λ_0	v', v''
8669·5	5, 0	8164·6	0, 0	7489·5	9, 1
8554·4	8, 2	7896·7	8, 1	7326·3	8, 0
8356·4	7, 1	7720·3	7, 0	7282·3	11, 2

TRIPLET BAND

Degraded to the red. Possibly due to a $^3\Pi - ^3\Pi$ transition. Heads 3850·7, 3853·7 and 3856·3.

Bi_2

References. G. M. Almy and F. M. Sparks, *P.R.*, **44**, 365 (1933)†.
G. Nakamura and T. Shidei, *Japan. J. Phys.*, **10**, 11 (1935)†.
S. P. Reddy and M. K. Ali, *J. Mol. Spec.*, **35**, 285 (1970)†.

VISIBLE SYSTEM, 7980–4500 Å, A — X

Occurrence. In microwave discharges, thermal emission, fluorescence and absorption.

Appearance. Degraded to the red. A very extensive system; Reddy and Ali record 270 bands 7980–4830 in emission.

The following are the strongest bands as listed by Nakamura and Shidei, in absorption:

λ	I	v', v''	λ	I	v', v''	λ	I	v', v''
5679·9	5	3, 3	5453·8	7	6, 1	5293·4	10	9, 0
5625·2	5	3, 2	5415·5	7	7, 1	5258·2	10	10, 0
5531·2	5	4, 1	5365·5	8	7, 0	5224·1	10	11, 0
5491·9	7	5, 1	5329·5	8	8, 0	5190·3	9	12, 0

VIOLET SYSTEM, 4200–4000 Å, D — A

Occurrence. In absorption at high temperature.

Appearance. Degraded to the red.

Strongest bands as observed by Almy and Sparks:

λ	I	v', v''
4150·7	4	0, 3
4128·0	5	0, 2
4105·5	5	0, 1
4064·5	2	1, 0

ULTRA-VIOLET SYSTEM, 2900–2600 Å, C — X

Occurrence. In absorption. This system appears to come up very easily and has been observed as an impurity, especially with cadmium.

Strong bands as recorded by Almy and Sparks:

λ	I	v', v''	λ	I	v', v''
2810·2	5	1, 6	2755·9	4	0, 1
2796·9	7	1, 5	2744·5	6	1, 1
2783·7	5	1, 4	2731·6	9	1, 0
2782·2	4	0, 3	2720·7	7	2, 0
2768·9	7	0, 2	2710·3	5	3, 0

FAR ULTRA-VIOLET SYSTEM, 2250–2000 Å

Weak diffuse bands observed in absorption; probably from ground state. Strong bands observed by Almy and Sparks, 2205·4, 2197·2, 2188·8, 2180·5, 2172·7, 2148·8, 2142·9, and 2135·8.

WEAK RED SYSTEMS

Reddy and Ali found three systems in a high-frequency discharge.
G — A, 8820–8030 Å. 16 bands degraded to the red.
H — A, 7050–6730 Å.   6 bands degraded to the violet.
I — A, 6570–6290 Å. 10 bands degraded to the violet.

BiBr

A number of systems have been observed in absorption.

λλ5438–4595. A weak system of red-degraded bands, strongest head 5246·5.

λλ4130–3862. Strongest system. Not clearly degraded either way; strong flutings from 4041 Å getting weaker to shorter Å.

λλ2968–2709. Strongest heads: 2856·0 (1, 0), 2814·7 (2, 0).

λλ2449–2336. Strongest heads: 2412·5, 2400·5, 2388·5 (0, 0), 2375·1, 2363·5, 2336·5.

λλ2350–2222. Strongest heads: 2350·6, 2306·2.

References. F. Morgan, *P.R.*, **49**, 41 (1936)†.

P. K. Sur and K. Majumdar, *Proc. Nat. Inst. Sci. India*, **20**, 235 (1954).

S. P. Singh, *Indian J. Pure Appl. Phys.*, **6**, 299 (1968).

BiCl

References. F. Morgan, *P.R.*, **49**, 41 (1936)†.

S. K. Ray, *Indian J. Phys.*, **16**, 35 (1942)†.

V. S. Rao and P. T. Rao, *Indian J. Phys.*, **39**, 65 (1965)†.

R. Yamdagni, *J. Mol. Spec.*, **35**, 149 (1970)†.

Occurrence. Three extensive systems have been observed; in emission from the flame surrounding a carbon arc containing $BiCl_3$ and in a high-frequency discharge; in absorption at $800–1000°C$.

BLUE-GREEN SYSTEM, 5660–4308 Å, A 0^+ − X 0^+

Appearance. Degraded to the red.

Wavelengths and intensities according to Ray are as follows:

λ	I	v', v''	λ	I	v', v''	λ	I	v', v''
5205·7	4	2, 10	4796·3	5	0, 3	4549·9	4	1, 0
5107·4	5	1, 8	4727·6	4	0, 2	4506·7	5	2, 0
5011·6	4	0, 6	4680·1	4	1, 2	4465·3	4	3, 0
4938·4	5	0, 5	4614·4	8	1, 1	4425·9	4	4, 0
4866·4	6	0, 4	4569·9	6	2, 1			

VIOLET SYSTEM, 4220–4000 Å, A′ 0^+ − X 0^+

Degraded to the red. A relatively strong $v'' = 0$ progression of single-headed bands. Strongest bands from Yamdagni, whose values of v' are 3 greater than those of Rao and Rao.

λ	I	v', v''	λ	I	v', v''	λ	I	v', v''
4323·7	9	2, 1	4197·0	8	4, 0	4102·6	9	7, 0
4286·8	8	3, 1	4164·1	10	5, 0	4073·8	9	8, 0
4251·4	9	4, 1	4132·8	9	6, 0	4046·9	10	9, 0
4231·1	8	3, 0						

ULTRA-VIOLET SYSTEM, 4000–3600 Å, B − X 0^+

Appearance. Degraded to the red.

Bands according to Ray:

BiCl (*contd.*)

λ	I	v′, v″		λ	I	v′, v″
3914·3	10	0, 0		3865·0	5	4, 4
3900·3	9	1, 1		3854·4	6	1, 0
3888·4	7	2, 2		3842·1	5	2, 1
3875·8	6	3, 3		3830·9	5	3, 2

In addition to these heads due to $BiCl^{35}$, some heads due to $BiCl^{37}$ were also observed.

BiF

References. H. G. Howell, *P.R.S.*, **155**, 141 (1936)†.
G. D. Rochester, *P.R.*, **51**, 486 (1937).
M. M. Patel and P. S. Narayanan, *Indian J. Pure Appl. Phys.*, **5**, 223 (1967)†.
K. M. Rao and P. T. Rao, *Indian J. Phys.*, **39**, 572 (1965)†.
A. K. Chaudhry *et al.*, *J. Phys.*, B, **1**, 1223 (1968).
B. S. Mohanty, D. K. Rai and K. N. Upadhya, *J. Phys.*, B, **1**, 523 (1968).

VISIBLE SYSTEM, A, 5700–4150 Å, $A(?^3\Sigma^-) - X_1 (?^3\Sigma^-)$

Occurrence. High-frequency discharge through BiF_3, and in absorption.
Appearance. Degraded to the red. Over 100 bands have been recorded.
The following are the main sequence heads from Howell:

λ	I	Sequence
4568·2	8	0, 2
4465·5	10	0, 1
4366·7	10	0, 0
4295·8	8	1, 0

SYSTEM B, 4000–3700 Å, $B - X_1$

Occurrence. In absorption and in transformer discharge.
Appearance. Degraded to the violet. Close triple-headed bands.
First heads of main bands from Patel and Narayanan:

λ	I	v′, v″		λ	I	v′, v″
3915·8	7	0, 1		3812·5	7	2, 2
3899·3	6	1, 2		3799·6	6	3, 3
3841·0	9	0, 0		3751·1	3	1, 0
3826·5	6	1, 1				

Bands of a weak red-degraded $A_3 - X_1$ system, with (0, 0) head at 3833·9, overlap this B system.

ULTRA-VIOLET SYSTEMS, C

Occurrence. In active nitrogen and in high-frequency and hollow-cathode discharges through BiF_3.
Appearance. All degraded to shorter wavelengths. These weak bands have been provisionally analysed into several systems and sub-systems.
3250–3000. $C (^3\Pi_1) - X_3 (^1\Sigma^+)$ 3158·5 (0, 1), 3106·8 (0, 0), 3049·1 (1, 0).
$C (^3\Pi_0) - X_3 (^1\Sigma^+)$ 3105·2 (0, 0).

BiF (*contd.*)

2850–2650.   C $(^3\Pi_2) - X_2$ $(^1\Delta_2)$ 2744·5 (0, 1), 2704·4 (0, 0).

C $(^3\Pi_1) - X_2$ $(^1\Delta_2)$ 2743·2 (0, 1), 2703·2 (0, 0).

2350–2250. $^1\Pi - X_1$ $(^3\Sigma^-)$ 2338 (0, 3), 2284·5 (0, 1), 2258 (0, 0).

BiH

Reference.   A. Heimer, *Z.P.*, **95**, 328 (1935)†.

M. A. Khan and Z. M. Khan, *Proc. Phys. Soc.*, **88**, 211 (1966).

Occurrence.   The first three band systems described below have been obtained in a bismuth arc in H_2 at reduced pressure and in discharge tubes. The 3059 Å system has been observed in absorption.

4698 Å SYSTEM, B $0^+ - X\,0^+$

Bands slightly degraded to the violet. Strongest system:

Origins

λ	v'. v''	
4361	1, 0	
4697	1, 1	
4698	0, 0	(0, 0) P branch closes up to a weak head near 4736 Å just beyond the strong Bi line 4722 Å.
5071	1, 2	
5089	0, 1	

6118 Å SYSTEM, B $0^+ - A\,1$

Bands slightly degraded to the violet.

λ	*Origins* v', v''	*Heads*
6118	0, 0	6117·5 Q
6792	1, 2	6792·2 Q
6814	0, 1	6813·5 Q

4842 Å SYSTEM, $D^1\Sigma - C^1\Sigma$

Bands slightly degraded to the red. Weak system.

λ	*Origins* v', v''
4842	0, 0
5171	0, 1

3059 Å SYSTEM, $E^1\Sigma - X\,0^+$

Single-headed bands, degraded to the red. R heads 3219·6 (0, 0), 3108·7 (1, 1), 3059·1 (0, 0) and 2958·8 (1, 0).

BiI

References.   F. Morgan, *P.R.*, **49**, 41 (1936)†.

P. T. Rao, *Indian J. Phys.*, **23**, 379 (1949)†.

S. P. Singh, *Indian J. Pure Appl. Phys.*, **6**, 445 (1968).

BiI (*contd.*)

Bands $\lambda\lambda 4308$–4164 have been observed in absorption and in a high frequency discharge; the direction of degradation is uncertain, and appearance of three or four groups of narrow bands is best judged from Rao's photographs; (0, 0) head at 4271·2.

A weaker system $\lambda\lambda 5900$–5800 occurs in the high frequency discharge. Four red-degraded sequences with heads at $\lambda\lambda 5859\cdot 9$ (0, 1), 5802·2 (0, 0), 5737·8 (1, 0) and 5676·6 (2, 0).

Another system B $-a$ is reported by Singh.

BiO

Occurrence. The main system is the visible and infra-red occurs readily in arcs, flames and in absorption. The 3500 Å system has been obtained in a discharge and in absorption in a furnace. Other systems have been reported in absorption only, and as they are predissociated may be difficult to obtain in emission.

References. O. Scari, *Acta. Phys. Hungar.*, **6**, 73 (1956)†.
N. K. Bridge and H. G. Howell, *Proc. Phys. Soc.*, A, **67**, 44 (1954)†.
W. J. M. Gissane and R. F. Barrow, *Proc. Phys. Soc.*, **85**, 1048 (1965)†.
R. F. Barrow, W. J. M. Gissane and D. Richards, *P. R. S.*, **300**, 469 (1967).

This molecule is close to case c coupling; transitions are indicated by Ω values, with probable equivalent case a or b types indicated in parentheses.

MAIN SYSTEM, A $\frac{1}{2}(^2\Pi_{\frac{1}{2}}) - X_1\ \frac{1}{2}(^2\Pi_{\frac{1}{2}})$

Appearance. Single-headed bands, degraded to the red.

There are a fairly large number of bands, and it is difficult to assign intensities because of varying plate sensitivity. The following appear prominent in Scari's photographs; those marked with an asterisk are especially strong. The $v'' = 0$ progression is strong in absorption.

λ	v', v''	λ	v', v''	λ	v', v''	λ	v', v''
8282·7	0, 3*	7184	1, 1	6412·6	3, 0	6037·7	5, 0
7950·2	1, 3	7115·2	4, 3	6378·5	6, 2	6020·1	8, 2
7845	5, 5*	7026·4	3, 2	6297·8	5, 1	5869·2	6, 0*
7748·3	3, 4	6936·5	2, 1*	6218·6	4, 0	5712·1	7, 0
7550	1, 2	6708·8	3, 1*	6193·7	7, 2	5564·5	8, 0
7277·3	2, 2*	6495·8	4, 1				

3500 Å SYSTEM, B $\frac{1}{2}(^4\Sigma^-) - X_1\ \frac{1}{2}(^2\Pi_{\frac{1}{2}})$

Appearance. Degraded to the red. Long progressions. Strongest heads:

λ	I	v', v''	λ	I	v', v''	λ	I	v', v''
3576·7	5	0, 0 R_1	3381·0	10	2, 0 R_1	3268·0	3	1, 0 R_2
3460·5	4	2, 1 R_1	3330·4	9	3, 1 R_1	3238·4	6	5, 0 R_1
3434·4	10	1, 0 R_1	3283·0	8	4, 0 R_1	3173·2	3	3, 0 R_2

3268 Å SYSTEM, C $\frac{3}{2}(^2\Delta_{\frac{3}{2}}) - X_1\ \frac{1}{2}(^2\Pi_{\frac{1}{2}})$

A weak (0, 0) band, degraded to the red, with origin at 3268·4.

3063 Å SYSTEM, D $\frac{1}{2}(^2\Pi_{\frac{1}{2}}) - X_1\ \frac{1}{2}(^2\Pi_{\frac{1}{2}})$

A progression of diffuse (predissociated) bands, degraded to the red. Approximate band origins 3197 (0, 2), 3129 (0, 1), 3063·6 (0, 0).

BiO (*contd.*)

2600 Å SYSTEMS

Originally reported as a doublet system by Bridge and Howell, it is provisionally divided into two systems by Barrow *et al.* All bands are degraded to shorter wavelengths and form marked sequences.

$E - X_1 \frac{1}{2}$						$F - X_1 \frac{1}{2}$		
λ	I	v', v''	λ	I	v', v''	λ	I	v', v''
2637·5	6	0, 1	2585·8	7	1, 1	2480·9	3	2, 0
2631·7	4	1, 2	2541·0	8	1, 0	2476·0	2	1, 2
2590·7	10	0, 0	2493·7	3	2, 0	2440·1	2	0, 0

BiS

Reference.   R. F. Barrow, O. V. Stobart and H. Vaughan, *Proc. Phys. Soc.*, **90**, 555 (1967).

A system of 16 bands, mostly in the $v'' = 0$ progression, observed in absorption. Degraded to the red. Strongest heads: 6659·6 (6, 0), 6534·8 (7, 0), 6414·6 (8, 0), 6300·2 (9, 0), 6190·5 (10, 0), 6085·4 (11, 0).

Bands reported by Sur are at least partially due to S_2.

BiSe

Occurrence.   In absorption by mixed Bi and Se in furnace.

Appearance.   Two systems of bands, with marked sequences, degraded to shorter wavelengths.

References.   C. B. Sharma, *Proc. Phys. Soc.*, A, **67**, 935 (1954).
R. Yamdagni, *Indian J. Pure Appl. Phys.*, **8**, 51 (1970).

The following are the strongest heads of Sharma's systems:

System I				*System II*		
λ	I	v', v''		λ	I	v', v''
2848·5	8	0, 2		2275·8	9	0, 2
2826·3	10	0, 1		2262·3	10	0, 1
2805·2	9	0, 0		2249·0	10	0, 0
2782·0	10	1, 0		2233·2	10	1, 0
2759·3	10	2, 0		2217·8	8	2, 0

Another system, 5300–4300 Å, of red-degraded bands has been reported in absorption by Yamdagni. Strongest heads probably 4998·2 (1, 2) and 4956·1 (2, 2).

BiTe

Occurrence.   In absorption by mixed Bi and Te in furnace.

Appearance.   Five systems, all degraded to shorter wavelengths. Systems I and V have not been analysed. II and III show a long $v' = 0$ progression. IV shows marked sequences.

Reference.   C. B. Sharma, *Proc. Phys. Soc.*, A, **67**, 935 (1954).

The following are the outstanding heads, with intensities:

System I: λλ2926·1 (8), 2908·8 (8), 2875·0 (10), 2871·4 (9), 2854·5 (10), 2835·5 (9).

BiTe (*contd.*)

System II				System III				System IV		
λ	I	v', v"		λ	I	v', v"		λ	I	v', v"
2429·3	7	0, 4		2355·9	7	0, 2		2339·8	8	0, 2
2417·4	8	0, 3		2348·7	7	2, 3		2325·3	8	1, 2
2405·6	10	0, 2		2344·5	7	0, 1		2317·3	3	0, 0
2393·8	10	0, 1		2335·5	8	1, 1		2314·2	8	1, 1
2382·1	9	0, 0		2324·3	8	1, 0		2311·6	10	2, 2
								2303·1	10	1, 0

System V, λλ2244·0 (7), 2235·0 (10), 2228·4 (7).

Br_2

The spectrum of bromine is complex and difficult to identify from wavelength measurements; comparison of spectrograms is better. This is because the small values of ω for many excited states lead to numerous bands; the nearly equal abundance of isotopes Br^{79} and Br^{81} causes heads to be split into three components and thus to appear diffuse. The main absorption systems have been studied in great detail, but the analysis of some ultra-violet systems is uncertain, as is the relation between systems in emission and absorption.

MAIN (ABSORPTION) SYSTEMS, $B^3\Pi_{0^+} - X^1\Sigma$ AND $A^3\Pi_1 - X^1\Sigma$

Occurrence. Readily in absorption by bromine vapour. Also in emission from the heated vapour by Uchida, in the flame of Br_2 burning in H_2 by Kitagawa, in the flame of ethyl bromide by Vaidya, and in fluorescence. Clyne and Coxon found emission of the $A^3\Pi_1 - X^1\Sigma$ component between 10000 and 6400 Å in a bromine afterglow.

Appearance. Degraded to longer wavelengths. Waves of a very closely spaced bands extending from 5100 Å to the near infra-red. In absorption there is a continuum in the blue, commencing at the short-wave limit of the band system at 5107 Å.

References. H. Kuhn, *Z.P.*, **39**, 77 (1926).
O. Darbyshire, *P.R.S.*, **159**, 93 (1937)†.
Y. Uchida, *Inst. Phys. Chem. Res. Tokyo Sci. Papers*, No. 651, 71 (1936).
T. Kitagawa, *Proc. Imp. Acad. Tokyo*, **11**, 262 (1935).
W. M. Vaidya, *Proc. Indian Acad. Sci.*, **7A**, 321 (1938).
J. A. Coxon, *J. Mol. Spec.*, **37**, 39 (1971)†; **41**, 548 (1972).
M. A. A. Clyne and J. A. Coxon, *J. Mol. Spec.*, **23**, 258 (1967)†.

A full table of wavelengths of the absorption bands without intensities is given by Kuhn. The following are the wavelengths (averaged from the above references) of the bands which have been observed in emission by most of the above authors. Most of these bands occur in absorption also.

λλ6546, 6472, 6415, 6364, 6342, 6312, 6291, 6263, 6239, 6220, 6189, 6168, 6120, 6071, 5957, 5942, 5864, 5826, 5752, 5725, 5603, 5588.

EMISSION BANDS, λλ4200–2000

Occurrence. Discharge tube containing bromine.
Appearance. Diffuse.
Reference. P. Venkateswarlu, *Proc. Indian Acad. Sci.*, A, **25**, 138 (1947)†.

Br$_2$ *(contd.)*

Wavelengths and intensities are given for sixty-seven diffuse maxima. Strongest:

λ	I	λ	I	λ	I	λ	I
4224·5	4	3336·6	10	2753·6	8	2526·9	6
3932·5	3	2923·8	8	2732·4	7	2510·9	6
3597·8	4	2900·4	10*	2709·8	7	2494·2	6
3549·4	10	2872·5	8	2638·9	6	2478·8	7
3366·8	6	2780·6	7	2623·1	6		

* This band, around 2900 Å, has been observed by Coleman and Gaydon in flames containing Br$_2$.

EMISSION IN ARGON, λλ3150–2970, 2950–2670, 2660–2590

Occurrence. Uncondensed transformer discharge through bromine vapour and argon.

References. P. Venkateswarlu and R. D. Verma, *Proc. Indian Acad. Sci.*, **46**, 251 and 416 (1957)†.

R. D. Verma, *Proc. Indian Acad. Sci.*, **47**, 196 (1958).

λλ3150–2970. About 90 bands are listed; most of them are not clearly degraded. The system origin is at 3067 Å, and the following are the strongest, with intensities: λλ3122·5 (8), 3119·9 (8), 3118·4 (8), 3110·7 (10), 3106·9 (8), 3105·1 (8), 3091·5 (7).

λλ2950–2670. About 170 bands are listed; they are shaded to the red, and have been analysed into a system with origin at 2786 Å. The following are the strongest heads, with intensities: λλ2924·2 (9), 2923·6 (7), 2920·0 (8), 2914·2 (10), 2913·9 (10), 2893·7 (7), 2881·8 (7).

λλ2660–2590. Some 34 bands are listed, with a possible analysis indicating emission to very high vibrational levels of the $^3\Pi_u$ state; the heads are not sharp. Strongest, with intensities: λλ2659·2 (3), 2652·1 (3), 2647·5 (3), 2648·8 (3), 2645·3 (5), 2640·5 (7), 2638·5 (3), 2634·1 (3).

FLASH-PHOTOLYSIS (ABSORPTION) BANDS, 2900–2400 Å

Reference. A. G. Briggs and R. G. W. Norrish, *P.R.S.*, **276**, 51 (1963)†.

Transient absorption bands were obtained following flash-photolysis of mixtures of Br$_2$ and inert gases. They are probably the absorption counterpart of Venkateswarlu and Verma's emission bands. A long progression of 50 bands, spaced at around 10 Å, occurs between 2872·4 (1, 0) and 2411·4 (50, 0).

FAR ULTRA-VIOLET EMISSION, H – X$^1\Sigma^+$

Reference. P. B. V. Haranath and P. T. Rao, *J. Mol. Spec.*, **2**, 428 (1958)†.

About 60 red-degraded bands between 2100 and 1850 Å were obtained in a discharge tube source. Other systems at still shorter wavelengths were also found.

EMISSION BANDS 6700–5000 Å

Bands found by Uchida and Ota are probably due to Br$_2^+$ (see below), but the assignment is not certain.

Br$_2^+$

References. Y. Uchida and Y. Ota, *Japan. J. Phys.*, **5**, 59 (1928)†.

Uchida and Ota observed a number of bands, degraded to the red, in an uncondensed discharge through Br$_2$ and made a provisional analysis into two systems. Haranath and Rao observed about

Br_2^+ (*contd.*)

300 bands and modified the vibrational assignments of both systems. They suggested that the emitter was probably Br_2^+. The following measurements are from Uchida and Ota with our estimates of intensity from the published photograph.

λ	I	λ	I	λ	I	λ	I
6646	3	6392·8	6	6027·8	8	5586·1	9
6579·0	4	6372·1	5	6004·6	10	5532·4	7
6540·7	4	6332·8	6	5945·7	9	5529·2	6
6519·7	4	6282·0	6	5880·6	10	5428·6	4
6475·6	7	6217·0	7	5819·7	9	5382·6	4
6455·1	7	6144·8	8	5758·3	10	5134·2	3
6435·9	8	6083·1	8	5699·9	10	5100·7	3
6421·2	5	6074·6	8	5644·3	9		

BrCl

MAIN SYSTEM, 8830–5640 Å, $^3\Pi_0 - X^1\Sigma^+$

Occurrence. In absorption through heated $Cl_2 + Br_2$ with 6 m path, and in emission from chemiluminescent reactions $Br + Cl + M$ and $Br + ClO_2$.

Appearance. Degraded to the red. Single-headed, apart from isotope splitting. Long progressions. In emission 49 bands 8830–6750, and in absorption 24 bands 6968–5640 have been recorded.

References. M. A. A. Clyne and J. A. Coxon, *P.R.S.*, **298**, 424 (1967)†; *J. Mol. Spec.*, **50**, 142 (1974)†.

The following are among the stronger heads of $BrCl^{35}$ in emission: λλ8250, 7985, 7736, 7498, 7384, 7275, 7166, 7061, 6962, 6872, 6682, 6499, 6324, 6254, 6091, 6032, 5978 and 5932; the last three are also prominent in absorption.

ULTRA-VIOLET EMISSION

Occurrence. Emission from condensed or high-frequency discharges through mixed Br_2 and Cl_2.

Appearance. Six diffuse bands, λλ3600–3520, 3390–3300, 3190–3070, 2950–2825, 2750–2710 and 2605–2550.

Reference. P. B. V. Haranath and P. T. Rao, *Indian J. Phys.*, **31**, 368 (1957)†.

BrF

References. R. A. Durie, *P.R.S.*, **207**, 388 (1951)†.
P. H. Brodersen and J. E. Sicre, *Z.P.*, **141**, 515 (1955).
M. A. A. Clyne, J. A. Coxon and L. W. Townsend, *J. Chem. Soc. Faraday Trans. II*, **68**, 2134 (1972)†.

Occurrence. In flame of Br_2 burning with fluorine, in absorption and probably in discharge through BrF_3. Also in $Br + F$ reaction using mixed flows from discharges through $CF_4 + Ar$ and $Br_2 + O_2$; singlet oxygen molecules are involved in the excitation process (see Clyne *et al.*).

Appearance. The main system shows rather evenly spaced single-headed bands, degraded to the red. A weaker progression of a second system underlies the main system.

MAIN SYSTEM, $0^+ \rightleftharpoons X\ ^1\Sigma^+$

The strong bands are listed below. Intensities are for emission, from Durie. Values of v', v'' and most λ are from Brodersen and Sicre. Clyne *et al.* list 80 bands, 4734–9711 Å.

BrF (*contd.*)

λ	I	v', v''	λ	I	v', v''	λ	I	v', v''
4772·1	2	9, 0	5212·2	6	5, 1	5923·9	10	2, 3
4828·5	2	8, 0	5308·1	8	6, 2	6026	8	3, 4
4891·9	3	7, 0	5396·3	9	5, 2	6153	9	2, 4
4930·0	5	9, 1	5491·5	9	4, 2	6293	10	1, 4
4987·8	4	8, 1	5593·5	9	3, 2,	6405	10	2, 5
5055·4	5	7, 1	5693·7	9	4, 3	6554	8	1, 5
5129·4	6	6, 1	5804·1	10	3, 3			

WEAKER SYSTEM, $1 - X^1\Sigma^+$

The following are probably the strongest heads: $\lambda\lambda 5578\cdot1$ (2, 0), $5499\cdot3$ (3, 0), $5424\cdot1$ (4, 0), $5359\cdot8$ (5, 0), $5307\cdot5$ (6, 0).

BrO

ETHYL BROMIDE FLAME BANDS

Occurrence. In flame of ethyl bromide and when Br_2 is added to the oxy-hydrogen flame. They have also been observed in absorption following flash photolysis of $Br_2 + O_2$.

Appearance. Degraded to the red. The $v' = 0$ progression is conspicuous in emission and the $v'' = 0$ in absorption.

References. W. M. Vaidya, *Indian Acad. Sci. Proc.*, A, **1**, 321 (1938).
E. H. Coleman and A. G. Gaydon, *Disc. Faraday Soc.*, **2**, 166 (1947)†.
R. A. Durie and D. A. Ramsay, *Canad. J. Phys.*, **36**, 35 (1958)†.

The following are the emission bands from Coleman and Gaydon, with revised v'' numbering:

λ	I	v', v''	λ	I	v', v''	λ	I	v', v''
4856	2	1, 11	4349	4	2, 8	4109	4	2, 6
4817	3	0, 10	4270	10	0, 6	4069	5	1, 5
4673	8	0, 9	4225	2	2, 7	4029	3	0, 4
4533	10	0, 8	4186	3	1, 6	3999	3	2, 5
4398	10	0, 7	4147	6	0, 5	3958	3	1, 4

The following are among the strongest absorption bands:

λ	I	v', v''	λ	I	v', v''	λ	I	v', v''
3433·8	9	3, 0	3288·6	10	6, 0	3168·5	10	9, 0
3383·4	10	4, 0	3247·7	10	7, 0	3132·6	9	10, 0
3333·2	10	5, 0	3207·9	10	8, 0	3100·4	8	11, 0

C_2

The strongest system is the well-known Swan system in the green; this is due to a $^3\Pi - ^3\Pi$ transition, and two other triplet systems are known (Fox-Herzberg and Ballik-Ramsay). The ground state of C_2 is a singlet lying 610 cm.$^{-1}$ below the lowest triplet level, and five singlet systems are known; that found by Mulliken is most readily observed, although the Phillips system in the less accessible near infra-red is probably stronger. The Deslandres-d'Azambuja one is fairly strong.

We have retained the old notation but denoted the lower state of the Swan bands as $X'\,^3\Pi_u$ and the lowest singlet as $x^1\Sigma_g^+$ (formerly $a^1\Sigma_g^+$). Some recent papers use capitals for the singlets and lower case letters for triplets.

C_2 *(contd.)*

SWAN SYSTEM, $A^3\Pi_g - X'^3\Pi_u$

Occurrence. These bands are of very frequent occurrence in sources containing carbon. They are especially strongly developed in the greenish inner cone of a Bunsen or Meker burner flame, and in discharge tubes with high current density through hydrocarbon and other organic vapours. They also occur in a carbon arc, in active nitrogen reacting with organic vapours, in discharge tubes through He + CO, and in both emission and absorption in shock tubes and furnaces. They have been reported in reactions between alkali metals and organic halides (e.g. Na + CCl_4). They are a prominent feature of comets and carbon-type stars.

Appearance. Degraded to the violet. Marked sequences of single-headed bands. See Plate 9.

References. R. C. Johnson, *Phil. Trans. Roy. Soc.*, A, **226**, 157 (1927).

D. C. Tyte, S. H. Innanen and R. W. Nicholls, *Identification Atlas of Mol. Spectra*, **5**, York Univ. (1967)†.

J. G. Phillips, *Astrophys. J.*, **108**, 434 (1948).

E. D. Bugrim, A. I. Lyutyi, V. S. Rossikhin and I. L. Tsikora, *Opt. Spectrosc.*, **19**, 292 (1965).

K. S. Kini and M. I. Savadatti, *J. Phys.*, B, **2**, 307 (1969).

The following are all the main heads:

λ	I	v', v''	λ	I	v', v''	λ	I	v', v''
6677·3	1	2, 5	5923·4	1	5, 7	4737·1	9	1, 0
6599·2	1	3, 6	5635·5	8	0, 1	4715·2	8	2, 1
6533·7	2	4, 7	5585·5	8	1, 2	4697·6	7	3, 2
6480·5	2	5, 8	5540·7	6	2, 3	4684·8	4	4, 3
6442·3	2	6, 9	5501·9	4	3, 4	4680·2	1	6, 5
6191·2	3	0, 2	5470·3	2	4, 5	4678·6	2	5, 4
6122·1	4	1, 3	5165·2	10	0, 0	4382·5	2	2, 0
6059·7	3	2, 4	5129·3	6	1, 1	4371·4	4	3, 1
6004·9	3	3, 5	5097·7	1	2, 2	4365·2	5	4, 2
5958·7	2	4, 6						

Tail Bands

Some weak 'tail bands', degraded to the red have been observed by Phillips and by Bugrim *et al.*, $\lambda\lambda$4996·7 (13, 12), 4911·0 (12, 11), 4836·1 (11, 10), 4770·1 (10, 9), 4255·0 (11, 8), 4197·1 (10, 7) and 4147·9 (9, 6). The bands with origins at 4734 (9, 8), 4395 (8, 6) and 4123 (8, 5) are headless. Kini and Savadatti record 12 more weak heads.

High Pressure Bands

These occur in condensed discharges through CO at relatively high pressures (10 to 100 torr) and some special sources. They were initially treated as a separate system, but are now known to be part of the Swan system with $v' = 6$ selectively excited, probably by the reaction $C + C_2O = C_2 (A^3\Pi) + CO$. They may occur without the ordinary Swan bands being present.

They are degraded to the violet, and under small dispersion usually appear double headed, perhaps due to anomalous rotational intensity distribution, but with larger dispersion the shorter wavelength head becomes less definite.

References. R. C. Johnson and R. K. Asundi, *P.R.S.*, **124**, 668 (1929).

A. Fowler, *Mon. Not. R. Astr. Soc.*, **70**, 484 (1910)†.

J. G. Fox and G. Herzberg, *P.R.*, **52**, 638 (1937).

C. Kunz, P. Harteck and S. Dondas, *J. Chem. Phys.*, **46**, 4157 (1967).

C_2 (contd.)

In the following table the longer wavelength heads are by Johnson and Asundi, and the shorter, where given, by Fowler.

λ	I	v', v''	λ	I	v', v''	λ	I	v', v''
7852·5	4	6, 11	5434·9	5	6, 7	4093	2	6, 3
—			5413	2		—		
7083·2	6	6, 10	5030	2	6, 6	3619·5	1	6, 1
—			5015	1		—		
6442·3	8	6, 9	4680·2	15	6, 5	3419	1	6, 0
6420	4		4663	6		—		
5899·3	10	6, 8	4368·8	7	6, 4			
5878	5		4353	3				

FOX-HERZBERG SYSTEM, $B^3\Pi_g - X'\,^3\Pi_u$

Occurrence. Weakly condensed discharge through helium containing benzene vapour.
Appearance. Shaded to red.
Reference. J. G. Fox and G. Herzberg, *P.R.*, **52**, 638 (1937).
J. G. Phillips, *Astrophys. J.*, **110**, 73 (1949)†.

λ	v', v''	λ	v', v''	λ	v', v''
*3283	0, 6	2772·1	1, 3	2527·9	3, 2
*3129	0, 5	2731·5	0, 2	2486·3	2, 0
**2987	0, 4	2698·8	2, 3	2429·9	3, 0
2896·4	1, 4	2656·3	1, 2	2378·2	4, 0
**2855	0, 3	2589·0	2, 2		

* Probably fairly strong.   ** Strong band.

BALLIK-RAMSAY, FAR INFRA-RED SYSTEM, $A'\,^3\Sigma_g^- - X'\,^3\Pi_u$

Occurrence. Emission from carbon-tube furnace. The system is likely to be strong in most sources but is too far in the infra-red to have been recorded so far from flames, etc.
Appearance. Marked sequences. Degraded to longer wavelengths.
Reference. E. A. Ballik and D. A. Ramsay, *Astrophys. J.*, **137**, 61 (1963).
Sequence heads, 24745 (0, 1), 17675 (0, 0), 14075 (1, 0) and 11724 Å (2, 0).

MULLIKEN SYSTEM, 2313 Å, $d^1\Sigma_u^+ - x^1\Sigma_g^+$

Occurrence. Emission in carbon arc, oxy-acetylene flame and discharge tubes containing hydrocarbons. Also in absorption.
References. R. S. Mulliken, *Z. Elektrochem.*, **36**, 603 (1930).
J. G. Fox and G. Herzberg, *P.R.*, **52**, 638 (1937).
O. G. Landsverk, *P.R.*, **56**, 769 (1939).
G. Messerle, *Z. Naturforsch.*, **23 A**, 470 (1968).

The headless (0, 0) band with single P and R branches spreading out from the origin at 2312·7 Å is by far the strongest. Other weaker bands have been recorded. Origins:

C_2 (contd.)

λ_0	v', v''
2414·8	0, 1
2316·8	3, 3
2315·4	2, 2
2314·0	1, 1
2312·7	0, 0

PHILLIPS NEAR INFRA-RED SYSTEM, $b^1\Pi_u - x^1\Sigma_g^+$

Occurrence. Originally found in heavy-current discharges, but now known to be a strong feature in most systems including arcs and flames.

Appearance. Degraded to longer wavelengths with clear sequences. Strong Q branches and weaker P and R branches.

References. J. G. Phillips, *Astrophys. J.*, **107**, 389 (1948)†.

E. A. Ballik and D. A. Ramsay, *Astrophys. J.*, **137**, 84 (1963).

R heads:

λ	v', v''	λ	v', v''
15484·1	0, 1	8750·8	2, 0
12070·2	0, 0	8108·2	5, 2
10147*	1, 0	7907·7	4, 1
8980·5	3, 1	7714·6	3, 0

* Band origin

DESLANDRES-D'AZAMBUJA SYSTEM, $c^1\Pi_g - b^1\Pi_u$

Occurrence. Condensed discharge through CO, CO_2 or C_2H_2, or through argon containing H_2 using carbon electrodes. In spark through liquid alcohol. In carbon arc in H_2 under high temperature conditions.

Appearance. Degraded to shorter wavelengths. Single headed.

References. R. C. Johnson, *Nature*, **125**, 89 (1930).

G. H. Dieke and W. Lochte-Holtgreven, *Z.P.*, **62**, 767 (1930).

G. Herzberg and R. B. Sutton, *Canad. J. Res.*, **40**, 74 (1940).

G. Messerle and L. Krauss, *Z. Naturforsch.*, **22A**, 2015 (1967).

This is a weak system. Intensities are relative to the strongest band of this system.

λ	I	v', v''	λ	I	v', v''	λ	I	v', v''
4102·3	9	0, 1	3852·2	10	0, 0	3587·6	7	3, 2
4068·1	6	1, 2	3825·6	5	1, 1	3399·7	5	2, 0
4041·8	3	2, 3	3607·3	8	1, 0	3398·1	5	3, 1
4026·9	1	3, 4	3592·9	7	2, 1			

Tail Bands

Herzberg and Sutton obtained the following from an uncondensed discharge through He + C_6H_6. They are degraded to the red:

3689·0 (6, 5), 3617·9 (5, 4), 3599·3 (4, 3), 3431·9 (5, 3).

MESSERLE-KRAUSS BANDS, $c'^1\Pi_g - b^1\Pi_u$

This is a weak system, observed mainly by its perturbations of overlapping Deslandres-d'Azambuja bands. The following are band origins. Degraded to the red:

C_2 (*contd.*)

λ_0	I	v', v''		λ_0	I	v', v''
3779·6	weak	4, 6		3586·0	strong	0, 2
3691·5	weak	3, 5		3405·7	strong	1, 2
3672·7	mod.	5, 6		3396·1	strong	0, 1
3627·0	strong	2, 4				

FREYMARK BANDS, $e^1\Sigma_g^+ - b^1\Pi_u$

Occurrence. Discharge through acetylene and probably in other sources.

Appearance. Degraded to shorter wavelengths. Apparently single headed, with clear sequences.

Reference. H. Freymark, *Ann. Phys. (Lpz.)*, **8**, 221 (1951)†.

Heads (λ_h) or origins (λ_0):

λ_h	I	v', v''		λ_h	I	v', v''		λ_0	I	v', v''
2218·2	9	0, 1		2072·4	6	1, 0		2087·1	5	4, 3
2216·6	7	1, 2		2075·6	6	2, 1		2096·9	4	5, 4
2142·9	10	0, 0		2081·2	5	3, 2				
2142·6	1	1, 1								

C_2^+

MEINEL SYSTEM, 2490 Å, $^2\Sigma_g^- - {}^2\Pi_u$

Occurrence. Absorption following a flash discharge through C_2H_2.

Appearance. Slightly degraded to the red. The only known band, the (0, 0), has a Q head at 2490·5, with P and R branches extending away from this.

Reference. H. Meinel, *Canad. J. Phys.*, **50**, 158 (1972)†.

The assignment to C_2^+ is probable but not quite certain; C_2^- is a possibility.

C_2^-

HERZBERG-LAGERQVIST SYSTEM, 6000–4800 Å, $^2\Sigma - {}^2\Sigma$

Occurrence. In emission and in absorption from flash discharges through CH_4, and in a shock tube.

Appearance. Marked sequences of single-headed bands. Degraded to the violet.

References. G. Herzberg and A. Lagerqvist, *Canad. J. Phys.*, **46**, 2363 (1968)†.
W. S. Cathro and J. C. Mackie, *J. Chem. Soc. Faraday Trans. II*, **69**, 237 (1973)†.

P heads with own estimates of intensity:

λ	I	v', v''		λ	I	v', v''		λ	I	v', v''
5984·8	4	0, 1		5363·3	6	1, 1		4833·8	5	3, 2
5912·7	5	1, 2		4902·0	6	1, 0		4804·5	3	4, 3
5844·6	2	2, 3		4866·4	6	2, 1		4779·1	2	5, 4
5415·9	10	0, 0								

C_3

THE 4050 Å COMET-HEAD GROUP

This group was at one time attributed to CH_2 but later work using deuterium and C^{13} proved that the emitter is C_3.

C₃ (*contd.*)

Occurrence. Originally observed in the heads of comets, the group was later observed in various laboratory sources, including discharges through flowing hydrocarbons or between carbon electrodes in $H_2 + Xe$, or in a graphite hollow-cathode discharge. Also in rich hydrocarbon flames, hydrocarbon/fluorine flames, in shock waves and in absorption and emission in a carbon-tube furnace. The most detailed work has been done on flash-photolysis absorption (and fluorescent emission) of diazomethane.

Appearance. A complex group of narrow red-degraded bands. The appearance depends a lot on the dispersion used, rotational structure being observed at large dispersion. The main head is at 4049·77 Å and is always prominent. Another head at 4072·4 is usually quite clear. The wavelength measurements in comets (Swings), flames (Kiess and Bass) and flash photolysis (Gausset *et al.*) do not agree at all well.

References. P. Swings, *Mon. Not. R. Astr. Soc.*, **103**, 92 (1943).
A. E. Douglas, *Astrophys. J.*, **114**, 466 (1951)†.
N. H. Kiess and A. M. Bass, *J. Chem. Phys.*, **22**, 569 (1954)†.
K. Clusius and A. E. Douglas, *Canad. J. Phys.*, **32**, 319 (1954).
L. Gausset, G. Herzberg, A. Lagerqvist and B. Rosen, *Astrophys. J.*, **142**, 45 (1965)†.

Gausset *et al.* list about 80 red-degraded heads between 4100 and 3440 Å in flash photolysis of diazomethane. The following are extracted from this list:

λ	I	λ	I	λ	I	λ	I
4098·5	1	3990·8	6	3935·8	3	3793·5	3
4072·4	6	3983·1	4	3925·9	3	3656·8	3
4049·8	10	3970·9	4	3914·5	3	3619·7	3
4038·4	4	3965·9	3	3879·4	3	3524·9	3
4018·3	4	3949·1	4				

The C_3 molecule is linear and a full analysis has been made. The transition is $^1\Pi_u - {}^1\Sigma_g^+$.

CBr

Occurrence. Absorption following flash photolysis of organic bromides.
References. R. N. Dixon and H. W. Kroto, *Trans. Faraday Soc.*, **59**, 1484 (1963)†.
W. J. R. Tyerman, *Spectrochim. Acta.*, **26a**, 1215 (1970).

3052 Å SYSTEM, $A^2\Delta - X^2\Pi$

Appearance. Double-headed bands, degraded to shorter wavelengths, but Q_1 heads are diffuse due to predissociation.

Heads from Dixon and Kroto:

λ	v', v''	remarks
3060·0	1, 1 Q_2	weak head
3052·3	0, 0 Q_2	strong head
3023·0	1, 1 Q_1	weak maximum
3014·5	0, 0 Q_1	strong maximum

2500 Å SYSTEM, PROBABLY $B^2\Sigma^+ - X^2\Pi$

Two diffuse bands at 2526 and 2496 observed by Tyerman.

CCl

There is a fairly strong band system near 2800 Å, and a system of diffuse bands. A weaker system has been observed at 2368 Å.

Occurrence. Uncondensed discharge through flowing CCl_4 vapour. The 2368 Å system occurs best at a low current density. The 2800 Å system also occurs in flames. Also in flash photolysis.

References. R. K. Asundi and S. M. Karim, *Proc. Indian Acad. Sci.*, A, **6**, 328 (1937).
P. Venkateswarlu, *P.R.*, **77**, 79 (1950).
R. F. Barrow, G. Drummond and S. Walker, *Proc. Phys. Soc.*, A, **67**, 186 (1954).
R. D. Gordon and G. W. King, *Canad. J. Phys.*, **39**, 252 (1961).
I. E. Ovcharenko, Y. Y. Kuzyakov and V. M. Tatevskii, *Opt. Spectrosc.*, **19**, 294 (1965).
W. J. R. Tyerman, *Spectrochim. Acta*, **26a**, 1215 (1970).

CONTINUOUS BANDS

Strong wide bands, maxima 4600, 3070 and 2580. Less strong bands 3348, 2430 and 2300. Diffuse heads were reported by Tyerman at 2311, 2307 and 2304 Å.

2800 Å SYSTEM, $A^2\Delta - X^2\Pi$

Appearance. Complex groups of bands, degraded to shorter wavelengths.

The following data are from Barrow *et al.* Alternative analyses have been proposed by Gordon and King and by Ovcharenko *et al.*

λ	I	heads	λ	I	heads	λ	I	heads
2931·9	1	0, 2 P_1	2849·2	4	0, 1 P_2	2778·9	7	0, 0 Q_2
2927·4	1	0, 2 Q_1	2846·0	5	0, 1 Q_2	2777·6	10	1, 1 Q_2
2919·5	1	0, 2 P_2	2794·2	4	0, 0 P_1	2724·0	1	1, 0 Q_1
2916·3	1	0, 2 Q_2	2789·8	6	0, 0 Q_1	2721·5	2	3, 2 Q_1
2861·5	3	0, 1 P_1	2788·3	10	1, 1 Q_1	2713·9	2	1, 0 Q_2
2857·1	5	0, 1 Q_1	2786·7	5	2, 2 Q_1	2711·5	1	3, 2 Q_2
2855·5	4	1, 2 Q_1	2782·3	8	0, 0 P_2			

2368 Å SYSTEM

Appearance. A single group of bands, degraded to the red.

λ	I
2367·8	10
2383·9	9
2400·6	5
2434·0	0

CF

There are two systems in the ultra-violet.

Occurrence. In discharge (especially of valve oscillator type) through flowing CF_4 or other fluorocarbon vapour. The $A - X$ system also occurs in flames of hydrocarbons burning with fluorine or with ClF_3. Also in absorption in carbon-tube furnace at 2400° K.

References. E. B. Andrews and R. F. Barrow, *Proc. Phys. Soc.*, **64**, 481 (1951)†.
Yu. Ya. Kuzyakov and V. M. Tatevskii, *Optika i Spektrosk.*, **5**, 699 (1959).
D. E. Mann, H. P. Broida and B. E. Squires, *J. Chem. Phys.*, **22**, 348 (1954).

CF (*contd.*)

T. L. Porter, D. E. Mann and N. Acquista, *J. Mol. Spec.*, **16**, 228 (1965)†.
P. K. Carroll and T. P. Grennan, *J. Phys.*, B, **3**, 865 (1970)†.

$A^2\Sigma - X^2\Pi$ SYSTEM

Appearance. Double double-headed bands, degraded to shorter wavelengths. The (0, 1), (0, 2) and (0, 3) are strongest.

All heads of the (0, 2) and the $^0P_{12}$ and P_1 heads of other strong bands.

λ	v', v''	λ	v', v''	λ	v', v''
2640·8	0, 4 $^0P_{12}$	2474·7	0, 2 P_1	2327·8	0, 0 P_1
2635·6	0, 4 P_1	2473·9	0, 2 Q_1	2308·7	1, 1 $^0P_{12}$
2558·2	0, 3 $^0P_{12}$	2404·1	0, 1 $^0P_{12}$	2304·5	1, 1 P_1
2553·3	0, 3 P_1	2399·6	0, 1 P_1	2242·2	1, 0 $^0P_{12}$
2479·4	0, 2 $^0P_{12}$	2332·0	0, 0 $^0P_{12}$	2238·2	1, 0 P_1
2478·4	0, 2 P_2				

$B^2\Delta - X^2\Pi$ SYSTEM

Appearance. Degraded to the red. Double double-headed bands, but the first head $^sR_{21}$ is rather weak. The (0, 1) band is strongest.

All heads of (0, 1) and R_1 heads of other main bands.

λ	v', v''	λ	v', v''	λ	v', v''
2192·1	0, 3 R_1	2081·8	0, 1 Q_2	2075·7	0, 1 $^sR_{21}$
2134·2	0, 2 R_1	2080·7	0, 1 R_2	2024·0	0, 0 R_1
2083·4	1, 2 R_1	2078·5	0, 1 R_1	1978·8*	1, 0 R_1

* This is λ_{vac}

CF⁺

Bands initially attributed to CF⁺ are now known to be due to BF (R. D. Verma, *Canad. J. Phys.*, **40**, 1865 (1962)).

CF₂

Occurrence. In discharge through flowing CF_4, and in absorption by CF_4 in carbon tube furnace at 1900° K. Also in a shock-tube and in flash photolysis.

Appearance. A rather large number of narrow red-degraded bands. Under small dispersion the complex rotational structure makes them appear slightly diffuse. The arrangement appears fairly regular, and a vibrational analysis has been made.

References. P. Venkateswarlu, *P.R.*, **77**, 676 (1950)†.
J. L. Margrave and K. Wieland, *J. Chem. Phys.*, **21**, 1552 (1953).
C. W. Mathews, *Canad, J. Phys.*, **45**, 2355 (1967).

The following are the strongest heads observed in emission by Venkateswarlu:

CF$_2$ (*contd.*)

λ	I	λ	I	λ	I
3214·1	6	2866·1	6	2675·5	6
3197·5	4	2852·5	6	2652·4	6
3053·7	5	2799·8	8	2628·5	9
3038·1	5	2774·2	5	2595·0	9
3022·8	5	2761·2	5	2550·6	8
2921·3	6	2749·1	8	2518·6	9
2893·5	7	2711·3	9	2487·8	9
		2688·1	5	2457·6	7

The following are the strongest bands, observed by Margrave and Wieland in absorption:

λ	I	λ	I
2118·7	3	2062·5	5
2101·7	3	2045·3	7
2099·9	4	2042·8	6
2083·3	3	2028·4	7
2079·5	5	2011·3	8
2063·5	6	2008·0	7

CH

Bands of CH are readily excited during the combustion of hydrocarbons and in electrical discharges where carbon and hydrogen are present. They are also observed in many astrophysical sources. Three systems are known in the regions of 4300 Å, 3900 Å and 3143 Å respectively. Their intensities decrease in the order in which they are given, the third usually being much the weakest.

4300 Å SYSTEM, $A^2\Delta - X^2\Pi$

Occurrence. In sources where carbon and hydrogen are present together such as flames of hydrocarbons, the carbon arc in hydrogen, discharge tubes under a great variety of conditions and in active nitrogen when a hydrocarbon is introduced. It is also observed in emission from the heads of comets and in absorption in the Sun's atmosphere. Difficult to detect in absorption, but has been recorded in flash photolysis and the reaction zone of flames. Also in stars and inter-stellar space.

Appearance. Degraded to the violet. Usually the (0, 0) band is the only one to appear. The (1, 1) and (2, 2) are overlaid by this (0, 0); a line-like feature at 4324 Å is the piled up Q branch of the (2, 2). The (0, 0) shows strong Q heads and the lines of the P branch can be traced to about λ4385. Both branches consist of narrow doublets. See Plate 4.

References. L. Gerö, *Z.P.*, **118**, 27 (1941)†.
N. H. Kiess and H. P. Broida, *Astrophys. J.*, **123**, 166 (1956)†.
C. E. Moore and H. P. Broida, *J. Res. Nat. Bur. Stand.*, A, **63**, 19 (1959).
A. M. Bass and H. P. Broida, *Nat. Bur. Stand. Monograph*, No. 24 (1961).

v', v''	Q Heads	I	P Heads	I
0, 1	4890·0	2	4941·2	1
1, 2	4857·6	0	4913·5	0
2, 2	4324	6		
0, 0	4314·2	10	(4385)	5

CH (*contd.*)

3900 Å SYSTEM, $B^2\Sigma - X^2\Pi$

Occurrence. Similar to the 4300 Å system.

Appearance. Degraded to the red. The $(0, 0)$ band has a very open rotational structure. Other bands are obtained rather more readily than for the 4300 system, but rotational structure of bands with $v' = 1$ is curtailed by predissociation. The $(2, 1)$ band is usually absent, due to predissociation, but has been observed by Durie. See Plate 4.

References. C. E. Moore and H. P. Broida, *J. Res. Nat. Bur. Stand.*, A, **63**, 19 (1959).
R. A. Durie, *Proc. Phys. Soc.*, A, **65**, 125 (1952)†.
A. M. Bass and H. P. Broida, *Nat. Bur. Stand. Monograph*, No. 24 (1961).

v', v''	Q Heads	I	R Heads	I
1, 2			4495·5	1
1, 1	4036	0	4025·3	2
0, 0	3889·0	5	3871·3	6
1, 0	3635	0	3627·2	2

3143 Å SYSTEM, $C^2\Sigma^+ - X^2\Pi$

Occurrence. Similar to the above, but favoured by a higher temperature. More readily observed in absorption.

Appearance. The $(0, 0)$ and $(1, 1)$ bands have been reported. The branches consist of doublets. The Q branches are at first degraded to the violet, then at the sixteenth member form a second head and turn to the red. This last head is the most intense.

References. T. Hori, *Z.P.*, **59**, 91 (1930)†.
T. Heimer, *Z.P.*, **78**, 771 (1932).

v', v''	Q Heads		
1, 1	3156·6 maximum		
0, 0	3144·9 deg. V	3144·1 deg. V	3143·4 deg. R

CH⁺

DOUGLAS-HERZBERG SYSTEM, $A^1\Pi - X^1\Sigma$

Occurrence. Discharge, especially in hollow cathode, through helium with a trace of benzene or acetylene.

Appearance. Open rotational structure, with single P, Q and R branches. Degraded to the red; heads not well developed.

References. A. E. Douglas and G. Herzberg, *Canad. J. Res.*, A, **20**, 71 (1942)†.
A. E. Douglas and J. R. Morton, *Astrophys. J.*, **131**, 1 (1960)†.

As the heads are not well developed, we list the Q(1) lines as well; these are close to the origins. No intensities are available, but the $(0, 0)$ and $(1, 0)$ bands are probably strong.

v', v''	R Heads	Q(1)	v', v''	R Heads	Q(1)
0, 1	4775·9	4794·0	1, 0	3954·4	3962·1
1, 1	4433·8	4444·4	3, 1	3967·1	3972·5
0, 0	4225·3	4237·6	2, 0	3743·7	3749·2
2, 1	4171·1	4178·4	4, 1	3806·1	3810·6

CH⁺ (*contd.*)

Absorption lines at $\lambda\lambda 4232\cdot58$, $3957\cdot72$ and $3745\cdot33$ in the spectra of certain stars have been identified with the R (0) lines of the (0, 0), (1, 0) and (2, 0) bands of CH^+ and attributed to absorption in interstellar space.

3503 Å SYSTEM, $B^1\Delta - A^1\Pi$

Occurrence. Excitation of hydrocarbons by a beam of positive ions.
Reference. M. Carré, *Physica*, **41**, 63 (1969)†.

It is difficult to extract data from this paper but there appear to be three bands 3500–3400. Heads, degraded to shorter wavelengths are at 3503 (0, 0) and 3468 (1, 1). This system overlaps the following one.

3490 Å SYSTEM, $b^3\Sigma - a^3\Pi$

Occurrence and *Reference* as above.
Degraded to the red. Probable heads 3490 (0, 0), 3270 (1, 0).

Bands at 2264 and 2367 Å, previously attributed to CH^+, have been shown, by Feast, to be due to HgH^+.

CH₂, Methylene Radical

Occurrence. Flash photolysis of diazomethane, CH_2N_2.
References. G. Herzberg, *P.R.S.*, **262**, 291 (1961).
G. Herzberg and J. W. C. Johns, *P.R.S.*, **295**, 107 (1966)†.

There is a strong singlet system in the vacuum ultra-violet at 1415 Å and a weaker triplet system 9200–4900 Å. This is a many-line structure with few features for identification. There are weak violet-degraded Q heads at 8192, 7708, 7316, 6953, 6532, 6226, 5906 and 5375 Å. There are also signs of other very weak singlet bands 3600–3300 with the most definite at 3470 Å.

The 4050 Å comet-head group was at one time attributed to CH_2 but is now known to be due to C_3.

CH₃, Methyl radical

Occurrence. In absorption following flash photolysis of mercury dimethyl, methyl iodide, acetaldehyde, etc., and in absorption by low-pressure flames of methane, methyl ether, acetone or acetaldehyde.
Appearance. Diffuse double bands, shaded slightly to the red.
References. G. Herzberg and J. Shoosmith, *Canad. J. Phys.*, **34**, 523 (1956)†.
A. G. Gaydon, G. N. Spokes and J. van Suchtelen, *P.R.S.*, **256**, 323 (1960).
R. Harvey and P. F. Jessen, *Nature, Phys. Sci.*, **241**, 102 (1973).

The strongest band of the quartz region has maxima at 2163·6 and 2157·6 Å. There is another weak band at 2202 and 2188 Å. The strongest bands are in the vacuum ultra-violet ($\lambda\lambda$1502·9, 1498·9, 1496·7, 1385·6, 1382·6).

C_2H_2, Acetylene

ULTRA-VIOLET ABSORPTION

The absorption shows bands 2400–2200 Å getting stronger to shorter wavelengths and superposed on some apparently continuous background. Bands are degraded to the red and are mostly close double-headed.

   References.   S. C. Woo, T. K. Liu, T. C. Chu and W. Chih, *J. Chem. Phys.*, **6**, 240 (1938). K. K. Innes, *J. Chem. Phys.*, **22**, 863 (1954)†.

   The following are a selection of heads, partly from Innes and partly from the list of about 250 by Woo *et al.*; intensities from Woo *et al.*:

λ	I	λ	I	λ	I	λ	I
2377·8	1	2289·0	5	2234·0	8	2158·9	7
2372·1	0	2286·0	6	2209·2	8	2140·5	10
2341·4	3	2268·5	5	2206·0	10	2121·4	9
2320·8	4	2256·3	5	2188·9	10	2114·6	7
2315·0	4	2238·4	7	2185·6	10	2096·1	8
2310·1	3	2237·2	7				

INFRA-RED ABSORPTION (VIBRATION-ROTATION SPECTRUM)

   Reference.   K. Hedfeld and R. Mecke, *Z.P.*, **64**, 151 (1930)†.

   Acetylene shows an absorption band with origin at 7887 Å, and maxima of intensity at 7874 and 7901 Å. There are also bands, which are presumably weaker, at 7956 and 8622 Å.

ULTRA-VIOLET EMISSION

About 40 bands, mostly degraded to the red, have been observed in an oscillator discharge through flowing acetylene. They probably form part of the main absorption system of C_2H_2 although C_2H is also considered as a possible alternative emitter. The following are the strongest bands: λλ2869, 2723, 2693, 2642, 2580·4, 2565·6, 2526·7, 2522·1, 2499·4, 2493·5, 2478·9, 2466·3, 2441·1, 2419·9.

   Reference.   P. Dyne, *Canad. J. Phys.*, **30**, 79 (1952)†.

C_2H_4, Ethylene

   Diffuse absorption bands, commencing near the limit of the quartz region and becoming stronger in the vacuum region. Maxima: λλ2069, 2032, 2000, 1970, 1935, etc.

   References.   C. P. Snow and C. B. Alsopp, *Trans. Faraday Soc.*, **30**, 93 (1934)†. P. G. Wilkinson and R. S. Mulliken, *J. Chem. Phys.*, **23**, 1895 (1955).

C_2H_5, Ethyl Radical

   A diffuse absorption band with maxima at 2242 and 2228 Å in low-pressure flames of ethane or ether is provisionally assigned to the ethyl radical.

   Reference.   A. G. Gaydon, G. N. Spokes and J. van Suchtelen, *P.R.S.*, **256**, 323 (1960).

C_3H_3, Propargyl Radical

   Reference.   D. A. Ramsay and P. Thistlewaite, *Canad. J. Phys.*, **44**, 1381 (1966)†.

C$_3$H$_3$, Propargyl Radical (*contd.*)

Diffuse bands 2900–3450 observed in flash photolysis of various compounds (methyl acetylene, propargyl bromide, etc.) attributed to $CH_2-C=CH$. Maxima of strongest 3320, 3217 and 3119 Å.

C$_3$H$_5$, Allyl Radical

Reference.   C. L. Currie and D. A. Ramsay, *J. Chem. Phys.*, **45**, 488 (1966)†.

Diffuse bands 4100–3700 observed in flash photolysis of various allyl compounds. The strongest band (0, 0) is at 4083 Å and there are others (short-wave limit of diffuse structure) at 4024, 3937, 3921, 3887 and 3752 Å.

C$_4$H$_2$, Diacetylene

Reference.   S. C. Woo and T. C. Chu, *J. Chem. Phys.*, **5**, 786 (1937).

About 700 narrow line-like bands have been observed between 2970 and 2068 Å in absorption by diacetylene vapour. The bands become stronger but more diffuse at shorter wavelengths. The following bands, with intensities, are probably most prominent: $\lambda\lambda 2810\cdot6$ (4), 2809·8 (5), 2806·2 (5), 2755·7 (5), 2703·8 (7), 2650·8 (5), 2650·2 (7), 2618·9 (6), 2618·0 (6), 2608·2 (5), 2600·9 (5), 2600·1 (5), 2559·8 (5), 2468·3 (6), 2433·0 (10), 2422·4 (7), 2408·2 (7), 2404·2 (7), 2395·0 (8), 2313·8 (9), 2282·9 (6), 2205·5 (6).

C$_4$H$_2$$^+$

SCHÜLER'S T SPECTRUM

Occurrence.   In low current discharges through a variety of organic vapours (benzene, naphthalene, acetylene, etc.) usually in a carrier gas such as helium.

Appearance.   Four main groups of bands of complex structure, and some weaker bands. Some heads are degraded to the red and show rotational fine structure.

References.   H. Schüler and L. Reinebeck, *Z. Naturforsch.*, **6a**, 160 and 170 (1951); **7a**, 285 (1952).

J. H. Callomon, *Canad. J. Phys.*, **34**, 1046 (1956)†.

Calloman has shown from isotopic studies and detailed analysis that the emitter is the diacetylene ion $C_4H_2^+$. The following are his measurements and designations, with intensities reduced to a scale of 10:

λ	I	design.	λ	I	design.	λ	I	design.
6502·2	0	F	5912·9	1	D$_2$	5330·4	5	B$_2$
6038·7	1	D$_5$	5696·4	3	C$_3$	5299·0	3	B$_1$
6026·1	2	D$_4$	5626·4	2	C$_2$	5142·0	1	A$'$
6001·2	2	D$_3$	5591·5	3	C$_1$	5067·9	10	A
5957·7	1	D$_2''$	5551·6	0	C$_0$	4856·0	1	Z

C$_5$H$_5$

Absorption bands at 3378·4 or 3380·3 and 3342·3 Å, degraded to the red, have been observed in flash photolysis and attributed to the cyclopentadienyl radical.

References.   B. A. Thrush, *Nature*, **178**, 155 (1956)†.

R. Engleman and D. A. Ramsay, *Canad. J. Phys.*, **48**, 964 (1970)†.

C_6H_5, Phenyl Radical

Reference. G. Porter and B. Ward, *P.R.S.*, **287**, 457 (1965)†.

About 50 bands 5300–4400 Å have been observed in flash photolysis of benzene and benzene derivatives. They are degraded to the red and close double-headed. First heads of strong bands:

λ	I	λ	I	λ	I
5287·3	6	4906·9	10	4631·1	9
5132·2	4	4830·2	10	4512·7	7
5048·0	10	4701·1	10	4448·5	5

C_6H_6, Benzene

EMISSION SPECTRUM

Occurrence. The bands are most clearly produced by a Tesla coil discharge through benzene vapour, but can also be observed in an ordinary uncondensed discharge through flowing benzene vapour.

Appearance. Degraded to the red. Evenly spaced groups of bands similar in general appearance to the sequences of diatomic molecules.

References. J. B. Austin and I. A. Black, *P.R.*, **35**, 452 (1934)†.
R. K. Asundi and M. R. Padhye, *Nature*, **156**, 368 (1945).

Austin and Black have published a good photograph and a long list of wavelengths, but no estimates of intensities; the following are probably the outstanding heads, the intensities being our estimates from the photograph:

λ	I	λ	I	λ	I	λ	I
2903·7	4	2822·6	4	2751·2	7	2673·5	5
2900·6	3	2820·0	4	2739·1	10	*2667·4	10
*2898·1	4	*2812·3	7	*2736·5	8	2657·5	3
2837·9	6	2810·6	3	2689·2	6	2613·6	8
2832·9	3	2764·9	5	2684·6	6	2608·7	6
2828·1	5	2757·7	4	2678·6	9	*2602·6	9

* Head of group.

ABSORPTION SPECTRUM

Occurrence. Absorption by benzene vapour. The bands are also observed in absorption by liquid benzene and in solution.

Appearance. Degraded to the red. Evenly spaced groups of bands, usually three strong bands to each group. They become diffuse in the heated vapour or in solution. See Plate 10.

References. V. Henri, *J. Phys. Radium*, **3**, 18 (1922).
W. F. Radle and C. A. Beck, *J. Chem. Phys.*, **8**, 507 (1940).

The following are the principal heads as observed by Henri for the vapour; no intensities are recorded and the estimates given below are our own from Henri's description and from the extinction coefficients in solution:

C₆H₆ (*contd.*)

λ	I	λ	I	λ	I	λ	I
2667·1	1	2539·0		2428·5		*2363·5	3
2602·9		*2528·6	10	2425·3		2324·4	
2599·8		2483·7		*2415·9	5	2313·1	
*2589·0	9	2480·8		2375·2		2275·2	
2541·7		*2471·0	9	2372·5			

* Head of group.

The following are the approximate wavelengths of the heads of the strongest groups of absorption bands in various solvents, etc.:

H₂O	.	.	.	2674	2594	2535	2477	2426	2372
CH₃OH	.	.	.	2682	2605	2543	2484	2430	2375
C₂H₅OH	.	.	.	2684	2606	2545	2485	2433	2377
Hexane	.	.	.	2686	2607	2547	2487	2435	2378
CCl₄	.	.	.	—	2618	2558	2498	2439	—
Liquid benzene	.	.	2691	2611	2552	2492	2436	2384
Benzene vapour	.	.	2667	2589	2528	2471	2416	2363

FLUORESCENCE SPECTRUM

Occurrence. Fluorescence of benzene in solution.

Reference. V. Henri (see above).

This system and the emission spectrum and also the absorption spectrum are apparently all parts of the same system, the red end appearing stronger in emission and the shorter wavelength bands in absorption. The fluorescence spectrum is similar to the emission spectrum apart from the shift to longer wavelengths due to the action of the solvent. The following are the heads as recorded by Henri for solution in pentane:

λ	I	λ	I
3005	7	2766	10
2917	8	2701	8
2847	9	2659	6

C₇H₇, Benzyl Radical

References. H. Schüler and J. Kusjakow, *Spectrochim. Acta*, **17**, 356 (1961)†.
B. Ward, *Spectrochim. Acta*, **24 A**, 813 (1968)†.
A. N. Singh and I. S. Singh, *Indian J. Pure Appl. Phys.*, **7**, 349 (1969).

Emission bands around 3050 and 4500 Å occur with transformer discharges through toluene, ethyl benzene and other benzene derivatives. These bands, and also another system at 2580 Å, also occur in absorption following flash photolysis of toluene, etc.

4500 Å SYSTEM

The following measurements are from Singh and Singh who observed about 150 heads. The 4477 band is strongest and those at 4465, 4492, 4674 and 4688 are well shown on Schüler and Kusjakow's plate:

C₇H₇ (*contd.*)

λ	I	λ	I	λ	I	λ	I
4888·1	7	4634·1	8	4496·4	8	4479·2	8
4688·0	8	4331·6	7	4492·3	9	4477·2	10
4674·0	9	4568·3	6	4487·5	8	4469·7	6
4653·7	7	4510·6	7	4482·5	8	4465·0	7

3050 Å SYSTEM

The following measurements of red-degraded heads are from Ward:

λ	I	λ	I	λ	I
3067·3	5	3055·1	8	2964·1	6
3061·9	7	3051·6	10	2948·6	5
3058·5	7	2967·3	5	2933·5	5

C₁₀H₈, Naphthalene

Reference. D. P. Craig, J. M. Hollas, M. F. Redies and S. C. Wait, *Phil. Trans. Roy. Soc.*, A, **253**, 543 (1961)†.

Red-degraded absorption bands. Strongest 3080·0; other strong bands 3035·7 and 3014·9 Å.

CHCl

Reference. A. J. Merer and D. N. Travis, *Canad. J. Phys.*, **44**, 525 (1966)†.

The chloromethylene radical has been observed in absorption following flash photolysis of HCClBr₂. There is a long progression 8200–5500 with most features degraded to the violet. Q heads, 7610 (0 1 0, 0 0 0), 7144 (0 2 0, 0 0 0), 6735·7 (0 3 0, 0 0 0) and 6368·3 (0 4 0, 0 0 0).

CHF

Reference. A. J. Merer and D. N. Travis, *Canad. J. Phys.*, **44**, 1541 (1966)†.

Complex bands 6000–4300 in absorption following flash photolysis of HCFBr₂. Heads are mostly degraded to the red but the K structure is degraded to the violet. The (0 0 0, 0 0 0) band is at 5791 with heads at 5733, 5703 and 5668. The (0 1 0, 0 0 0) band is at 5460 Å.

CHN see HCN

CHNO, Isocyanic acid

Reference. Sho Chow Woo and Ta Kong Liu, *J. Chem. Phys.*, **3**, 544 (1935).

In absorption by vapour, diffuse bands λλ2565, 2545, 2525, 2513, 2505, 2495, 2477, 2465, 2445, 2434, 2415, 2400, 2385, 2370, 2357, 2345. There is continuous absorption below 2240 Å.

CH₃NO₂, Methyl nitrite

Reference. H. W. Thompson and C. H. Purkis, *Trans. Faraday Soc.*, **32**, 674 and 1466 (1936)

In absorption by the vapour there are diffuse bands in the near ultra-violet and continuous absorption further in the ultra-violet. Strongest bands with our intensity estimates: λλ3651 (4), 3508 (9), 3390·2 (10), 3284·1 (8), 3187·1 (5).

$C_2H_5NO_2$, Ethyl nitrite

References. See methyl nitrite.

In absorption there are diffuse bands in the near ultra-violet and continuous absorption at shorter wavelengths. Strong bands with our intensity estimates λλ3689 (4), 3549 (10), 3429 (8), 3316 (4), 3221 (2).

CHNS

Reference. G. C. Chaturvedi *et al.*, *J. Mol. Spec.*, **39**, 242 (1971)†.

In absorption isothiocyanic acid shows two progressions of bands 3400–2800, and two others around 2600 and 2300 Å. Diffuse red-degraded bands 3066, 3023, 2981 and 2941 are relatively strong.

CHO

VAIDYA'S HYDROCARBON FLAME BANDS

Occurrence. In inner cones of hydrocarbon and other flames, especially in low-temperature flames. First observed in C_2H_4/air and known as 'Ethylene flame bands'. Also in short-wave fluorescence of formaldehyde and high-frequency discharge through methyl formate.

Appearance. Degraded to the red. Fairly definite heads but complex rotational structure. Most of the bands fall into two progressions denoted A_0 and A_1 below, with some weaker bands in progressions A_2 and A_3.

References. W. M. Vaidya, *P.R.S.*, **147**, 513 (1934)†.

W. M. Vaidya, *Proc. Phys. Soc.*, A, **64**, 428 (1951).

A. G. Gaydon, *The Spectroscopy of Flames*, Chapman and Hall (1974)†.

G. A. Hornbeck and R. C. Herman, *Nat. Bur. Stand. Circular*, No. 523, 9 (1954)†.

R. N. Dixon, *Trans. Faraday Soc.*, **65**, 3141 (1969)†.

Table compiled from the above references. Intensities by Vaidya. The bands of the B progression only occur in hotter flames.

λ	I	Prog.	λ	I	Prog.	λ	I	Prog.	λ	I	Prog.
4088·2	3	A_1	3587·5	8	A_1	3186·0	9	A_1	2751·5	4	—
3986·0	2	A_0	3538·7	2	A_2	3114·8	9	A_0	2714·7	5	A_1
3823·3	4	A_1	3501·0	8	A_0	3013·7	8	A_1	2704·5		B
3802·7		B	3472·5		B	3001·5		B	2658·1	4	A_0
3729·3	5	A_0	3457·2	1	A_2	2947·4	7	A_0	2639·8	1	A_2
3697·7		B	3417·3	3	—	2857·8	6	A_1	2618·0	3	—
3674·6	1	A_2	3376·3	10	A_1	2796·1	5	A_0	2584·1	3	A_1
3635·6	3	—	3298·2	10	A_0	2780·4		B	2531·8	2	A_0
			3215·5	7	—	2773·8	1	A_2	2516·8	1	A_2

FLASH PHOTOLYSIS BANDS 4500–7500 Å

Occurrence. Observed in absorption following the flash photolysis of acetaldehyde, glyoxal, etc.

Appearance. Degraded to the red. Some bands show double heads and apparently simple rotational structure. Some bands are diffuse.

Reference. G. Herzberg and D. A. Ramsay, *P.R.S.*, **233**, 34 (1956)†.

CHO (*contd.*)

R heads	Q heads	v_1', v_2', v_3'	I	R heads	Q heads	v_1', v_2', v_3'	I
7551·6	7560·9	0, 6, 0	4	5570·0	–	0, 8, 1	3
6766·3	6774·1	0, 8, 0	7	5195·6	5201·0	0, 14, 0	5
6436	–	0, 9, 0	5d	5148·2	5152·7	0, 10, 1	4
6138·0	6144·7	0, 10, 0	10	4833·3	4838·3	0, 16, 0	4
5906/5860	–	0, 11, 0	5d	4791·1	4796·5	0, 12, 1	3
5624·0	5629·2	0, 12, 0	8				

CH_2O, Formaldehyde

ABSORPTION SPECTRUM

References. V. Henri and S. A. Schou, *Z.P.*, **49**, 774 (1928)†.
J. C. D. Brand, *J. Chem. Soc.*, 858 (1956).

The absorption shows complex but well resolved fine structure, with some heads degraded to the red. Under small dispersion identification is best made by comparison of spectrograms. See Plate 10. These bands occur during the slow combustion of hydrocarbons. The following wavelengths of red-degraded heads are from Henri and Schou:

λ	I	λ	I	λ	I	λ	I
3387	4	3057	3	2948	6	2787	7
3288	6	3051	5	2931	10	2756	5
3198	3	3033	7	2874	6	2747	5
3164	9	2978·9	6	2839	9	2667	1

EMELÉUS'S COOL FLAME SPECTRUM; FORMALDEHYDE FLUORESCENCE

Occurrence. In the cool flame of ether, acetaldehyde, hexane and other organic substances. Also in fluorescence by formaldehyde and in a Tesla discharge.

Appearance. A number of narrow approximately equally spaced bands; probably degraded to the red. See Plate 8.

References. H. J. Emeléus, *J. Chem. Soc.*, 2948 (1926)†.
J. C. D. Brand, *J. Chem. Phys.*, **19**, 377 (1951); *J. Chem. Soc.*, 858 (1956).
A. G. Gaydon, *Spectroscopy of Flames*, Chapman and Hall (1974)†.

The following are based on Brand's measurements, but with Gaydon's estimates of intensity:

λ	I	λ	I	λ	I
5227	1	4434	8	3846·5	8
5097	1	4347	8	3777–63	5
4947	1	4240–20	10	3698–79	8
4821	3	4121	8	3544	4
4695–73	5	4044	6	3405	2
4569–51	8	3952	10		

CH_3O

Reference. D. W. G. Style and J. C. Ward, *Trans. Faraday Soc.*, **49**, 999 (1953).

Bands believed to be due to the methoxy radical have been observed in the short-wave fluorescence of methyl nitrite, dimethyl carbonate, methyl formate or methyl chloroformate.

CH₃O (*contd.*)

Diffuse narrow bands, possibly shaded to the red. They have been arranged into three progressions, and Professor Style has kindly provided the following wavelengths; intensities are our estimates from his photograph.

λ	I	λ	I	λ	I	λ	I
4205	1	3886	4	3609	6	3361	2
4134	1	3837	8	3570	6	3322	4
4072	4	3777	7	3514	10	3282	2
4036	4	3736	2	3978	6	3249	1
3985	7	3698	8	3438	6	3214	1
3910	4	3642	10	3399	7		

A band at 2064 Å found in flash photolysis is due to HNO.

CHOOH, Formic Acid

Reference. B. Sugarman, *Proc. Phys. Soc.*, **55**, 429 (1943)†.

The absorption spectrum shows a number of diffuse bands without well-defined heads in the range λλ2600–2260. Below λ2260 the absorption becomes continuous at least to λ1900.

Centres of bands:

λ	I	λ	I	λ	I
2500·0	1	2394·7	5	2340·3	10
2461·2	4	2391·2	4	2335·8	6
2443·4	4	2382·2	9	2325·2	9
2420·8	6	2377·2	6	2318·9	7
2414·3	4	2372·8	5	2302·9	8
2408·4	4	2361·5	9	2284·9	10
2398·3	10	2354·0	6	2273·3	5

C₂H₄O, Acetaldehyde

Occurrence. Absorption by acetaldehyde vapour.

Appearance. Strong absorption with maximum at 2900 Å, with a complex structure of discrete bands from 3400 Å to 3200 Å, after which they become diffuse and finally merge into continuum around 2800 Å. Some of the bands appear to be shaded to the red.

References. S. A. Schou, *Jour. de Chim. Phys.*, **26**, 27 (1929).

K. K. Innes and L. E. Giddings, *J. Mol. Spec.*, **7**, 435 (1961).

The following are the limits of the outstanding bands as taken from Schou's list:

λ	I	λ	I	λ	I	λ	I
3399·3	3	3341·7	3	3296·2	6	3254·4	6
3381·1		3328·9		3289·9		3247·3	
3376·6	2	3320·0	6	3281·3	5	3241·1	6
3363·0		3314·7		3274·4		3234·1	
3359·0	4	3305·1	4	3267·6	7	3231·4	5
3344·5		3299·9		3258·3		3228·7	

C_2H_4O, Acetaldehyde (*contd.*)

λ	I	λ	I	λ	I	λ	I
3222·1	5	3207·3	6	3190·9	9	3177·5	8
3216·9		3202·0		3180·2		3172·2	
3215·6	6	3199·0	7			Diffuse bands extend	
3212·7		3196·1				to 2800, after which	
						absorption is con-	
						tinuous.	

C_2H_5CHO, Propionaldehyde

Occurrence. Absorption by the vapour.

Appearance. Complex system of diffuse bands from 3400 Å, merging into a continuum at 3250 Å, this continuum extending to about 2500 Å.

Reference. S. A. Schou, *Jour. de Chim. Phys.*, **26**, 39 (1929).

The following are the limits of the strongest bands:

λ	I	λ	I	λ	I
3370·8	4	3322·0	8	3276·7	8
3363·9		3316·4		3272·3	
3343·6	5	3298·1	9	3269·0	7
3339·6		3294·2		3262·0	
3336·1	6	3288·7	8	3258·7	8
3332·3		3284·8		3248·9	
3331·4	6				
3324·9					

CH_2CHCHO, Propenal

Reference. J. C. D. Brand, J. H. Callomon and J. K. G. Watson, *Disc. Faraday Soc.*, **35**, 175 (1963)†.

The strongest band maxima in absorption are at 3865·6, 3821·2, 3794·1, 3685–76 and 3665·8 Å.

C_3H_6O, Acetone

Occurrence. Absorption by the vapour.

Reference. E. J. Bowen and H. W. Thompson, *Nature*, **133**, 571 (1934).

Acetone shows continuous absorption from 3200 Å to 2400 Å, with a maximum at about 2800 Å. At low pressure this continuum breaks up into four groups each of about 25 diffuse bands; these groups have maxima at 3150, 2900, 2710 and 2570 Å.

C_6H_5OH, Phenol

References. J. Kahane-Paillous, *J. Chim. Phys.*, **57**, 1058 (1960).
H. D. Bist, J. C. D. Brand and D. R. Williams, *J. Mol. Spec.*, **21**, 76 (1966).

C_6H_5OH, Phenol (contd.)

The 2750 Å system (2900–2500 Å) occurs in absorption and in emission in a Schüler tube. In absorption the strongest red-degraded head is at 2750·0, with other strong heads 2692·1, 2681·1, 2680·9, 2678·2, 2657·0 and 2636·6. In emission the strongest head is 2813·9 with others at 2994·2, 2849·8, 2829·8, 2828·3 and 2790·7.

C_6H_5CHO, Benzaldehyde

Occurrence. Absorption by benzaldehyde vapour.
Reference. M. Hemptinne, *J. Phys. Radium*, **9**, 357 (1928)†.

The following maxima of the narrow headless bands are taken from Hemptinne's published spectrogram. Intensities (on a scale of 5) are our estimates from the spectrogram, and some of the wavelengths are also taken from this.

λ	I	λ	I	λ	I
2851	1	2766*	4	2716	1
2841	5	2746	3	2709	2
2806	2	2735	3	2696	2
2777	2	2726	1	2677	2

* A misprint in the wavelength recorded has been corrected.

CHOCHO, Glyoxal

References. A. G. Gaydon, *Trans. Faraday Soc.*, **43**, 36 (1947)†.
J. Paldus and D. A. Ramsay, *Canad. J. Phys.*, **45**, 1380 (1967)†.
G. N. Currie and D. A. Ramsay, *Canad. J. Phys.*, **49**, 317 (1971)†.

Characteristic narrow bands, mostly degraded to the violet, are shown by glyoxal vapour in absorption, in emission in a Tesla discharge, and in fluorescence. The intensities I_a, I_t and I_f are for these three sources respectively. Wavelengths by Gaydon, from small dispersion studies. Paldus and Ramsay have examined the rotational structure.

λ	I_a	I_t	I_f	λ	I_a	I_t	I_f	λ	I_a	I_t	I_f
5209	–	7	3	4784	–	6	(7)	4605	3	5	7
4946	–	5	5	4777	1	9	10	4555	10	10	8
4906	–	5	6	4752	–	6	7	4532	7	7	5
4793	–	5	(7)	4670	2	7	7	4404	6	1	6
								4280	8	0	3

Currie and Ramsay have also studied the 4875 Å band of *cis*-glyoxal, which occurs in fluorescence and weakly in absorption.

CHOF

References. L. E. Giddings and K. K. Innes, *J. Mol. Spec.*, **6**, 528 (1961)†.
G. Fischer, *J. Mol. Spec.*, **29**, 37 (1969).

Formyl fluoride shows absorption 2700–2000 Å, strongest at the shorter wavelengths. The following band centres may be prominent, 2551, 2482, 2356, 2300, 2285, 2246, 2233, 2183 Å.

CN

CN is a particularly stable radical and its spectrum is dominant in many sources, and frequently occurs as an impurity. It is also important in astrophysics. The Violet system is most easily observed, and the Red system, with strongest bands in the near infra-red is also important. Four weaker systems in the near ultra-violet were discovered by Douglas and Routly, and weak inter-combination systems have been observed by Le Blanc and Carroll.

RED SYSTEM, $A^2\Pi - X^2\Sigma$

References. F. A. Jenkins, Y. K. Roots and R. S. Mulliken, *P.R.*, **39**, 16 (1932)†.
G. Herzberg and J. G. Phillips, *Astrophys. J.*, **108**, 163 (1948)†.
F. J. LeBlanc, *J. Chem. Phys.*, **48**, 1980 (1968).
B. Brocklehurst *et al.*, *Identification Atlas of Molecular Spectra*, 8, York Univ. Ontario (1971)†.
S. P. Davis and J. G. Phillips, *The Red System of CN*, Univ. California Press (1963).

Occurrence. In the carbon arc in air, in shock heated $CO + N_2$, in flames of C_2N_2 and of organic fuels burning with N_2O, in discharge tubes containing nitrogen and carbon compounds, and especially strongly when organic vapours such as C_2H_2 or $HCCl_3$ react with active nitrogen. Also in carbon-type stars and the sun.

Appearance. Degraded to longer wavelengths. Triple-headed bands with the R_2, R_1 and Q_1 heads of about equal strength and a much weaker $^sR_{21}$ head in front. See Plate 9.

The following table gives all four heads of five of the strongest bands and the first main head (R_2) of all bands except a few of the very weak ones. Wavelengths are from Davis and Phillips, or from summarized data by Brocklehurst *et al.* Estimates of intensity are difficult because of varying plate sensitivity over the large wavelength range; values given refer to a discharge tube. In flames, bands with low v' in the infra-red are strong. In active nitrogen bands with v' from 5 to 9 are strong, and in active nitrogen/CH_4 LeBlanc records bands with v' up to 25.

λ	I	v', v''	λ	I	v', v''	λ	I	v', v''
19644	2	0, 2	7624·1	5	7, 4	5729·9	7	6, 1
15050	1	2, 3	7437·3	8	6, 3	5728·1	1	11, 5
14545	3	1, 2	7259·1	8	5, 2	5606·9	2	5, 0
14074	5	0, 1	7088·7	6	4, 1	5597·9	3	10, 4
11247	2	1, 1	6925·8	2	3, 0	5508·0	*	18, 10
10996		0, 0 Q_1	6791·9	2	8, 4	5473·3	5	9, 3
10963		0, 0 R_1	6631·3	9	7, 3	5354·1	4	8, 2
10925	10	0, 0 R_2	6502·4		6, 2 Q_1	5239·3	3	7, 1
10872		0, 0 $^sR_{21}$	6494·1		6, 2 R_1	5156·0	1	11, 4
9381·1	4	2, 1	6478·5	10	6, 2 R_2	5135·4	*	19, 10
9189·5		1, 0 Q_1	6466·5		6, 2 $^sR_{21}$	5129·6	1	6, 0
9168·4		1, 0 R_1	6432·2	1	11, 6	5043·5	1	10, 3
9140·6	7	1, 0 R_2	6332·2	9	5, 1	5010·1	*	18, 9
9106·8		1, 0 $^sR_{21}$	6278·9	1	10, 5	5003·9	1	14, 6
8709·2	3	6, 4	6192·0	4	4, 0	4935·6	2	9, 2
8484·6	4	5, 3	6132·4	3	9, 4	4891·9	1	13, 5
8270·8	5	4, 2	6106·0	*	18, 11	4832·4	1	8, 1
8067·1	8	3, 1	5992·6	6	8, 3	4784·6	1	12, 4
7908·6		2, 0 Q_1	5877·7		7, 2 Q_1	4733·3	0	7, 0
7894·1		2, 0 R_1	5871·4		7, 2 R_1	4681·9	1	11, 3
7872·7	10	2, 0 R_2	5858·2	9	7, 2 R_2	4488·9	0	9, 1
7850·2		2, 0 $^sR_{21}$	5849·3		7, 2 $^sR_{21}$	4373·7	0	12, 3

* Observed in active nitrogen/CH_4.

CN (*contd.*)

VIOLET SYSTEM, $B^2\Sigma - X^2\Sigma$

Occurrence. In carbon arc in air, in discharge tubes containing nitrogen and carbon compounds, and when carbon compounds are introduced into active nitrogen. These bands occur very frequently as impurities in discharge tube sources, and in arcs between carbon poles. Also in shock tubes, flames, in many stellar sources, and in absorption.

Appearance. In arc sources three strong sequences with heads at λ4216, 3883, and 3590, degraded to the violet. In active nitrogen the heads of the 4216 and 3590 sequences are less marked, and 'tail' bands are observed; these are degraded to the red. See Plate 9.

References. Main system. W. Jevons, *P.R.S.*, **112**, 407 (1926)†.
B. Brocklehurst *et al.*, *Identification Atlas of Molecular Spectra*, **9**, York Univ., Ontario, (1972)†.
Tail bands. F. A. Jenkins, *P.R.*, **31**, 539 (1928)†.
        M. W. Feast, *Proc. Phys. Soc.*, A, **62**, 121 (1949).
        A. E. Douglas and P. M. Routly, *Astrophys. J.* Supplement **1**, 295 (1955).

Main System. Intensities are our own estimates. Degraded to the violet.

λ	I	v', v''	λ	I	v', v''	λ	I	v', v''
4606·1	1	0, 2	4197·2	8	1, 2	3861·9	8	2, 2
4578·0	2	1, 3	4181·0	7	2, 3	3854·7	6	3, 3
4553·1	2	2, 4	4167·8	6	3, 4	3850·9	4	4, 4
4531·9	2	3, 5	4158·1	5	4, 5	3590·4	8	1, 0
4514·8	2	4, 6	4152·4	4	5, 6	3585·9	7	2, 1
4502·2	2	5, 7	3883·4	10	0, 0	3583·9	6	3, 2
4216·0	9	0, 1	3871·4	9	1, 1			

Tail Bands. These bands show heads degraded to the red. They only occur when the system is very strongly developed, as in active nitrogen, or the carbon arc. Intensities are for a carbon arc on an increased scale, based on 10 for the (11, 11) band. In many bands the heads are weak or even absent; we list heads, λ_h, and origins, λ_o from above references.

λ_h	λ_o	I	v', v''	λ_h	λ_o	I	v', v''
4255·6	4258·8		18, 18	3424·7	3426·5		16, 13
4078·7	4083·2		15, 15	3404·8	3411·7	7	10, 8
4050·5	4052·8		19, 18	3385·9	3387·1		19, 15
4029·3	4034·6	6	14, 14	3380·3	3389·5	6	9, 7
3986·4	3988·8		18, 17	3359·1	3371·6	4	8, 6
3984·6	3991·1	9	13, 13	3340·6	3357·4	4	7, 5
3944·7	3953·3	9	12, 12	3322·3	3347·1	4	6, 4
3909·5	3920·8	10	11, 11	3296·3	3339·9	3	5, 3
	3894·1	(10)	10, 10	3203·5	3208·9	4	10, 7
	3697·1		12, 11	3180·2	3186·7	4	9, 6
3658·1	3665·3	5	11, 10	3159·9	3168·0	4	8, 5
3628·9	3638·4	7	10, 9	3149·1	3150·8		14, 10
3603·0	3616·6	7	9, 8	3142·6	3152·9	4	7, 4
3501·4	3504·8		13, 11	3127·6	3141·1	3	6, 3
3465·3	3469·1	6	12, 10	3114·3	3131·0	2	5, 2
3433·0	3437·8	6	11, 9				

CN (*contd.*)

DOUGLAS AND ROUTLY'S SYSTEMS

Occurrence. In hollow cathode of discharge through He containing a trace of C_2N_2.

References. A. E. Douglas and P. M. Routly, *Astrophys. J.*, Supplement 1, 295 (1955).

B. L. Jha and D. R. Rao, *Proc. Indian Acad. Sci.*, A, **63**, 316 (1966).

B. L. Lutz, *Astrophys. J.*, **164**, 213 (1971).

$H^2\Pi - B^2\Sigma$. Red-degraded bands with heads at 3025·6 (0, 1) and 2842·9 (0, 0). They have a similar appearance to bands of the red system.

$F^2\Delta - A^2\Pi$. Double-headed bands (separation about 6 Å), degraded to the red. First (R_{21}) heads; intensities from Jha and Rao.

λ	I	v', v''	λ	I	v', v''	λ	I	v', v''
2640·5	2	2, 9	2435·3	4	2, 7	2291·4	4	0, 4
2613·1	2	1, 8	2409·1	4	1, 6	2229·3	3	1, 4
2534·5	4	2, 8	2383·5	4	0, 5	2204·9	4	0, 3
2507·8	4	1, 7	2342·0	2	2, 6	2123·6	2	0, 2
2462·5	1	3, 8	2316·4	4	1, 5	2047·9	–	0, 1

$D^2\Pi - A^2\Pi$. A weak system of red-degraded bands 2200–3000 Å. The bands have open structure, and the heads are not obvious.

λ origins	v', v''	λ origins	v', v''	λ origins	v', v''
3042·2	0, 7	2571·1	1, 4	2408·4	2, 3
2898·1	0, 6	2512·1	2, 4	2355·2	3, 3
2691·6	1, 5	2466·0	1, 3	2261·9	3, 2

$D^2\Pi - X^2\Sigma$. Weak bands, degraded to the red without obvious heads. Only two of the seven bands have been measured.

λ origins	v', v''
2879·5	0, 10
2737·3	0, 9

LE BLANC'S SYSTEM, $B^2\Sigma - A^2\Pi$

Reference. F. J. LeBlanc, *J. Chem. Phys.*, **48**, 1841 (1968).

Occurrence. Trace of methane in active nitrogen.

Appearance. Degraded to the violet. Bands show two main heads.

This is a weak system. Intensities are our estimates based on LeBlanc's data.

λ	I	v', v''	λ	I	v', v''	λ	I	v', v''
4939·4	5	7, 6 P_2	4327·1	4	4, 1 P_2	4239·0	10	7, 4 P_2
4925·1		7, 6 Q_1	4316·2		4, 1 Q_1	4228·4		7, 4 Q_1
4684·3	5	4, 2 P_2	4292·4	7	5, 2 P_2	4221·1	5	8, 5 P_2
4671·3		4, 2 Q_1	4281·5		5, 2 Q_1	4198·9	9	9, 6 Q_1
4638·2	5	5, 3 P_2	4263·0	6	6, 3 P_2			
4625·5		5, 3 Q_1	4252·2		6, 3 Q_1			

CARROLL'S SYSTEM, $E^2\Sigma - A^2\Pi$

References. P. K. Carroll, *Canad. J. Phys.*, **34**, 83 (1956)†.

B. L. Lutz, *Astrophys. J.*, **164**, 213 (1971).

CN (*contd.*)

Occurrence. Hollow cathode discharge through He with C_2N_2.
Appearance. Close triple-headed. Degraded to the red.
First heads:

λ	v', v''	λ	v', v''
2244·0	1, 4	2157·9	0, 2
2161·0	1, 3	2078·8	0, 1
2241·9	0, 3	2004·4	0, 2

CN⁺

Occurrence. Hollow cathode of discharge in He containing a trace of C_2N_2.
References. A. E. Douglas and P. M. Routly, *Astrophys. J.*, **119**, 303 (1954)†.
B. L. Lutz, *Astrophys. J.*, **163**, 131 (1971).

2180 Å SYSTEM $f\,^1\Sigma - a\,^1\Sigma$.

Appearance. Headless bands with single P and R branches spreading from a clear region at the origin.

λ origins	I	v', v''
2180·6	10	0, 0
2153·5	7	1, 1
2129·9	4	2, 2
2109·3	3	3, 3

3185 Å SYSTEM $c\,^1\Sigma - a\,^1\Sigma$.

Appearance. Degraded to the red; single-headed.
The following are the R heads, with our own estimates of intensity:

λ	I	v', v''	λ	I	v', v''
3263·3	6	1, 1	3019·6	6	3, 1
3185·1	5	0, 0	2952·5	6	2, 0
3092·4	3	4, 2	2915·7	5	4, 1
3063·3	7	1, 0			

2660 Å SYSTEM, $d\,^1\Pi - b\,^1\Pi$

Appearance. Degraded to the red. Single-headed.

λ	v', v''	λ	v', v''
2843·2	1, 2	2677·9	2, 1
2782·0	0, 1	2660·1	0, 0
2717·5	1, 1	2639·6	3, 1
2716·9	4, 2	2600·3	1, 0

2652 Å SYSTEM, $f\,^1\Sigma - b\,^1\Pi$

Degraded to shorter wavelengths. Heads 2274·5 (0, 1) and 2652·6 (0, 0).

CN₂

References. G. Herzberg and D. N. Travis, *Canad. J. Phys.*, **42**, 1658 (1964)†.
H. W. Kroto, *Canad. J. Phys.*, **45**, 1439 (1967)†.

CN$_2$ (*contd.*)

A $^3\Pi - {}^3\Sigma$ system has been obtained in flash photolysis of diazomethane by Herzberg and Travis, and also in emission from a flame of active nitrogen with organic compounds by Jennings and Linnett. Strongest heads 3294·5 (deg. R), 3290·1 (deg. R) and 3286·3 (deg. V).

A $^1\Pi - {}^1\Delta$ system, 3337–3300 Å, with strongest head at 3327·3 (deg. R) has been found by Kroto in flash photolysis of cyanogen azide.

C$_2$N

References.   A. J. Merer and D. N. Travis, *Canad. J. Phys.*, **43**, 1795 (1965)†; **44**, 353 (1966).

Observed in flash photolysis of HC(CN)N$_2$. An A$^2\Delta - X^2\Pi$ system has 24 violet-degraded heads, 4708·0, 4699·0, etc. There is a weak B$^2\Sigma - X^2\Pi$ system with strongest head at 4457·1. A system C$^2\Sigma - X^2\Pi$ shows 13 violet-degraded heads, with first head at 3752·9 Å.

C$_2$N$_2$, Cyanogen

Two groups of absorption bands have been observed for cyanogen gas. The stronger lies below 2300 Å, and the weaker, requiring a path length of 1 to 5 m at S.T.P., is in the region 3100–2400 Å; bands of this system also occur in emission from discharges and afterglows (Meyer *et al.*). Progress has been made with analysis into several systems.

References.   Soo-Chow Woo and R. M. Badger, *P.R.*, **39**, 932 (1932).

Soo-Chow Woo and Ta-Kong Liu, *J. Chem. Phys.*, **5**, 161 (1937).

J. H. Callomon and A. B. Davey, *Proc. Phys. Soc.*, **82**, 335 (1963)†.

J. A. Merer, D. H. Stedman and D. W. Setser, *J. Mol. Spec.*, **44**, 206 (1972)†.

Strong bands of main system as observed by Sho-Chow Woo and Badger; most of the bands are shaded to longer wavelengths:

λ	I	λ	I	λ	I
2237·7	3	2164·5	4	2107·4	7
2226·4	5	2145·6	4	2093·1	10
2200·0	5	2137·7	4	2054·6	7
2188·2	9	2125·0	7	2035·2	7
				2007·0	10

The weaker group contains a $^3\Sigma_u^+ - {}^1\Sigma_g^+$ system with origin at 3002 Å. Narrow bands are degraded to the red. In absorption the following are relatively strong (Woo and Liu):

λ	I	λ	I	λ	I
3002·7	3	2759·2	3	2614·0	4
2831·1	3	2678·5	4	2540·3	5
2828·6	5	2675·6	6	2507·0	7

In emission there are heads of red-degraded band groups at 3825·8, 3489·6, 3229·1 and 3002·7 Å.

CNO and CNS, see NCO and NCS

CO

Many band systems are known for CO. The electronic levels involved are singlet and triplet. Some of the systems are very frequently obtained as impurities in the spectra from discharge tubes. The bands most commonly encountered are the Ångström bands in the visible, the Third Positive system in the near ultra-violet and the Fourth Positive system in the far ultra-violet and vacuum regions. A trace of CO in one of the rare gases often gives the Triplet system.

A valuable summary of the spectra of CO is given by P. H. Krupenie, U. S. National Standard Reference Data Series, N.S.R.D.S. – N.B.S.5 (1966).

THE ÅNGSTRÖM SYSTEM, $B^1\Sigma - A^1\Pi$

Occurrence. Readily obtained with CO or CO_2 in the positive column of an uncondensed discharge. The glass of a new discharge tube usually produces enough CO to give these bands and those of the Third Positive system.

Appearance. A progression of bands with strong single heads degraded to the violet. See Plate 2.

References. R. C. Johnson and R. K. Asundi, *P.R.S.*, **123**, 560 (1929).

E. Hulthén, *Ann. Phys. (Lpz.)*, **71**, 41 (1923).

O. Jasse, *C.R. Acad. Sci. (Paris)*, **182**, 692 (1926).

The following are the heads usually observed:

λ	I	v', v''	λ	I	v', v''	λ	I	v', v''
6620·3	7	0, 5	5610·2	10	0, 3	4835·3	10	0, 1
6299	2	1, 6	5399	2	1, 4	4510·9	10	0, 0
6079·9	9	0, 4	5198·2	10	0, 2	4393·1	8	1, 1
5818	2	1, 5	5016	1	1, 3	4123·6	7	1, 0

THE HERZBERG SYSTEM, $C^1\Sigma - A^1\Pi$

Occurrence. From CO in discharge tubes and controlled electron sources with conditions favourable for the production of the Ångström bands.

Appearance. The bands are degraded to the violet and are very similar in appearance to the Ångström bands.

References. O. S. Duffendack and G. W. Fox, *Astrophys. J.*, **65**, 214 (1927).

R. C. Johnson and R. K. Asundi, *P.R.S.*, **123**, 560 (1929).

G. Herzberg, *Z.P.*, **52**, 815 (1929).

λ	I	v', v''	λ	I	v', v''
5705·9	1	0, 7	4380·3	7	0, 3
5318·4	1	0, 6	4124·8	7	0, 2
4972·8	2	0, 5	3893·1	7	0, 1
4661·3	5	0, 4	3680·9	4	0, 0

FOURTH POSITIVE SYSTEM, $A^1\Pi - X^1\Sigma$

Occurrence. The system appears very readily in the positive column of discharges through carbon monoxide or carbon dioxide and other organic vapours if oxygen is present. It is emitted weakly by a carbon arc. In the reaction zone of hydrocarbon flames, especially C_2H_2/O_2, these bands are weakly excited to $v' = 9$ by a chemiluminescent reaction $O + C_2O = CO^* + CO$ (Fontijn and Johnson). The short-wave end of the system, in the vacuum ultra-violet, occurs strongly in absorption.

CO (*contd.*)

Appearance. Degraded to the red. An extensive system of apparently single-headed bands extending from 2800 Å to 1000 Å. See Plate 2.

References. R. S. Estey, *P.R.*, **35**, 309 (1930).

L. B. Headrick and G. W. Fox, *P.R.*, **35**, 1033 (1930)†.

D. N. Read, *P.R.*, **46**, 571 (1934).

A. Fontijn and S. E. Johnson, *J. Chem. Phys.*, **59**, 6193 (1973).

The following measurements of the bands of wavelength greater than 2000 Å are by Estey; these values are about $\frac{1}{2}$ Å smaller than the measurements by Read or Headrick and Fox, whose measurements are, however, chiefly concerned with the shorter wavelength end of the spectrum.

λ	I	v', v''	λ	I	v', v''	λ	I	v', v''
2799·7	9 ?	9, 22	2492·9	8	10, 20	2247·2	7	8, 16
2785·4	8 ?	4, 18	2483·8	3	6, 17	2238·3	9	4, 13
2742·6	6	11, 23	2463·2	10	9, 19	2221·5	10	7, 15
2740·0	4	7, 20	2458·0	2	5, 16	2215·8	3	3, 12
2712·1	4	6, 19	2433·9	9	8, 18	2196·8	10	6, 14
2698·3	6	13, 24	2424·1	5	11, 20	2173·0	9	5, 13
2684·0	3	5, 18	2407·6	7	7, 17	2150·2	8	4, 12
2680·8	5	9, 21	2394·2	3	10, 19	2137·0	5	7, 14
2662·9	4	12, 23	2393·1	4	13, 21	2128·3	8	3, 11
2661·5	4	15, 25	2381·6	6	6, 16	2113·1	9	6, 13
2659·6	4	4, 17	2365·5	5	9, 18	2107·2	7	2, 10
2630·0	6	11, 22	2356·5	4	5. 15	2089·9	10	5, 12
2598·3	4	10, 21	2337·9	7	8, 17	2086·9	1	1, 9
2594·5	1	16, 25	2332·5	3	4, 14	2067·6	10	4, 11
2567·8	5	9, 20	2311·5	8	7, 16	2046·3	10	3, 10
2556·0	3	12, 22	2286·1	7	6, 15	2025·8	9	2, 9
2538·6	4	8, 19	2273·9	3	9, 17	2011·8	8	5, 11
2521·8	3	14, 23	2272·3	1	12, 19	2005·8	5	1, 8
2509·9	8	7, 18	2261·7	9	5, 14			

$E^1\Pi - A^1\Pi$ SYSTEM

Occurrence. In ordinary discharge through CO_2.

Appearance. Three single-headed bands, degraded to shorter wavelengths.

Reference. R. Kepa and M. Rytel, *Acta Phys. Polonica*, A, **39**, 629 (1971)†.

Confirmed by isotope shifts. Heads:

λ	v', v''
3959·8	0, 2
3745·7	0, 1
3548·0	0, 0

THE THIRD POSITIVE AND 5B BANDS, $b^3\Sigma - a^3\Pi$

These bands fall into two progressions which at one time were throught to belong to two different systems. Some papers therefore refer to one progression as the Third Positive Bands and the other as the 5B Bands; other papers refer to all the bands as the Third Positive System.

Occurrence. With CO in discharge tubes under a wide range of conditions. A mere trace of CO gives the bands strongly in the positive column with an uncondensed discharge.

CO (*contd.*)

Appearance. A progression of five strong bands with five close subheads forming the Third Positive group, consisting of the strongest members of the $(0, v'')$ progression, and a weaker progression of bands of a similar type forming the 5B group. Degraded to the violet. See Plate 2.

References. G. H. Dieke and J. W. Mauchly, *P.R.*, **43**, 12 (1933).

O. S. Duffendack and G. W. Fox, *Astrophys. J.*, **65**, 214 (1927).

B. S. Beer, *Z.P.*, **107**, 73 (1937).

	3rd Pos.				5B	
λ	I	v', v''		λ	I	v', v''
2833·1	10	0, 0		2665·3	8	1, 0
2977·4	9	0, 1		2793·1	2	1, 1
3134·4	8	0, 2		2930	1	1, 2
3305·7	7	0, 3		3079·9	5	1, 3
3493·3	6	0, 4		3242·1	6	1, 4
3699	2	0, 5		3419·2	5	1, 5
				3612·7	5	1, 6
				3825·1	2	1, 7

THE ASUNDI BANDS, $a'\,^3\Sigma - a\,^3\Pi$

Occurrence. In the positive column of a discharge tube containing CO.

Appearance. Bands degraded to the red with complex structure; probably containing five heads.

References. R. K. Asundi, *P.R.S.*, **124**, 277 (1929).

L. Gerö and K. Lörinczi, *Z.P.*, **113**, 449 (1939).

The following measurements of the first heads are from Asundi who used small dispersion and recorded only two extreme heads for each band; values of v' have been corrected.

λ	I	v', v''	λ	I	v', v''	λ	I	v', v''
8592	2	4, 0	7116·5	3	11, 3	6366·9	5	11, 2
8222·5	1	6, 1	6990·2	2	8, 1	6244·0	5	8, 0
7833·9	3	5, 0	6804·0	8	10, 2	6105·2	5	10, 1
7552·5	1	7, 1	6685·7	7	7, 0	5861·0	6	9, 0
7314·0	2	9, 2	6513·5	9	9, 1	5749·1	6	11, 1
7210·4	5	6, 0						

THE TRIPLET BANDS, $d^3\Delta - a^3\Pi$

Occurrence. Obtained by Merton and Johnson in a wide-bore tube containing He and H_2 with a trace of CO using an uncondensed discharge. Weakly in oxy-acetylene flames.

Appearance. With moderate dispersion the bands, which are degraded to the red, show a well-marked triplet structure. See Plate 2.

References. T. R. Merton and R. C. Johnson, *P.R.S.*, **103**, 383 (1923)†.

L. Gerö and F. Szabó, *Ann. Phys. (Lpz.)*, **35**, 597 (1939).

R. Herman and L. Herman, *J. Phys. Radium*, **9**, 160 (1948)†.

The following wavelengths are from Herman and Herman, with revised vibrational assignments; intensities from Merton and Johnson.

CO (*contd.*)

λ	I	v', v''	λ	I	v', v''	λ	I	v', v''
6464·6			4996·9			4494·4		
6433·1	10	0, 0	4979·0	6	6, 1	4478·8		12, 3
6401·0			4959·0			4462·9		
6037·0			4935·5			4405·0		
6010·5	8	1, 0	4917·2	2	8, 2	4390·9		7, 0
5982			4897·5			4374·0		
5670·5			4823·5			4343·8		
5647·6	6	2, 0	4806·7	8	5, 0	4328·7		11, 2
5624			4787			4314·1		
5554·1			4764·8			4227·2		
5532·5	5	4, 1	4747·5	5	7, 1	4213·7		8, 0
5508			4729·1			4198·9		
5351·2			4602·6			4201·5		
5330·5	5	3, 0	4586·4	7	6, 0	4188·4		10, 1
5308			4571			4174·6		
5258·3			4556·5			4182·5		
5238·4	5	5, 1	4541·0	5	8, 1	4171		12, 2
5216			4524·0			4157		
5070·9			4520·7			4036·4		
5052·7	8	4, 0	4505·5		10, 2	4023·7		13, 2
5033			4488·4			4011		

THE CAMERON BANDS, $a\,^3\Pi - X\,^1\Sigma$

Occurrence. Cameron obtained these bands in a wide-bore discharge tube filled with neon using an uncondensed discharge but with low intensity. Hansche using a continous-wave oscillator to excite CO in a 12-litre flask finds that the Cameron bands reach a maximum intensity at a pressure between 0·003—0·002 torr. May also be obtained in absorption.

Appearance. Degraded to the red. Five close heads to each band.

References. W. H. B. Cameron, *Phil. Mag.*, **1**, 405 (1926).

G. Herzberg, *Z.P.*, **52**, 815 (1929).

G. E. Hansche, *P.R.*, **57**, 289 (1940).

Y. Tanaka, A. S. Jursa and F. le Blanc, *J. Chem. Phys.*, **26**, 862 (1957).

K. N. Rao, *Astrophys. J.*, **110**, 304 (1949)†.

The following wavelenths are mostly from Cameron and are for the first (R_3) head. For the (0, 0) band all heads are from Tanaka *et al.*

λ	I	v', v''	λ	I	v', v''	λ	I	v', v''
2575·3	8	4, 8	2409·2	7	2, 5	2063·7	1	0, 0 R_1
2553·3	6	3, 7	2388·8	7	1, 4	2062·5	4	0, 0 Q_2
2531·9	4	2, 6	2369·0	3	0, 3	2061·9	5	0, 0 R_2
2510·9	6	1, 5	2277·0	1	1, 3	2061·0	1	0, 0 Q_3
2492·9*	4	0, 4	2257·7	1	0, 2	2060·0	2	0, 0 R_3
2451·8	6	4, 7	2154·6	2	0, 1	1989·3	3	1, 0
2430·3	3	3, 6						

* Second head; first missing.

CO (*contd.*)

THE 3A BANDS, $c\,^3\Pi - a\,^3\Pi$

Occurrence. In discharge tube containing CO. Schmid and Gerö obtained them with high intensity in a discharge tube containing neon with carbon electrodes.

Appearance. Degraded to the violet with five heads close together. Under low dispersion they appear double-headed.

References. R. Schmid and L. Gerö, *Nature*, **139**, 928 (1937).

R. K. Asundi, *P.R.S.*, **124**, 277 (1929).

L. Gerö, *Z.P.*, **109**, 210 (1938).

S. G. Tilford, *J. Chem. Phys.*, **50**, 3126 (1969).

λ	I	v', v''
2711·3	3	0, 4
2596·9	4	0, 3
2489·9	5	0, 2
2389·7	5	0, 1
2295·9	4	0, 0

OTHER BANDS OF CO

Knauss Bands. Four bands were obtained in an electrodeless discharge through CO. The bands were degraded to the violet and assigned to the transition $C^1\Sigma - a\,^3\Sigma$.

Reference. H. P. Knauss and J. C. Cotton, *P.R.*, **36**, 1099 (1930).

The approximate wavelengths are:

λ	v', v''
3253·	0, 3
3138	0, 2
3028	0, 1
2925	0, 0

Kaplan Bands. Three bands were obtained in a long atomic hydrogen tube. Each band apparently contained six heads degraded to the violet and resembled the bands of the Third Positive and 3A systems. Wavelengths given by Kaplan were 2750 Å, 2630 Å, 2518 Å.

Reference. J. Kaplan, *P.R.*, **35**, 1298 (1930).

Herman Bands. These were initially thought to be part of the Triplet system, but are now assigned to a separate system, $e\,^3\Sigma^- - a\,^3\Pi$. It occurs also in a flame of carbon suboxide reacting with atomic oxygen. Triple-headed bands, degraded to the red. The following are the known heads:

λ	v', v''	λ	v', v''	λ	v', v''
5428·3		4885·0		4454·5	
5414·5	2, 0	−	4, 0	4445·5	6, 0
5402·5		−		4437	
5140·3		4657·2		4270·8	
5128·1	3, 0	−	5, 0	−	7, 0
5116·2		−		−	

References. R. Herman and L. Herman, *J. Phys. Radium*, **9**, 160 (1948).

R. F. Barrow, *Nature*, **189**, 480 (1961).

CO⁺

There are two strong systems, known as the First Negative bands of carbon and the Comet-tail system. Two weaker systems are known by the names of their discoverers, Baldet-Johnson and Marchand-D'Incan-Janin systems.

FIRST NEGATIVE SYSTEM, $B^2\Sigma - X^2\Sigma$

Occurrence. In the cathode glow of discharge tubes containing CO or CO_2, especially in hollow cathode. Also in discharges through helium containing a trace of CO. These bands, like most systems of CO, are very frequent impurities in discharge tubes, especially at low pressures. They also occur in an electron beam.

Appearance. Degraded to red. Single-headed bands forming fairly obvious sequences. See Plate 2.

References. R. C. Johnson, *P.R.S.*, **108**, 343 (1925)†.
H. Biskamp, *Z.P.*, **86**, 33 (1933).
K. N. Rao, *Astrophys. J.*, **111**, 50 (1950)†.

The following values are from Biskamp's measurements. A few weak bands below 2000 Å are omitted.

λ	I	v', v''	λ	I	v', v''	λ	I	v', v''
3152·7	1	8, 13	2607·2	8	2, 5	2222·7	4	5, 4
3107·5	2	7, 12	2577·7	10	1, 4	2220·3	0	8, 6
3064·0	3	6, 11	2550·3	7	0, 3	2214·5	5	1, 1
3023·0	2	5, 10	2530·8	1	4, 6	2189·8	10	0, 0
2984·2	2	4, 9 8, 12	2504·6	10	3, 5	2185·1	4	7, 5
2947·6	1	3, 8	2474·2	10	2, 4	2164·3	5	3, 2
2938·5	1	7, 11	2445·8	10	1, 3	2154·1	4	6, 4
2913·2	1	2, 7	2419·4	8	0, 2	2137·8	6	2, 1
2897·2	3	6, 10	2412·4	4	4, 5	2123·8	3	5, 3
2882·2	2	1, 6	2381·5	5	3, 4	2112·4	8	1, 0
2874·5	0	9, 12	2362·5	1	6, 6	2095·3	5	4, 2
2858·1	4	5, 9	2352·5	6	2, 3	2091·0	4	7, 4
2820·8	5	4, 8	2325·2	9	1, 2	2067·9	1	3, 1
2785·8	5	3, 7	2299·6	10	0, 1	2067·8	1	9, 5
2752·9	6	2, 6	2298·2	3	4, 4	2061·0	3	6, 3
2745·1	1	6, 9	2293·7	1	10, 8	2042·3	4	2, 0
2722·3	7	1, 5	2268·6	3	3, 3	2034·3	2	8, 4
2707·9	3	5, 8	2255·7	1	9, 7	2032·3	1	5, 2
2693·9	2	0, 4	2254·3	2	6, 5	2004·7	0	4, 1
2672·4	7	4, 7	2240·4	4	2, 2	2003·1	0	7, 3
2638·8	8	3, 6						

COMET-TAIL SYSTEM, $A^2\Pi - X^2\Sigma$

Occurrence. In discharge tubes containing CO or CO_2 at relatively very low pressure, in electron beam through CO at low pressure, in discharge tubes containing helium with a trace of CO, and in the tails of comets.

References. M. F. Baldet, *C. R. Acad. Sci. (Paris)*, **180**, 271 (1925).
M. F. Baldet, *C. R. Acad. Sci. (Paris)*, **180**, 820 (1925).
T. R. Merton and R. C. Johnson, *P.R.S.*, **103**, 383 (1923)†.

CO⁺ (*contd.*)

R. C. Johnson, *P.R.S.*, **108**, 343 (1925).
D. Coster, H. H. Brons and H. Bulthuis, *Z.P.*, **79**, 787 (1932).
K. N. Rao, *Astrophys. J.*, **111**, 306 (1950)†.

The measurements and intensities as listed below are average values compiled from the above references; the list given by Baldet is the most complete. Except for the two strongest bands only the two R heads are given here. The corresponding Q heads lie from 5 (for bands in the red) to 1·5 Å to the red of the R heads. The Q heads are rather stronger than the R heads, but are usually masked by the overlapping lines of the R branch. The R_2 head is always a little stronger than the R_1 head. The values of v' given previously have been reduced by 3 to make them consistent with the analysis of the Baldet-Johnson system.

λ	I	v', v''
6405	0	R_2
6354		R_1
6238·7	7	0, 2 R_2
6189·4		R_1
6015	0	R_2
5970		R_1
5900·4	1	2, 3 R_2
5856·5		R_1
5806	0	R_2
5764		R_1
5693·6	3	1, 2 R_2
5652·6		R_1
5499·9	6	0, 1 R_2
5461·4		R_1
5072·1	5	1, 1 R_2
5039·7		R_1
4910·9	3	0, 0 R_2
4879·5		R_1
4865·8	1	3, 2 R_2
4836·6		R_1
4711·2	5	2, 1 R_2
4683·4		R_1
4565·8	8	1, 0 R_2
4539·4		R_1

λ	I	v', v''
4518·0	3	4, 2 R_1
4403·3	2	3, 1 R_2
4378·9		R_1
4274·3	10	2, 0 Q_2
4272·0		R_2
4252·4		Q_1
4248·9		R_1
4244·1	1	5, 2 R_1
4151·9	1	7, 3 R_2
4130·4		R_1
4138·9	2	4, 1 R_2
4117·3		R_1
4019·7	9	3, 0 Q_2
4017·7		R_2
3999·6		Q_1
3997·3		R_1
3908·0	2	5, 1 R_2
3888·6		R_1
3795·8	8	4, 0 R_2
3778·8		R_1
3705·3	4	6, 1 R_2
3688·1		R_1

λ	I	v', v''
3600·8	6	5, 0 R_2
3584·2		R_1
3525·6	2	7, 1 R_2
3510·3		R_1
3427·9	4	6, 0 R_2
3413·3		R_1
3366·1	2	8, 1 R_2
3351·7		R_1
3314·2	1	10, 2 R_2
3300·7		R_1
3273·9	2	7, 0 R_2
3260·4		R_1
3222·4	1	9, 1 R_2
3209·7		R_1
3180·3	1	11, 2 R_2
3168·1		R_1
3135·5	1	8, 0 R_2
3123·2		R_1
3093·3	0	10, 1 R_2
3081·5		R_1

CO^+ (*contd.*)

BALDET-JOHNSON SYSTEM; $B^2\Sigma - A^2\Pi$

Occurrence. In discharge through helium containing a trace of CO, and in electron beam through CO at low pressure. This is an intercombination system between the initial levels of the First Negative and the Comet-tail systems and occurs under similar conditions.

Appearance. Degraded to shorter wavelengths. Double double-headed bands.

References. R. C. Johnson, *P.R.S.*, **108**, 343 (1925)†.

M. F. Baldet, *C.R. Acad. Sci. (Paris)*, **178**, 1525 (1924).

The measurements and intensities listed below are averaged values from the above references:

λ	I	v', v''		λ	I	v', v''
4236·2	3	0, 1 P_1		3729·7	3	1, 0 P_1
4231·6	8	Q_1		3724·9	8	Q_1
4212·9	7	P_2		3711·2	9	P_2
4209·1	8	Q_2		3707·4	9	Q_2
				3515·8	2	2, 0 P_1
4201·5	1	1, 2 Q_1		3511·7	7	Q_1
4182·6	1	P_2		3500·4	3	P_2
4179·1	1	Q_2		3496·7	4	Q_2
3977·7	4	0, 0 P_1		3331·9	1	3, 0 P_1
3973·5	9	Q_1		3329·0	1	Q_1
3957·0	7	P_2		3317·9	1	P_2
3953·6	10	Q_2		3314·8	1	Q_2

MARCHAND-D'INCAN-JANIN SYSTEM, $C^2\Delta - A^2\Pi$

Occurrence. Microwave discharge through CO + Ne.

Appearance. Double-headed bands, degraded to the red in the region 3000–2100 Å.

Reference. J. Marchand, J. D'Incan and J. Janin, *Spectrochim. Acta*, A, **25**, 605 (1969)†.

The following are heads of most of the bands with our own provisional estimates of intensity from the published photo. Values of v' are uncertain, but n is probably 0.

λ	I	v', v''	λ	I	v', v''	λ	I	v', v''
2806·2	0	$n+1$, 5	2569·5	4	$n+0$, 2	2409·1	6	$n+1$, 1
2789·7			2555·2			2396·9		
2728·6	–	$n+2$, 5	2529·7	7	$n+2$, 3	2383·0	4	$n+0$, 0
2713·2			2516·5			2370·7		
2698·1	4	$n+1$, 4	2473·6	10	$n+0$, 1	2323·2	2	$n+1$, 0
2682·6			2460·3			2311·9		
2596·3	2	$n+1$, 3	2438·2	9	$n+2$, 2			
2582·0			2426·0					

CO_2

In flames there is a complex system of narrow bands, the Carbon Monoxide Flame Bands of CO_2, but no spectrum of neutral CO_2 is readily excited in a discharge. CO_2 is also transparent throughout the visible and ultra-violet to beyond 1700 Å. The vibration-rotation 'Venus bands' extend into the visible in absorption by very long paths.

CARBON MONOXIDE FLAME BANDS, $^1B_2 - X^1\Sigma^+$

Occurrence. In the flame of carbon monoxide burning in air, oxygen or nitrous oxide, and weakly in the outer cone of Bunsen flames. The bands have also been observed in the afterglow of a heavy-current discharge through carbon dioxide.

Appearance. A great number of narrow bands, diffuse under small dispersion, on a strong continuous or apparently continuous background. Bands are clearest between 5000 and 3500 Å with maximum intensity around 4200 Å, but the system can be extended from the green to beyond the OH band at 3064 Å. The bands, when seen under larger dispersion are narrow with complex rotational structure; some heads are shaded to shorter wavelengths and occur in pairs. This is part of the absorption system which lies below 1700 Å and is due to a transition from a triangular $^1B_2$ excited state to the linear $^1\Sigma$ ground state. See Plate 9.

References. A. Fowler and A. G. Gaydon, *P.R.S.*, **142**, 362 (1933)†.
A. G. Gaydon, *The Spectroscopy of Flames*, Chapman and Hall, London (1974)†.
R. N. Dixon, *P.R.S.*, **275**, 431 (1963)†.

The bands are favoured, relatively to the continuum, by low temperature or low pressure, and are clearest in the afterglow. The following heads of maxima of the narrow bands, as measured with small dispersion, may assist identification, which is, however, best done by comparison of plates.

λ	λ	λ	λ	λ
5430	4980	4674	4411	4104
5318	4933	4646	4344	4093
5276	4893	4567	4335	4045
5169	4798	4553	4260	4035
5129	4768	4527	4156	3912

VENUS ABSORPTION BANDS

Occurrence. Absorption by very long paths of CO_2 (up to 5500 m) and in the planet Venus.

Appearance. Degraded to longer wavelengths. Single P and R branches.

References. W. S. Adams and T. Dunham, *Publ. Astronom. Soc. Pacific*, **44**, 243 (1932)†.
G. Herzberg and L. Herzberg, *J. Opt. Soc. Amer.*, **43**, 1037 (1953)†.

This is part of the vibration-rotation spectrum of ground-state CO_2. The strongest bands are in the far infra-red (4·4 and 2·8 μm). These heads, in the photographable region, are from Herzberg and Herzberg.

λ	I	λ	I	λ	I
12177·3	100	10361·8	8	7882·8	0·1
12030·4	100	8688·9	1	7820·1	0·3
10487·6	20	8735·9	0·05	7158·2	0·02

CO_2^+

FOX, DUFFENDACK AND BARKER'S SYSTEM, $A^2\Pi - X^2\Pi$

Occurrence. In the negative glow of discharges through streaming carbon dioxide; the bands appear strongly in a hollow cathode and have also been produced with a trace of carbon dioxide in the presence of helium or neon. The bands have been observed as an impurity in the spectrum of what was thought to be pure oxygen. The spectrum has been studied most completely when excited by a beam of electrons through streaming carbon dioxide at low pressure; as observed in this way the bands are fairly free from the many systems due to CO.

Appearance. Narrow bands degraded to the red. The system extends from 2800 to nearly 5000 Å. At the ultra-violet end the appearance is relatively simple, the bands forming marked groups which resemble sequences, but the longer wavelength end is very confused and presents few definite features. See Plate 2.

References. H. D. Smyth, *P.R.*, **38**, 2000 (1931)†.
S. Mrozowski, *P.R.*, **60**, 730 (1941); **72**, 682 and 691 (1947).
J. W. C. Johns, *Canad. J. Phys.*, **42**, 1004 (1964).

The following measurements are by Fox, Duffendack and Barker (after Smyth) and the intensities are by Smyth for electron beam excitation. Only the strong bands are reproduced here. Bands marked with an asterisk are at the head of characteristic groups which are prominent in the spectrum as obtained from discharge tubes.

λ	I	λ	I	λ	I	λ	I
4159·5	5	3761·4	4	3400·9	5	3155·2	5
4137·6	6	3691·8	6	3394·5	4	3149·5	4
4120·8	6	3674·1	5	3388·9	4	3139·2	5
4107·9	5	*3661·6	5	3377·5	8	3136·7	5
4070·7	5	3621·0	7	*3370·0	8	3134·6	4
4048·9	5	3565·5	5	3284·3	3	*3132·9	4
3960·9	7	3562·2	6	3269·9	5	3063·5	4
3890·4	4	3551·4	6	3264·6	5	3058·3	4
3870·5	7	3545·9	7	3253·9	6	3048·6	5
3853·2	4	3533·8	4	*3246·9	6	3034·2	4
3838·8	5	3510·8	3	3164·9	3	2874·3	5
3774·6	4	*3503·2	3				

BANDS λλ2883 AND 2896, $A^2\Sigma^+ - X^2\Pi$

Occurrence. These bands are very persistent and occur in almost all discharge tubes containing carbon dioxide or even carbon monoxide. They are favoured by freshly streaming gas and relatively energetic excitation, *i.e.*, negative glow or hollow cathode. They often occur as an impurity.

Appearance. Two strong narrow bands of almost line-like sharpness, each being a close doublet intensity maxima at about λλ2897·5, 2895·1, 2884·0 and 2881·8. There are also weaker heads at λλ2890·5, 2877·5 and 2874·8. See Plate 2.

References. R. Schmid, *Z.P.*, **83**, 711 (1933)†.
F. Bueso-Sanllehi, *P.R.*, **60**, 556 (1941)†.

C_2O

8580 Å SYSTEM, $A^3\Pi - X^3\Sigma$

Reference. C. Devillers and D. A. Ramsay, *Canad. J. Phys.*, **49**, 2839 (1971)†.

C_2O (contd.)

Violet-degraded bands, 9000–5000 Å, have been observed in flash photolysis of C_3O_2 and pulsed radiolysis of CO. Outstanding heads:

λ	λ	λ	λ
8607·6	8557·8*	7699·8	7298·4
8585·5*	7751·6	7688·4	7282·1*
8579·9	7740·3	7318·4	6652·0
8569·9	7734·9	7303·1*	6374·2

* Relatively strong.

C_3O_2

Occurrence. In absorption by carbon suboxide vapour.

Appearance. Narrow diffuse bands, evenly spaced in the region 3380–3250 Å and occurring in pairs on a complex banded background in the region 3250–2910 Å.

Reference. H. W. Thompson and N. Healey, *P.R.S.*, **157**, 331 (1936).

The following are the maxima of the strong bands, $\lambda\lambda$3350, 3332, 3316, 3302, 3292, 3277, 3251, 3175, 3166, 3136, 3127, 3092, 3082, 3047, 3038, 3015, 3006, 2994, 2987, 2955 and 2946.

$COCl_2$, Phosgene

References. V. Henri and O. R. Howell, *P.R.S.*, **128**, 192 (1930)†.
D. C. Moule and P. D. Foo, *J. Chem. Phys.*, **55**, 1262 (1971)†.

In absorption there are waves of narrow red-degraded bands 9 to 12 Å apart from 3050 to 2380 Å. With a path length of 1 m bands below 2500 appear at 1 torr pressure, spreading to 2850 at 4 torr and to 3050 at 1 atm. There is a mention of emission systems with maxima 3050 and 5100 Å.

The following are the strongest ($I = 9$ or 10) of the 270 bands listed by Henri and Howell, $\lambda\lambda$2899·4, 2863·7, 2816·7, 2809·8, 2803·0, 2783·5, 2776·9, 2770·7, 2764·0, 2751·4, 2739·9, 2739·1, 2720·0.

COS ?

EMISSION

Occurrence. Emission and absorption in discharge through carbonyl sulphide vapour.

Appearance. A regularly spaced group of narrow bands. Fowler and Vaidya described them as degraded to the violet, but Akriche *et al.* show them degraded to the red.

References. A. Fowler and W. M. Vaidya, *P.R.S.*, **132**, 310 (1931).
J. Akriche, L. Herman and R. Herman, *J. Quant. Spectrosc. Rad. Transfer*, **4**, 863 (1964)†.

The following are the strongest heads, from Akriche *et al.*, $\lambda\lambda$3076·9, 3041·2, 3006·4, 2972·3, 2939·4, 2907·7, 2876·9, 2846·4.

ABSORPTION

Reference. W. Lochte-Holtgreven and C. E. H. Bawn, *Trans. Faraday Soc.*, **28**, 698 (1932).
Absorption band extending from a sharp edge at 2550 ± 20 Å, towards shorter wavelengths.

COS^+

Occurrence. Excitation of carbonyl sulphide by an electron beam.

Appearance. 10 bands, due to a transition $A^2\Pi - X^2\Pi$, 4320–3180 Å. They occur in pairs and are degraded to the red.

Reference. M. Horani, S. Leach, J. Rostas and G. Berthier, *J. Chim. Phys.*, **63**, 1015 (1966)†.

λ	I	λ	I	λ	I
4316·9	9	3693·6	8	3208·9	0
4272·2	9	3659·8	8	3183·0	0
3984·8	10	3436·5	5		
3846·2	10	3407·4	5		

CP

Two systems with the same initial level are known.

Occurrence. In discharge tubes containing argon, phosphorous and an organic vapour.

References. H. Barwald, G. Herzberg and L. Herzberg, *Ann. Phys. (Lpz.)*, **20**, 569 (1934)†.
A. K. Chaudhry and K. N. Upadhya, *Indian J. Phys.*, **43**, 83 (1969).

SYSTEM A, NEAR ULTRA-VIOLET, $B^2\Sigma - X^2\Sigma$

Appearance. Degraded to the red. Single-headed bands.

The following are the strong bands. Intensities have been increased to a scale of 10.

λ	I	v', v''	λ	I	v', v''	λ	I	v', v''
4014·8	8	1, 4	3459·2	10	0, 0	3190·2	8	3, 0
3957·1	6	0, 3	3363·5	10	1, 0	3111·8	6	4, 0
3777·9	8	0, 2	3320·1	6	3, 1	3038·6	5	5, 0
3612·4	10	0, 1	3273·7	6	2, 0	3054·8	6	8, 2
3508·2	8	1, 1	3235·3	6	4, 1	2969·1	6	6, 0

SYSTEM B, BLUE, $B^2\Sigma - A^2\Pi$

Appearance. Degraded to the red. Double double-headed bands.

The following are the strong heads. Intensities have been raised to a scale of 10 for the strongest band of System A.

λ	I	v', v''	λ	I	v', v''	λ	I	v', v''
4653·0	4	2, 2 Q_2	4551·3	4	0, 0 R_2	4454·7	6	3, 2 Q_1
4619·1	4	2, 2 Q_1	4524·6	8	0, 0 Q_1	4438·7	4	2, 1 R_2
4605·3	4	1, 1 Q_2	4517·5	3	0, 0 R_1	4434·2	4	2, 1 Q_2
4572·0	4	1, 1 Q_1	4502·2	4	4, 3 Q_1	4407·7	4	2, 1 Q_1
4557·3	8	0, 0 Q_2	4486·3	5	3, 2 Q_2	4392·8	3	1, 0 Q_2

CS

MAIN SYSTEM, $A^1\Pi - X^1\Sigma$

Occurrence. In discharges through CS_2, in flames containing sulphur, in a carbon arc fed with sulphur, and in absorption following flash photolysis of CS_2. A frequent impurity.

Appearance. Degraded to the red. Close double-headed bands. The sequences are fairly obvious and the head of the (0, 0) band at 2576 Å is usually very outstanding. See Plate 9.

CS (*contd.*)

References. L. C. Martin, *P.R.S.*, **89**, 127 (1913)†.

W. Jevons, *P.R.S.*, **117**, 351 (1928).

A. Lagerqvist, H. Westerlund, C. V. Wright and R. F. Barrow, *Ark. Fys.*, **14**, 387 (1958)†.

R. F. Barrow, R. N. Dixon, A. Lagerqvist and C. V. Wright, *Ark. Fys.*, **18**, 543 (1960)†.

The following are the R heads of all the prominent bands as measured by Jevons. The Q heads lie from 0·5 to 2 Å to the red of the R heads. Intensities are for a discharge through CS_2.

λ	I	v', v''	λ	I	v', v''	λ	I	v', v''
2852·3	2	0, 3	2677·0	6	1, 2	2523·2	7	2, 1
2836·8	2	5, 7	2662·6	9	0, 1	2511·2	3	6, 4
2819·5	3	4, 6	2638·9	2	4, 4	2507·3	4	1, 0
2801·5	5	3, 5	2621·6	7	3, 3	2493·7	6	5, 3
2785·7	5	2, 4	2605·9	8	2, 2	2477·0	4	4, 2
2769·2	3	1, 3	2589·6	6	1, 1	2473·4	3	8, 5
2754·7	7	0, 2	2575·6	10	0, 0	2460·2	5	3, 1
2743·9	3	5, 6	2572·7	5	5, 4	2454·3	1	7, 4
2726·7	4	4, 5	2555·8	5	4, 3	2444·8	3	2, 0
2708·9	6	3, 4	2538·7	8	3, 2	2436·0	1	6, 3
2693·2	7	2, 3	2530·0	3	7, 5	2418·4	0	5, 2

The upper state $A^1\Pi$ is perturbed by several other states, causing irregular rotational structure. In flash photolysis some extra heads due to transitions to these perturbing states were found by Barrow *et al.*; 2588·9 $a^*\Pi - X^1\Sigma$, 2572·9 $a'^3\Sigma^+ - X^1\Sigma$, and 2504·8 $e^3\Sigma^- - X^1\Sigma$.

TEWARSON AND PALMER'S SYSTEM, $a^3\Pi - X^1\Sigma$

Occurrence. Emission from chemiluminescent reactions of $CSCl_2$, etc. with potassium vapour and reactions of metastable argon atoms with CS_2, COS, etc.

Appearance. Degraded to the red. Close double-headed (separation 2 to 4 Å) with a weaker third head further to the red.

References. A. Tewarson and H. B. Palmer, *J. Mol. Spec.*, **27**, 246 (1968)†.

G. W. Taylor and D. W. Setser, *J. Mol. Spec.*, **44**, 108 (1972)†.

First heads of strong bands, mainly from Tewarson and Palmer, with own estimates of intensity.

λ	I	v', v''	λ	I	v', v''	λ	I	v', v''
3837·1	6	2, 3	3661·5	7	2, 2	3523·1	6	3, 2
3816·1	4	1, 2	3642·0	6	1, 1	3500·6	8	2, 1
3795·8	2	0, 1	3621·2	10	0, 0	3480·0	5	1, 0

Taylor and Setser also note possible triplet-triplet bands in emission around 5300 and 6375–6200 Å.

CS_2

Occurrence. Absorption by carbon disulphide vapour.

Appearance. A very complex system of headless bands.

References. E. D. Wilson, *Astrophys. J.*, **69**, 34 (1929).

L. N. Liebermann, *P.R.*, **60**, 496 (1941).

A. E. Douglas and I. Zanon, *Canad. J. Phys.*, **42**, 627 (1964)†.

CS$_2$ (contd.)

B. Kleman, *Canad. J. Phys.*, **41**, 2034 (1963)†.
A. G. Gaydon, G. H. Kimbell and H. B. Palmer, *P.R.S.*, **279**, 313 (1964)†.

The following are the strongest maxima, with intensities reduced to a scale of 10 from Wilson's measurements.

λ	I	λ	I	λ	I	λ	I
3346·0	1	3227·4	3	3154·0	5	3080·7	4
3321·6	2	3214·3	7	3150·9	6	3056·8	2
3301·3	2	3204·4	9	3144·0	8	3036	2
3274·8	8	3189·5	10	3126·6	6	3023	2
3260·4	2	3181·5	6	3119·3	6	3009·4	1
3250·6	5	3170·2	6	3100·4	4	2993·4	1
3235	8	3161·9	5	3092·5	5		

This system becomes stronger and extends to at least 4000 Å on heating. There is another strong system (see Douglas and Zanon) further in the ultra-violet, 2300–2050 Å.

CS$_2^+$

Occurrence. Hollow cathode discharge through streaming CS$_2$ vapour.
Reference. J. H. Callomon, *P.R.S.*, **244**, 220 (1958)†.

SYSTEM A

Two strong narrow bands at 2819 and 2855 Å, and a number of much weaker irregularly spaced bands. Some of the heads are degraded to the red. The following are the most prominent heads:

λ	I	λ	I
2907·1	1	2843·5	1
2873·3	1	2819·2	10
2869·1	1	2817·9	2
2854·7	10	2817·4	1
2852·2	2	2808·6	1

SYSTEM B

This is a complex system of numerous narrow bands, very like the carbon monoxide flame spectrum, extending from the near ultra-violet to the infra-red. It shows rotational structure under high dispersion. Assignment to CS$_2^+$ is provisional. No wavelengths are available; Callomon shows a photograph.

C$_3$S$_2$

Reference. L. C. Robertson and J. A. Merritt, *J. Chem. Phys.*, **56**, 5428 (1972).

Absorption by carbon subsulphide shows a strong system 2530–2200 Å, with strong maxima at 2402·4, 2382·0, 2361·6, 2341·2, and a weaker system 5300–4300 with strongest heads 4854·2 and 4735·2 Å.

CSe

MAIN SYSTEM, $A^1\Pi - X^1\Sigma$

Occurrence. High-frequency discharge through selenium vapour in a quartz tube on which carbon has been deposited.

Appearance. Degraded to the red. Some of the bands show double heads (separation about 5 Å).

Reference. B. Rosen and M. Désirant, *C.R. Acad. Sci. (Paris)*, **200**, 1659 (1935).

The following are the stronger (presumably the R) heads of the bands observed:

λ	v', v''	λ	v', v''	λ	v', v''
3053·3	2, 4	2963·9	2, 3	2861·6	1, 1
3038·3	1, 3	2948·0	1, 2	*2844·5	0, 0
3021·1	0, 2	2931·0	0, 1	2779·1	1, 0

* The (0, 0) band is stated to be the strongest; the weaker Q head is at 2848·9 Å.

4100 Å SYSTEM, $a^3\Pi - X^1\Sigma$

Occurrence. In Schuler-type discharge.

Appearance. Degraded to the red. The $^3\Pi_{0^+}$ bands are single-headed and the $^3\Pi_1$ are close double-headed.

Reference. J. Lebreton, G. Bosser and L. Marsigny, *J. Phys.*, B, **6**, L 226 (1973).

R heads:

λ	v', v''		λ	v', v''	
4357·0	1, 2	$^3\Pi_{0^+} - {}^1\Sigma$	4001·3	1, 0	$^3\Pi_{0^+} - {}^1\Sigma$
4334·1	0, 1	$^3\Pi_{0^+} - {}^1\Sigma$	4275·4	0, 1	$^3\Pi_1 - {}^1\Sigma$
4172·5	0, 0	$^3\Pi_{0^+} - {}^1\Sigma$	4096·0	0, 0	$^3\Pi_1 - {}^1\Sigma$

CSe_2

References. A. B. Callear and W. J. R. Tyerman, *Trans. Faraday Soc.*, **61**, 2395 (1965)†.
G. W. King and K. Srikamenan, *J. Mol. Spec.*, **31**, 269 (1969)†.

In absorption there is a strong system 2400–2200 Å, a moderately strong one 4000–3500, and weaker systems 4500–4300 and below 2100 Å.

Ca_2

Calcium, heated in a furnace, shows a 'many-line' system of bands, strongly degraded to the violet, due to the transition $A^1\Sigma^+ - X^1\Sigma^+$, in the region 6000–4600 Å. In emission there are bands 5147–4823 Å. The 4226 Å line of Ca is broadened asymmetrically. Bands in the orange, observed in an arc, are probably due to CaCl.

References. S. Weniger, *Proc. Colloq. Late Type Stars* (Ed. M. Hack) p.25, Trieste (1966).
G. V. Kovalenk and V. A. Sokolov, *Izv. Sib. Otd. Akad. Nauk., S.S.S.R., Ser. Khim. Nauk*, **9**, 118 (1967).
W. J. Balfour and R. F. Whitlock, *Canad. J. Phys.*, **53**, 472 (1975)†.

CaBr

There are two overlapping systems in the red and orange-red, one in the violet and two in the near ultra-violet.

CaBr (*contd.*)

RED SYSTEM, $A^2\Pi - X^2\Sigma$

Occurrence. In absorption and when calcium bromide is introduced into a flame; the bands do not appear readily in an arc.

Appearance. Marked close sequences degraded to shorter wavelengths.

References. K. Hedfeld, *Z.P.*, **68**, 610 (1931).

O. H. Walters and S. Barratt, *P.R.S.*, **118**, 120 (1928).

The following measurements and analysis are by Hedfeld. The intensities I_a and I_f are for absorption and for emission in a flame respectively, the former being by Walters and Barratt, who also report a band (intensity 5) at 6106·6.

λ	I_a	I_f	Sequence	λ	I_a	I_f	Sequence
6399·0		0	0, 1 P_1	6258·8	0	5	0, 0 P_2
6390·5	0	0	0, 1 Q_1	6252·9	10	10	0, 0 Q_2
6370·9		0	0, 1 P_2	6176·8		0	1, 0 P_1
6364·8	0	0	0, 1 Q_2	6168·8	0	0	1, 0 Q_1
6286·0	0	4	0, 0 P_1	6150·6		0	1, 0 P_2
6277·7	10	10	0, 0 Q_1	6145·0	0	0	1, 0 Q_2

ORANGE-RED SYSTEM, 6130–6000 Å, $B^2\Sigma - X^2\Sigma$

This was reported as a band at 6106 Å, intensity 5, by Walters and Barratt. R. E. Harrington (after Rosen *et al.*) records violet-degraded absorption bands in red-degraded sequences at 6104·7 (0, 0) and 6002·4 (1, 0).

VIOLET SYSTEM, $C^2\Pi - X^2\Sigma$

Occurrence. In absorption and in a flame.

Appearance. Close sequences of bands degraded to the red.

Data from Rosen *et al.* based on a thesis by E. Harrington. Strong sequence heads:

λ	Sequence	λ	Sequence
4040·7	0, 2 R_1	*3950·4	0, 0 R_1
4005·3	0, 2 Q_2	*3915·9	0, 0 R_2
3995·2	0, 1 R_1	3969·7	1, 0 R_1
3959·9	0, 1 R_2	3876·1	1, 0 R_2

* Strongest heads.

ULTRA-VIOLET SYSTEMS

Occurrence. In high-frequency discharge. Also noted in absorption by Walters and Barratt.

Appearance. Both systems are degraded to shorter wavelengths and show marked sequences.

References. Y. P. Reddy and P. T. Rao, *Indian J. Pure Appl. Phys.*, **6**, 181 (1968)†.

S. G. Shah, *Indian J. Pure Appl. Phys.*, **8**, 118 (1970).

3300–3100 Å, $D^2\Sigma - X^2\Sigma$

Sequence heads:

λ	v', v''	λ	v', v''
3292·6	0, 3	3170·1	1, 0
3262·3	0, 2	3138·1	2, 0
3232·5	0, 1	3106·8	3, 0
3203·0	0, 0		

CaBr (*contd.*)

3025–2900 Å, $E^2\Sigma - X^2\Sigma$

Sequence heads:

λ	I	v', v''		λ	I	v', v''
3019·1	2	0, 3		2943·9	10	0, 0
2993·8	4	0, 2		2916·7	6	1, 0
2968·7	10	0, 1				

CaCl

Occurrence. When $CaCl_2$ is introduced into an arc or flame.
Also in absorption, frequently as an impurity.

References. K. Hedfeld, *Z.P.*, **68**, 610 (1931).
A. E. Parker, *P.R.*, **47**, 349 (1935).
E. Morgan and R. F. Barrow, *Nature*, **185**, 754 (1960).
M. Schütte, *Z. Naturforsch.*, A, **9**, 891 (1954).

There are three strong systems, in the red, orange, and ultra-violet respectively. A few weak fragmentary systems are also reported by Parker.

RED SYSTEM, λλ6361–6047, $A^2\Pi - X^2\Sigma$

Appearance. Degraded to violet. Marked close sequences.
Strongest sequence heads:

λ	I	v', v''		λ	I	v', v''
6353·5	2	0, 1 Q_1		6193·4	5	0, 0 P_2
6325·8	2	0, 1 Q_2		6184·9	10	0, 0 Q_2
6224·9	5	0, 0 P_1		6076·6	2	1, 0 Q_1
6211·6	10	0, 0 Q_1		6051·6	2	1, 0 Q_2

ORANGE SYSTEM, λλ6067–5810, $B(?\ ^2\Sigma) - X^2\Sigma$

Appearance. Degraded to red. Close sequences.
λ5934·0 head of (0, 0) sequence, intensity 10.
λ5809·9   ”   ” (1, 0)   ”   ”   4.

ULTRA-VIOLET SYSTEM, 4023–3644 Å, $C(?\ ^2\Pi) - X^2\Sigma$

Appearance. Degraded to red. Close sequences.
No intensities given. Heads of sequences:

λ	Sequence		λ	Sequence
3828·1	0, 1 Q_1		3764·2	0, 0 Q_2
3816·9	0, 1 Q_2		3763·5	0, 0 R_2
3775·0	0, 0 Q_1		3728·0	1, 0 Q_1
3774·4	0, 0 R_1		3727·4	1, 0 R_1
			3717·3	1, 0 Q_2

CaF

Occurrence. When calcium fluoride is put in carbon arc or a flame. These bands often occur as impurities in arc spectra and have been recommended for use in analytical work as a test for the presence of fluorine. Also in absorption.

References. S. Datta, *P.R.S.*, **99**, 436 (1921)†.
R. C. Johnson, *P.R.S.*, **122**, 161 (1929).
A. Harvey, *P.R.S.*, **133**, 336 (1931)†.
C. A. Fowler, *P.R.*, **59**, 645 (1941)†.
S. C. Prasad and M. K. Narayan, *Indian J. Phys.*, **43**, 205 (1969).
K. V. Subbaram and D. R. Rao, *Indian J. Phys.*, **43**, 312 (1969)†.

There are three systems in emission, usually known as the orange, green and ultra-violet systems. The two former appear strongly, but the ultra-violet bands are weak and usually masked by CaO bands. C. A. Fowler has obtained three additional systems in absorption in the ultra-violet.

ORANGE SYSTEM, 6300–5830 Å, $A^2\Pi - X^2\Sigma$

There are long sequences of heads of about equal intensity. The (0, 1) and (0, 0) sequences are degraded to the violet, the separation between successive heads being about 4·5 Å for the (0, 1) sequence and 2 Å for the (0, 0). The (1, 0) sequence is piled up on itself and shows a head at 5830 Å, degraded to the red. See Plate 1.

The heads of the sequences only are listed:

λ	I	Sequence	λ	I	Sequence
6285·3	4	(0, 1) Q_{12}	6050·8	6	(0, 0) P_2
6256·6	4	(0, 1) Q_2	6036·8	6	(0, 0) Q_2
6086·9	5	(0, 0) P_{12}	5830	5	(1, 0) Deg. to red
6064·4	10	(0, 0) Q_{12}			

GREEN SYSTEM, $B^2\Sigma - X^2\Sigma$

Strong (0, 0) sequence of double-headed bands degraded to the red. The heads are separated by about 1·8 Å, and successive bands of the sequence by about 6 Å. There is a weaker (1, 0) sequence of similar appearance. See Plate 1.

λ	v', v''	λ	v', v''
5291·0	(0, 0) R_2	5145·4	(1, 0) R_2
5292·9	(0, 0) R_1	5146·4	(1, 0) R_1
5296·8	(1, 1) R_2	5151·9	(2, 1) R_2
5298·6	(1, 1) R_1	5152·8	(2, 1) R_1
etc.		etc.	

ULTRA-VIOLET SYSTEMS

3310 Å, $C^2\Pi - X^2\Sigma$

Marked sequences, degraded to the red. Bands within each sequence are separated by about 10 Å. Weaker R_1 and R_2 heads lie about 1 and 3 Å to shorter wavelengths. The following are Q_1 heads from Fowler's formula; intensities for absorption:

λ	I	Sequence
3442·2	2	0, 2
3375·1	7	0, 1
3310·0	10	0, 0
3258·5	5	1, 0

CaF (*contd.*)

3245 Å, $D^2\Sigma - X^2\Sigma$

Appearance. Degraded to the violet. Fowler obtained absorption over the range $\lambda\lambda 3245{-}3081$. The following wavelengths have been calculated from Fowler's formula for the heads. The intensities where given are those observed by Fowler.

λ	I	v', v''	λ	I	v', v''	λ	I	v', v''
3372·2		0, 2	3238·8	8	1, 1	3167·3	9	3, 2
3307·9		0, 1	3232·2	6	2, 2	3115·6	1	2, 0
3300·4		1, 2	3178·9	6	1, 0	3110·5	5	3, 1
3245·4	10	0, 0	3173·1	8	2, 1	3055·2		3, 0

2926 Å, $E^2\Sigma - X^2\Sigma$

Appearance. Degraded to the violet. Fowler obtained absorption over the range $\lambda\lambda 3035{-}2754$. Wavelengths calculated from Fowler's formula for the heads with his observed intensities:

λ	I	v', v''	λ	I	v', v''	λ	I	v', v''
3028·8		0, 2	2921·2	4	1, 1	2821·1	4	2, 0
2976·9	4	0, 1	2872·4	8	1, 0	2717·5	6	3, 1
2926·2	10	0, 0	2868·2	7	2, 1	2814·0	7	4, 2

2659 Å, $F^2\Pi - X^2\Sigma$

Appearance. Degraded to the violet, region $\lambda\lambda 2700{-}2554$.

P_1 heads obtained from Fowler's formula:

2700·9 (6) (0, 1), 2659·1 (10) (0, 0), 2612·2 (6) (1, 0), 2567·5 (1) (2, 0).

The P_2 heads are 0·1 Å and the Q_2 heads 1·0 Å to the violet of the P_1 heads.

CaH

Eleven systems have now been reported. The strongest are A and B. The weak ultra-violet systems F to K have been studied by Khan and colleagues in absorption.

References. E. Hulthén, *P.R.*, **29**, 97 (1927).

B. Grundström and E. Hulthén, *Nature*, **125**, 634 (1930).

B. Grundström, *Z.P.*, **69**, 235 (1931)†; **75**, 302 (1932); **95**, 574 (1935).

W. W. Watson and R. L. Weber, *P.R.*, **48**, 732 (1935).

M. A. Khan, *Proc. Phys. Soc.*, **80**, 593 (1962)†; **87**, 569 (1966)†.

M. A. Khan and M. K. Afridi, *J. Phys.*, B, **1**, 260 (1968).

M. A. Khan and S. S. Hasnain, *Nuovo Cimento*, B, **18**, 384 (1973)†.

6946 Å, A-SYSTEM, $A^2\Pi - X^2\Sigma$

Bands are degraded to the violet. Obtained in Ca arcs in H_2 at various pressures. Also in absorption in a furnace by mixtures of Ca vapour and H_2. Identified in the sun-spot spectrum.

v', v''	λ *Origins*	v', v''	λ *Origins*	λ *Heads*	
0, 1	7613	2, 3	7484		
1, 2	7571	0, 0	6946	6942·6 Q_1	7028 P_1
0, 1	7567	1, 1	6930	6928·6 Q_1	7006 P_1
2, 3	7531	0, 0	6908	6919·8 Q_2	7035 P_2
1, 2	7525	1, 1	6891	6902·6 Q_2	7005 P_2

CaH (*contd.*)

6346 Å, B-SYSTEM, B $^2\Sigma - $ X $^2\Sigma$

Bands degraded to the violet. Occurrence similar to A-System.

v', v''	Origins	Heads	
2, 2	6358		
1, 1	6352		
0, 0	6346	6389·3 P_1	6382·1 P_2

3534 Å, C-SYSTEM, C $^2\Sigma - $ X $^2\Sigma$

Bands degraded to the violet forming fairly sharp P heads. Complete system observed at high pressures of hydrogen (3–4 atms.).

v', v''	P Heads	v', v''	P Heads
0, 1	3696·6	1, 0	3367·6
0, 0	3533·6	2, 1	3356·3
1, 1	3515·4	3, 2	3346·5
2, 2	3498·1	4, 3	3337·6
3, 3	3482·0	5, 4	3330·8

4500 Å, D-SYSTEM, D $^2\Sigma - $ X $^2\Sigma$

Bands degraded to the red. System less intense than A and B. The R heads very weak.

v', v''	Origins	v', v''	Origins
0, 3	5301	1, 1	4473
0, 2	4988	1, 0	4235
1, 2	4732	2, 0	4059
0, 1	4702		

4900 Å, E-SYSTEM, E$^2\Pi - $ X $^2\Sigma$

Weak system slightly degraded to the violet. Observed in arc in hydrogen at low pressure with water-cooled copper cathode and anode of metallic calcium.

v', v''	Origin	Heads	
0, 0	4900	4898·1 Q_1	4899·1 Q_2
1, 1		4915·7 Q_1	4916·6 Q_2
2, 2		4934·8 Q_1	4935·8 Q_2

2720 Å, F-SYSTEM, F$^2\Sigma - $ X$^2\Sigma$

Single-headed bands, degraded to shorter wavelengths, observed in absorption. (0, 0) band at 2720·2, and weaker (0, 1) at 2815·9 Å.

2883 Å, G-SYSTEM, G$^2\Sigma - $ X$^2\Sigma$

Single-headed. Degraded to shorter wavelengths. (0, 0) at 2882·4 and weaker (0, 1) at 2752·2 Å.

SYSTEMS H, I, J AND K

All weak single-headed bands, degraded to shorter wavelengths, observed in absorption. Transitions H$^2\Sigma$, I$^2\Sigma$, J$^2\Sigma$ and K$^2\Sigma$ to X$^2\Sigma$.

CaH (*contd.*)

λ	v', v''	System
2852·5	0, 0	K
2851·8	0, 0	J
2580	0, 0	H
2535	0, 0	I

CaI

There are two overlapping systems in the red and further systems in the violet and near ultra-violet.

RED BANDS

Occurrence. In absorption and in flames, arcs and high-frequency discharges.
Appearance. Degraded to the violet with marked close sequences.
References. K. Hedfeld, *Z.P.*, **68**, 610 (1931).
O. H. Walters and S. Barratt, *P.R.S.*, **118**, 120 (1928).
P. S. Murty, Y. P. Reddy and P. T. Rao, *J. Phys.*, B, **3**, 425 (1970).
L. K. Khanna and V. S. Dubey, *Indian J. Pure Appl. Phys.*, **11**, 375 (1973).

These bands were originally analysed into three overlapping systems, but are now attributed to two sub-systems of $A^2\Pi - X^2\Sigma$ and a weaker $B^2\Sigma - X^2\Sigma$. The following list of main Q heads of sequences is compiled from the above references, intensities I_a and I_f being for absorption and emission in a flame. Some sequences of sub-systems A_1 and A_2 ($A^2\Pi - X^2\Sigma$) have weak P heads in front.

λ	I_a	I_f	Sequence	λ	I_a	I_f	Sequence
6513·7	1	0	0, 1 A_1	6315·4	8	0	1, 0 A_1
6488·7	1	0	0, 1 A_2	6291·9	6	0	1, 0 A_2
6412·9	10	6	0, 0 A_1	6265·1	8	0	?
6388·8	10	9	0, 0 A_2	6239·8	—	—	0, 0 B
6361·3	6	3					

VIOLET SYSTEM, C $(?\ ^2\Pi) - X^2\Sigma$

Occurrence. In absorption and in flames and high frequency discharge.
Appearance. Degraded to the red. Double-headed with clear sequences.
References. P. S. Murty, Y. P. Reddy and P. T. Rao, *J. Phys.*, B, **3**, 425 (1970)†.
R. C. Maheshwari, M. M. Shukla and I. D. Singh, *Indian J. Pure Appl. Phys.*, **9**, 327 (1971)†.

Outstanding sequence heads from Murty *et al.*; intensities for absorption:

λ	I	Sequence	λ	I	Sequence
4331·9	6	0, 1 R_1	4211·7	8	0, 0 R_2
4288·6	10	0, 0 Q_1	4211·0	6	0, 0 $^S R_{21}$
4287·9	8	0, 0 R_1	4171·8	4	1, 0 $^S R_{21}$
4246·4	5	1, 0 R_1	4133·6	2	2, 0 $^S R_{21}$

ULTRA-VIOLET SYSTEM, D $(?\ ^2\Sigma) - X^2\Sigma$

Reference. L. K. Khanna and V. S. Dubey, *Indian J. Pure Appl. Phys.*, **11**, 286 (1973).

This is a weak system observed in absorption and in an arc. Over 70 heads, degraded to shorter wavelengths, have been measured and sequence heads are listed below, but agreement with earlier

CaI (*contd.*)

observations by Walters and Barratt (who found heads at 3215 and 3186 Å) is poor and the intensity distribution within the sequences is peculiar:

λ	I	Sequence
3273·0	1	0, 2
3247·6	2	0, 1
3222·8	3	0, 0
3196·6	1	1, 0

CaO

There has been considerable difficulty in the assignments and analyses of spectra associated with calcium. Strong orange and green systems in arcs are given under 'calcium oxide'. Strong orange and green flame bands are assigned to CaOH. There are definite singlet systems of CaO in the extreme red, the blue and the violet; analyses proposed by Meggers, Mahanti, Brodersen, Rosen and Lagerqvist do not, in general, agree; near equality of ω' and ω'' and complex perturbations create difficulty. In the infra-red there may be another system, although this may be part of the extreme red system. It is not certain that the state designated $X\,^1\Sigma$ is really the true ground state.

Occurrence. All these systems of CaO have been obtained in an arc, either of calcium salts on carbon poles, or between calcium electrodes. Some have also been observed in a hollow cathode discharge, and in absorption in a shock tube.

EXTREME RED SYSTEM, $A^1\Sigma - X^1\Sigma$

Appearance. Degraded to longer wavelengths, with marked sequences.

References. W. F. Meggers, *J. Res. Nat. Bur. Stand.*, **10**, 669 (1933).

M. Hultin and A. Lagerqvist, *Ark. Fys.*, **2**, 471 (1951)†.

L. Brewer and R. Hauge, *J. Mol. Spec.*, **25**, 330 (1968).

This list of the stronger heads is mainly from Hultin and Lagerqvist, but includes Meggers's old infra-red system, with intensity scale adjusted, and Brewer and Hauge's data.

λ	I	v', v''	λ	I	v', v''	λ	I	v', v''
10615·3	–	0, 3	9807·4	8	4, 6	8153·0	9	1, 0
10599·7	–	1, 4	9775·4	5	5, 7	7724·0	2	5, 3
10533·3	1	3, 6	9741·3	2	6, 8	7721·1	6	4, 2
10491·4	1	4, 7	9700·5	4	7, 9	7715·5	8	3, 1
10444·7	2	5, 8	9229	8	0, 1	7712·2	8	2, 0*
10396·6	2	6, 9	9215·4	2	3, 4	7687·8		
10340·0	2	7, 10	9199·1	1	5, 6	7326	7	5, 2
10289·0	2	8, 11	8652·2	10	0, 0	7318·5	7	4, 1
9835·0	8	3, 5	8167·3	7	2, 1	7308·3	1	3, 0

* Double-headed because of a perturbation.

BLUE SYSTEM, $B^1\Pi - X^1\Sigma$

Appearance. Degraded to the red.

References. P. C. Mahanti, *P.R.*, **42**, 609 (1932)†.

A. Lagerqvist, *Ark. Fys.*, **8**, 83 (1954).

The following are the strongest bands observed by Mahanti, with vibrational assignments of a few bands by Lagerqvist:

CaO (*contd.*)

λ	I	λ	I	λ	I	v′, v″
4519·1	3	4366·7	5	4126·0	3	
4505·0	4	4351·2	5	4104·1	4	1, 3
4425·8	3	4240·8	3	4084·3	5	0, 2
4403·9	6	4221·9	5	3973·9	3	0, 1
4384·8	6	4205·1	6	3872·9	2	
				3773·2	2	1, 0

ULTRA-VIOLET SYSTEM, $C^1\Sigma - X^1\Sigma$

Appearance. Degraded to the red.

References. P. C. Mahanti, *P.R.*, **42**, 609 (1932)†.

A. Lagerqvist, *Ark. Fys.*, **8**, 83 (1954).

The following are the strongest bands from Mahanti:

λ	I	v′, v″	λ	I	v′, v″
3753·2	3	0, 3	3494·7	3	1, 1
3676·5	3	1, 3	3475·0	6	0, 0
3656·6	4	0, 2	3409·1	4	1, 0
3583·7	3	1, 2	3346·7	3	2, 0
3564·0	5	0, 1	3287·4	3	3, 0

FIELD, CAPELLE AND JONES'S FLAME SYSTEM, $A'^1\Pi - X^1\Sigma$

Reference. R. W. Field, G. A. Capelle and C. R. Jones, *J. Mol. Spec.*, **54**, 156 (1975).

A long (v′, 0) progression of red-degraded bands, 7520–5500 Å, obtained in the flame of metallic calcium vapour burning in N_2O or O_3. The following are probably strong and free from overlapping λλ6772·0 (12, 0), 6558·5 (13, 0), 6003·0 (16, 0), 5842·0 (17, 0); the (14, 0) and (15, 0) are masked.

Calcium Oxide

The strong band systems in the orange and green obtained with calcium salts in an arc in air were until recently attributed to CaO, but it now appears possible that the emitter is a polyatomic oxide, such as Ca_2O_2. The flame bands, in closely the same spectrum region, are due to CaOH and are not the same as the arc bands.

Occurrence. In arc in air or oxygen, in King furnace, probably in exploding wires.

Reference. A. G. Gaydon, *P.R.S.*, **231**, 437 (1955)†.

ORANGE SYSTEM

Appearance. The orange system consists of numerous bands, with heads shaded in both directions. Even under very large dispersion the rotational structure is too complex to permit analysis. See Plate 1.

The following are the most prominent features, being rough measures from our plates taken in the second order of a 20-ft. concave grating. The letters R, V or M indicate direction of degradation of the head.

Calcium Oxide (*contd.*)

λ		I	λ		I	λ		I
*6362	M	4	6183	V	6	6065	R ?	5
*6344	M	4	6097	V	10	6056	R	5
*6318	M	2	6092	V	6	6041	R	3
6281	M	3	6088	R ?	5	6006	V	8
6278	V	4	6075·5	V	5	6003	R	8
6262	M	8	6069	V	7	5983	R	8
6258·5	V	9						

* Dr G. H. Newson (personal communication) has suggested that these 'bands' are really auto-ionized lines of Ca.

GREEN SYSTEM

Appearance. A diffuse banded structure, apparently without rotational fine structure, extending from 5473 Å to about 5560 Å. Under small dispersion the appearance is of a diffuse band with head at 5473 Å degraded to the red. Under large dispersion the head is less obvious, but there are maxima of intensity at about λλ5506, 5498, 5496, 5488, 5484, 5476 and 5473; the head at λ5498 is relatively strong and appears to be shaded to the red. See Plate 1.

CaOH

The strong coloration of flames containing calcium salts is due to bands in the orange-red and green. These were formerly attributed to CaO or Ca_2, but James and Sugden studied them in flames of various composition and showed that the emitter was CaOH. Gaydon confirmed this by isotope displacement, using deuterium. The bands also occur in an arc at reduced pressure in water vapour. Bands of a calcium oxide lie in a similar region.

References. C. G. James and T. M. Sugden, *Nature*, **175**, 333 (1955).
A. G. Gaydon, *P.R.S.*, **231**, 437 (1955)†.
L. Brewer and R. Hauge, *J. Mol. Spec.*, **25**, 330 (1968).

The orange-red band is very diffuse and has a maximum at 6230 Å; it is accompanied by much weaker diffuse bands at 6038 and 6415 Å.

The green band in the region 5570–5530 Å shows some structure, with narrow line-like maxima as given below; intensities on a scale of 5.

λ	I	λ	I	λ	I	λ	I
5559·1	1	5552·4	3	5546·4	4	5541·2	1
5557·8	2	5551·2	1	5545·9	2	5540·8	2
5556·4	0	5550·4	1	5544·9	3	5539·7	5
5555·2	1	5549·0	2	5544·4	3	5538·6	5
5554·3	4	5548·5	2	5543·0	4	5537·7	1
5553·4	3	5547·4	5	5542·6	5		

CbO see NbO

CaS

RED SYSTEM, $A^1\Sigma^+ - X^1\Sigma^+$

Occurrence. Absorption at $1900°$ C.

Appearance. Degraded to the red. Single-headed.

Reference. R. C. Blues and R. F. Barrow, *Trans. Faraday Soc.*, **65**, 646 (1969).

λ	v', v''	λ	v', v''
6804·9	1, 2	6578·3	0, 0
6782·8	0, 1	6428·3	2, 1
6600·5	1, 1	6406·1	1, 0

Cd$_2$

Several papers have appeared on the emission, absorption and fluorescence spectra of cadmium and there has been some controversy on the origin of the various bands which have been observed. The bands and continua produced by the Cd_2 molecule have been listed by Cram.

References. S. W. Cram, *P.R.*, **46**, 205 (1934).

W. R. S. Garton, *Proc. Phys. Soc.*, A, **64**, 430 (1951)†.

λ2124

A narrow band between 2140 and 2110 Å has been observed in emission (in discharge tubes, especially ring discharges), absorption and fluorescence.

λ2212

A narrow band has been observed in absorption and Garton has obtained it strongly in emission from an arc.

λ2288

A broad band with maximum at 2288 Å has been observed in emission, absorption and fluorescence; flutings are superposed on the red side of this. The limits and regions of the flutings are roughly:

Source			Limits	Region of Fluting
Emission	.	.	2191–3050	–
Absorption	.	.	2212–3050	2590–2825
Fluorescence	.	.	2260–3050	2700–3050

λ3178

Narrow band at 3178 in emission, absorption and fluorescence.

λ3261

Band overlapping Cd line at 3261, in emission, absorption, and fluorescence.

BLUE REGION

Broad band observed in emission and fluorescence.

Emission	.	.	4058–5400 Å
Fluorescence	.	.	3800–5000 Å, maximum at 4000 Å

CdBr

Occurrence. In discharges, including high-frequency, through $CdBr_2$ vapour. Also in fluorescence (Oeser).

References. K. Wieland, *Helv. Phys. Acta*, **2**, 46 (1929).

E. Oeser, *Z.P.*, **95**, 699 (1955).

C. Ramasastry, *Indian J. Phys.*, **23**, 453 (1949)†.

A. B. Darji, *Indian J. Pure Appl. Phys.*, **8**, 240 (1970).

VISIBLE SYSTEM, 8100–3500 Å, B ($?^2\Sigma$) – X ($?^2\Sigma$)

Appearance. Numerous bands on a continuous background, which has a maximum at 8000 Å. The red-degraded bands are strongest 4900–3850. Ramasastry lists about 50, of which the strongest are λλ4570·4, 4548·4, 4525·6, 4504·4, 4461·7, 4195·9 and 4175·6.

C SYSTEM, 3298 Å, C ($?^2\Pi$) – X

Violet-degraded (0, 0) sequence at 3298·1 and a weak (0, 1) band at 3323·7.

D SYSTEM, 3247–3064 Å, D – X

Appearance. Bands degraded to shorter wavelengths.

Heads of strongest sequences:

λ	I	Sequence
3223·4	4	0, 2
3199·9	8	0, 1
3176·6	9	0, 0
3151·3	7	1, 0
3126·5	2	2, 0

Bands 3551–3407 Å reported by Walter and Barratt are due to TlBr.

CdCl

References. J. M. Walter and S. Barratt, *P.R.S.*, **122**, 201 (1929).

K. Wieland, *Helv. Phys. Acta*, **2**, 46 (1929).

E. Oeser, *Z.P.*, **95**, 699 (1935).

S. D. Cornell, *P.R.*, **54**, 341 (1938).

H. G. Howell, *P.R.S.*, **182**, 95 (1943).

C. Ramasastry, *Indian J. Phys.*, **21**, 265 (1947)†.

M. M. Patel and S. P. Patel, *Indian J. Pure Appl. Phys.*, **4**, 388 (1966)†.

BANDS 8700–3300 Å, $A^2\Sigma - X^2\Sigma$

Diffuse bands, mostly degraded to the red, on a continuum with an intensity maximum at 8500 Å have been obtained with $CdCl_2$ in low-pressure discharges (Wieland). Bands 4750–4050 Å have been measured by Ramasastry, and by Patel and Patel, and analysed. Strong heads from Patel and Patel:

λ	I	v', v''	λ	I	v', v''	λ	I	v', v''
4638·6	4	0, 14	4453·9	9	0, 11	4308·6	7	1, 9
4607·2	5	1, 14	4424·7	9	1, 11	4288·4	7	7, 11
4575·8	6	0, 13	4394·6	8	0, 10	4283·3	7	2, 9
4545·7	7	1, 13	4366·4	8	1, 10	4252·0	7	1, 8
4514·0	10	0, 12	4340·0	8	2, 10	4233·4	7	7, 10
4484·3	10	1, 12	4315·5	7	3, 10	4178·9	6	7, 9

CdCl (*contd.*)

BANDS λλ3400–3300. A set of bands degraded to shorter wavelengths has been observed in this region by Wieland. They were not obtained by Ramasastry.

BANDS λλ3181–3115 and λλ3104–3018. These bands have been observed in absorption by Walter and Barratt and by Oeser. Howell assigns them to the transition $^2\Pi - {}^2\Sigma$, ground state.
    Strongest absorption bands recorded by Walter and Barratt:

λ	I	λ	I	λ	I
3181	5	3163	3	3066	4
3174	2	3074	2	3060	3
3172	5	3072	5	3054	2

    These bands were also observed in absorption by Oeser but were not obtained in emission by Ramasastry.

BANDS λλ2240–2185. Weak bands degraded to the red have been observed in high frequency discharges by Cornell and by Ramasastry. Transition probably $^2\Sigma - {}^2\Sigma$
    Strongest bands from Ramasastry:

λ	I	v', v''	λ	I	v', v''	λ	I	v', v''
2191·9	4	1, 0	2219·9	6	0, 1	2239·9	5	1, 3
2205·0	10	0, 0	2223·6	5	1, 2	2252·5	1	0, 3
2207·4	8	1, 1	2236·1	4	0, 2	2256·3	2	1, 4

BANDS λλ2163–1774. Twelve absorption bands have been observed by Oeser.

CdF

    The absorption spectrum of the vapour produced by heating CdF_2 in a carbon-tube electric furnace has been investigated by C. A. Fowler. With low dispersion, covering the range 7000–1950 Å, narrow regions of absorption were observed near λ2800 at a temperature of 1350°C. These increased in strength with rising temperature up to 1600°C., but at higher temperatures were covered by continuous absorption spreading from shorter wavelengths. High dispersion showed six bands degraded to the red, spaced almost equally, with intensities decreasing toward shorter wavelengths. The two strongest have very diffuse heads; the remainder increase in sharpness toward the ultra-violet. The third, fourth and fifth bands have weak satellite heads. A narrow continuum with maximum at λ2824 and a further group of three closely spaced heads apparently not belonging to the main system were also reported.
    Reference.   C. A. Fowler, *P.R.*, **62**, 141 (1942)†.
    Fowler gives the following wavelengths for the band heads of CdF in absorption together with tentative values of $v'-v''$:

Main Heads		Satellite Heads		Unclassified Heads
λ	$v'-v''$	λ	$v'-v''$	λ
2923·5	0			2788·2
2880·6	1			2786·5
2838·5	2	2837·2	2	2784·9
2797·1	3	2796·0	3	
2756·3	4	2755·5	4	
2716·2	5			

CdH

Reference. E. Svensson, *Z.P.*, **59**, 333 (1930).

4500 Å SYSTEM, A $^2\Pi - X^2\Sigma$

Bands degraded to the violet, each with P, Q and R branches. The system is easily obtained in discharges where cadmium vapour is mixed with hydrogen. See Plate 4.

$^2\Pi_{1/2} - {}^2\Sigma$				$^2\Pi_{3/2} - {}^2\Sigma$			
v', v''	Origins	P Heads	Q Heads	v', v''	Origins	P Heads	Q Heads
0, 4	5624			0, 2	4835		
0, 3	5368	5368·7	5359·3	1, 3	4693		
1, 4	5146	5146·1	5141·2	0, 1	4571		
0, 2	5080	5081·7	5071·1	1, 2	4470		
1, 3	4926			0, 0	4300	4313·3	4297·6
0, 1	4791	4791·1*	4777·4	1, 1	4247		
1, 2	4683			1, 0	4026	4026·7	4008·5
0, 0	4500	4509·0	4491·3	2, 1	3980	3980·4	3965·6
1, 1	4437			2, 0	3789		
1, 0	4198	4198·6	4177·3				

* Line-like head.

4037 Å SYSTEM, $B^2\Sigma - X^2\Sigma$

Degraded to the red. A long progression of bands with double P and R branches. The (0, 0) band was originally thought to be at 3520 Å, but additional bands listed by Rosen *et al.* (from Svensson's thesis) are included here. R heads:

λ	v', v''	λ	v', v''	λ	v', v''
4037	0, 0	3524·3	4, 0	3174·3	8, 0
3882	1, 0	3422·2	5, 0	3106	9, 0
3747	2, 0	3332·2	6, 0	3042·9	10, 0
3629	3, 0	3249·8	7, 0	2985·6	11, 0

2483 Å SYSTEM, $C^2\Sigma - X^2\Sigma$

Occurrence. In absorption through a discharge in heated Cd + H_2.
Appearance. Degraded to shorter wavelengths. Diffuse due to predissociation.
Reference. M. A. Khan, *Proc. Phys. Soc.*, **80**, 1264 (1962)†.
Heads $\lambda\lambda$2483·1 (0, 0), 2390·1 (1, 0), 2382·7 (2, 1), 2376·7 (3, 2).

Svensson also reports a red-degraded band at 4933·6 Å, attributed to a $^2\Sigma - A^2\Pi$ transition. It is not mentioned by later authors.

CdH$^+$

References. E. Svensson and F. Tyrén, *Z.P.*, **85**, 257 (1933).
E. Bengtsson and R. Rydberg, *Z.P.*, **57**, 648 (1929).

2341 Å SYSTEM, $^1\Sigma - {}^1\Sigma$

Extensive system of singlet bands degraded to the red. System occurs with cadmium arc in hydrogen at low pressure; and in discharges through a mixture of hydrogen and cadmium vapour where ionization is favoured, such as hollow cathode or high-frequency discharge.

CdH⁺ (*contd.*)

λ	I	v', v''		λ	I	v', v''
2239·8	4	3, 1		2481·4	6	2, 3
2275·2	5	1, 0		2538·2	8	0, 2
2340·9	8	0, 0		2558·5	8	1, 3
2437·9	10	0, 1		2574·8	7	2, 4
2461·3	8	1, 2		2587·3	4	3, 5

CdI

The bands fall into four groups, but the second and third of these are probably part of the same $^2\Pi - X^2\Sigma$ system.

References.   K. Wieland, *Helv. Phys. Acta*, **2**, 46 and 77 (1929).

E. Oeser, *Z.P.*, **95**, 699 (1935).

T. S. Subbaraya, N. A. N. Rao and B. N. Rao, *Proc. Indian Acad. Sci.*, **5**, 372 (1935)†.

H. G. Howell, *P.R.S.*, **182**, 95 (1943).

C. Ramasastry and K. R. Rao, *Indian J. Phys.*, **20**, 100 (1946).

M. M. Patel, S. P. Patel and A. B. Darji, *Indian J. Pure Appl. Phys.*, **5**, 526 (1967)†; **6**, 342 (1968)†.

BANDS $\lambda\lambda6600-3600$, B $(?^2\Sigma) - X^2\Sigma$

Obtained in discharge tubes containing CdI_2, and in fluorescence. The band structure is clearer when excited in presence of excess of inert gas. The bands are diffuse and are degraded to the red on a continuous background with maximum 6500 Å. The following are given as intensity 10 or 9 by Patel *et al.* who list 110 bands.

4611·5 (1, 13), 4577·5 (1, 12), 4558·5 (0, 11), 4538·6 (4, 12), 4524·0 (0, 10), 4518·9 (3, 11), 4509·1 (1, 10), 4505·5 (4, 11), 4484·9 (3, 10).

BANDS $\lambda\lambda3600-3480$, C $(?^2\Pi_{\frac{1}{2}}) - X^2\Sigma$

Obtained in discharges and in fluorescence. Degraded slightly to the red. Sequence heads from Patel *et al.*:

λ	I	Sequence
3563·4	4	0, 1
3541·0	10	0, 0
3518·0	3	1, 0
3495·3	2	2, 0

BANDS $\lambda\lambda3500-3250$, D $(?^2\Pi_{\frac{3}{2}}) - X^2\Sigma$

Occurs in absorption, in fluorescence and in discharges. Degraded slightly to shorter wavelengths. Wieland measured 90 heads and still more have been recorded since. Heads of sequences:

λ	I	Sequence
3404·8	5	0, 1
3384·8	7	0, 0
3362·2	5	1, 0
3340·8	3	2, 0

BANDS $\lambda\lambda2550-2350$, E $(?^2\Sigma) - X^2\Sigma$

A number of weak bands, degraded to the red, observed by Wieland (27 bands) and attributed to CdI_2, were later assigned to CdI by Ramasastry and Rao who measured 56 bands, but with

CdI (*contd.*)

considerable differences in wavelength and intensity.

The following are probably relatively strong, λλ2460·8 (1, 8), 2451·7 (1, 7), 2441·7 (1, 6), 2435·3 (2, 6), 2425·5 (2, 5), 2415·6 (2, 4), 2399·6 (3, 3), 2389·7 (3, 2), 2385·5 (2, 1).

CdIn

Reference.   C. Santaram and J. G. Winans, *J. Mol. Spec.*, **16**, 309 (1965)†.

Diffuse bands 4102 and 4511 Å and some other systems were observed in a high-frequency discharge.

System B → A. Degraded to the violet.

5544 (0, 0), 5308 and 5226 Å.

System D → C. 5760·4 (0, 0), degraded violet.

5615·7 degraded red.

System E.   5735·4 (0, 1), 5701·3 (0, 0).

CdS

Continuous absorption 3150–2600 and below 2350 Å.

Reference.   P. K. Sen Gupta, *P.R.S.*, **143**, 438 (1933).

CdTl and CḋTl₂

Bands of $CdTl_2$ around 6400 Å and of CdTl at 4777 Å (deg. violet) have been reported in a high-frequency discharge.

Reference.   C. Santaram, V. K. Vaidyan and J. G. Winans, *J. Phys.*, B, **4**, 133 (1971)†.

CeO

Occurrence.   Cerium oxide or chloride in arc, and in absorption.

Appearance.   All bands are degraded to longer wavelengths and sequences are prominent.

References.   W. W. Watson, *P.R.*, **53**, 639 (1938).

L. L. Ames and R. F. Barrow, *Proc. Phys. Soc.*, **90**, 869 (1967).

Watson divided the bands into five overlapping systems. These are, according to Ames and Barrow, components of two systems, $a^3\Gamma - x^3\Phi$ and $b^3\Gamma - x^3\Phi$.

SYSTEM A

Single-headed bands:

λ8396   head of (0, 1) sequence.

λ7879·3 (1, 1) band.

λ7831·8 head of strong (0, 0) sequence.

λ7380   head of (1, 0) sequence.

SYSTEMS B AND C

These overlap and appear as a double-headed system.

λ7716   head of (0, 1) sequence of System B.

λ7347   (1, 1) band of System C.

λ7297·2 (0, 0) sequence of System C.

CeO (*contd.*)

λ7275·5 (1, 1) band of System B.

λ7235·8 (0, 0) sequence of System B.

λ6847   (1, 0) sequence of System B.

SYSTEMS D AND E

λ4863·2 (0, 0) sequence of System D.

λ4791·7 (0, 0) sequence of System E.

λ4683   (1, 0) sequence of System D.

λ4614   (1, 0) sequence of System E.

Cl_2

ABSORPTION SPECTRUM, $A^3\Pi_0 - X^1\Sigma^+$

References. A. E. Douglas, C. K. Møller and B. P. Stoicheff, *Canad. J. Phys.*, **41**, 1174 (1963)†.
M. A. A. Clyne and J. A. Coxon, *J. Mol. Spec.*, **33**, 381 (1970)†.

Chlorine gas shows a strong continuous absorption extending from the blue to around 2500 Å and having a maximum around 3300 Å.

In greater thicknesses (1 m atmosphere) there is also a banded absorption extending from the convergence limit at 4780 Å to about 6000 Å. Full measurements of individual lines of 30 bands are given by Douglas *et al.*, and the system has been extended in heated Cl_2 by Clyne and Coxon. Bands are strongly degraded to the red and heads are not obvious under large dispersion, but under small dispersion Clyne and Coxon's photographs show relatively prominent heads at λλ5317·8 (8, 1), 5272·0 (9, 1), 5229·7 (10, 1), 5212·5 (7, 0), 5191·2 (11, 1), 5165·4 (8, 0), 5156·0 (12, 1), 5122·0 (9, 0), 5082·2 (10, 0).

EMISSION SPECTRA

References. A. Elliott and W. H. B. Cameron, *P.R.S.*, **158**, 681 (1937).
W. H. B. Cameron and A. Elliott, *P.R.S.*, **169**, 463 (1939).
P. B. V. Haranath and P. T. Rao, *J. Mol. Spec.*, **2**, 428 (1958).
P. Venkateswarlu and B. N. Khanna, *Proc. Indian Acad. Sci.*, A, **49**, 117 (1959).
B. N. Khanna, *Proc. Indian Acad. Sci.*, A, **49**, 293 (1959).
L. W. Bader and E. A. Ogryzlo, *J. Chem. Phys.*, **41**, 2926 (1964)†.

Early work is well summarized by Elliott and Cameron, but bands 6000–3800, strongest in the blue violet, which occur readily in discharge tubes are now assigned to Cl_2^+. The following are the main banded regions and continua.

Orange-red, $A^3\Pi_0 - X^1\Sigma^+$

Observed (by Kitagawa) in flame of Cl_2 burning with H_2 and in the Cl_2 afterglow by Bader and Ogryzlo. The latter's photograph shows the following red-degraded heads as relatively strong: λλ6633 (8, 8), 6562 (9, 8), 6496 (10, 8), 6347 (9, 7) 6143 (9, 6), 6072·8 (7, 5), 6009·1 (8, 5), 5950·5 (9, 5).

2600–2390 Å

About 180 red-degraded bands observed in active nitrogen (Strutt and Fowler) and in a transformer discharge through Ar + Cl_2 by Venkateswarlu and Khanna. An analysis has been made. Relatively strong bands λλ2568·6, 2565·1, 2560·5, 2557·3, 2550·5, 2544·9, 2543·5.

Cl$_2$ (*contd.*)

2365–2239 Å

About 120 bands, mostly degraded to shorter wavelengths, were observed by Khanna in a transformer discharge through Ar + Cl$_2$. An analysis has been given. Relatively strong heads λλ2333·8, 2329·3, 2328·9, 2326·5, 2321·1, 2310·9, 2306·8, 2304·4, 2298·9, 2294·6.

Continua

Diffuse bands and continua occur under various conditions. The strongest are at 3070 and 2570 Å, occurring in discharges at relatively high pressures, in Tesla discharges, high-frequency discharges and in active nitrogen.

Cl$_2^+$

Occurrence. Readily in various types of discharge through Cl$_2$.

Appearance. Degraded to the red. Numerous bands between 6000 and 3800 Å with few heads showing clearly above the rotational structure. Most of the bands have single P and R branches.

References. P. B. V. Haranath and P. T. Rao, *Indian J. Phys.*, **32**, 401 (1958)†.
V. V. Rao and P. T. Rao, *Canad. J. Phys.*, **36**, 1557 (1958)†.
F. P. Huberman, *J. Mol. Spec.*, **20**, 29 (1966)†.

The bands apparently form three sub-systems of a $^2\Pi - {}^2\Pi$ transition, but the upper state is strongly perturbed and analyses differ. Huberman has made a rotational analysis of 38 bands. The following heads, from Haranath and Rao, are of bands which appear strongly in their studies and also those of Huberman.

λ	I	λ	I	λ	I	λ	I
5080·8	7	4655·1	8	4405·4	5	4140·5	6
4795·0	8	4613·6	6	4316·2	5	4112·7	5
4751·0	8	4549·1	8	4194·4	6	4064·4	5
4682·5	8	4487·0	5	4178·7	5	3862·6	5

Cl$_2$CS

Thiocarbonyl chloride (or thiophosgene) shows a weak absorption system with origin at 5340 Å, and a still weaker system near 5700 Å.

Reference. J. C. D. Brand *et al.*, *Trans. Faraday Soc.*, **61**, 2365 (1965).

ClF

Occurrence. In absorption.

Appearance. Degraded to red. Strong $v'' = 0$ progression, converging to a limit, with continuous absorption beyond this.

Transition. $^3\Pi_0 - {}^1\Sigma^+$.

References. A. L. Wahrhaftig, *J. Chem. Phys.*, **10**, 248 (1942).
H. J. Schumacher, H. Schmitz and P. H. Brodersen, *An. Assoc. Quim. Argentina*, **38**, 98 (1950)†.
W. Stricker and L. Krauss, *Z. Naturforsch.*, A, **23**, 1116 (1968).

The following wavelengths are from the above references, with our estimates of intensity from the photograph of Schumacher *et al.*

ClF (contd.)

λ	I	v', v''	λ	I	v', v''	λ	I	v', v''
5231·2	1	4, 1	4961·0	5	5, 0	4703·4	8	11, 0
5183·6	0	8, 2	4942·9	5	9, 1	4685·1	7	12, 0
5159·1	3	5, 1	4906·4	3	10, 1	4671·9	5	13, 0
5102·6	2	3, 0	4901·2	6	6, 0	4661·7	3	14, 0
5094·3	4	6, 1	4848·0	8	7, 0	4654·3	2	15, 0
5037·2	4	7, 1	4801·2	10	8, 0	4649·9	1	16, 0
5028·0	4	4, 0	4761·3	10	9, 0	4648·2	0	17, 0
4985·8	5	8, 1	4727·6	10	10, 0			

ClF$_3$

Reference.   H. Schmitz and H. J. Schumacher, *Z. Naturforsch.*, A, **2**, 363 (1947).

Absorption bands increasing in strength towards shorter wavelengths. $\lambda\lambda$3412, 3291, 3210, 3100, 3009, 2908, 2807, 2706, 2630, 2578, 2545, 2575, 2500, 2485, 2460, 2409, 2404, 2360, 2305, 2296, 2231.

ClO

Occurrence.   In emission from oxy-hydrogen flame to which Cl_2 has been added, and in absorption following flash photolysis of $Cl_2 + O_2$.

Appearance.   Degraded to the red. In absorption there is a strong $v'' = 0$ progression; in emission the $v' = 0$, 1 and 2 progressions are strong. Many of the bands are slightly diffuse due to predissociation.

Transition. $A^2\Pi - X^2\Pi$. Most of the bands belong to the $^2\Pi_{\frac{3}{2}} - ^2\Pi_{\frac{3}{2}}$ sub-system.

References.   G. Pannetier and A. G. Gaydon, *Nature*, **161**, 242 (1948)†.

G. Porter, *Disc. Faraday Soc.*, **9**, 60 (1950).

R. A. Durie and D. A. Ramsay, *Canad. J. Phys.*, **36**, 35 (1958)†.

The following are the emission bands, from Pannetier and Gaydon:

λ	I	v', v''	λ	I	v', v''	λ	I	v', v''
4502	2	4, 14	4154	4	2, 10	3841	10	0, 6
4459	3	3, 13	4114	9	1, 9	3761	9	1, 6
4417	3	2, 12	4078	6	0, 8	3729·5	7	0, 5
4373	3	1, 11	3991	8	1, 8	3652·5	6	1, 5
4335	1	3, 12	3957	7	0, 7	3625	3	0, 4
4283	7	2, 11	3874	10	1, 7	3587	2	2, 5
4241	5	1, 10						

The following are the absorption bands of the main $^2\Pi_{\frac{3}{2}} - ^2\Pi_{\frac{3}{2}}$ sub-system from Porter:

λ	I	v', v''	λ	I	v', v''	λ	I	v', v''
3034·5	1	4, 0	2796·0	8	11, 0	2671·2	5	18, 0
2993·0	2	5, 0	2771·6	10	12, 0	2661·0	5	19, 0
2954·3	3	6, 0	2749·5	8	13, 0	2652·5	5	20, 0
2918·0	5	7, 0	2729·4	7	14, 0	2645·8	3	21, 0
2884·0	5	8, 0	2711·1	7	15, 0	2640·6	2	22, 0
2851·8	6	9, 0	2695·0	6	16, 0	2636·3	2	23, 0
2822·4	8	10, 0	2682·5	5	17, 0			

ClO (*contd.*)

The following weaker bands of the $^2\Pi_{\frac{1}{2}} - ^2\Pi_{\frac{1}{2}}$ sub-system have been observed in absorption by Porter: $\lambda\lambda 2975\cdot0$ (5, 0), $2937\cdot6$ (6, 0), $2902\cdot0$ (7, 0), $2866\cdot7$ (8, 0), $2835\cdot7$ (9, 0), $2807\cdot0$ (10, 0), $2781\cdot4$ (11, 0), $2758\cdot0$ (12, 0), $2735\cdot3$ (13, 0).

ClO_2

Occurrence. Absorption by chlorine dioxide.

Appearance. Degraded to the red. This is, for a triatomic molecule, a remarkably regular system of well-defined single-headed bands. The transition is $^2A_2 - ^2B_1$, with origin at 4750 Å, and the strongest bands fall on a $(v_1', 0, 0-0, 0, 0)$ progression extending to 3200 Å, with strongest bands around 4000 Å. Beyond 3850 they become diffuse. Bands occur in groups consisting of one strong band and two or three weaker ones to shorter wavelengths.

References. C. F. Goodeve and C. P. Stein, *Trans. Faraday Soc.*, **25**, 738 (1929)†.
Z. W. Ku, *P.R.*, **44**, 376 (1933)†.
R. W. Redding and J. C. D. Brand, *J. Mol. Spec.*, **29**, 93 (1969).

Heads for which an analysis is given, with quantitative measurements of intensity on a scale $I(000-000) = 1\cdot00$, are by Redding and Brand, and others from the older references.

λ	I	v_1'	v_2'	v_3',	v_1''	v_2''	v_3''	λ	λ
4756·9	1	0	0	0	0	0	0	3869·7	3291·2
4601·9	6	1	0	0	0	0	0	3772·5	3226·0
4458·0	21	2	0	0	0	0	0	3680·5	3163·5
4324·2	38	3	0	0	0	0	0	3594·1	3105·7
4199·4	68	4	0	0	0	0	0	3511·4	3050·9
4082·5	110	5	0	0	0	0	0	3434·0	2999·4
3972·8	155	6	0	0	0	0	0	3360·5	2953·5

ClOO radical

In the flash photolysis of Cl_2 (1·8 torr) mixed with O_2 (1 atm.) a short lived intermediate, believed to be the peroxy radical, has been observed. Absorption extends from 2700 to 2250 Å, with maximum around 2500 Å.

Reference. E. D. Morris and H. S. Johnston, *J. Amer. Chem. Soc.*, **90**, 1918 (1968).

CoBr

Occurrence. Cobalt bromide in heavy-current or high-frequency discharge.

Appearance. A number of systems, all degraded to the red with narrow bands forming clear sequences. These systems are probably components of transitions of high multiplicity.

References. S. V. K. Rao and P. T. Rao, *Indian J. Phys.*, **36**, 609 (1962)†.
N. V. K. Rao and P. T. Rao, *Curr. Sci., India*, **37**, 608 (1968)†.
N. V. K. Rao and P. T. Rao, *Indian J. Pure Appl. Phys.*, **10**, 389 (1972)†.

The following are the main sequence heads of the various systems:

CoBr (*contd.*)

λ	I	Sequence	λ	I	Sequence	λ	I	Sequence
7271	–	0, 0 N$_2$	5357·0	5	1, 0 G	4518·8	10	0, 0 C
7224	–	0, 0 N$_1$	4829·5	10	0, 0 F	4493·9	10	0, 0 B
5587·9	10	0, 0 I	4736·5	10	0, 0 E	4457·7	5	1, 0 C
5577·6	10	0, 0 H	4671·3	4	1, 0 E	4433·6	4	1, 0 B
5497·9	5	1, 0 I	4602·8	10	0, 0 D	4389·2	4	0, 1 A
5477·2	3	1, 0 H	4539·7	5	1, 0 D	4337·5	5	0, 0 A
5448·5	8	0, 0 G						

CoCl

Occurrence. Cobalt chloride in heavy-current or high-frequency discharge.

Appearance. A large number of systems have been reported. Most show well developed sequences of narrow red-degraded bands.

References. P. Mesnage, *C. R. Acad. Sci. (Paris)*, **201**, 389 (1935).

K. R. More, *P.R.*, **54**, 122 (1938).

V. G. Krishnamurthy, *Indian J. Phys.*, **26**, 177 (1952).

S. V. K. Rao and P. T. Rao, *Indian J. Phys.*, **35**, 556 (1961)†.

The following is a composite list of sequence heads. Systems are named A to O but some are probably related sub-systems. System M, observed by Krishnamurthy, may be due to a $^5\Pi - {}^5\Sigma$ transition and each sequence shows 5 well-spaced Q heads covering about 50 Å.

λ	I	Sequence	λ	I	Sequence	λ	I	Sequence
7882·4	7	0, 0 O	5041·9	9	0, 0 L	4685·8	8	0, 0 G
7597·2	4	1, 0 O	5031·4	10	0, 0 K	4677·6	3	1, 0 I
7171·2	4	0, 0 N$_2$	5010·9	9	0, 0 J	4619·7	8	0, 0 F
7117·0	3	0, 0 N$_1$	4929·7	5	1, 0 K	4541·3	5	0, 0 E
6941·2	4	1, 0 N$_2$	4910·6	7	1, 0 J	4529·3	–	0, 0 D
6892·8	4	1, 0 N$_1$	4775·8	5	0, 1 G	4506·8	8	0, 0 C
5804·8	5	0, 1 M	4767·6	7	0, 0 I	4462·5	–	0, 0 B
5667·8	10	0, 0 M	4708·9	5	0, 1 F	4353·0	7	0, 0 A
5542·6	5	1, 0 M	4701·5	10	0, 0 H			

CoH

Occurrence. In King furnace and in absorption in shock-heated gases.

References. A. Heimer, *Z.P.*, **104**, 448 (1937).

R. E. Smith, *P.R.S.*, **332**, 113 (1973)†.

The main system, $^3\Phi_4 - {}^3\Phi_4$ or $\Omega = 4$ to $\Omega = 4$, shows widely spaced P and R branches and a few strong Q lines. Degraded to the red.

v', v''	R heads	Q heads
0, 0	4477·9	4495·0
1, 1	4194·5	4206·3

There is also a weaker (0, 0) band of another system or sub-system with R head at 4547·9 Å.

CoO

Reference. L. Malet and B. Rosen, *Bull. Soc. Roy. Sci. Liege*, 382 (1945).

Red-degraded bands observed in an arc between cobalt electrodes.

λ	I	λ	I	λ	I	λ	I
9125	4	8204	2	7440	2	6651	3
8985	3	7935	4	7220	2	6532	4
8705	4	7685	4	6910	5	6340	2
8494	4	7500	2	6670	3	6305	1
8247	2						

CrBr

Occurrence. Discharge through chromium bromide in a quartz tube.

Appearance. Not stated. Photograph is poor.

Reference. V. R. Rao, *Curr. Sci. India*, **18**, 338 (1949).

The following bands are listed as intensity 2: λλ6301·4, 6296·8, 6280·7, 6275·4, 6272·0, 6267·0, 6238·1, 6198·8. No analysis has been made.

CrCl

Occurrence. Heavy-current discharge through tube containing chromium chloride.

Appearance. Five groups of bands of complex structure.

Transition. Probably $^6\Pi - {}^6\Sigma$.

Reference. V. R. Rao and K. R. Rao, *Indian J. Phys.*, **23**, 508 (1949).

The (0, 0) band shows numerous heads 6225–6140 Å. The following are the strongest heads, with intensities: λλ6451·7 (3), 6315·0 (4), 6309·9 (3), 6298·5 (3), 6261·8 (3), 6259·4 (3), 6205·0 (4), 6195·3 (3), 6193·0 (3), 6172·5 (5), 6162·6 (5), 6158·6 (5), 6140·5 (6), 6116·5 (4), 6103·7 (8), 6096·0 (5), 6087·8 (4), 6047·8 (4), 6040·4 (4), 6028·9 (4), 6008·2 (4), 5998·0 (4).

Bands around 2600 Å reported by Mesnage appear to be due to AlCl.

CrF

Occurrence. Heavy-current discharge through quartz tube containing chromium fluoride.

Appearance. A strong group of bands between 4230 and 4470 Å is possibly due to the transition $A^6\Pi - X^6\Sigma$, and four weak groups 3830–4050 Å are possibly due to $B^6\Sigma - X^6\Sigma$. There are numerous heads, mostly degraded to the red.

Reference. B. K. Durgavathi and V. R. Rao, *Indian J. Phys.*, **28**, 525 (1954)†.

The following are strong heads of the first system:

λ	I	v', v''	λ	I	v', v''	λ	I	v', v''
4229·6	6		4264·0	3	1, 1	4431·1	5	0, 2
4235·3	8	0, 0	4368·5	4	1, 2	4462·7	3	—
4241·5	9		4401·0	5	—	4467·2	3	—

The following are strong heads of the second system:

CrF (*contd.*)

λ	I	v', v''	λ	I	v', v''	λ	I	v', v''
3841·7	6	0, 0	3860·2	3	4, 4	3938·5	3	—
3846·6	2	1, 1	3899·2	2	—	3975·4	2	—
3850·2	3	2, 2	3921·6	5	0, 1	3984·4	7	—
3854·4	2	3, 3	3933·8	6	3, 4	3987·4	2	—

CrH

There are band systems in the infra-red and near ultra-violet.

Occurrence. In a King furnace containing Cr and H_2. The ultra-violet system was first obtained in a high tension arc between Cr electrodes in a hydrogen flame. The ultra-violet systems have also been found in absorption in shock-heated gases.

References. A. G. Gaydon and R. W. B. Pearse, *Nature*, **140**, 110 (1937)†.

B. Kleman and B. Liljeqvist, *Ark. Fys.*, **9**, 345 (1955)†.

B. Kleman and U. Uhler, *Canad. J. Phys.*, **37**, 537 (1959)†.

S. O'Connor, *Proc. Roy. Irish Acad.*, A, **65**, 15 (1967)†.

INFRA-RED SYSTEM, $^6\Sigma^+ - ^6\Sigma^+$

Appearance. Marked sequences of apparently single-headed bands, degraded to longer wavelengths.

λ	I	v', v''
6889·6	1	2, 0
7641·7	4	1, 0
8611·1	10	0, 0
8696	3	1, 1
9968·5	3	0, 1

ULTRA-VIOLET SYSTEMS

Appearance. A strong band at 3675 and a weaker band at 3290 Å, both with complex structure.

The following are the heads, intensities and direction of degradation:

λ	I	λ	I
3706	5 V	3280	2 R
3678	10 V	3290	5 R
		3298	4 V
		3318	3 V
		3324	1 V

CrO

Occurrence. Chromium salts in carbon arc, flame or shock tube.

Appearance. Degraded to red. Marked sequences.

Transition. Probably $^5\Pi - ^5\Pi$.

References. C. Ghosh, *Z.P.*, **78**, 521 (1932)†.

M. Ninomiya, *J. Phys. Soc. Japan*, **10**, 829 (1955)†.

The following are the strongest heads:

CrO (*contd.*)

λ	I	v′, v″	λ	I	v′, v″	λ	I	v′, v″
6891·5	5	2, 4	6394·3	9	0, 1	5564·1	5	2, 0
6829·7	6	1, 3	6051·6	10	0, 0	5416·5	4	4, 1
6771·8	6	0, 2	5794·4	8	1, 0	5356·4	3	3, 0
6451·5	7	1, 2	5623·5	5	3, 1	5168·2	3	4, 0

CrS

Occurrence. Thermal emission in a furnace at 2200° C.

Appearance. Degraded to the red. Clear sequences.

Reference. A. Monjazeb and H. Mohan, *Spectrosc. Lett.*, **6**, 143 (1973).

Main heads, λλ4633·2 (0, 3), 4508·4 (0, 2), 4388·7 (0, 1), 4273·6 (0, 0), 4183·4 (1, 0), 4097·6 (2, 0), 4015·1 (3, 0), 3937·5 (4, 0).

Cs₂

Several systems have been observed in absorption, and some of them in fluorescence. The character and relative strengths of the various systems are best seen from the photographs shown by Loomis and Kusch.

References. R. Rompe, *Z.P.*, **74**, 175 (1932).

F. W. Loomis and P. Kusch, *P.R.*, **46**, 292 (1934)†.

P. Kusch and M. M. Hassel, *J. Mol. Spec.*, **25**, 205 (1968); **32**, 181 (1969)†.

The following are the most important features in absorption:

A strong system around 10500 Å.

Diffuse bands, degraded to longer wavelengths, in the near infra-red, heads λλ8733, 8346, 8262 and 8170.

The most intense system is near 7690 Å. The photograph by Kusch and Hessel shows a strong red-degraded head at about 7691 Å. Rompe recorded heads (with intensities) at 7680 (0), 7670 (1), 7640 (2), 7610 (3) and 7540 (5).

A weak system of ten sharp heads, degraded to the red, 7400–7230.

A strong system with heads degraded to the violet at λλ7078, 7075, 7072 and diffuse bands 7185 and 7128.

A sharp head, degraded to the red, at λ6250 appears at rather high temperature. Kusch and Hessel record 342 heads of this system, some of these being degraded to the red and others to the violet. Also in fluorescence.

A system of moderate intensity around 4800 Å.

Bands associated with 4555, 4593 Cs doublet.

A group of heads of moderate intensity at λλ3959, 3953, 3947, 3941, degraded to shorter wavelengths.

λ3920, a weak head degraded to shorter wavelengths.

Bands associated with 3889, 3877 Cs doublet.

CsBr

Occurrence. In absorption in a furnace.

Reference. R. F. Barrow and A. D. Caunt, *P.R.S.*, **219**, 120 (1953).

CsBr (*contd.*)

A continuum and regularly spaced diffuse bands which have a maximum at about 2750 Å and extend to longer wavelengths with increasing temperature, the longest wavelength recorded being 3582 Å.

CsCd

Barratt has studied the absorption spectra of mixed metallic vapours of the alkali metals with zinc, cadmuim and mercury. In most cases he observed diffuse bands, with a sharp head on the long wavelength side, extending to the violet as a weak continuum.

For CsCd he observed bands at λλ5316 and 5228.

Reference.   S. Barratt, *Trans. Faraday Soc.*, **25**, 758 (1929).

CsCl

Occurrence.   In absorption in a furnace.

Reference.   R. F. Barrow and A. D. Caunt, *P.R.S.*, **219**, 120 (1953).

A continuum and regularly spaced diffuse bands which have a maximum at about 2447 Å and extend to longer wavelengths with increasing temperature, the longest wavelength recorded being 3350 Å.

CsF

Occurrence.   In absorption in a furnace.

Reference.   R. F. Barrow and A. D. Caunt, *P.R.S.*, **219**, 120 (1953).

A continuum and regularly spaced diffuse bands which have a maximum at about 2095 Å and extend to longer wavelengths with increasing temperature, the longest wavelength recorded being 2711 Å.

CsH

5500 Å SYSTEM, $A^1\Sigma^+ - X^1\Sigma^+$

Reference.   G. M. Almy and M. Rassweiler, *P.R.*, **53**, 890 (1938).

A many-lined system, analogous to those of the other alkali hydrides, obtained in absorption from a mixture of hydrogen and cæsium vapour at about 550° C.

Origins of the strongest bands:

v', v''	λ	v', v''	λ	v', v''	λ	v', v''	λ
0, 2	6110·8	2, 0	5402·6	6, 0	5138·2	10, 0	4887·1
1, 2	6031·4	3, 0	5336·3	7, 0	5073·6	11, 0	4827·6
1, 1	5740·4	4, 0	5269·9	8, 0	5010·0	12, 0	4769·8
2, 1	5667·8	5, 0	5203·8	9, 0	4947·8	13, 0	4713·3

3640–3470 Å SYSTEM, $B^1\Sigma^+ - X^1\Sigma^+$

Reference.   U. Ringström, *J. Mol. Spec.*, **36**, 232 (1970).

A progression of 11 absorption bands. Strongly degraded to the red, but heads not prominent. Origins λλ3582·2, 3571·7, 3560·8, 3549·7, 3538·5, 3527·2, 3504·2, 3481·6, 3471·3.

CsHg

Diffuse bands, degraded to the violet, observed in absorption by Barratt (see CsCd). Heads λλ5343, 5222, 5112, 4932 and 4817. Mention is also made of sharp bands with heads at λλ4991, 4984 and 4975.

CsI

Occurrence. In absorption in a furnace.
Reference. R. F. Barrow and A. D. Caunt, *P.R.S.*, **219**, 120 (1953).
A continuum and regularly spaced diffuse bands which have a maximum at about 3240 Å and extend to longer wavelengths with increasing temperature, the longest wavelength recorded being 4366 Å.

CsKr

Reference. J. A. Gwinn, P. M. Thomas and J. F. Kielkopf, *J. Chem. Phys.*, **48**, 568 (1968).
Diffuse satellite bands, due to molecule formation, accompany the Cs resonance lines when excited in a discharge through Cs vapour in krypton or xenon.

CsMg

A band with maximum at 5706 Å and a violet-degraded band at 4839·2 were observed by Barratt (see CsCd).

CsXe see CsKr

CsZn

Diffuse bands, degraded to the violet, observed in absorption by Barratt (see CsCd). Heads λλ5163 and 5126.

Cu₂

Occurrence. Emission and absorption in a King furnace.
Appearance. Two systems, in green and blue, both degraded to red with fairly obvious sequences.
References. B. Kleman and S. Lindqvist, *Ark. Fys.*, **8**, 333 (1954)†.
D. S. Pešić and S. Weniger, *C. R. Acad. Sci.* (*Paris*), B, **273**, 602 (1971).

GREEN SYSTEM, $A^1\Pi_u - X^1\Sigma_g^+$
The following are the strongest heads, with own estimates of intensity:

λ	I	v', v"	λ	I	v', v"
4856·3	4	1, 0	4965·9	4	0, 1
4873·4	4	2, 1	4982·7	8	1, 2
4901·5	10	0, 0	5000·1	7	2, 3
4919·0	8	1, 1			

Weaker bands extend to 5738 Å.

Cu₂ (*contd.*)

BLUE SYSTEM, $B^1\Pi_u - X^1\Sigma_g^+$

Strongest heads; own estimates of intensity:

λ	I	v', v''	λ	I	v', v''
4497·7	4	2, 0	4612·5	8	3, 3
4546·4	7	1, 0	4653·4	10	0, 1
4596·9	10	0, 0	4710·7	5	0, 2
4606·9	8	2, 2			

CuAg see AgCu

CuBi

Reference.   Y. Lefebvre and R. Houdart, *C. R. Acad. Sci. (Paris)*, B, **270**, 1485 (1973)†.
In emission in a King furnace at 1900° C. Violet-degraded bands 6400–6150; (0, 0) at 6276 Å. Red-degraded bands 5200–5000; (0, 1) at 5123·4, (0, 0) at 5072·7.

CuBr

Occurrence.   Four systems have been observed in absorption, and in emission from flames and microwave discharges.
Appearance.   All systems are degraded to the red and show marked sequences.
References.   R. Ritschl, *Z.P.*, **42**, 172 (1927).
P. R. Rao and K. V. S. R. Apparao, *Proc. Indian Acad. Sci.*, A, **60**, 57 (1964)†.
The following measurements are by Ritschl; only the strong bands are given. Intensities, which have been increased to a scale of 10, are for absorption.

System A, $A^1\Pi - X^1\Sigma$			System B, $B^1\Pi - X^1\Sigma$			System C, $C^1\Sigma - X^1\Sigma$		
λ	I	v', v''	λ	I	v', v''	λ	I	v', v''
5032·2	3	0, 2	4461·9	4	0, 2	4379·3	6	0, 2
4954·7	4	0, 1	4400·9	4	0, 1	4320·5	4	0, 1
4883·4	4	1, 1	4341·1	10	0, 0	4262·8	8	0, 0
4879·3	8	0, 0	4288·6	7	1, 0	4210·2	6	1, 0
4810·4	2	1, 0	4237·8	4	2, 0	4159·3	5	2, 0

SYSTEM D, $D^1\Sigma - X^1\Sigma$

Sequence heads from Rao and Apparao λλ4015·2 (0, 2), 3965·3 (0, 1), 3917·1 (0, 0), 3874·7 (1, 0), 3833·6 (2, 0).

CuCl

Occurrence.   Six systems have been observed in flames, in fluorescence and in absorption, and also in discharge tubes. They also appear when CuCl is introduced into active nitrogen, and in an arc. The bands frequently occur as impurities in flames, especially the CO flame.
Appearance.   All systems are degraded to the red and form clear sequences.
References.   R. Ritschl, *Z.P.*, **42**, 172 (1927).
A. G. Gaydon, *P.R.S.*, **182**, 199 (1943)†.
P. R. Rao, R. K. Asundi and J. K. Brody, *Canad. J. Phys.*, **40**, 42 and 1443 (1962)†.

CuCl (*contd.*)

The following measurements of sequence heads are mostly from Ritschl, with a few more accurate measurements from Rao *et al.* Intensities refer to absorption (from Ritschl).

System A, $A^1\Pi - X^1\Sigma$			*System B*, $B^1\Pi - X^1\Sigma$			*System C*, $C^1\Sigma - X^1\Sigma$		
λ	I	v', v''	λ	I	v', v''	λ	I	v', v''
5380	2	0, 1	4982·2	4	0, 1	4946·1	4	0, 1
5262·3	4	0, 0	4881·5	8	0, 0	4846·9	8	0, 0
5152	2	1, 0	4789·0	5	1, 0	4755·7	5	0, 0

System D, $D^1\Pi - X^1\Sigma$			*System E*, $E^1\Sigma - X^1\Sigma$			*System F*, $F^1\Pi - X^1\Sigma$		
λ	I	v', v''	λ	I	v', v''	λ	I	v', v''
4515·9	1	0, 2	4493·0	4	0, 2	4089·4	8	0, 2
4433·8	6	0, 1	4412·0	6	0, 1	4021·8	10	0, 1
4353·2	9	0, 0	4333·2	10	0, 0	3955·9	8	0, 0
4281·3	7	1, 0	4259·4	8	1, 0	3897·0	6	1, 0
4211·0	4	2, 0	4188·6	6	2, 0			
			4119·9	5	3, 0			

CuF

References.   R. Ritschl, *Z.P.*, **42**, 172 (1927).
L. H. Woods, *P.R.*, **64**, 259 (1943)†.

Occurrence.   Three systems have been observed in absorption. The same three systems have also been obtained by Woods from a copper hollow cathode containing gaseous HF at 0·1 torr pressure. System C has also been observed in an arc. The bands have been obtained weakly by heating CuF_2 in active nitrogen and may occur in flame sources.

Appearance.   Marked sequences with vibrational structure degraded to the violet but rotational structure degraded to the red.

Strong bands as observed by Ritschl. Intensities, which are for absorption, have been increased to a scale of 8.

System A, $A^1\Pi - X^1\Sigma$			*System B*, $B^1\Sigma - X^1\Sigma$			*System C*, $C^1\Pi - X^1\Sigma$		
λ	I	v', v''	λ	I	v', v''	λ	I	v', v''
5694·3	6	0, 0	5061·1	7	0, 0	5086·4	2	0, 1
5685·7	6	1, 1	5052·3	6	1, 1	4932·0	8	0, 0
5677·2	5	2, 2	4901·3	5	1, 0	4926·8	6	1, 1
						4781·9	4	1, 0

CuGa

Twelve red-degraded bands have been observed in emission from a King furnace, 6860–6470 Å. 6656·1 (0, 1), 6559·5 (0, 0), 6497·0 (1, 0).

Reference.   M. Biron, *C. R. Acad. Sci.* (*Paris*), B, **271**, 1096 (1970).

CuH

Seven systems have been reported, of which the strongest is at 4280 Å. All systems are degraded to the red and show very open rotational structure. They all occur in absorption in a King

CuH (*contd.*)

furnace, and the 4280 system occurs readily in emission in most sources containing copper and hydrogen. See Plate 4.

References. A. Heimer and T. Heimer, *Z.P.*, **84**, 222 (1933).

T. Heimer, *Z.P.*, **95**, 321 (1935).

U. Ringström, *Ark. Fys.*, **32**, 211 (1966)†.

4280 Å SYSTEM, $A^1\Sigma^+ - X^1\Sigma^+$

v', v''	Origins	R Heads	v', v''	Origins	R Heads
2, 3	4745·3	4734·1	0, 0	4288·6	4279·6
1, 2	4701·5	4689·0	2, 1	4067·6	4061·9
0, 1	4661·9	4648·4	1, 0	4011·5	4005·4
2, 2	4387·4	4399·0	2, 0	3780·5	3776·3
1, 1	4336·2	4327·7	3, 0	3586·2	3583·0

3804 Å SYSTEM, $B^3\Pi_0 - X^1\Sigma^+$

P and R branches, degraded to the red.

v', v''	Origins	R Heads	v', v''	Origins	R Heads
1, 4	4800·9		1, 1	3848·4	3843·8
1, 2	4133·5	4127·4	0, 0	3803·8	3798·6
0, 1	4094·6	4087·4	2, 1	3660·8	
2, 2	3917·9	3914·3	1, 0	3590·4	

3688 Å SYSTEM, C $1 - X^1\Sigma^+$

v', v''	Origins	R Heads	Q Head
0, 1	3960·6	–	–
0, 0	3688·0	3684·0	3689·2
1, 0	3500·8	3497·4	–
2, 0	3350·2	3347·7	–

3576 Å SYSTEM, c $1 - X^1\Sigma^+$

v', v''	Origins	R Heads	Q Head
0, 1	3830·8	–	–
0, 0	3575·1	3572·2	3576·3
1, 0	3406·6	3404·4	

$b\Delta_2 - X^1\Sigma^+$ SYSTEM

The (1, 0) band has an origin at 3363 Å.

2239 Å SYSTEM, $D^1\Pi - X^1\Sigma^+$

Reference. B. Grundström, *Z.P.*, **98**, 128 (1935)†.

v', v''	Origins	R Heads	Q Heads
0, 0	2239	2228	2239·2
1, 1	2242	2234	2242

CuH (*contd.*)

$E^1\Sigma^+ - X^1\Sigma^+$ SYSTEM

Reference. U. Ringström, *Canad. J. Phys.*, **46**, 2291 (1968).

This is a 'many line' system of 22 headless bands, strongly degraded to the red, in the region 2740–2270 Å.

The following are some R(0) lines, $\lambda\lambda2512\cdot0$ (5, 1), $2481\cdot8$ (6, 1), $2412\cdot1$ (8, 1), $2405\cdot3$ (5, 0), $2371\cdot9$ (6, 0), $2278\cdot2$ (9, 0).

CuI

Occurrence. In flame spectra and on introduction of copper iodide vapour into active nitrogen. Also in microwave discharges and in absorption.

Appearance. Degraded to red. Single-headed (apart from isotope shift). Close sequences.

Transition. There are four systems, A, C, D and E all involving transitions to the ground state. The B system has now been shown to be due to Cu_2 (B. Kleman and S. Lindqvist, *Ark. Fys.*, **8**, 333 (1954)).

References. R. S. Mulliken, *P.R.*, **26**, 1 (1925).

R. Ritschl, *Z.P.*, **42**, 172 (1927).

P. R. Rao and K. V. S. R. Apparao, *Canad. J. Phys.*, **44**, 2241 (1966)†.

K. P. R. Nair and K. N. Upadhya, *Canad. J. Phys.*, **44**, 1267 (1966)†.

R. K. Pandey, K. N. Upadhya and B. S. Mohanty, *Indian J. Phys.*, **42**, 154 (1968)†.

The strongest bands listed by Mulliken (in active nitrogen) are given in the following tables. Intensities have been increased to a scale of 10.

SYSTEM A, $A^1\Pi - X^1\Sigma$

λ	I	v', v''	λ	I	v', v''	λ	I	v', v''
5477·1	4	3, 8	5312·4	5	2, 5	5101·9	3	2, 2
5402·2	5	3, 7	5297·5	6	1, 4	5072·8	10	0, 0
5386·5	6	2, 6	5283·8	4	0, 3	5034·6	3	2, 1
5371·5	5	1, 5	5226·1	5	1, 3	5019·7	7	1, 0
5357·4	3	0, 4	5212·2	7	0, 2	4983·9	4	3, 1
5328·5	3	3, 6	5141·1	7	0, 1	4968·3	5	2, 0

SYSTEM C, $C^1\Sigma - X^1\Sigma$

λ	I	v', v''
4687·8	3	0, 2
4630·6	5 ?	0, 1
4575·1	4	0, 0
4527·9	4	1, 0

SYSTEM D, $D^1\Pi - X^1\Sigma$

λ	I	v', v''	λ	I	v', v''
4514·8	3	0, 3	4369·9	3	1, 1
4462·2	4	0, 2	4359·9	4	0, 0
4410·8	5	0, 1	4320·0	3	1, 0

CuI (*contd.*)

SYSTEM E, $E^1\Sigma - X^1\Sigma$

λ	I	v', v''	λ	I	v', v''	λ	I	v', v''
4419·1	3	2, 7	4315·4	3	1, 4	4214·6	4	0, 1
4413·7	2	1, 6	4309·6	4	0, 3	4174·6	4	1, 1
4369·9	3	2, 6	4280·1	2	3, 5	4168·5	1	0, 0
4364·0	4	1, 5	4261·7	5	0, 2	4129·4	2	1, 0
4358·3	3	0, 4	4227·0	4	2, 3	4091·3	3	2, 0

CuO

Occurrence. Copper arc and Cu salts in flame or active nitrogen. Also in exploded copper wires. Most of the bands have been recorded in flash absorption.

Appearance. The spectrum is very complex. The strongest bands, wide doublets in the orange red, are shown in Plate 10, and the whole spectrum by Antić-Jovanović *et al.* Most bands are degraded to the red but a few weaker ones shade to the violet. Vibrational structure is sometimes degraded in the opposite direction so that sequence heads are complex. Early analyses are mostly wrong and even now there is some lack of agreement with frozen-matrix studies.

References. A. Antić-Jovanović, D. S. Pešić and A. G. Gaydon, *P.R.S.*, **307**, 399 (1968)†. A. Lagerqvist and U. Uhler, *Z. Naturforsch.*, **226**, 551 (1967). O. Appleblad and A. Lagerqvist, *J. Mol. Spec.*, **48**, 607 (1973).

ORANGE SYSTEM, $A^2\Sigma - X^2\Pi$

Antić-Jovanović *et al.* list about 80 heads. The following are the first heads of groups which appear prominent with small dispersion:

λ	I	v', v''	λ	I	v', v''	λ	I	v', v''
6401·5	7	2, 3 Q_2	6161·6	9	1, 1 Q_2	5958·4	4	5, 4 Q_2
6392·5	6	0, 1 R_2	6146·8	7	2, 2 R_2	5940·4	5	4, 3 R_2
6377·7	4	3, 4 R_2	6059·3	10	0, 0 R_1	5847·7	7	2, 1 R_1
6294·1	6	1, 2 R_1	6045·0	8	1, 1 ^sR	5841·5	4	3, 2 ^sR

GREEN REGION

The arc spectrum shows a number of weak bands, including a little group of heads between 5237 and 5228 Å, and heads $\lambda\lambda 5344$, 5312·2, 5308, 5279, 5274 and 5235·7. The 5312·2 and 5235·7 heads are probably components of the (0, 0) band of another transition $B^2\Sigma - X^2\Pi$. Bands of CuOH also occur in the green.

BLUE BANDS, $C^2\Pi - X^2\Pi$

Again structure is complex. Antić-Jovanović *et al.* made a vibrational analysis of some features treating them as part of a $^2\Sigma - {}^2\Pi$ transition; Appleblad and Lagerqvist give the same vibrational analysis but argue for a $^2\Pi - {}^2\Pi$ transition. Assignments, where given, are based on their analysis.

λ	I	v', v''	λ	I	v', v''	λ	I	v', v''
4916·7	5	0, 1 i Q	4710·8	7	0, 0 ii Q	4525	5	—
4883·4	5	0, 1 ii R	4687·4	5	0, 0 ii R	4518	5	—
4862·9	5	1, 2 ii Q	4635·0	6	1, 0 i Q	4464	7	—
4855·1	5	0, 1 ii Q	4583·8	6	1, 0 ii Q	4457	8	—
4828·9	4	0, 1 ii R	4532	5	—	4453	8	—
4769·9	6	0, 0 i Q						

CuO (*contd.*)

VIOLET REGION

There are a few weak bands in the violet and near ultra-violet. A red-degraded band at 4182 Å has been analysed by Lagerqvist and Uhler as the (1, 0) band of a transition $D^2\Pi_{\frac{3}{2}} - X^2\Pi_{\frac{3}{2}}$.

CuOH

The green coloration of flames containing copper compounds has been shown to be due to the formation of CuOH. There is a strong diffuse band 5350–5550 Å and a weaker band 6150–6250 Å. Singh has given measurements of structure in the flame bands, although he provisionally attributed them to Cu_2. Antić-Jovanović and Pesić list 80 features; strongest maxima $\lambda\lambda$5458·7, 5457, 5434, 5418·6, 5388, 5374·6, 5371, 5356·5, 5324, 5246.

References. E. M. Bulewicz and T. M. Sugden, *Trans. Faraday Soc.*, **52**, 1481 (1956).
N. L. Singh, *Proc. Indian Acad. Sci.*, A, **25**, 1 (1947)†.
A. M. Antić-Jovanović and D. S. Pešić, *Bull. Chem. Soc. Belgrade*, **34**, 5 (1970)†.

CuS

Reference. M. Biron, *C. R. Acad. Sci. (Paris)*, **258**, 4228 (1964).

Two sub-systems of double-headed bands, degraded to the red, were obtained in emission from a King furnace. 23 bands were found. R heads of main sequences:

Sub-system A		Sub-system B	
λ	v', v''	λ	v', v''
5704·9	0, 1	5848·7	0, 1
5574·6	0, 0	5712·1	0, 0
5463·0	1, 0	5594·9	1, 0

CuSe

Reference. J. C. Joshi, *J. Mol. Spec.*, **8**, 79 (1962)†.

Observed in thermal emission. Two systems of red-degraded bands in marked sequences. Sequence heads:

$A - X$			$B - X$		
λ	I	v', v''	λ	I	v', v''
6347·8	2	0, 2	5765·9	3	0, 2
6231·1	5	0, 1	5668·4	5	0, 1
6116·3	5	0, 0	5574·1	6	0, 0
6025·6	3	1, 0	5498·7	3	1, 0
5929·9	2	2, 0	5427·5	1	2, 0

CuTe

References. R. C. Maheshwari and D. Sharma, *Proc. Phys. Soc.*, **81**, 898 (1963).
Y. Lefebvre and J. L. Bocquet, *J. Phys.*, B, **8**, 1322 (1975)†.

The main $A - X$ system consists of 31 red-degraded bands; sequence heads 6465·7 (0, 2), 6364·0 (0, 1), 6263·3 (0, 0), 6187·0 (1, 0), 6113·0 (2, 0).

There is also a weak $B - X$ system with (0, 0) band at 5647 Å.

DyO

Reference.   R. Herrmann and C. T. J. Alkemade, *Flame Photometry*, Interscience (1963), translated by P. T. Gilbert.

About 40 bands, some degraded to the red, occur in arcs and flames and are listed in collected data. Strongest, $\lambda\lambda$5868, 5856, 5834, 5729, 5705, 5694, 5273, 5263.

ErO

About 50 bands appear in arcs and flames and are listed under erbium oxide by Herrmann and Alkemade (see DyO above). Strongest bands, with direction of degradation, $\lambda\lambda$5613(V), 5596(M), 5563(R), 5552(M), 5545(M), 5528(M), 5067(R).

EuOH

Diffuse bands occur in flames and are provisionally attributed to europium hydroxide (see DyO above). Strongest maxima $\lambda\lambda$7020, 6840, 6620, 6550, 6470, 6230.

F_2

Occurrence.   Discharge through F_2, especially high-voltage low-frequency with flowing gas.
References.   T. L. Porter, *J. Chem. Phys.*, **48**, 2071 (1968).
H. G. Gale and G. S. Monk, *Astrophys. J.*, **59**, 125 (1924)†; **69**, 77 (1929)†.

7510–4220 Å, $C^1\Sigma^+ - B^1\Pi$

Appearance.   Strongly degraded to the red. Strong perturbations with some bands having diffuse lines. The main bands form a $v' = 0$ progression.

R heads from Porter, with intensities on a scale of 100:

λ	I	v', v''	λ	I	v', v''	λ	I	v', v''
7509·4	2	0, 7	6480·4	25	1, 6	5398·0	10	0, 2
6982·5	4	0, 6	6102·8	70	0, 4	5388·5	10	1, 3
6931·9	5	1, 7	5731·4	25	0, 3	5092·3	3	1, 2
6517·2	100	0, 5	5715·6	4	1, 4	4823	1	1, 2

5853 Å, $C^1\Sigma^+ - B'^1\Pi$

Under similar conditions to the main system Porter found bands at 5852·6 ($I = 8$) and 5495·7 (1).

F_2^+

Reference.   T. L. Porter, *J. Chem. Phys.*, **48**, 2071 (1968).

In discharges through flowing fluorine some 2900 lines 5090–4220 Å have been arranged into 36 branches of a 'many-line' red-degraded system. Widely spaced (20 Å) double-headed bands, but heads not outstanding. R_2 heads:

λ	v', v''	λ	v', v''	λ	v', v''	λ	v', v''
5524·1	0, 9	4974·9	2, 8	4735·0	2, 7	4513·2	2, 6
5234·7	0, 8	4969·7	0, 7	4726·0	0, 6	4401·2	1, 5
5099·7	1, 8	4847·7	1, 7	4615·6	1, 6	4308·1	0, 5

FeBr

VISIBLE SYSTEMS

Occurrence. In high frequency or heavy current D.C. discharge through $FeBr_3$ vapour.

Appearance. Several groups of narrow bands in the region 5720–6410 Å have been arranged into four systems. Bands assigned to systems II, III and IV are degraded to the violet. The two pairs of bands of system I are degraded to the red.

References. S. P. Reddy, *J. Sci. Industr. Res.*, B, **18**, 188 (1959)†.

P. Mesnage, *C. R. Acad. Sci. (Paris)*, **204**, 761 (1937).

The following are all heads with intensity ≥ 4, with analysis proposed by Reddy. Direction of degradation, R or V is indicated after the wavelength, and the head, P or Q, after the v', v'' assignment.

λ	I	Syst.	v', v''		λ	I	Syst.	v', v''	
6399·7 V	5	IV	3, 5	Q	5959·3 V	8	II	0, 0	
6302·4 V	7	IV	0, 1	P	5957·1 V	10	II	1, 1	
6299·4 V	7	IV	0, 1	Q	5953·5 V	6	II	2, 2	
6293·2 V	4	IV	1, 2	Q	5949·5 V	5	II	3, 3	
6187·5 V	8	IV	0, 0	P	5852·8 V	7	II	1, 0	
6182·7 V	7	IV	0, 0	Q	5849·8 V	7	II	2, 1	
5977·7 V	10	III	0, 0	P	5846·2 V	6	II	3, 2	
5975·6 V	10	III	0, 0	Q	5842·5 V	5	II	4, 3	
5971·4 V	8	III	1, 1	P	5829·7 R	10	I	1, 2	
5969·4 V	8	III	1, 1	Q	5822·4 R	7	I	0, 1	
5966·2 V	7	III	2, 2	P	5729·6 R	8	I	1, 1	
5964·1 V	7	III	2, 2	Q	5622·8 R	5	I	0, 0	

ULTRA-VIOLET SYSTEM, 3760–3650 Å

Reference. N. V. K. Rao and P. T. Rao, *J. Phys.*, B, **3**, 725 (1970)†.

A high-frequency discharge shows 27 heads, degraded to shorter wavelengths, probably due to a $^4\Pi - {}^4\Sigma$ transition. The (0, 0) band has 10 heads of which the most prominent are at 3711·9, 3702·0 and 3694·1 Å.

FeCl

Occurrence. In high-frequency or high current-density discharges through $FeCl_2$.

References. P. Mesnage, *C. R. Acad. Sci. (Paris)*, **201**, 389 (1935).

E. Miescher, *Helv. Phys. Acta*, **11**, 463 (1938).

W. Müller, *Thesis, Basel* (1943)†.

S. P. Reddy and P. T. Rao, *J. Mol. Spec.*, **4**, 16 (1960)†.

V. V. K. Rao and P. T. Rao, *Indian J. Pure Appl. Phys.*, **9**, 102 (1971)†.

Seven systems have been analysed. Systems II, III and IV probably have the same lower state.

SYSTEM I, 3447 Å

Degraded violet. Observed by Mesnage and Miescher. Transition probably $^6\Pi - {}^6\Sigma$. Miescher gives the following heads of the (0, 0) sequence, λλ3447·6, 3443·5, 3437·9, 3431·5, 3424·0, 3415·6. Among the other heads the following may be prominent, λλ3463·4, 3412·8, 3397·3, 3381·9, 3374·8, 3366·8.

FeCl (*contd.*)

SYSTEM II, 3582 Å

Degraded violet. Observed by Müller and attributed to a $^4\Pi - {}^4\Sigma$ transition.
Heads of (0, 1) sequence $\lambda\lambda 3634\cdot1, 3630\cdot2, 3625\cdot6, 3618\cdot2$.
Heads of (0, 0) sequence $\lambda\lambda 3582\cdot8, 3579\cdot2, 3574\cdot6, 3567\cdot3$.
Heads of (1, 0) sequence $\lambda\lambda 3529\cdot0, 3525\cdot5, 3521\cdot3, 3514\cdot0$.

SYSTEM III, 4696–5002 Å

Four sequences of complex red-degraded bands, attributed by Reddy and Rao to a $^4\Pi - {}^4\Sigma$
transition. Most prominent heads:

λ	I	v', v''	λ	I	v', v''
4696·8	4	2, 0 R_4	4862·5	6	0, 0 R_4
4774·5	4	1, 0 S_4	4865·0	10	0, 0 Q_4
4778·1	6	1, 0 R_4	4880·4	10	0, 0 Q_3
4796·1	10	1, 0 Q_3	4960·7	5	0, 1 Q_4
4808·7	8	1, 0 Q_2			

SYSTEM IV, 5752 Å

Two groups of violet-degraded bands, possibly due to a $^4\Sigma - {}^4\Sigma$ transition. All heads from Reddy
and Rao.

λ	I	v', v''
5885·2	7	0, 1
5879·6	7	1, 2
5752·4	10	0, 0
5748·2	8	1, 1

SYSTEM V, 6064 Å

Two sequences of single-headed bands, degraded to the violet. Strongest heads from Reddy and
Rao.

λ	I	v', v''	λ	I	v', v''
6063·9	10	0, 0	5989·6	10	1, 0
6060·3	8	1, 1	5986·5	10	2, 1
6056·0	8	2, 2	5983·2	10	3, 2

SYSTEM VI, 6474 Å

Three sequences of single-headed bands, degraded to the violet. Strongest heads from Reddy and
Rao.

λ	I	v', v''	λ	I	v', v''
6542·5	7	2, 3	6474·1	10	0, 0
6535·6	7	3, 4	6467·5	8	1, 1
6527·3	7	4, 5	6461·0	8	2, 2
6520·8	7	5, 6	6390·2	7	1, 0
6513·2	7	6, 7	6382·5	7	2, 1

α SYSTEM, 3250–3150 Å

Red-degraded bands, probably due to a $^4\Pi - {}^4\Sigma$ transition, observed by Rao and Rao. The strongest
individual head, listed as intensity 10, is at 3205·7 (0, 0 R_1) but the more prominent R_3 heads of
the complex multi-headed sequences are:

FeCl (*contd.*)

λ	I	v', v''
3243·6	3	0, 1
3199·3	6	0, 0
3157·3	1	1, 0

FeF

Rosen *et al.* (*Spectroscopic data relative to diatomic molecules*, (1970)) report unpublished data from R. F. Barrow *et al.* Absorption and emission in a King furnace shows bands, degraded to shorter wavelengths, with sequence heads $\lambda\lambda$3379·7 (0, 1), 3305·5 (0, 0) and 3236·5 (1, 0). There is also a red-degraded sequence at 3118 Å.

FeH

References. R. E. Smith, *P.R.S.*, **332**, 113 (1973).
P. K. Carroll and P. McCormack, *Astrophys. J.*, **177**, L33, (1972)†.

Headless bands with open but very complex rotational structure, strongest 5370–5200 and 5000–4850 Å, have been observed in absorption by shock-heated iron dust in H_2 + Ar. P. K. Carroll and P. McCormack have also found the bands in absorption and emission in a furnace and in sun-spot spectra. Maxima at 5320 and 4920 Å; also a head (deg. R) at 8690·8 Å.

FeO

Strong bands in the orange and infra-red are emitted by the flame of an iron arc in air and when iron carbonyl is introduced into a flame. These systems, and another weak system in the blue, are also emitted from exploding wires. The orange bands also occur in absorption following flash heating of an iron surface. These orange bands have been analysed into two systems, A and B; the bands in the blue can be arranged into another system C; these three systems have a common lower level. The infra-red bands, system D, probably have the same initial level as system A.

References. A. G. Gaydon, *PhD. Thesis, London*, (1937)†.
A. Delsemme and B. Rosen, *Bull. Soc. Roy. Sci., Liege*, 70 (1945)†.
L. Malet and B. Rosen, *Bull. Inst. Roy. Colon. Belg.*, 377 (1945)†.
R. F. Barrow and M. Senior, *Nature*, **223**, 1359 (1969).

ORANGE BANDS, SYSTEMS A AND B

The general appearance is complex. Most of the bands are degraded to the red, but some features (indicated by M) appear to be maxima of headless structures. In the following list, bands given with intensities are from Gaydon, while others are from Delsemme and Rosen. See Plate 9.

λ	I	v', v''	λ	I	v', v''	λ	I	v', v''
6651·5		A 1, 4 i	6278·9		A 1, 3 ii	5934·8		A 1, 2 ii
6596·6		A 1, 4 ii	6218·9	10	A 0, 2 i	5919	4M	
6566·7	2	A 0, 3 i	6180·5	9	A 0, 2 ii	5911	4M	
6524·1	2	A 0, 3 ii	6109·9	9	B 1, 2	5903·0	6	A 0, 1 i
6445·2		B 1, 3	6097·3	9M		5868·1	9	A 0, 1 ii
6430·3		B 0, 2	6084·7		B 0, 1	5819·2	6	B 2, 2
6295·9		A 1, 3 i	5974·6	6	A 1, 2 i	5807·4	2	B 1, 2

FeO (*contd.*)

λ	I	v', v''	λ	I	v', v''	λ	I	v', v''
5789·8	9	B 0, 0	5614·0	6	A 0, 0 i	5430	2	
5678·9		A 1, 1 i	5582·8	6	A 0, 0 ii	5408·6		A 1, 0 i
5646·6	6	A 1, 1 ii	5543·2	2	B 2, 1	5382·1		A 1, 0 ii
5624·1	4		5531·4	4		5289·5		B 2, 0
5621·3	4		5527·9		B 1, 0			

BLUE SYSTEM, C

From Malet and Rosen. Degraded red.

λ	I	v', v''	λ	I	v', v''
4929	1	1, 3	4544	3	1, 1
4730	4	1, 2	4478	4	0, 0
4659	3	0, 1	4386	2	
4604	1	2, 2			

INFRA-RED SYSTEM, D

Degraded to the red. The following measurements are by Malet and Rosen, with intensities, where given, by Gaydon whose observations cover only part of the region. Gaydon's measurements come about 5 Å less than those given here. The vibrational analysis is not very convincing.

λ	I	v', v''	λ	I	v', v''	λ	I	v', v''
9408			8578	6	1, 2	7527		2, 1
9333		1, 3	8302	8	3, 3	7428		
9258			8230	8		7265		
9200			8137*	8		7022		2, 0
9088		0, 2	8112	10	2, 2	6830		
8864			7775			6700		3, 0
8790		2, 3	7690		3, 2			

* From Gaydon

Some of the bands not classified in the above list may fit into two other fragmentary systems.

A. M. Bass and W. S. Benedict (*Astrophys. J.*, **116**, 652 (1952)) have made measurements of the infra-red bands under small dispersion, and have proposed an analysis.

ULTRA-VIOLET, λλ2430–2410

A. B. Callear and R. G. W. Norrish (*P.R.S.*, **259**, 304 (1960)†), have observed a group of bands in absorption following flash photolysis of explosive mixtures containing $Fe(CO)_5$; no measurements are available.

FeOH

Two diffuse groups of bands have been observed in the ultra-violet by Gaydon when $Fe(CO)_5$ is introduced into a hydrogen flame and in shock tube excitation of $Fe(OH)_2$. The emitter is probably FeOH. The first group lies in the region 3530–3580 Å and the second 3630–3675 Å, with a diffuse head, degraded to shorter wavelengths, at 3675 Å.

Ga$_2$

Reference. D. S. Ginter and M. L. Ginter, *J. Phys. Chem.*, **69**, 2480 (1965).

Bands 5143–4648 Å observed in emission and absorption in a King furnace. Most are degraded to the red, but some to the violet and a few are headless. Red-degraded bands (listed to 0·1 Å) λλ4993·2, 4975·5, 4971·5, 4942·5, 4892·2, 4852·2, 4804·2, 4767·4, 4719·2, 4684·2, 4648·4.

GaBr

Occurrence. In high-frequency discharge and in absorption.
References. A. Petrikaln and J. Hochberg, *Z.P.*, **86**, 214 (1933)†.
E. Miescher and M. Wehrli, *Helv. Phys. Acta*, 7, 331 (1934).

SYSTEM A, λλ3616–3452

Not clearly degraded either way, but Miescher and Wehrli note the following red-degraded R heads:

λ	I	v', v''
3582·8	8	0, 1
3549·3	9	0, 0
3516·1	4	1, 0
3484·0	6	2, 0

SYSTEM B, λλ3568–3439

Not clearly degraded either way. (0, 0) band at λ3503·3.

SYSTEM C, λλ2874–2667

Diffuse bands. 2679 (0, 0), 2720 (0, 1), 2754 (0, 2).

GaCl

Occurrence. In high-frequency discharge and in absorption.
References. A. Petrikaln and J. Hochberg, *Z.P.*, **86**, 214 (1933)†.
E. Miescher and M. Wehrli, *Helv. Phys. Acta*, 7, 331 (1934)†.
F. K. Levin and J. G. Winans, *P.R.*, **84**, 431 (1951)†.

SYSTEM A, λλ3469–3253, A$^3\Pi_0$ – X$^1\Sigma^+$

Degraded to shorter wavelengths. Strongest bands:

λ	I	v', v''
3426·5	3	0, 1
3384·4	6	0, 0
3340·2	3	1, 0

SYSTEM B, λλ3430–3220, B$^3\Pi_1$ – X$^1\Sigma^+$

Degraded to shorter wavelengths. Strongest bands:

λ	I	v', v''
3388·0	3	0, 1
3346·8	6	0, 0
3303·9	3	1, 0

GaCl (*contd.*)

SYSTEM C, $\lambda\lambda 2700-2483$, $C^1\Pi - X\,^1\Sigma^+$

Degraded to the red. Strongest bands:

λ	I	v', v''
2536·5	6	0, 2
2513·3	8	0, 1
2490·6	10	0, 0
2483	8	1, 0

GaCl$_2$

W. Wenk (*Dissertation, Basel*, 1941) has observed absorption continua at 2275, 2130, 1990 and 1735 Å.

GaF

Occurrence. In absorption in furnace, and in emission from hot hollow-cathode discharge containing metallic gallium and AlF$_3$.

References. D. Welti and R. F. Barrow, *Proc. Phys. Soc.*, A, **65**, 629 (1952).

R. F. Barrow, J. A. T. Jacquest and E. W. Thompson, *Proc. Phys. Soc.*, A, **67**, 528 (1954).

R. F. Barrow, P. G. Dodsworth and P. B. Zeeman, *Proc. Phys. Soc.*, A, **70**, 34 (1957).

A $^3\Pi_0 - X\,^1\Sigma$ SYSTEM

Appearance. Degraded to the violet; single-headed. Two sequences.

Strongest bands, with intensities for absorption:

λ	I	v', v''
3018·4	5	0, 0
3015·1	3	1, 1
3960·5	2	1, 0
2957·6	2	2, 1

B $^3\Pi_1 - X\,^1\Sigma$ SYSTEM

Appearance. Degraded to the violet. Close double heads and marked sequences.

The following are the Q heads (except for (0, 0)) of strongest bands, with intensities for absorption.

λ	I	v', v''	λ	I	v', v''
3044·8	0	0, 1	2981·5	2	2, 2
2989·8	10	0, 0 P	2931·2	3	1, 0
2988·8	10	0, 0 Q	2928·2	2	2, 1
2985·1	5	1, 1	2925·5	3	3, 2

C $^1\Pi - X\,^1\Sigma$ SYSTEM

Appearance. Double-headed bands, mostly degraded to the red.

The following are the heads (P, Q or R) of the most prominent bands. The direction of degradation, R or V, is indicated after the wavelength. Intensities of Q heads are for absorption.

GaF (*contd.*)

λ	I	v', v''	λ	I	v', v''	λ	I	v', v''
2169·4 V		0, 2 P	2121·7 R	9	2, 2 Q	2110·2 R		0, 0 R
2168·6 V	4	0, 2 Q	2121·0 R		2, 2 R	2094·6 R	7	2, 1 Q
2144·3 R	8	1, 2 Q	2116·6 R	10	1, 1 Q	2094·0 R		2, 1 R
2142·7 R		1, 2 R	2115·4 R		1, 1 R	2089·4 R	6	1, 0 Q
2140·3 V	7	0, 1 Q	2112·4 R	10	0, 0 Q	2067·9 R	4	2, 0 Q

GaH

Occurrence. In absorption and emission in King furnace.
References. H. Neuhaus, *Ark. Fys.*, **14**, 551 (1959).
M. L. Ginter and K. K. Innes, *J. Mol. Spec.*, 7, 64 (1961)†.
P. C. Poynor, K. K. Innes and M. L. Ginter, *J. Mol. Spec.*, **23**, 237 (1967).

$a\,^3\Pi_0 - X\,^1\Sigma^+$ BAND

Slightly degraded to the violet.

λ *origin*	v', v''
5763·5	0, 0

$a\,^3\Pi_1 - X\,^1\Sigma^+$ BAND

Shaded to the violet; double-headed.

λ	v', v''
5697·7	0, 0 P
5671·3	0, 0 Q

$a\,^3\Pi_2 - X\,^1\Sigma^+$

A weak (0, 0) band at 5582·1, largely masked by the $^3\Pi_1 - X\,^1\Sigma$ transition.

$a\,^3\Pi_{0^-} - X\,^1\Sigma^+$

Weak bands, of a forbidden transition, with violet-degraded Q branches only. Origins 5844·8 (2, 2), 5765·3 (0, 0), 5308·4 (1, 0).

$A\,^1\Pi - X\,^1\Sigma^+$

Weak diffuse red-degraded double-headed bands. R heads 4826·5, 4502·0 and 4208·6.

Bands with open structure in the region 2400–2160 Å have also been reported by W. R. S. Garton (*Proc. Phys. Soc.*, A, **64**, 509 (1951)).

GaI

Occurrence. In high-frequency discharges and in absorption.
Reference. E. Miescher and M. Wehrli, *Helv. Phys. Acta*, 7, 331 (1934)†.
There is a $^3\Pi - X\,^1\Sigma$ system 4180–3810 Å. It is very weakly degraded but most heads shade to the violet. The $A\,^3\Pi_0 - X\,^1\Sigma^+$ sub-system has prominent P heads 3944·1 (0, 1), 3911·4 (0, 0), 3892·9 (3, 2), 3887·5 (2, 1), 3882·8 (1, 0). The $B\,^3\Pi_1 - X\,^1\Sigma^+$ sub-system shows bands of the (0, 0) sequence from 3880·4 to 3862·2 Å. There is a continuum near 3065 Å associated with a $C\,^1\Pi - X\,^1\Sigma$ transition.

GaO

Occurrence. Gallium in copper arc in air, and in a heavy-current discharge. Also in absorption.

References. M. L. Guernsey, *P.R.*, **46**, 114 (1934)†.

V. Raziunas, G. J. Macur and S. Katz, *J. Chem. Phys.*, **39**, 1161 (1963).

The bands belong to a $^2\Sigma - {}^2\Sigma$ transition, probably involving the ground state.

(0, 1) sequence degraded to red.

λ	I	v', v''
4006·9	7	0, 1
4004·8	8	1, 2 etc.

(0, 0) sequence degraded only slightly to red.

λ	I	v', v''
3889·3	10	0, 0

(1, 0) sequence degraded to violet.

λ	I	v', v''
3778·5	9	1, 0
3779·4	9	2, 1

GdO

Occurrence. Gadolinium chloride in oxy-hydrogen flame or Gd salts in arc.

Appearance. A large number of systems, mostly with well-developed sequences, usually degraded to the red. The long-wave end of the spectrum is rather confused.

References. G. Piccardi, *Gazz. Chim. Ital.*, **63**, 887 (1933)†.

C. B. Suárez and R. Grinfeld, *J. Chem. Phys.*, **53**, 1110 (1970).

The heads have been arranged into ten systems by Piccardi. Suárez and Grinfeld (who list about 200 heads, without intensities) confirm many of these systems, but have found a new system α and have reanalysed III, VI, VII and VIII into only three systems denoted β, γ and δ. System V is doubtful. The following wavelengths are from Suárez and Grinfeld; our selection of prominent heads and estimates of intensity, where given, are based on Piccardi's photograph. All heads are degraded to the red except the three violet-degraded heads indicated by V.

λ	I	System	λ	I	System	λ	I	System
6220·8	6	IX 1, 1	5807·6V	9	IV 0, 0	4816·6	5	I 1, 2
6211·6	4	X 0, 0	5698·6	6	β 1, 1	4798·5	5	I 0, 1
6200·8	7	IX 0, 0	5680·8	8	β 0, 0	4633·3	8	I 1, 1
6182·6	–	δ 0, 0	5664·3	–	α 0, 0	4615·6	10	I 0, 0
5926·5V	–	γ 0, 0	4909·8	6	II 1, 1	4480·5	4	I 2, 1
5819·1V	9	IV 1, 1	4892·2	7	II 0, 0	4462·6	4	I 1, 0

GeBr

VISIBLE SYSTEM, $A^2\Sigma - X^2\Pi_{\frac{1}{2}}$

Occurrence. High-frequency discharge.

Appearance. About 40 red-degraded bands, arranged rather irregularly.

Reference. K. B. Rao and P. B. V. Haranath, *J. Phys.*, B, **2**, 1385 (1969)†.

No intensities are available but the following appear strong in the photograph:

GeBr (*contd.*)

λ	v', v''		λ	v', v''
5292·6	0, 7		4943·1	1, 3
5163·6	1, 6		4641·9	5, 1
5114·6	2, 6		4604·5	6, 1
4989·6	0, 3		4567·5	7, 1

VIOLET SYSTEM, $B^2\Delta - X^2\Pi$

Occurrence. Oscillator discharge through $GeBr_4$ vapour.

Appearance. Degraded to the red. Two sub-systems.

Reference. E. B. Andrews and R. F. Barrow, *Proc. Phys. Soc.*, A, **63**, 957 (1950)†.

Strongest heads:

λ	I	v', v''	λ	I	v', v''	λ	I	v', v''
3664·9	4	1, 0 i	3810·4	7	1, 0 ii	3898·0	6	1, 2 ii
3688·7	6	0, 0 i	3829·1	6	2, 1 ii	3926·1	8	0, 2 ii
3728·9	5	0, 1 i	3837·5	10	0, 0 ii	3942·9	8	1, 3 ii
3770·3	4	0, 2 i	3881·2	10	0, 1 ii	3971·5	4	0, 3 ii

NEAR ULTRA-VIOLET SYSTEM, $A'^2\Sigma^+ - X^2\Pi$

Occurrence. Discharge through $GeBr_4$ vapour and in absorption following flash photolysis of the vapour.

Appearance. Degraded to shorter wavelengths. Bands 3259–2946 Å.

Reference. W. Jevons, L. A. Bashford and H. V. A. Briscoe, *Proc. Phys. Soc.*, **49**, 532 (1937).

Strongest bands $\lambda\lambda$3068·8 (0, 3), 3041·3 (0, 2), 3014·7 (0, 1), 3006·4 (1, 2), 2988·0 (0, 0), 2980·2 (1, 1), 2954·2 (1, 0).

FAR ULTRA-VIOLET SYSTEMS

Occurrence. Absorption following flash photolysis of $GeBr_4$.

Reference. G. A. Oldershaw and K. Robinson, *Trans. Faraday Soc.*, **67**, 2499 (1971)†.

Five overlapping systems, all degraded to shorter wavelengths. The following are relatively strong:

λ	System	v', v''	λ	System	v', v''	λ	System	v', v''
2413·0	C i	1, 0	2250·2	D	0, 0	2195·1	E	2, 0
2392·8	C i	2, 0	2229·5	E	0, 0	2175·8	F	0, 0
2373·5	C i	3, 0	2211·9	E	1, 0	2101·2	G	0, 0
2352·0	C i	4, 0						

GeCl

VISIBLE SYSTEM, 6360–3800 Å

Occurrence. In high-frequency and Schuler-type discharges.

References. P. Deschamps, A. F. Robert and G. Pannetier, *J. Chim. Phys.*, **65**, 1084 (1968)†. K. B. Rao and P. B. V. Haranath, *J. Phys.*, B, **2**, 1080 (1969)†.

Analyses differ, and wavelengths by Deschamps *et al.* are systematically greater than those by Rao and Haranath by up to 2 Å. They agree in attributing bands to two sub-systems of a transition to the ground $^2\Pi$ state. About 100 heads are listed in these papers. They are degraded to the red.

GeCl (*contd.*)

The following may be relatively strong and serve for identification:
λλ6208, 6066, 5928, 5851, 5724, 5602, 5484, 5371, 5091, 5029, 4931, 4503, 4454, 4422, 4376, 4329, 4242, 4159.

3216–2848 Å SYSTEM, $B^2\Sigma - X^2\Pi$

Occurrence. Discharge through $GeCl_4$ and also absorption following flash photolysis.
Reference. W. Jevons, L. A. Bashford and H. V. A. Briscoe, *Proc. Phys. Soc.*, **49**, 532 (1937).
This is probably the strongest system. Bands are degraded to shorter wavelengths. Strongest heads; first sub-system 3098·0 (0, 2), 3059·8 (0, 1), 3022·4 (0, 0); second sub-system 3007·0 (0, 2), 2971·2 (0, 1), 2936·0 (0, 0), 2891·2 (1, 0).
There is also continuous emission 2660–2510 Å.

3500 Å SYSTEM, $A'^2\Delta - X^2\Pi$

Occurrence. In discharge through $GeCl_4$ and when excited by active nitrogen.
Reference. R. F. Barrow and A. Lagerqvist, *Ark. Fys.*, **1**, 221 (1949)†.
Two sub-systems with red-degraded sequence heads at 3500 and 3392, but some individual heads degrade the other way. Main heads with direction of degradation:

λ	I	v', v''	λ	I	v', v''
3500·3	7 V	0, 0 ii	3386·1	3 M	?
3501·5	10 R	0, 0 ii	3392·1	5 V	0, 0 i
3511·0	9 R	1, 1 ii	3392·7	9 R	0, 0 i
3521·1	7 R	2, 2 ii	3401·1	5 R	1, 1 i
3571·0	4 R	2, 3 ii	3440·0	2 R	0, 1 i

ULTRA-VIOLET SYSTEMS, $C - X^2\Pi$ AND $C' - X^2\Pi$

Occurrence. Absorption following flash photolysis of $GeCl_4$.
Reference. G. A. Oldershaw and K. Robinson, *Trans. Faraday Soc.*, **66**, 532 (1970)†.
All bands are degraded to shorter wavelengths. Main heads:

$C - X^2\Pi_{\frac{3}{2}}$		$C - X^2\Pi_{\frac{1}{2}}$		$C' - X^2\Pi$	
λ	v', v''	λ	v', v''	λ	v', v''
2201·8	1, 0	2179·5	0, 0	2339·6	1, 0
		2156·5	1, 0	2312·9	2, 0
		2134·2	2, 0	2287·2	3, 0

There is also a band at 2102 Å.

$GeCl_2$

J. W. Hastie, R. Hauge and J. L. Margrave (*J. Mol. Spec.*, **29**, 152 (1969)) report 30 bands, 3301–3140 Å, in absorption and in emission in a microwave discharge. A. Tewarson and H. B. Palmer (*J. Mol. Spec.*, **22**, 117 (1967)†) and C. M. Pathak and H. B. Palmer (*J. Mol. Spec.*, **31**, 170 (1969)) report complex band systems 6660–5600 and 4900–4100 Å in a flame of $GeCl_4$ reacting with potassium.

GeF

Strong systems lie in the visible and near ultra-violet; six weaker systems, between 2400 and 1900 Å, and two intercombination systems in the infra-red have been reported.

Occurrence. In discharge through GeF_4 vapour, or in hollow cathode discharge with GeF_4 in helium.

References. E. B. Andrews and R. F. Barrow, *Proc. Phys. Soc.*, A, **63**, 185 (1950)†.
R. F. Barrow, D. Butler, J. W. C. Johns and J. L. Powell, *Proc. Phys. Soc.*, **73**, 317 (1959).
R. W. Martin and A. J. Merer, *Canad. J. Phys.*, **51**, 125 (1973)†; **52**, 1458 (1974)†.

VISIBLE SYSTEM, λλ5200–3900, $A^2\Sigma^+ - X^2\Pi$

Appearance. Two sub-systems of close double-headed bands, degraded to the red. Strongest bands:

λ	I	v', v''	λ	I	v', v''	λ	I	v', v''
4773·8	6	0, 2 i	4436·8	9	0, 1 ii	4256·8	4	3, 0 i
4628·9	8	0, 1 i	4410·7	7	1, 0 i	4235·5	7	1, 0 ii
4569·3	6	0, 2 ii	4332·0	6	2, 0 i	4163·0	6	2, 0 ii
4492·0	7	0, 0 i	4310·7	8	0, 0 i	4094·0	4	3, 0 ii

NEAR ULTRA-VIOLET SYSTEM, λλ3100–2600, $B^2\Sigma^+ - X^2\Pi$

Appearance. Two sub-systems of close double-headed bands. Degraded to shorter wave-lengths. Strongest bands:

λ	I	v', v''	λ	I	v', v''
3045·7	4	0, 2 ii	2904·9	6	0, 1 i
2986·1	6	0, 1 ii	2862·0	5	1, 0 ii
2974·1	4	1, 2 ii	2850·2	9	0, 0 i
2961·2	4	0, 2 i	2787·4	6	1, 0 i
2928·3	10	0, 0 ii	2777·7	4	2, 1 i

$a^4\Sigma^- - X^2\Pi$ SYSTEM

A weak system underlies the above stronger one. Martin and Merer have analysed it. Red-degraded heads of the (1, 0) band are at 2867·3 $^PQ_{23}$, 2865·0 Q_{33} and 2792·4 Q_{22}.

FAR ULTRA-VIOLET SYSTEMS

All bands are degraded to shorter wavelengths and systems are double because of the splitting in the $X^2\Pi$ lower level. The following are the two (0, 0) bands of each:

$C' - X$	2349·5	2301·0	$E - X$	2184·7	2142·7
$C - X$	2321·4	2273·5	$F - X$	2154·9	2112·7
$D - X$	2263·5	2218·4	$G - X$	2060·5	2023·4

INFRA-RED

Barrow *et al.* report weak intercombination systems $B^2\Sigma - A^2\Sigma$ between 8500 and 7400 Å and $E^2\Pi - B^2\Sigma$ between 8700 and 8550; there is a head at 8626, degraded to longer wavelengths, and a line-like maximum at 8629 Å.

GeF$_2$

R. Hauge, V. M. Khanna and J. L. Margrave (*J. Mol. Spec.*, **27**, 143 (1968)) report 27 red-degraded bands of a complex system 2424–2225 Å in absorption.

GeH

Three systems have been reported.

RED SYSTEM, $^4\Sigma - X\,^2\Pi$

Occurrence. In emission from King furnace.
Appearance. Complex headless structure, with apparent origins at 6504 and 6154 Å; these are believed to be the two (0, 0) sub-bands.
Reference. B. Kleman and E. Werhagen, *Ark. Fys.*, **6**, 399 (1953).

VIOLET SYSTEM, $^2\Delta - X\,^2\Pi$

Occurrence. In emission from King furnace.
Appearance. Degraded to the red. Rather complex structure.
References. B. L. Kleman and E. Werhagen, *Ark. Fys.*, **6**, 359 (1953)†.
L. Klynning and B. Lindgren, *Ark. Fys.*, **32**, 575 (1966)†.

The following heads have been observed:

λ	I	v', v''		λ	I	v', v''
3718·6	1	1, 0 $^SR_{21}$		3874·4	5	0, 0 $^SR_{21}$
3727·8	2	1, 0 R_1		3897·6	10	0, 0 R_1
3728·8	2	1, 0 $^RQ_{21}$		4015·3	6	0, 0 R_2
3845·5	1	1, 0 R_2		4037·2	10	0, 0 $^QR_{12}$
3854·4	2	1, 0 $^QR_{12}$		4135·7	1	1, 1 R_1
				4147·7	2	1, 1 $^QR_{12}$

ULTRA-VIOLET SYSTEM

Occurrence. In absorption in King furnace.
Appearance. Degraded to the red.
Transition. Probably $^2\Delta - X\,^2\Pi$ or $^2\Sigma - X\,^2\Pi$.
Reference. R. F. Barrow, G. Drummond and W. R. S. Garton, *Proc. Phys. Soc.*, A, **66**, 191 (1953).

The following provisional assignments of heads have been made:

λ	I	v', v''		λ	I	v', v''
2454·2	8	0, 0 ^SR		2519·0	10	0, 0 Q_2
2464·0	10	0, 0 R_1		2565·9	2	0, 1 ^SR
2510·5	9	0, 0 R_2		2580·1	4	0, 1 R_1

GeI

References. G. A. Oldershaw and K. Robinson, *Trans. Faraday Soc.*, **64**, 2256 (1968)†.
A. Chatalic, P. Deschamps and G. Pannetier, *J. Chim. Phys.*, **67**, 1567 (1970).

VISIBLE SYSTEM, $A^2\Sigma - X^2\Pi_{\frac{1}{2}}$

About 45 red-degraded bands, 6000–4500 Å, were obtained in a high-frequency discharge by Chatalic *et al.*; no details are available.

GeI (*contd.*)

ULTRA-VIOLET SYSTEM, $A^2\Sigma - X^2\Pi$

About 25 bands, degraded to shorter wavelengths and forming a prominent $v'' = 0$ progression were obtained in absorption following flash photolysis of GeI_4. Analyses in above papers differ and values of v' are uncertain. Data from Oldershaw and Robinson:

λ	v', v''		λ	v', v''
3059·2	2, 0		2976·5	5, 0
3031·0	3, 0		2950·0	6, 0
3003·3	4, 0		2924·3	7, 0

GeO

The main system lies in the near ultra-violet. There is a weaker system in the blue-green and two others in the far ultra-violet.

Occurrence. The main and ultra-violet systems have been found in arcs and discharges through $GeCl_4 + O_2$ and in absorption. The blue-green system occurs in emission in a furnace.

References. W. Jevons, L. A. Bashford and H. V. A. Briscoe, *Proc. Phys. Soc.*, **49**, 543 (1937).
G. Drummond and R. F. Barrow, *Proc. Phys. Soc.*, A, **65**, 277 (1952)†.
R. F. Barrow and H. C. Rowlinson, *P.R.S.*, **224**, 374 (1954)†.
D. P. Tewari and H. Mohan, *J. Mol. Spec.*, **39**, 290 (1971).
A. A. N. Murty, Y. P. Reddy and P. T. Rao, *Indian J. Pure Appl. Phys.*, **10**, 834 (1972)†.

MAIN SYSTEM, $\lambda\lambda 3319-2261$

Appearance. Degraded to the red. The $v' = 0$ progression is prominent in emission, and the $v'' = 0$ in absorption.

Transition. $D\,^1\Pi - X\,^1\Sigma$, more recently called $A\,^1\Pi - X\,^1\Sigma$.

The following are the R heads of strong bands, in emission:

λ	I	v', v''	λ	I	v', v''	λ	I	v', v''
3048·8	5	0, 5	2804·2	10	0, 2	2614·4	5	1, 0
2989·9	5	1, 5	2779·7	6	2, 3	2571·8	5	2, 0
2963·0	7	0, 4	2730·0	8	0, 1	2531·1	5	3, 0
2908·1	8	1, 4	2683·0	8	1, 1	2492·2	5	4, 0
2881·7	10	0, 3	2659·4	3	0, 0			

BLUE-GREEN SYSTEM, $B - X^1\Sigma$

Tewari and Mohan report 16 red-degraded bands. Strongest:

λ	I	v', v
5014·8	4	0, 1
4779·4	5	0, 0
4741·3	3	2, 1
4553·3	4	1, 0

ULTRA-VIOLET SYSTEMS

Murty *et al.* report 21 red-degraded bands of an $E - X^1\Sigma$ system in a radio-frequency discharge. Strongest:

GeO (*contd.*)

λ	I	v', v''	λ	I	v', v''	λ	I	v', v''
2362·5	5	1, 8	2264·0	7	1, 6	2195·3	5	0, 4
2312·6	5	1, 7	2241·8	7	0, 5	2171·7	6	1, 4
2289·8	6	0, 6	2217·2	7	1, 5	2127·6	7	1, 3

In absorption, bands at 2065·2 (3, 2), 2024·4 (3, 1) and 2004·5 (4, 1) are strong. Another system $F - X$ is also obtained in absorption 1532–1434 Å.

GeS

Two band systems, both degraded to the red, have been observed in absorption by Shapiro, Gibbs and Laubengayer. System A has also been obtained in emission by Barrow from a heavy-current uncondensed discharge through a mixture of sulphur, germanium oxide and aluminium.

References. C. V. Shapiro, R. C. Gibbs and A. W. Laubengayer, *P.R.*, **40**, 354 (1932).
R. F. Barrow, *Proc. Phys. Soc.*, **53**, 116 (1941)†.
G. Drummond and R. F. Barrow, *Proc. Phys. Soc.*, A, **65**, 277 (1952)†.

SYSTEM A, $\lambda\lambda 3750$–2709, $D - X\,^1\Sigma$

Appearance. Degraded to the red. The bands are evenly spaced and form short sequences. Heads due to $Ge\,^{70}S$, $Ge\,^{72}S$, and $Ge\,^{76}S$ as well as those due to the most abundant molecule $Ge\,^{74}S$, were observed.

Strongest bands:

λ	I	v', v''	λ	I	v', v''	λ	I	v', v''
3574·1	e	2, 10	3275·3	3	0, 4	3014·4	7	1, 0
3506·3	e	2, 9	3216·1	4	0, 3	2982·1	2	2, 0
3485·2	e	1, 8	3158·8	3	0, 2	2949·3	7	3, 0
3440·9	e	2, 8	3103·0	6	0, 1	2917·9	6	4, 0
3419·8	e	1, 7	3067·5	8	1, 1	2887·6	7	5, 0
3356·6	e	1, 6	3048·9	5	0, 0	2858·2	4	6, 0
3336·3	1	0, 5	3033·1	8	2, 1			

The letter e in the intensity column indicates that the band has been observed only in emission.

SYSTEM B, $\lambda\lambda 2782$–2146, $E - X\,^1\Sigma$

Appearance. Degraded to the red. A very long $v'' = 0$ progression is the main feature in absorption.

The following are among the strongest bands; these wavelengths, by Drummond and Barrow, are about 1 Å greater than those of Shapiro *et al.*

λ	v', v''	λ	v', v''	λ	v', v''
2658·5	0, 2	2539·6	2, 0	2414·0	9, 0
2654·6	2, 3	2520·4	3, 0	2397·3	10, 0
2637·2	1, 2	2501·5	4, 0	2381·4	11, 0
2616·0	2, 2	2483·0	5, 0	2366·4	12, 0
2597·6	1, 1	2465·3	6, 0	2351·4	13, 0
2577·7	2, 1	2448·0	7, 0	2336·5	14, 0
2557·5	3, 1	2430·8	8, 0	2321·9	15, 0

GeSe

Two systems, denoted $D - X$ and $E - X$ have been observed in absorption; the former has been obtained in emission in a discharge tube, as well.

Reference. G. Drummond and R. F. Barrow, *Proc. Phys. Soc.*, A, **65**, 277 (1952)†.

D − X SYSTEM, $\lambda\lambda 3411-2954$

Appearance. Degraded to the red. Long $v' = 0$ and $v'' = 0$ progressions.
Strongest absorption bands:

λ	I	v', v''	λ	I	v', v''	λ	I	v', v''
3472·2	1*	0, 5	3306·4	7*	1, 2	3220·5	8	1, 0
3427·2	2*	0, 4	3291·7	6	0, 1	3193·1	10	2, 0
3380·9	2*	0, 3	3262·9	9	1, 1	3166·4	9	3, 0
3336·0	4*	0, 2	3234·9	10	2, 1	3140·3	8	4, 0

* Relatively strong in emission.

E − X SYSTEM, $\lambda\lambda 2995-2685$

Appearance. Degraded to the red.
Strongest bands: $\lambda\lambda 2822\cdot8$ (4, 2), 2807·8 (3, 1), 2806·6 (5, 2), 2791·3 (4, 1), 2775·3 (5, 1), 2760·0 (4, 0), 2759·9 (6, 1), 2744·8 (5, 0), 2744·7 (7, 1), 2729·5 (6, 0).

GeTe

Occurrence. Two systems, denoted $D - X$ and $E - X$ have been observed in absorption. The former also occurs in emission from a discharge containing GeO_2, Al and Te.

References. R. F. Barrow and W. Jevons, *Proc. Phys. Soc.*, **52**, 534 (1940)†.
G. Drummond and R. F. Barrow, *Proc. Phys. Soc.*, A, **65**, 277 (1952)†.

D − X SYSTEM, $\lambda\lambda 3829-3267$

Appearance. Degraded to the red. Long $v'' = 0$ and $v' = 0$ progressions.
The following are prominent heads, with intensities for absorption:

λ	I	v', v''	λ	I	v', v''	λ	I	v', v''
3739·2	5	0, 3	3622·1	8	1, 1	3553·0	9	2, 0
3694·7	6	0, 2	3609·1	5	0, 0	3526·2	9	3, 0
3664·9	7	1, 2	3594·0	8	2, 1	3499·9	10	4, 0
3651·1	6	0, 1	3580·5	7	1, 0	3474·2	9	5, 0

E − X SYSTEM, $\lambda\lambda 3354-2827$

Appearance. Degraded to the red. Very long progressions. There is still some doubt about the complete vibrational analysis. The following heads appear prominent on the photograph: $\lambda\lambda 3232\cdot3$ (1, 2), 3216·2 (0, 1), 3199·6 (1, 1), 3183·7 (0, 0), 3167·5 (1, 0), 3152·5 (2, 0), 3136·0 (3, 0), 3121·0 (4, 0).

H_2

Occurrence. In discharge tubes containing hydrogen or water vapour. The system frequently appears when a discharge tube is first evacuated, especially if it has metal electrodes which occlude hydrogen.

H₂ *(contd.)*

Appearance. The hydrogen molecular spectrum, or 'secondary' spectrum of hydrogen as it is often called, has few characteristic features, as the rotational structure is so open that there are no heads or close groups of lines to form anything resembling the usual band structure. The system is strongest in the orange but extends throughout the whole visible spectrum. Identification is rendered easier by the almost invariable presence of the strong hydrogen atomic lines, H_α 6562·79, H_β 4861·33, H_γ 4340·47 Å. See Plate 4.

References. H. G. Gale, G. S. Monk and K. O. Lee, *Astrophys. J.*, **67**, 89 (1928).

O. W. Richardson, *Molecular Hydrogen and its spectrum*, Oxford Univ. Press, 1934.

E. W. Foster and O. W. Richardson, *P.R.S.*, **217**, 433 (1953).

G. H. Dieke, *J. Mol. Spec.*, **2**, 494 (1958).

S. P. S. Porto and N. Januzzi, *J. Mol. Spec.*, **11**, 378 (1963).

The following are the strongest lines of the spectrum, listed as intensities, 8, 0, or 10 by Gale, Monk and Lee.

λ	λ	λ	λ	λ	λ
8349·52	6135·39	5878·50	5505·52	5039·82	4709·54
8273·26	6121·79	5849·32	5499·58	5030·37	4683·82
8164·64	6098·22	5836·13	5495·96	5015·07	4662·81
7524·64	6095·96	5822·76	5481·08	5013·04	4661·40
7253·28	6080·78	5812·59	5459·60	5011·19	4631·85
7195·66	6069·99	5806·10	5434·82	5007·99	4627·99
7168·81	6063·28	5775·05	5425·89	5003·40	4617·53
6428·11	6031·90	5736·88	5419·89	4973·31	4582·59
6399·47	6027·98	5731·92	5401·05	4934·24	4579·99
6327·06	6021·27	5728·55	5388·17	4928·79	4575·88
6299·42	6018·29	5689·19	5303·10	4928·64	4572·71
6285·39	6002·82	5655·75	5291·60	4873·01	4568·13
6238·39	5994·06	5634·81	5272·30	4856·55	4554·16
6224·81	5975·44	5612·54	5266·04	4849·30	4498·11
6201·18	5949·89	5597·64	5196·37	4822·94	4212·50
6199·39	5938·62	5552·53	5084·84	4763·84	4205·10
6182·99	5931·37	5537·47	5055·09	4723·03	4177·12
6161·60	5888·17	5518·47	5041·63	4719·04	4171·31
					4069·63
					4066·88

Lines of these triplet systems, with a different intensity distribution, have also been obtained in absorption following a flash discharge through hydrogen by G. Herzberg, *Science of Light*, **16**, 14 (1963)†

HBr

Reference. J. G. Stamper and R. F. Barrow, *J. Phys. Chem.*, **65**, 250 (1961).

A many-line spectrum 2175–2080 Å with maximum around 2150 has been observed in a radio-frequency discharge. Branches have been picked out but there are no clear heads and no full analysis has been made.

HBr⁺

Occurrence. High-frequency electrodeless or hollow-cathode discharge through HBr gas.
Appearance. Open rotational structure. Degraded to the red.
Transition. $A\,^2\Sigma - X\,^2\Pi$.
References. F. Norling, *Z.P.*, **95**, 179 (1935).
R. F. Barrow and A. D. Caunt, *Proc. Phys. Soc.*, A, **66**, 617 (1953).

It is rather difficult to extract the data from the publications. The following are the probable heads and assignments. Norling observed the strong (0, 0) and (1, 0) bands and Barrow and Caunt the weaker (0, 1) and (1, 1).

λ	v', v''	λ	v', v''	λ	v', v''
3269·3	1, 0 $^Q R_{12}$	3581·5	1, 0 $^S R_{21}$	3761·0	0, 0 Q_2
3271·8	1, 0 Q_2	3582·0	1, 0 R_1	3910·1	1, 1 $^Q R_{12}$
3417·6	0, 0 $^S R_{21}$	3710·6	0, 1 $^S R_{21}$	3910·9	1, 1 Q_2
3420·5	0, 0 R_1	3716·1	0, 1 R_1	4124·3	0, 1 $^Q R_{12}$
3537·1	1, 1 $^S R_{21}$	3760·3	0, 0 $^Q R_{12}$	4125·2	0, 1 Q_2
3542	1, 1 R_1				

HCCl see CHCl

HCF see CHF

HCN

Reference. L. Lindholm, *Z.P.*, **108**, 454 (1938).

The infra-red vibration-rotation absorption spectrum of hydrocyanic acid extends into the visible. There are weak bands with centres at 8563 and 7912 (with a possible head at 7880), and very weak ones at 8608, 7961, 6814 and 6428 Å.

HCO see CHO

HCOF see CHOF

HCP

Reference. J. W. C. Johns, H. F. Shurvell and J. K. Tyler, *Canad. J. Phys.*, **47**, 893 (1969)†.

About 80 absorption bands, 4100–2350 Å, have been assigned to 7 systems. They are degraded to the red with strongest R heads at λλ2825·7, 2785·5, 2751·7, 2740·1, 2711·2, 2709·4, 2670·8, 2641·2.

HCl

VIBRATION-ROTATION BANDS

Reference. E. Lindholm, *Ark. Mat. Astr. Fys.*, B, **29**, No. 15 (1943).

With long absorption paths the far infra-red vibration-rotation bands extend into the photographic region. The bands may also occur in emission from flames. Heads of the (4, 0) and (5, 0) bands lie at 9152 and 7463 Å (degraded to longer wavelengths).

HCl (*contd.*)

ULTRA-VIOLET λλ2375−1980

Reference.   J. K. Jacques and R. F. Barrow, *Proc. Phys. Soc.*, **73**, 538 (1959).

A 'many-line' system of headless bands occurs in a hollow cathode discharge. The transition is $V\,{}^1\Sigma - X\,{}^1\Sigma$ but no data are available.

HCl⁺

References.   M. Kulp, *Z.P.*, **67**, 7 (1931).

W. D. Shearley and C. W. Mathews, *J. Mol. Spec.*, **47**, 420 (1973).

3500 Å SYSTEM, $A^2\Sigma - X^2\Pi$

Occurrence.   In discharges, especially hollow-cathode and microwave.

Appearance.   Bands are double and possess P, Q and R branches degraded to the red. The strongest bands form a $v'' = 0$ progression.

Main Q heads and intensities from Shearley and Mathews. R heads from Kulp.

		$^2\Sigma - {}^2\Pi_{\frac{1}{2}}$		$^2\Sigma - {}^2\Pi_{\frac{3}{2}}$	
v', v''	*I*	*R heads*	*Q heads*	*R heads*	*Q heads*
0, 1	4	3955·7	3964·7	3853·6	3863·3
0, 0	5	3591·6	3885·5	3507·3	3514·5
1, 0	7	3405·8	3410·9	3330·0	3335·6
3, 1	4		3380·9		3307·2
2, 0	10	3245·9	3250·0	3177·3	3181·7
3, 0	6		3110·8		3048·2
4, 0	4		2989·5		2931·7
5, 0	2		2883·3		2829·6

Continua.   Kulp records also a weak continuum with a maximum of intensity at 3000 Å and a second stronger continuum with a maximum at 2570−2580 Å..

HF

VIBRATION-ROTATION BANDS

Reference.   S. M. Naudé and H. Venlager, *Proc. Phys. Soc.*, A, **63**, 470 (1950).

The open-structured vibration-rotation bands in the far infra-red extend into the photographic region with very long absorption paths. The (3, 0) and (4, 0) heads lie at λλ8667 and 6686·4. The (3, 0) band has been obtained in emission from flames.

ULTRA-VIOLET SYSTEM, 2670−2000, $B\,{}^1\Sigma^+ - X\,{}^1\Sigma^+$

Occurrence.   In hollow-cathode discharge, especially with helium as carrier.

Appearance.   This is a 'many-line' system, without heads. The rotational structure is strongly degraded to the red.

References.   J. W. C. Johns and R. F. Barrow, *P.R.S.*, **251**, 504 (1959).

G. Di Lonardo and A. E. Douglas, *Canad. J. Phys.*, **51**, 434 (1973).

There are no outstanding features. The following are origins of the strongest bands:

HF (*contd.*)

λ origin	v', v''	λ origin	v', v''	λ origin	v', v''
2553·2	1, 17	2353·8	0, 14	2198·4	1, 13
2544·3	0, 16	2293·1	1, 14	2154·5	0, 12
2473·6	1, 16	2254·1	0, 13	2103·5	1, 12
2451·5	0, 15				

HNCO, Isocyanic acid see CHNO

HF⁺

Reference. S. Gewurtz, H. Lew and P. Flainek, *Canad. J. Phys.*, **53**, 1097 (1975).

Bands 4830–3580 Å have been observed in a low-pressure discharge. They are strongly degraded to the red, with weak heads close to the origins. Origins,

λ_0	v', v''	λ_0	v', v''	λ_0	v', v''
4682·5	2, 2	4340·7	1, 1	3703·0	2, 0
4599·3	0, 1	4055·8	0, 0	3600·9	3, 0
4520·3	3, 2	3853·3	1, 0		

HNF

Reference. C. M. Woodman, *J. Mol. Spec.*, **33**, 311 (1970)†.

A progression of 7 complex bands 5000–3800 Å has been found in absorption following flash photolysis of HNF_2. Bands have many heads; the $^PQ_1$ (degraded to the violet) and $^RQ_0$ (shaded to the red) are relatively prominent. Strongest bands:

$(v_1$	v_2	$v_3)'$,	$(v_1$	v_1	$v_3)''$	$^PQ_1$	$^RQ_0$
0	0	0	0	0	0	4967·6	4957·0
0	1	0	0	0	0	4715·9	4707·0
0	2	0	0	0	0	4492·7	4482·1
0	3	0	0	0	0	4293·5	4282·9

HNO

FAR RED SYSTEM

References. F. W. Dalby, *Canad. J. Phys.*, **36**, 1336 (1958)†.
M. J. Y. Clement and D. A. Ramsay, *Canad. J. Phys.*, **39**, 205 (1961)†.
J. L. Bancroft, J. M. Hollas and D. A. Ramsay, *Canad. J. Phys.*, **40**, 322 (1962)†.
P. N. Clough *et al.*, *Chem. Phys. Lett.*, **23**, 155 (1973).

In absorption following flash photolysis of various mixtures and of nitro compounds, complex bands, with some sharp red-degraded heads, are found. The (0, 0, 0-0, 0, 0) is strongest and covers the region 7718–7341 Å, with prominent heads λλ7379·3, 7416·4, 7452·7, 7520 and 7542. There are weaker bands 7095–7061 and 6860–6715.

In emission in the H + NO reaction, the (0, 0, 0-0, 0, 0) band is again strong with main maximum around 7625 Å. Less strong bands are centred around 9120±40, 8650±80, 7965±70 and 6925±60. The earlier analysis may require revision (Clough *et al.*).

FAR ULTRA-VIOLET BANDS

Reference. A. B. Callear and P. M. Wood, *Trans. Faraday Soc.*, **67**, 3399 (1971)†.

HNO (*contd.*)

In the flash photolysis of H_2 + NO and other substances, bands have been observed with centres at 2072 (I = 10), 2042 (–), 2001 (2) and 1980 (3).

HNO_2

Occurrence. In absorption by mixed NO, N_2O and H_2O. In self absorption by explosive flames of CO, N_2O and NO.

Appearance. A fairly regular system of narrow diffuse bands.

References. E. H. Melvin and O. R. Wulf, *J. Chem. Phys.*, **3**, 755 (1935)†.

D. M. Newitt and L. E. Outridge, *J. Chem. Phys.*, **6**, 752 (1938).

G. Porter, *J. Chem. Phys.*, **19**, 1278 (1951)†.

The following table of band maxima is from Porter. Some of the bands found by Newitt and Outridge are due to NO_2.

λ	I	λ	I	λ	I
3844	3	3417	7	3277	2
3680	9	3387·5	4	3203	0
3542·5	10	3306	3	3177	0
3509	2				

H_2O

VIBRATION-ROTATION SPECTRUM

References. W. Baumann and R. Mecke, *Z.P.*, **81**, 445 (1933)†.

K. Freudenberg and R. Mecke, *Z.P.*, **81**, 465 (1933).

T. Kitagawa, *Proc. Imp. Acad. Tokyo*, **12**, 281 (1936)†.

A. G. Gaydon, *P.R.S.*, **181**, 197 (1942)†.

A. G. Gaydon, *The Spectroscopy of Flames*, Chapman and Hall, London, (1974).

The fundamentals at 1·87, 2·66 and 6·26 μm are responsible for absorption and also for the main emission from flames in the infra-red. Weaker overtone bands at shorter wavelengths, even into the visible region, are responsible for absorption over long path-lengths, e.g. in atmospheric absorption of the solar spectrum, and also for emission from hot flames such as oxy-hydrogen; they occur especially in diffusion flames of O_2 burning in an atmosphere of H_2.

In absorption, heads are not prominent, but in emission, red-degraded heads may be observed, especially for bands in the infra-red; in the visible, intensity maxima occur due to bunching of lines of complex rotational structure.

The following are the origins (λ_0) and heads (λ_h) of the main bands <10 000 Å.

λ_0	λ_h	$(\nu_1\ \nu_2\ \nu_3)'$			$(\nu_1\ \nu_2\ \nu_3)''$			λ_0	λ_h	$(\nu_1\ \nu_2\ \nu_3)'$			$(\nu_1\ \nu_2\ \nu_3)''$		
9419·6	9277	2	0	1	0	0	0	6513·8	6457·5	3	1	1	0	0	0
9061·8	8916	0	0	3	0	0	0	6314·4	6165·7	1	1	3	0	0	0
8227·4	8097	2	1	1	0	0	0	5943·1	–	3	2	1	0	0	0
7956·4	–	0	1	3	0	0	0	5915·9	–	4	0	1	0	0	0
7228·2	7164·5	3	0	1	0	0	0	5714·2	–	2	0	3	0	0	0
6981·9	6919·0	1	0	3	0	0	0								

In emission from flames, the following maxima or red-degraded heads are recorded. Bands <7000 Å are from Kitagawa, and >7000 from Gaydon, with intensities based on the latter's observations.

H₂O (*contd.*)

λ	I	λ	I	λ	I	λ	I
9669	7*	8916	7*	6468·0	5	5988·8	3
9610	4	8097	8*	6457·5	4	5948·8	3
9559	4	7299	5	6377·1	4	5923·8	2
9485	3	7164·5	6*	6321·6	4	5900·2	3
9440	3	6922·0	2*	6255·1	4	5880·2	3
9333	7	6628·6	4	6220·0	3	5861·6	2
9277	10*	6574·5	4	6202·6	4	5806·9	2
9183	4	6516·8	5	6181·5	2	5715·3	1
9129	4	6490·4	5	6165·7	4	5683·3	1
8974	3						

* Outstanding head, degraded to longer λ.

FAR ULTRA-VIOLET ABSORPTION

References. S. Leifson, *Astrophys. J.*, **63**, 73 (1926)†.
G. Rathenau, *Z.P.*, **87**, 32 (1933)†.

Water vapour has a fairly sharp cut-off at 1800 Å. There is a strong region of absorption 1780 to 1610 Å, and another banded region of strong absorption between 1300 and 1400 Å.

LIQUID WATER

Liquid water shows infra-red absorption bands at about 7750 and 9850 Å and cuts off the ultra-violet beyond 1800 Å.

—OH HYDROXYL

Compounds with an OH— group, *i.e.*, alcohols, all show a strong infra-red absorption band around 9500 Å.

Reference. R. M. Badger and S. H. Bauer, *J. Chem. Phys.*, **4**, 711 (1936).

H₂O⁺

References. P. A. Weniger *et al.*, *Astrophys. J.*, **190**, L 43 (1974)†.
H. Lew and I. Heiber, *J. Chem. Phys.*, **58**, 1246 (1973).

Occurrence. In a low-pressure discharge which uses an electron beam from a tungsten filament, and in the tails of comets. A report of bands in absorption is doubtful.

Appearance. A long progression of bands (7540–4270 Å) with alternating (Σ and Π) structure. In low-temperature sources each band has a number (around 10) of narrow maxima. The following table gives up to three of the strongest maxima for each band, based mainly on observations in comet Kohoutek, presumably a low-temperature fluorescence spectrum.

$(v_1\ v_2\ v_3)'$			$(v_1\ v_2\ v_3)''$			λ *narrow maxima*		
0	5	0	0	0	0	7468		
0	6	0	0	0	0	6987	7040	
0	7	0	0	0	0	6542	6594	6686
0	8	0	0	0	0	6158	6199	6210
0	9	0	0	0	0	5799	5827	5914
0	10	0	0	0	0	5480	5489	5521
0	11	0	0	0	0	5194		

H_2O_2

Hydrogen peroxide vapour shows continuous absorption in the ultra-violet. The absorption commences around 3700 Å, increases slowly in strength to 3000 Å, and then more rapidly to the limit of observations at 2150 Å.

Reference.   H. C. Urey, L. H. Dawsey and F. O. Rice, *J. Amer. Chem. Soc.*, **51**, 1371 (1929).

HPO see PHO

HS see SH

H_2S, H_2Se, H_2Te

Reference.   C. F. Goodeve and N. O. Stein, *Trans. Faraday Soc.*, **27**, 393 (1931).

Hydrogen sulphide, selenide and telluride show continuous absorption in the ultra-violet commencing respectively at about 2700, 3400 and 4000 Å and strengthening to shorter wavelengths.

H_2S^+

References.   R. N. Dixon *et al.*, *Molec. Phys.*, **22**, 977 (1971)†.
G. Duxbury, M. Horani and J. Rostas, *P.R.S.*, **331**, 109 (1972)†.

Occurrence.   Controlled electron beam.

Appearance.   Bands of complex structure, usually with a central red-degraded Q head. The main bands form a progression and alternate in structure (Σ and Π type).

The following heads (s = strong, m = moderate and w = weak) are from Dixon *et al.*:

λ	I	$(v_1\ v_2\ v_3)'$			$(v_1\ v_2\ v_3)''$			λ	I	$(v_1\ v_2\ v_3)'$			$(v_1\ v_2\ v_3)''$		
4962·4	m	0	3	0	0	1	0	4507·2	s	0	4	0	0	0	0
4818·5	m	0	2	0	0	0	0	4421·3	m	0	7	0	0	2	0
4755·7	m	0	4	0	0	1	0	4336·4	m	0	5	0	0	0	0
4692·4	s	0	3	0	0	0	0	4172·5	m	0	6	0	0	0	0
4616·9	m	0	6	0	0	2	0	4012·0	w	0	7	0	0	0	0

HS_2

Reference.   R. K. Gosavi *et al.*, *Chem. Phys. Lett.*, **21**, 318 (1973).

Absorption bands 3800—3070 Å following flash photolysis of H_2S_2. Strongest red-degraded bands λλ3482·7, 3445·5, 3408·6, 3340·0, 3277·8.

HSi— For compounds of Si see SiHBr, etc.

He_2

Occurrence.   In a midly condensed discharge through helium at a pressure of a few centimetres of mercury.

Appearance.   A number of apparently irregularly spaced bands of open rotational structure in the visible and near ultra-violet. Most of the bands are degraded to longer wavelengths.

He₂ (*contd.*)

Transition. The helium bands can be arranged into a number of series resembling the Rydberg series of line spectra. A treatment of the analysis of the helium spectrum is beyond the scope of this work. The ground state of He_2 is unstable; the excited stable levels fall into two groups, singlets and triplets; the lowest stable levels in order from the ground state are $^3\Sigma_u$, $^1\Sigma_u$, $^3\Pi_g$, $^1\Pi_g$.

References. A very large number of papers have appeared on the molecular spectrum of helium. The data given below are based on the following early papers:

W. E. Curtis, *P.R.S.*, **89**, 146 (1913)†.

A. Fowler, *P.R.S.*, **91**, 208 (1915).

W. E. Curtis, *P.R.S.*, **101**, 38 (1922)†; **103**, 315 (1923).

W. E. Curtis and R. C. Long, *P.R.S.*, **108**, 513 (1925).

W. F. Meggers and G. H. Dieke, *J. Res. Nat. Bur. Stand*, **9**, 121 (1932).

D. Cuthbertson, *C. R. Acad. Sci.* (*Paris*), **236**, 1757 (1953).

M. L. Ginter, *J. Chem. Phys.*, **42**, 561 (1965).

A. B. Callear and R. E. M. Hedges, *Trans. Faraday Soc.*, **66**, 2921 (1970)†.

The following appear to be the most prominent heads which are degraded to the red; the intensities are our estimates from the published spectrograms:

λ	I	λ	I	λ	I	λ	I
9346·1	–	5133·2	2	4157·8	4	3634	3
9123·1	–	5108·2	1	4002·3	4	3462·4	1
8051·5	–	5056·1	3	3989·1	5	3356·4	4
6398·7	10	4648·5	10	3777	4	3348·0	3 ·
6310	3	4625·6	10	3676·5	7	3206·4	2
5862·1	6	4545·8	5	3665·0	5	3200·6	2

Degraded to the violet:

λ5733·0   9

In addition to the above well-marked heads there is a complex region of strong band structure λλ6250–5750 (there may appear to be a head at 5950 under low dispersion) and another from 4500–4400 Å with a head near 4393 when seen under low dispersion. There is also band structure from 4050–3900 Å.

Callear and Hedges have observed the 4648·5 Å band and some weaker ones in absorption following a pulsed discharge.

HeNe

In discharges through mixed helium and neon there is band structure between 4114–4063 and 4270–4235 Å with a red-degraded head at 4238·7 and a peak at 4262, with weaker bands 4217 and 4203 Å.

References. H. J. Oskam and H. M. Jongerius, *Physica*, **25**, 1092 (1958)†.

J. Fache, *C. R. Acad. Sci.* (*Paris*), **259**, 2195 (1964)†.

W. R. Henderson, F. A. Matsen and W. W. Robertson, *J. Chem. Phys.*, **43**, 1290 (1965).

HfBr

Reference. T. Savithry, D. V. K. Rao and P. T. Rao, *Physica*, **67**, 400 (1973)†.

Emission in a high-frequency discharge shows bands 6500–6100. The strong red-degraded (0, 0) sequence is at 6205·5, with weak (0, 1) at 6304 and (1, 0) at 6115 Å.

HfI

References.   T. Savithry, D. V. K. Rao and P. T. Rao, *Curr. Sci. (India)*, **40**, 516 (1971)†; **42**, 533 (1973)†.

Three systems have been obtained in a radio-frequency discharge.

System A, 7000–6700. Degraded to the red. 6866·4 (0, 0).

System B, 4890–4820, $^4\Pi - {}^4\Sigma$. Degraded to the red. Q heads of (0, 0) sequence 4874·6, 4867·8, 4830·6 and 4825·4.

System C, 5080–4820, $^4\Pi - {}^4\Sigma$. Sequences degraded to the red but some bands shade to the violet. Q heads of (0, 0) 5025·8, 4993·4, 4953·6 and 4932·1.

HfO

Occurrence.   In arc containing hafnium salts and in microwave discharge.

Appearance.   At least nine systems, all degraded to the red. Usually obvious sequences.

References.   W. F. Meggers, *J. Res. Nat. Bur. Stand.*, **1**, 151 (1928)†.

G. Edvinsson and C. Nylén, *Phys. Scripta*, **3**, 261 (1971)†.

The following wavelengths are from Edvinsson and Nylén; intensities, where given, are based on those of Meggers:

λ	I	System		λ	I	System
6021·1		$A^1\Sigma - X^1\Sigma$ (0, 0)		4102·3	9	$D^1\Pi - X^1\Sigma$ (1, 0)
5698·0	7	$B^1\Pi - X^1\Sigma$ (0, 0)		3970·0	7	$E^1\Pi - X^1\Sigma$ (0, 0)
5074·7	9	C (0, 0)		3839·6		$E^1\Pi - X^1\Sigma$ (1, 0)
4703·1		H (0, 0)		3654·2	5	$F^1\Sigma - X^1\Sigma$ (0, 0)
4434·4		$D^1\Pi - X^1\Sigma$ (0, 1)		3545·2		$F^1\Sigma - X^1\Sigma$ (1, 0)
4408·8		J (0, 0)		3327·8		$G^1\Sigma - X^1\Sigma$ (0, 0)
4252·1	10	$D^1\Pi - X^1\Sigma$ (0, 0)		3236·8		$G^1\Sigma - X^1\Sigma$ (1, 0)

Hg₂

Numerous papers have appeared on the emission, absorption and fluorescence bands and continua attributed to Hg_2. Good photographs of these have been published by Rayleigh.

References.   Lord Rayleigh, *P.R.S.*, **116**, 702 (1927)†; **119**, 349 (1928)†.

J. M. Walter and S. Barratt, *P.R.S.*, **122**, 201 (1929).

T. Mrozowska, *Acta Physica Polonica*, **2**, 81 (1933).

J. Okubo and E. Matuyama, *Tohoku Union Sci. Reports*, **22**, 383 (1933)†.

S. Mrozowski, *P.R.*, **76**, 1714 (1949).

λ2345 TO SHORTER WAVELENGTHS

Strong continuum commencing at about 2345 Å and fading out to shorter wavelengths with superposed diffuse bands observed in emission, absorption and fluorescence.

Maxima of bands:

λ	I
2342	10
2337	5
2333	1
2330	0

Hg₂ (*contd.*)

λ2540 TO RED

Strong continuum stretching from near the strong line λ2537 to the red with superposed diffuse bands.

In emission and fluorescence bands are observed 2659–3097 Å.
In absorption the superposed bands are clearest 2613–2943 Å.

λ3350

Broad continuum with maximum at 3350 Å observed in fluorescence.

λλ3650–4047 and λλ4078–4340

Emission bands have been observed in this region by Okubo and Matuyama.

λ4850

Broad continuum with maximum at 4850 Å observed in fluorescence.

Rosen *et al.* (*Spectroscopic Data Relative to Diatomic Molecules*, 1970) also record a number of systems of closely-spaced bands, including 12 systems observed by H. Takeyama.

Hg₂⁺

Occurrence. Mercury vapour in discharge tubes, especially with Tesla coil excitation.
Appearance. Diffuse bands around 2480 Å degraded to red.
With large dispersion each band is seen to be a sequence.
References. J. G. Winans, *P.R.*, **42**, 800 (1932)†.
S. Mrozowski, *P.R.*, **76**, 1714 (1949).

The assignment to Hg₂⁺ is not quite certain.
Bands as observed by Winans:

λ	I	v', v''	λ	I	v', v''	λ	I	v', v''
2525·4	3		2489·5	4	0, 2	2464	3	2, 0
2518·0	3		2482	5	0, 1	2458·0	3	3, 0
2509·4	2		2476·1	10	0, 0	2449·5	1	
2495·6	3	0, 3	2469·5	5	1, 0			

HgAr, Hg-rare gases

The mercury resonance line at 2537 Å may be broadened asymmetrically, sometimes developing weak band structure, in the presence of rare gases, due to incipient formation of molecules HgAr, HgNe, HgKr or HgXe.

HgBr

Occurrence. Transformer and high-frequency discharges through HgBr₂ vapour. Some of the ultra-violet bands also occur in absorption. Also in fluorescence of HgBr₂.
References. J. Lohmeyer, *Z. wiss. Photogr.*, **4**, 367 (1906).
K. Wieland, *Helv. Phys. Acta*, **2**, 46 (1929)†.
A. Terenin, *Z.P.*, **44**, 713 (1927).
K. Wieland, *Z.P.*, **77**, 157 (1932)†.

HgBr (*contd.*)

K. Wieland, *Helv. Phys. Acta*, **12**, 295 (1939); **19**, 408 (1946).

H. G. Howell, *P.R.S.*, **182**, 95 (1943).

V. G. Krishnamurthy, *Z.P.*, **152**, 242 (1958)†.

K. Wieland, *Z. Elektrochem.*, **64**, 761 (1960)†.

K. R. Rao and G. V. S. R. Rao, *Indian J. Phys.*, **18**, 281 (1944)†.

SYSTEM B, λλ5080–3200, B($?\,^2\Sigma$)$-$X$^2\Sigma$

Appearance. Complex system of line-like bands on a continuum with intensity maximum at about 5010 Å.

Strongest bands measured from end to end or at centre of the 'lines' (unpublished data from Wieland, see also Lohmeyer):

5076–71–66, 5056–48, 5042–34, 5024–22–20, 5009–05–02, 4999–95, 4988–84, 4976–72–69, 4959–55–52, 4940, 4878, 4865–62, 4848, 4834, 4820, 4775, 4761, 4747, 4734, 4719, 4705, 4691, 4678, 4666–62, 4652–50, 4637, 4626–22.

The same system, excited in the presence of an inert gas in large excess, shows a simple vibrational structure of bands degraded to the red (Wieland, 1939).

Prominent heads (unpublished data from Wieland):

λ	I	v', v''	λ	I	v', v''	λ	I	v', v''
5036·1	8	3, 26	4951·1	9	0, 20*	4879·1	7	0, 18
5017·7	10	0, 22*	4945·2	8	0, 20	4850·9	7	0, 17*
5010·1	9	0, 22	4918·0	8	0, 19*	4845·5	6	0, 17
4983·7	9	0, 21*	4912·4	7	0, 19	4817·9	7	1, 17*
4977·2	8	0, 21	4884·6	8	0, 18*	4784·5	6	1, 16*

* HgBr^{79}, all other heads HgBr^{81}.

SYSTEM C₁, λλ2940–2815

Wieland's system C has been split into three by Krishnamurthy. We designate them C_1, C_2 and C_3. These systems may be due to a transition $C^2\Pi_{\frac{1}{2}} - X^2\Sigma$. Degraded to the red. Strongest heads:

λ	I	v', v''	λ	I	v', v''
2906·3	5	0, 2	2868·0	6	1, 1
2898·8	6	1, 3	2853·2	7	1, 0
2890·8	7	0, 1	2846·5	6	2, 1
2883·5	7	1, 2	2839·9	5	3, 2
2875·3	8	0, 0	2817·9	5	4, 2

SYSTEM C₂, λλ2825–2777

Degraded to the red. Strongest heads.

λ	I	v', v''	λ	I	v', v''
2824·7	5	0, 2	2791·7	6	1, 1
2811·5	6	0, 1	2789·7	4	4, 5
2805·7	5	1, 2	2786·3	3	2, 2
2797·3	6	0, 0	2777·7	3	1, 0

SYSTEM C₃, λλ2777–2730

Diffuse bands, possibly shaded to the red. No complete analysis has been made. Strongest heads λλ2764·2, 2763·1, 2751·3, 2750·2.

HgBr (*contd.*)

SYSTEM D, $\lambda\lambda 2665-2471$, $D(?\,^2\Pi_{\frac{3}{2}}) - X^2\Sigma$

Degraded to shorter wavelengths. Heads of strong sequences:

λ	I	Sequence	λ	I	Sequence
2627·8	4	0, 3	2575·1	7	1, 0
2615·3	7	0, 2	2560·2	6	2, 0
2602·7	8	0, 1	2545·5	5	3, 0
2590·2	6	0, 0	2531·5	3	4, 0

SYSTEM E, 2470–2430 Å

Rao and Rao report 20 diffuse headless bands. An analysis has been made, but it is not quite certain. Strongest heads $\lambda\lambda 2467\cdot 3, 2457\cdot 0, 2448\cdot 1$.

HgBr$_2$

All emission bands previously ascribed to HgBr$_2$ by Wieland are now ascribed to HgBr. In absorption several continua in the far ultra-violet have been observed by Wieland, and a well-developed system at 1862–1813 Å of bands degraded to the red has been analysed by Wehrli.

References. K. Wieland, *Z.P.*, **76**, 801 (1932); **77**, 157 (1932).
M. Wehrli, *Helv. Phys. Acta*, **11**, 339 (1938)†.

HgCl

Three systems, B, C and D, have been observed in low-pressure discharge tubes. System B has also been observed in fluorescence of HgCl$_2$ and in chemiluminescence. In absorption only the strongest bands of system C have been observed above 1000° C.

References. J. Lohmeyer, *Z. Wiss. Photogr.*, **4**, 367 (1906).
A. Terenin, *Z.P.*, **44**, 713 (1927).
K. Wieland, *Helv. Phys. Acta*, **2**, 46 (1929)†; **10**, 323 (1937); **14**, 420 (1941)†; **19**, 408 (1946).
S. D. Cornell, *P.R.*, **54**, 341 (1938)†.
H. G. Howell, *P.R.S.*, **182**, 95 (1943).
V. G. Krishnamurthy, *Z.P.*, **150**, 287 (1958)†.

SYSTEM B, 5700–3000 Å, $B^2\Sigma - X^2\Sigma$

Appearance. Complex system of line-like bands on a continuum with pronounced intensity maximum at about 5550 Å. Strongest bands measured from end to end or at the centre of the 'lines' (unpublished data of Wieland, see also Lohmeyer):
5672–68 (3), 5654 (3), 5646 (4), 5624–20 (7), 5615–10 (8), 5588 (9), 5567–64 (9), 5559–55 (8), 5553 (9), 5540–38 (7), 5532–28 (8), 5518 (6), 5512 (5), 5497 (9), 5485 (7), 5457 (6), 5446 (7), 5424 (8), 5321 (8), 5304–01 (6), 5295 (7), 5272 (8), 5255–52 (8), 5232 (6).

The same system, with HgCl$_2$ excited in the presence of an inert gas in large excess, shows a simple vibrational structure of bands degraded to the red.

Prominent heads (Wieland, 1941):

λ	I	v', v''	λ	I	v', v''	λ	I	v', v''
5635·7	7	2, 25	5393·7	7	0, 19	5208·5	3	0, 16
5576·3	10	0, 22	5332·1	6	0, 18	5157·5	3	1, 16
5517·2	9	0, 21	5270·3	5	0, 17	5146·8	2	0, 15
5455·2	8	0, 20	5217·8	4	1, 17			

HgCl (*contd.*)

SYSTEM C, 2913–2700 Å, C($?\,^2\Pi_{\frac{1}{2}}$) − X$^2\Sigma$

Appearance. Two groups of bands, degraded to shorter wavelengths. Krishnamurthy notes strong heads at λλ2811, 2790, 2747, 2740, 2719, and has discussed the vibrational analysis.

Heads of prominent groups (Wieland, 1929):
2891·6 (3), 2870·0 (4), 2858 (2), 2812·2 (7), 2805·3 (6), 2790·5 (10), 2783·8 (9), 2742·1 (5), 2740·8 (8), 2721·3 (3), 2719·5 (6).

SYSTEM D, 2637–2380 Å, D($?\,^2\Pi_{\frac{3}{2}}$) − X$^2\Sigma$

Appearance. Bands degraded to shorter wavelengths.

Heads of strong sequences (Wieland, 1929):

λ	I	Sequence	λ	I	Sequence	λ	I	Sequence
2609·1	2	0, 5	2535·0	7	0, 1	2474·6	5	2, 0
2590·5	3	0, 4	2516·5	7	0, 0	2454·6	3	3, 0
2572·0	4	0, 3	2495·4	7	1, 0	2435·1	1	4, 0
2553·4	6	0, 2						

HgCl$_2$

Emission bands previously ascribed to HgCl$_2$ are now ascribed to HgCl. In absorption several continuous regions have been observed in the far ultra-violet by Wieland and a well-developed band system at λλ1731−1671, degraded to the red, has been analysed by Wehrli.

References. K. Wieland, *Z.P.*, **76**, 801 (1932); **77**, 157 (1932).
M. Wehrli, *Helv. Phys. Acta*, **11**, 339 (1938)†.

HgF

Occurrence. In high-frequency discharge.
References. H. C. Howell, *P.R.S.*, **182**, 95 (1943)†.
Y. K. S. C. Babu, P. T. Rao and B. R. Reddy, *Indian J. Pure Appl. Phys.*, **4**, 467 (1966)†.

2560 Å SYSTEM, C$^2\Pi_{\frac{1}{2}}$ − X$^2\Sigma$

Appearance. A very strong (0, 0) sequence and less strong (1, 0) and (0, 1) sequences degraded to shorter wavelengths.

Howell lists the Q heads of 35 bands, but without intensities. The following may be prominent:
λλ2624·2 (0, 2), 2591·6 (0, 1), 2590·9 (1, 2), 2559·8 (0, 0), 2559·1 (1, 1), 2527·0 (2, 1).

2325 Å SYSTEM, D$^2\Pi_{\frac{3}{2}}$ − X$^2\Sigma$

Appearance. The marked sequences are degraded to the red, but the rotational structure of most individual bands is degraded to shorter wavelengths.

The following are the most outstanding of the 43 heads listed by Howell:

λ	I	v′, v″	λ	I	v′, v″	λ	I	v′, v″
2355·5	5	2, 3 Q	2333·0	7	3, 3 Q	2326·2	8	0, 0 P
2353·3	5	1, 2 Q	2329·7	8	2, 2 Q	2325·6	10	0, 0 Q
2351·9	8	0, 1 Q	2327·3	8	1, 1 Q	2299·7	4	1, 0 Q

HgF (*contd.*)

2135 Å SYSTEM, $E^2\Sigma - X^2\Sigma$

Babu *et al.* report about 40 red-degraded heads 2230–2100 Å, but sequences are degraded to shorter wavelengths. Bands are close double-headed. R heads of strongest:

λ	I	v', v"	λ	I	v', v"	λ	I	v', v"
2224·3	5	0, 4	2192·6	7	2, 5	2170·7	6	2, 4
2202·0	9	0, 3	2179·5	10	0, 2	2157·2	9	0, 1
2196·8	8	1, 4	2174·8	8	1, 3	2135·1	7	0, 0

HgH

References. E. Hulthén, *Z.P.*, **32**, 32 (1925); **50**, 319 (1928).
R. Rydberg, *Z.P.*, **73**, 74 (1931).
T. L. Porter, *J. Opt. Soc. Amer.*, **52**, 1201 (1962).
D. M. Eakin and S. P. Davis, *J. Mol. Spec.*, **35**, 27 (1970)†.

4017 Å SYSTEM, $A^2\Pi - X^2\Sigma$

A widely spaced doublet system with bands degraded to the violet, each showing P, Q and R branches. Obtained in discharges through mercury vapour and hydrogen.

P Heads

	$^2\Pi_{\frac{1}{2}} - {}^2\Sigma$			$^2\Pi_{\frac{3}{2}} - {}^2\Sigma$	
v', v"	λ	I	v', v"	λ	I
0, 3	4520	2	0, 2	3785	2
0, 2	4394	4	0, 1	3647	3
0, 1	4219	7	0, 0	3500	5
0, 0	4017	10	1, 0	3274	3
1, 1	3900	2	2, 1	3200	2
1, 0	3728	4			

2950 Å SYSTEM, $B^2\Sigma - X^2\Sigma$

Bands with double P and R branches degraded to the red.

v', v"	Origins	R_1 Heads	R_2 Heads
0, 1	3059·8	3057·6	3057·8
0, 0	2951	2949·5	2949·7

2807 Å SYSTEM, $C^2\Sigma - X^2\Sigma$

Two bands are observed, degraded to the red. Double P and R branches.

v', v"	Origins	R_1 Head	R_2 Head
0, 1	2904		
0, 0	2808	2807·3	2807·0

2700 Å SYSTEM, $D^2\Sigma - X^2\Sigma$

Band with double P and R branches degraded to the red.

v', v"	Origin	R_2 Head
0, 0	2699	2696

HgH⁺

References.   T. Hori, *Z.P.*, **61**, 481 (1930)†.
S. Mrozowski and M. Szulc, *Acta. Phys. Polonica*, **6**, 44 (1937).

2264 Å SYSTEM, $A^1\Sigma - X^1\Sigma$

An extensive system of single-headed bands, degraded to the red. They occur in discharges where ionization is favoured, as in the hollow cathode. R heads:

λ	I	v', v''	λ	I	v', v''
2493·9	7	1, 3	2367·3	9	0, 1
2474·7	4	0, 2	2286·7	5	1, 1
2413·3	3	2, 3	2263·9	10	0, 0
2388·1	8	1, 2			

HgI

Nine systems have been observed with HgI_2 in discharge tubes, including high-frequency discharges. The most prominent of the systems, B, C and D, which involve the ground state, have also been observed in fluorescence of HgI_2. The blue-violet system B can also be excited with active nitrogen. In absorption, above 1000°C., the strongest bands of system C and, very faintly, those of system B have been observed.

References.   J. Lohmeyer, *Z. wiss. Photogr.*, **4**, 367 (1906).
R. S. Mulliken, *P.R.*, **26**, 1 (1925).
A. Terenin, *Z.P.*, **44**, 713 (1927).
K. Wieland, *Helv. Phys. Acta*, **2**, 46 (1929)†.
K. Wieland, *Z.P.*, **76**, 801, (1932)†.
T. S. Subbaraya, B. N. Rao and N. A. N. Rao, *Proc. Indian Acad. Sci.*, A, **5**, 365 (1937)†.
H. G. Howell, *P.R.S.*, **182**, 95 (1943).
K. R. Rao, M. G. Sastry and V. G. Krishnamurthy, *Indian J. Phys.*, **18**, 323 (1944).
K. R. Rao and V. R. Rao, *Indian J. Phys.*, **20**, 148 (1946).
C. Ramasastry and K. R. Rao, *Indian J. Phys.*, **21**, 143 (1947)†.
C. Ramasastry, *Indian J. Phys.*, **22**, 95 (1948)†.
V. G. Krishnamurthy, *Z.P.*, **160**, 438 (1960).
K. Wieland, *Z. Elektrochem.*, **64**, 761 (1960)†.

SYSTEM B, 4520–3600 Å, $B^2\Sigma - X^2\Sigma$

A complex system of line-like bands is superposed on a continuum with pronounced intensity maximum at about 4440 Å. When obtained by excitation of HgI_2 in excess of inert gas the structure is simpler and the following red-degraded heads have been observed (unpublished data by Wieland):

λ	I	v', v''	λ	I	v', v''	λ	I	v', v''
4476·1	6	1, 19	4391·6	7	1, 15	4277·4	4	3, 11
4455·4	10	1, 19	4373·3	6	1, 14	4257·4	3	4, 11
4440·2	9	1, 18	4355·4	6	1, 13	4238·8	3	4, 10
4423·3	8	1, 17	4316·2	5	2, 12	4219·3	2	5, 10
4411·6	10	0, 15	4295·8	5	3, 12	4200·5	2	5, 9

HgI (*contd.*)

SYSTEM C, 3095–2850 Å, $C(?^2\Pi_{\frac{1}{2}}) - X^2\Sigma$

Appearance. Bands degraded to shorter wavelengths.
Strongest heads (Wieland, 1932):

λ	I	v', v''	λ	I	v', v''	λ	I	v', v''
3083·6	3	0, 3	3051·5	6	2, 4	3028·0	5	1, 0
3072·4	6	0, 2	3049·5	7	0, 0	3007·4	4	2, 0
3061·8	8	1, 3	3041·4	5	(2, 3)	2998·5	4	3, 1
3061·0	10	0, 1	3031·8	3	4, 6	2979·0	4	4, 1

SYSTEM D, 2850–2640 Å, $D(?^2\Pi_{\frac{3}{2}}) - X^2\Sigma$

Appearance. Bands degraded to shorter wavelengths somewhat obscured by overlapping bands of systems C and E.

The (0, 0) band lies at 2754·4 Å (Rao, Sastry and Krishnamurthy, 1944).

SYSTEM E, 2700–2550 Å

Wieland (1932) recorded a number of broad bands in this region. Krishnamurthy has divided it into two systems 2650–2605 (strongest heads $\lambda\lambda$2633·3, 2629·8, 2625·1, 2621·7), and 2600–2557 ($\lambda\lambda$2577·7, 2575·0, 2570·1, 2554·2, 2548·4, 2537·4), but the bands do not show well in his published photograph.

SYSTEM F_1, 2534–2450 Å, $F_1 - X^2\Sigma$

Ramasastry has measured 54 heads of this system of bands degraded to the red. Strongest bands:
2487·6 (5), 2493·1 (5), 2500·6 (5), 2506·1 (6), 2509·9 (5), 2532·7 (6), 2533·8 (6).

SYSTEM F_2, 2437–2381 Å

Rao and Rao measured 34 closely-spaced bands degraded to the red.

SYSTEM F_3, 2350–2260 Å, $F_3 - X^2\Sigma$

Ramasastry has measured 84 heads degraded to the red. Strongest bands:
2321·7 (6), 2322·0 (5), 2325·6 (6), 2330·1 (6), 2331·1 (6), 2332·0 (5), 2334·8 (6), 2335·7 (5), 2336·9 (6), 2340·7 (10).

SYSTEM G, 2230–2165 Å, $G - X^2\Sigma$

Ramasastry and Rao (1947) have measured 59 bands degraded to the red. Strongest heads:
2213·6 (5), 2207·8 (6), 2197·7 (6), 2187·6 (6), 2183·4 (5), 2179·3 (5).

SYSTEM H, 2170–2110 Å, $H - X^2\Sigma$

Ramasastry and Rao (1947) have measured 22 bands degraded to the red. Strongest heads:
2150·4 (5), 2144·3 (5), 2139·0 (6), 2133·7 (5).

HgI_2

All emission bands previously ascribed to HgI_2 are probably due to HgI. In absorption there are several regions of continuum (Wieland) and a system of bands degraded to the red (Wehrli).

References. K. Wieland, *Z.P.*, **76**, 801 (1932); 77, 157 (1932).
M. Wehrli, *Helv. Phys. Acta*, **11**, 339 (1938)†.

HgIn

Reference.   C. Santaram and J. G. Winans, *J. Mol. Spec.*, **16**, 309 (1965)†.

In a high-frequency or Tesla discharge through mixed indium and mercury vapour broadening of the In resonance lines 4102 and 4511 Å occurs, due to molecule formation; there is a violet-degraded system at 5226 Å, with prominent heads:

λ	I	v', v''	λ	I	v', v''
5226·4	10	0, 0	5160·0	7	2, 1
5213·3	9	1, 1	5120·9	5	2, 0
5172·7	7	1, 0	5069·0	4	3, 0

Bands at 5760 and 5544 reported by Purbrich are due to CdIn.

HgO ?

Reference.   J. M. Walter and S. Barratt, *P.R.S.*, **122**, 201 (1929).

Walter and Barratt record absorption bands $\lambda\lambda 2943-2739$ using mercury vapour and oxygen. The assignment to HgO is uncertain.

HgS

Reference.   P. K. Sen Gupta, *P.R.S.*, **143**, 438 (1933–4).

Continuous absorption in three regions beginning at $\lambda\lambda 4450, 3100$ and 2250.

HgTl

Reference.   J. G. Winans and W. J. Pearce, *P.R.*, **74**, 1262 (1948).

In a brief note, intensity maxima at 6550, 5200, 4580 and 3810 Å are reported as observed in a Tesla discharge. Formulae indicate (0, 0) bands at 6539, 6517, 4419 and 3808.

HoF

Reference.   D. J. W. Robbins and R. F. Barrow, *J. Phys.*, B, **7**, L 234 (1974)†.

Red degraded bands in absorption.

A − X.   5400·0 (0, 1), 5229·2 (0, 0), 5088·0 (1, 0).

B − X.   4509·1 (2, 0), 4611·1 (1, 0), 4719·2 (0, 0).

HoO

Reference.   R. Herrmann and C. T. J. Alkemade, *Flame Photometry*, Interscience (1963).

About 60 bands 5880–5069 are reported in arcs and flames. The following are among the strongest; the direction of shading is indicated as R or V.

λ	I	λ	I	λ	I	λ	I
5849 V	6	5820 V	6	5696 V	10	5251 V	3
5935 V	6	5729 R	7	5660 −	10	5157 R	5
5831 V	6	5713 R	7	5657 V	7	5105 R	2

I$_2$

A large number of papers have been published on the spectrum of iodine. The most readily observed system is the well known visible absorption bands which are responsible for the violet colour of the vapour. All the band systems of iodine are composed of a very large number of narrow bands, and distinctive features are lacking.

VISIBLE SYSTEM, B 0_u^+–X$^1\Sigma_g^+$

Occurrence. This system has been studied principally in absorption 8400–4990 Å, the convergence limit. Bands 6700–5000 have been obtained in emission from high-frequency discharges through I$_2$, but they are suppressed by argon and other gases. Uchida observed the bands in emission from heated vapour. Also in fluorescence.

Appearance. Degraded to longer wavelengths. Numerous regularly spaced bands extending from the far red to the absorption limit at about 5000 Å. The system is too extensive to publish wavelengths, but a spectrogram is shown in Plate 10.

References. C. A. Goy and H. O. Pritchard, *J. Mol. Spec.*, **12**, 38 (1964).
S. M. Singh and J. Tellinghuisen, *J. Mol. Spec.*, **47**, 409 (1973).
Y. Uchida, *Inst. Phys. Chem. Res. Tokyo Sci. Papers*, No. 651, 71 (1936)†.

INFRA-RED SYSTEM, 9300–8300 Å, A$^3\Pi_{1u}$ – X$^1\Sigma_g^+$

Readily in absorption. Also in fluorescence and in emission in a discharge tube.

References. W. G. Brown, *P.R.*, **38**, 1187 (1931).
R. K. Asundi and P. Venkateswarlu, *Indian J. Phys.*, **21**, 101 (1947).

EMISSION SYSTEMS

A number of systems, diffuse bands and regions of continuous emission have been observed. Some of these occur in discharges through pure I$_2$ vapour and are suppressed by an inert diluent, while others do not occur well in the pure gas but are enhanced by argon or nitrogen. Most of them occur in fluorescence. The following are among the most important references.

References. A. E. Elliott, *P.R.S.*, **174**, 273 (1940)†.
P. Venkateswarlu, *P.R.*, **81**, 821 (1951).
K. Wieland and J. Waser, *P.R.*, **85**, 385 (1952).
L. Mathieson and A. L. G. Rees, *J. Chem. Phys.*, **25**, 753 (1956).
P. B. V. Haranath and P. T. Rao, *J. Mol. Spec.*, **2**, 428 (1958).
R. D. Verma, *J. Chem. Phys.*, **32**, 738 (1960).
P. B. V. Haranath and T. A. P. Rao, *Indian J. Phys.*, **34**, 123 (1960)†.
K. Wieland, J. B. Tellinghuisen and A. Nobs, *J. Mol. Spec.*, **41**, 69 (1972)†.
P. Venkateswarlu, *Canad. J. Phys.*, **48**, 1055 (1970).

There is some confusion about notation and identity of the systems, but the following are the best established.

(1) Diffuse bands obtained from pure I$_2$ at $\lambda\lambda$5100, 3524, 3481, 2880, 2868, 2857, 2847, 2836, 2825.

(2) A group of weak violet-degraded bands 4630–4440 reported by Elliott in I$_2$ + N$_2$ discharges.

(3) Red-degraded emission bands 4420–4000 Å in presence of a foreign gas. Mathieson and Rees assign them as E – B$^3\Pi_{0_u^+}$ although Venkateswarlu referred to them as D$^3\Sigma_g^-$ – B$^3\Pi_{0_u^+}$. Strongest bands (I = 9 or 10 from Wieland *et al.*): $\lambda\lambda$4298·4 (3, 26), 4293·2 (0, 22), 4290·4 (1, 23), 4277·2 (0, 21), 4260·9 (0, 20), 4244·4 (0, 19), 4227·6 (0, 18), 4210·7 (0, 17).

I_2 (*contd.*)

(4) A continuum near 3425 or 3416 Å is strong in flames containing iodine and in discharges through the pure gas. In the presence of argon this changes to a well-developed band system 3450–3000 Å; the red-degraded bands close up to give the appearance of a diffuse violet-degraded head at 3436. Mathieson and Rees denoted this system $D - X^1\Sigma_g^+$, while Venkateswarlu (1970) refers to it as $F^1\Sigma_g^+ - B^3\Pi_{0_u^+}$. Wieland *et al.* (1972) consider the 3425 band as separate and refer to it as system D.

(5) $\lambda\lambda 2785–2750$. Red-degraded bands observed in discharge in presence of foreign gas. Strongest heads: 2772·2, 2769·4, 2755·9. Called $F - B^3\Pi_{0_u^+}$.

(6) $\lambda\lambda 2740–2520$. Red-degraded bands observed in emission and fluorescence in presence of a foreign gas. Referred to as $G^3\Sigma_u^- - X^1\Sigma_g^+$ by Venkateswarlu, but as $F^1\Sigma_u^+ - X^1\Sigma_g^+$ by Wieland *et al.* Strongest heads: 2720·5 (5, 67), 2713·7 (6, 67), 2713·2 (4, 65), 2705·2 (3, 63).

(7) Bands 2370–1830 Å observed by Verma and others in discharges. Also in fluorescence and in absorption. Venkateswarlu refers to this as $H^1\Sigma_u^+ - X^1\Sigma_g^+$.

Venkateswarlu (1972) has also observed a large number of systems, including Rydberg series, below 2000 Å.

IBr

A number of systems have been reported. All consist of numerous closely-spaced red-degraded bands. There are absorption systems in the extreme red (6817–8000 Å), red (6240–6764 Å) and 5458–6186 Å. Uncondensed discharges through pure IBr give various patches of continuum between 5200 and 2300 Å; some bands occur in fluorescence and magnetic rotation. Better developed emission bands occur in discharges through IBr in argon.

References.   R. M. Badger and D. M. Yost, *P.R.*, **37**, 1548 (1931).

H. Cordes, *Z.P.*, **74**, 34 (1932).

W. G. Brown, *P.R.*, **42**, 355 (1932).

P. B. V. Haranath and P. T. Rao, *Indian J. Phys.*, **31**, 368 (1957)†.

P. Venkateswarlu and R. D. Verma, *Proc. Indian Acad. Sci.*, A, **47**, 150 and 161 (1958)†.

L. E. Selin, *Ark. Fys.*, **21**, 479 (1962)†.

M. A. A. Clyne and J. A. Coxon, *J. Mol. Spec.*, **23**, 238 (1967)†.

The following summary is mainly based on that by Venkateswarlu and Verma.

$\lambda\lambda 8000–6871$. Absorption, $A^3\Pi_1 - X^1\Sigma^+$ (Brown); emission by reaction Br + I (Clyne and Coxon).

$\lambda\lambda 6764–6240$. Absorption $B^3\Pi_{0^+} - X^1\Sigma^+$ (Brown).

$\lambda\lambda 6186–5458$. Absorption $B'\ 0^+ - X^1\Sigma^+$ (Brown).

$\lambda\lambda 5425–5360$. Emission in argon. $C - B'$. Strongest heads 5362·6 and 5373·8.

$\lambda\lambda 4520–4415$. Emission in argon. $E - B'$. Strongest heads 4479·0 and 4478·2.

$\lambda\lambda 4120–4010$. Emission in argon. $I - H$. Strongest heads 4086·5, 4076·2, 4073·5, 4069·8.

$\lambda\lambda 3915–3540$. Emission in argon. $D - A$. Nearly 300 bands.

Vacuum ultra-violet. Absorption systems $F - X$ and $G - X$.

ICl

ABSORPTION

Occurrence.   In absorption. Also in magnetic rotation and fluorescence.

Appearance.   Extensive systems of closely spaced red-degraded bands. In the near infra-red the bands are attributed to the transition $A^3\Pi_1 - X^1\Sigma^+$, while bands extending to the yellow green are

ICl (*contd.*)

due to transition $B^3\Pi_{0^+} - X^1\Sigma^+$ and $0^+ - X^1\Sigma^+$. Wavelengths and some photographs have been published by Curtis and Patkowski.

References.   O. Darbyshire, *P.R.*, **40**, 366 (1932).
G. E. Gibson and H. C. Ramsperger, *P.R.*, **30**, 598 (1927).
W. E. Curtis and J. Patkowski, *Phil. Trans. Roy. Soc.*, A, **232**, 395 (1934)†.
E. Hulthén, N. Johansson and V. Pilsater, *Ark. Fys.*, **14**, 31 (1958).

EMISSION

References.   R. K. Asundi and P. Venkateswarlu, *Indian J. Phys.*, **21**, 76 (1947).
P. B. V. Haranath and P. T. Rao, *Indian J. Phys.*, **31**, 156 (1957)†.
M. A. A. Clyne and J. A. Coxon, *P.R.S.*, **298**, 424 (1967)†.

In uncondensed discharges there is a continuum 5640–5120 and a series of diffuse bands between 4800 and 5120. In a condensed discharge about 300 bands occur between 4400 and 3800; the (0, 0) head is at 4196·2 Å.

Clyne and Coxon record 57 bands of the $A^3\Pi_1 - X^1\Sigma^+$ system between 8883 and 6038 Å in the recombination of free atoms I + Cl.

IF

Occurrence.   Flame of iodine reacting with fluorine. Reaction I + F in presence of O_2.
Appearance.   Degraded to the red. Strong $v' = 0$ and $v'' = 0$ progressions.
References.   R. A. Durie, *P.R.S.*, **207**, 388 (1951)†.
R. A. Durie, *Canad. J. Phys.*, **44**, 337 (1966)†.
M. A. A. Clyne, J. A. Coxon and L. W. Townsend, *J. Chem. Soc. Faraday Trans. II*, **68**, 2134 (1972)†.
J. W. Birks, S. D. Gabelnick and H. S. Johnston, *J. Mol. Spec.*, **57**, 23 (1975).

Over 40 bands have been recorded by Durie. The following are the strongest:

λ	I	v', v''	λ	I	v', v''	λ	I	v', v''
6798·6	9	1, 8	6031·1	10	0, 4	4961·3	8	3, 0
6728·0	9	0, 7	5825·5	10	0, 3	4868·2	8	4, 0
6549·3	10	1, 7	5691·5	9	1, 3	4780·3	8	5, 0
6481·2	10	0, 6	5631·5	9	0, 2	4697·4	8	6, 0
6249·3	10	0, 5	5505·8	9	1, 2	4619·3	7	7, 0

IO

METHYL IODIDE FLAME BANDS

Occurrence.   In flames containing iodide or methyl iodide. They are emitted most strongly in the region just above the inner cone. Bands with low v'' have also been obtained in absorption following flash photolysis of iodine/oxygen mixtures.
Appearance.   Degraded to the red. A fairly simple system.
Transition. $A^2\Pi - X^2\Pi$.
References.   E. H. Coleman, A. G. Gaydon and W. M. Vaidya, *Nature*, **162**, 108 (1948)†.
R. A. Durie and D. A. Ramsay, *Canad. J. Phys.*, **36**, 35 (1958).
R. A. Durie, F. Legay and D. A. Ramsay, *Canad. J. Phys.*, **38**, 444 (1960)†.

The following are the strongest heads. Most of the values are from Coleman *et al.*, but the more accurate values from Durie *et al.* are given where available, (*i.e.*, for the strongest bands).

IO (*contd.*)

λ	I	v', v''	λ	I	v', v''	λ	I	v', v''
6231·5	4	3, 11	5493·1	9	0, 5	4693·5	9	1, 1
6192·2	4	2, 10	5307·3	10	0, 4	4586·2	10	2, 1
5973·4	7	2, 9	5208·5	5	2, 5	4487·2	10	3, 1
5939	6	1, 8	5131·2	9	0, 3	4448·9	7	2, 0
5900	5	0, 7	5002·3	6	1, 3	4396·7	6	4, 1
5730	8	1, 7	4964·2	6	0, 2	4355·6	9	3, 0
5689·7	7	0, 6	4844·5	10	1, 2	4269·8	7	4, 0
5533	7	1, 6	4730·3	7	2, 2	4189·0	7	5, 0

Bands with $v' = 1, 4$ and 5 are diffuse, due to predissociation.

In$_2$

Occurrence. In absorption and fluorescence of indium vapour.
Reference. R. Wajnkranc, *Z.P.*, **104**, 122 (1936); **105**, 516 (1937).
The following are the chief features of the absorption spectrum:
Diffuse broad bands maxima λλ3818, 3734, 3680.
3548–3523 Å. A group of narrow bands degraded to the red.
3259 Å. A group of narrow bands to the red of the In line.
2340 Å. A group of narrow bands on the shorter wavelength side of the In line.

InBr

Three systems have been observed in emission in a high-frequency discharge and in absorption. Also in fluorescence of InBr$_2$ (W. Wenk).
References. A. Petrikaln and J. Hochberg, *Z.P.*, **86**, 214 (1933)†.
M. Wehrli and E. Miescher, *Helv. Phys. Acta*, 7, 298 (1934)†.
L. Lakshminarayana and P. B. V. Haranath, *Indian J. Phys.*, **44**, 504 (1970)†.

SYSTEM A, λλ3852–3641, $A^3\Pi_{0^+} - X^1\Sigma^+$

Some bands degraded slightly to shorter wavelengths. Strongest bands:

λ	I	v', v''
3789·8	4	0, 1
3758·5	8	0, 0
3727·2	4	1, 0

SYSTEM B, λλ3726–3568, $B^3\Pi_1 - X^1\Sigma^+$

Some bands degraded slightly to red. Strongest absorption bands:

λ	I	v', v''
3681·1	9	1, 2
3651·2	10	0, 0
3596·7	6	3, 1
3595·3	6	2, 0

SYSTEM C, λλ3083–2852

Single progression of diffuse bands, 2852 (0, 0), 2896 (0, 1), 2926 (0, 2), 2956 (0, 3), etc.

InBr$_2$

W. Wenk (*Dissertation, Basel*, 1941) has observed absorption continua at 2555, 2275, 2060 and 1895 Å.

InCl

Three systems have been observed in emission in a high-frequency discharge and in absorption in a furnace containing In and InCl$_3$. The 2672 Å system is a persistent impurity in absorption and has been erroneously attributed to In$_2$, InCd, HgIn and Bi$_2$ in the past. Also in fluorescence of InCl$_2$ (W. Wenk).

References. A. Petrikaln and J. Hochberg, *Z.P.*, **86**, 214 (1933)†.
M. Wehrli and E. Miescher, *Helv. Phys. Acta*, **7**, 298 (1934)†.
E. Miescher and M. Wehrli, *Helv. Phys. Acta*, **6**, 256 (1934).
P. Youngner and J. G. Winans, *J. Mol. Spec.*, **4**, 23 (1960)†.
H. M. Froslie and J. G. Winans, *P.R.*, **72**, 481 (1947)†.
Ashrafunnisa *et al.*, *Physica*, **73**, 421 (1974)†.

SYSTEM A, λλ3640–3471, $A^3\Pi_{0^+} - X^1\Sigma^+$

Degraded to shorter wavelengths. Strongest bands:

λ	I	v', v''	λ	I	v', v''
3599·2	6	0, 0	3554·0	4	2, 1
3596·5	2	1, 1	3514·6	2	2, 0
3556·2	5	1, 0	3513·0	3	3, 1

SYSTEM B, $B^3\Pi_1 - X^1\Sigma^+$

Degraded to shorter wavelengths. Q heads of strongest bands:

λ	I	v', v''
3499·0	8	0, 0
3458·5	5	1, 0
3456·3	4	2, 1
3419·3	3	2, 0

SYSTEM C, $C^1\Pi - X^1\Sigma^+$

Degraded to the red. Strongest bands at λλ2672·3 (0, 0), 2694·7 (0, 1), 2717·5 (0, 2), 2740·6 (0, 3) and λλ2661·3 (1, 0), 2683·7 (1, 1) in absorption only.

There are also continua around 2610 and 2100 Å in absorption at high pressure.

InCl$_2$

W. Wenk (*Dissertation, Basel*, 1941) has observed absorption continua at 2390, 2160, 1920 and 1818 Å.

InF

Occurrence. In absorption in furnace and in emission from hot hollow-cathode discharge containing metallic indium and AlF$_3$.

InF (*contd.*)

References. D. Welti and R. F. Barrow, *Proc. Phys. Soc.*, A, **65**, 629 (1952).
R. F. Barrow, J. A. T. Jacquest and E. W. Thompson, *Proc. Phys. Soc.*, A, **67**, 528 (1954).
R. F. Barrow, D. V. Glaser and P. B. Zeeman, *Proc. Phys. Soc.*, A, **68**, 962 (1955).

$A^3\Pi_0 - X^1\Sigma$ SYSTEM

Appearance. A weak system of two sequences of single-headed bands, degraded to the violet.
Heads: 3282·1 (0, 0), 3278·2 (1, 1), 3222·4 (1, 0), 3219·3 (2, 1), etc.

$B^3\Pi_1 - X^1\Sigma$ SYSTEM

Appearance. Degraded to the violet. Marked sequences of close double-headed bands.
The following are the Q heads of the prominent bands, with intensities for absorption:

λ	I	v', v''	λ	I	v', v''
3308·4	–	0, 2	3193·2	5	1, 1
3251·7	0	0, 1	3140·0	2	1, 0
3196·9	6	0, 0	3137·0	3	2, 1

$C^1\Pi - X^1\Sigma$ SYSTEM

Appearance. Double-headed bands, mostly degraded to the red. Marked sequences.
The following are outstanding Q heads as observed in absorption. The direction of degradation, R or V, is indicated after the wavelengths.

λ	I	v', v''	λ	I	v', v''
2366·6 V	1	0, 1	2337·3 V	10	0, 0
2347·1 R	4	2, 2	2313·1 R	1	1, 0
2341·8 R	5	1, 1	2290·4 R	1	2, 0

$D - X^1\Sigma$ SYSTEM

A few red-degraded bands have been tentatively assigned to a fourth system. Strongest heads: 2196·2 (4), 2192·5 (7), 2190·0 (9), 2188·9 (4), 2187·1 (10), 2185·8 (5).

InH

Occurrence. Arc in hydrogen between carbon and indium electrodes. Also in absorption.
References. B. Grundström, *Z.P.*, **113**, 721 (1939).
B. Grundström, *Nature*, **141**, 555 (1938).
H. Neuhaus, *Z.P.*, **150**, 4 (1958).
M. L. Ginter, *J. Mol. Spec.*, **11**, 301 (1963); **20**, 240 (1966).

6148 Å SYSTEM, $a^3\Pi_{0^+} - X^1\Sigma^+$

Barely degraded. Single P and R branches. Origins:

λ_0	v', v''	λ_0	v', v''
7043·9	3, 4	6190·9	1, 1
6861·1	2, 3	6148·5	0, 0
6459·4	3, 3	5785·8	2, 1
6286·9	2, 2	5688·6	1, 0

InH (*contd.*)

6166 Å SYSTEM, $a^3\Pi_{0^-} - X^1\Sigma^+$

This is a weak partially forbidden system, observed by Ginter (1966). Slightly degraded to the violet with Q branches only. Origins $\lambda\lambda 6741\cdot4$ (0, 1), $6213\cdot7$ (1, 1), $6166\cdot7$ (0, 0), $5708\cdot8$ (1, 0).

5913 Å SYSTEM, $b^3\Pi_1 - X^1\Sigma^+$

Headless bands, with single P, Q and R branches. In some the piled up Q branch, near the origin, is prominent. Origins:

λ_0	v', v''		λ_0	v', v''
5495·9	1, 0 deg. R		6071·0	2, 2 —
5913·8	0, 0 deg. V		6458·5	0, 1 deg. V.
5963·3	1, 1 deg. V			

4540 Å SYSTEM, $A^1\Pi - X^1\Sigma^+$

Strongly degraded to the red. Single P, Q and R branches. Origins and R heads.

λ_0	λ_h	v', v''	λ_0	λ_h	v', v''
4495·3		2, 0	4821·9		1, 1
4511·7		1, 0	4855·1	4850·8	0, 1
4540·7	4537·3	0, 0	5203·2	5197·7	0, 2

Garton (*Proc. Phys. Soc.*, A, **64**, 509 (1951)) has also reported structure 2500–2330 Å in absorption in a King furnace.

InI

Two overlapping systems, $A^3\Pi_{0^+} - X^1\Sigma^+$ and $B^3\Pi_1 - X^1\Sigma^+$, 4293–3948 Å with piled up (0, 0) sequences at 4098·5 (deg. violet) and 3993·4 (maximum) have been observed in a high-frequency discharge and in absorption in a furnace. Also in fluorescence (W. Wenk, see below). There is a continuum with maximum at about 3173 Å.

Reference. M. Wehrli and E. Miescher, *Helv. Phys. Acta*, 7, 298 (1934)†.

InI$_2$

W. Wenk (*Dissertation, Basel*, 1941) has observed absorption continua at 2640, 2465 and 2100 Å.

InO

Bands 4763–3847 have been observed in an arc. Most, but not all, heads are degraded to the red. The analysis is obviously wrong. See also InO⁺. Strong heads ($I = 9$ or 10) 4282·4, 4270·3, 4243·8, 4233·1, 4224·2, 4154·9.

References. M. L. Guernsey, *P.R.*, **46**, 114 (1934).
W. W. Watson and A. Shambon, *P.R.*, **50**, 607 (1936)†.

InO⁺

A red-degraded (0, 0) band, $^1\Pi - {}^1\Sigma$, in a discharge tube is reported at 3744·4 Å.

Reference. V. F. Shevel'kov, D. I. Kataev and A. A. Mal'tsev, *Vestnik. Moskov Univ. Ser II, Khim.*, **24**, 108 (1969).

IrC

References. K. Jansson, R. Scullman and B. Yttermo, *Chem. Phys. Lett.*, **4**, 188 (1969)†.
K. Jansson and R. Scullman, *J. Mol. Spec.*, **36**, 248 (1970)†.

Five systems of red-degraded bands have been observed in a King furnace. Strongest heads of each system:

$E^2\Delta_{\frac{3}{2}} - X^2\Delta_{\frac{3}{2}}$. 8273·9 (1, 1), 8217·1 (0, 0).

$E^2\Delta_{\frac{3}{2}} - X^2\Delta_{\frac{5}{2}}$. 6655·8 (1, 1), 6612·1 (0, 0).

$D^2\Phi_{\frac{7}{2}} - X^2\Delta_{\frac{5}{2}}$. 7509·1 (0, 1), 7022·9 (1, 1), 6960·4 (0, 0), 6540·5 (1, 0).

$K^2\Pi_{\frac{3}{2}} - X^2\Delta_{\frac{5}{2}}$. 5253·8 (1, 1), 5193·8 (0, 0).

$L^2\Phi_{\frac{7}{2}} - X^2\Delta_{\frac{5}{2}}$. 4801·2 (0, 0).

IrO

References. V. Raziunas, G. Macur and S. Katz, *J. Chem. Phys.*, **43**, 1010 (1965).
K. Jansson and R. Scullman, *J. Mol. Spec.*, **43**, 208 (1972)†.

Razuinas *et al.* note red-degraded bands at 6972·7, 5990·5 and 5856·7. Jansson and Scullman observed 15 bands, partly analysed into 3 sub-systems, 6400–4200 Å, in a hollow-cathode discharge. The following seem, from the description, likely to be strong or moderately strong:

λ	Assignment	λ	Assignment
5983·7	0, 1 $D-A'$	4765·4	0, 1 $H''-A''$
5677·9	0, 0 $D-A'$	4736·0	v, 1 $H'-A'$
5469·1	1, 0 $D-A'$	4631·9	
5201·7	0, 3 $H''-A''$	4570·7	0, 0 $H''-A''$
4975·2	0, 2 $H''-A''$	4447·7	
4944·6	v, 2 $H'-A'$	4338·7	

K_2

Strong band systems attributed to K_2 have been observed in the red, blue and violet, and weaker systems are reported in the ultra-violet.

FAR RED SYSTEM, 8850–7700 Å, $A^1\Sigma_u^+ - X^1\Sigma_g^+$

Occurrence. In absorption.

Appearance. Numerous bands in the region $\lambda\lambda 8840–7728$ degraded to longer wavelengths.

References. J. C. McLennan and D. S. Ainslie, *P.R.S.*, **103**, 304 (1923)†.
W. O. Crane and A. Christy, *P.R.*, **36**, 421 (1930).

Crane and Christy record a long list of heads, but no intensities. The following wavelengths may assist the identification: 8702·0 (0, 2), 8634·4 (0, 1), 8566·3 (0, 0), 8515·7 (1, 0), 8468·2 (2, 0). McLennan and Ainslie record maxima at $\lambda\lambda 8602$, 8547, 8492, 8447, 8407, 8375, 8309, 8266 and 8213.

NEAR RED SYSTEM, 6950–6250 Å, $B^1\Pi_u - X^1\Sigma_g^+$

Occurrence. In absorption, in fluorescence and in magnetic rotation.

Appearance. Numerous bands in the region $\lambda\lambda 6922–6280$, degraded to the red.

References. W. R. Fredrickson and W. W. Watson, *P.R.*, **30**, 429 (1927).
W. O. Crane and A. Christy, *P.R.*, **36**, 421 (1930).
F. W. Loomis and R. E. Nusbaum, *P.R.*, **39**, 89 (1932).

K$_2$ (*contd.*)

Strongest heads:

λ	I	v', v''	λ	I	v', v''
6629·8	5	1, 4	6473·9	10	1, 0
6622·8	6	0, 3	6443·2	8	2, 0
6583·4	9	0, 2	6413·0	7	3, 0
6544·1	8	0, 1	6383·7	5	4, 0
6512·6	5	1, 1			

BLUE SYSTEM, 4510–4220 Å, C$^1\Pi_u$ − X$^1\Sigma_g^+$

Occurrence. In absorption and in fluorescence.

Appearance. Degraded to the red.

References. H. Yamamoto, *Japan J. Phys.*, **5**, 153 (1929).

S. P. Sinha, *Proc. Phys. Soc.*, **60**, 436 (1948).

E. W. Robertson and R. F. Barrow, *Proc. Chem. Soc.*, 329 (1961).

Strongest heads:

λ	I	v', v''	λ	I	v', v''	λ	I	v', v''
4460·3	5	0, 6	4367·8	5	2, 2	4320·9	7	3, 0
4442·6	6	0, 5	4361·0	6	1, 1	4310·0	7	4, 0
4425·5	6	0, 4	4355·1	8	0, 0	4299·0	8	5, 0
4407·7	5	0, 3	4349·7	6	2, 1	4294·5	5	7, 1
4390·2	5	0, 2	4343·5	10	1, 0	4288·4	8	6, 0
4378·4	6	1, 2	4338·1	5	3, 1	4277·6	6	7, 0
4372·9	7	0, 1	4332·3	7	2, 0	4273·6	6	9, 1

VIOLET SYSTEM, 4165–3940 Å, D − X$^1\Sigma_g^+$

Occurrence. In absorption.

Appearance. Degraded to the red.

References. H. Yoshinaga, *Proc. Phys. Math. Soc. Japan*, **19**, 847 (1937).

S. P. Sinha, *Proc. Phys. Soc.*, **60**, 436 (1948).

Strongest heads:

λ	I	v', v''	λ	I	v', v''	λ	I	v', v''
4127·6	5	1, 5	4099·1	5	4, 5	4033·5	6	3, 0
4123·6	5	3, 6	4097·4	7	1, 3	4024·9	6	4, 0
4122·7	6	0, 4	4092·3	8	0, 2	4016·3	5	5, 0
4119·8	5	5, 7	4087·5	6	2, 3	4008·0	5	6, 0
4112·8	8	1, 4	4082·7	10	1, 2	3999·6	5	7, 0
4108·6	6	3, 5	4078·2	6	3, 3	3991·6	5	8, 0
4107·3	7	0, 3	4067·0	8	1, 1	3984·7	5	11, 1
4103·0	6	2, 4	4057·9	5	2, 1	3978·0	5	12, 1

ULTRA-VIOLET SYSTEM, 3924–3686 Å, E − X$^1\Sigma_g^+$

Occurrence. In absorption.

Appearance. A weak system of about 30 bands, degraded to the red.

Reference. S. P. Sinha, *Proc. Phys. Soc.*, A, **63**, 952 (1950).

K₂ (*contd.*)

Strongest heads:

λ	I	v', v"	λ	I	v', v"	λ	I	v', v"
3725·3	6	6, 0	3762·8	7	3, 1	3789·2	7	3, 3
3733·8	7	5, 0	3771·5	8	2, 1	3793·7	10	1, 2
3738·1	7	6, 1	3776·0	7	3, 2	3797·6	10	2, 3
3746·6	7	5, 1	3784·4	7	2, 2	3806·2	7	1, 3

OTHER BANDS, λλ3640–3420

Sinha (see above) has also observed a number of weak absorption bands in this region. Strongest: λλ3590·7, 3586·4, 3583·5, 3575·8, 3568·4.

KAr

A diffuse band at 7370 Å accompanies the potassium resonance lines in emission from rapidly compressed potassium vapour in argon.
Reference. G. T. Lalos and G. L. Hammond, *Astrophys. J.*, **135**, 616 (1962)†.

KBr

The heated vapour shows strong continuous absorption in the middle ultra-violet, merging into diffuse banded absorption in the near ultra-violet.
Reference. K. Sommermeyer, *Z.P.*, **56**, 548 (1929).

KCd

Diffuse bands, degraded to the violet, observed in absorption by Barratt (see CsCd). Heads: λλ4191, 4172.

KCs

Reference. J. M. Walter and S. Barratt, *P.R.S.*, **119**, 257 (1928).
Diffuse absorption band at λ5387.

KF

Occurrence. In absorption in a furnace.
Reference. R. F. Barrow and A. D. Caunt, *P.R.S.*, **219**, 120 (1953).
A continuum and regularly spaced diffuse bands which have a maximum at about 2140 Å and extend to longer wavelengths with increasing temperature, the longest wavelength recorded being 2913 Å.

KH

References. G. M. Almy and C. D. Hause, *P.R.*, **42**, 242 (1932).
T. Hori, *Mem. Ryojun Coll. Eng.*, **6**, 1, 33 (1933).
G. M. Almy and A. Beiler, *P.R.*, **61**, 476 (1942).
I. R. Bartky, *J. Mol. Spec.*, **20**, 299 (1966).

KH (*contd.*)

5100 Å SYSTEM, $^1\Sigma - {}^1\Sigma$, GROUND STATE

System of the many-lined type with weak R-heads degraded to the red. The system is readily obtained in absorption from a mixture of hydrogen and potassium vapour and in emission from a potassium arc in hydrogen or from discharges through a mixture of hydrogen and potassium vapour.

Origins of strongest bands:

λ_0	v', v''	λ_0	v', v''	λ_0	v', v''	λ_0	v', v''
5613·4	4, 2	5183·3	6, 1	4802·8	8, 0	4547·5	12, 0
5528·7	5, 2	5108·3	7, 1	4736·7	9, 0	4487·7	13, 0
5444·7	6, 2	5034·0	8, 1	4672·2	10, 0	4429·7	14, 0
5362·3	7, 2	4960·6	9, 1	4608·9	11, 0	4373·5	15, 0
5259·0	5, 1	4870·2	7, 0				

Diffuse bands, 4000–3000 Å, probably due to a $^1\Pi - {}^1\Sigma$ transition, have also been noted.

KHg

Diffuse bands, degraded to the violet, observed in absorption by Barratt (see CsCd). Heads: λλ6188, 6150, 4113 and 3988.

KI

The heated vapour shows strong continuous absorption in the near ultra-violet, merging into diffuse banded absorption in the violet.

Reference.   K. Sommermeyer, *Z.P.*, **56**, 548 (1929)†.

KMg

In absorption a red-degraded band at 6549·7 and a violet-degraded one at 4611 have been reported.

Reference.   S. Barratt, *P.R.S.*, **109**, 194 (1934).

KRb

Reference.   J. M. Walter and S. Barratt, *P.R.S.*, **119**, 257 (1928).
Diffuse absorption band at λ4959.

KZn

Diffuse band, degraded to the violet, observed in absorption by Barratt (see CsCd). Head: λ4147.

Kr₂ or Kr₂⁺

Regularly spaced bands, about 2 cm.$^{-1}$ apart, on the long-wave side of the krypton line 7601·5 Å, have been observed in a discharge at 10 atm.

Reference.   L. Herman and R. Herman, *Nature*, **195**, 1086 (1962)†.

KrF_2 ?

In a microwave discharge through krypton and lithium fluoride at $300°C$ diffuse bands, shaded to shorter wavelengths, were found; maxima 2485 and 2220 Å.

Reference. S. L. N. G. Krishnamachari, N. A. Narasimham and M. Singh, *Curr. Sci.* (*India*), **34**, 75 (1965)†.

KrO

Reference. C. D. Cooper, G. C. Cobb and E. L. Tolnas, *J. Mol. Spec.*, **7**, 223 (1961)†.

A system of red-degraded bands near the forbidden oxygen line λ5577 in a discharge through krypton at 1 atm. with trace of oxygen present. Heads λλ5561·6, 5543·1, 5521·2, 5510·5, 5501·6, 5494·8, 5489·4, 5482·3.

KrXe

Occurrence. Beam of fast electrons through mixed krypton and xenon.

Appearance. A strong band from a head at 4950 Å, extending to 4600 Å.

Reference. W. Friedl, *Z. Naturforsch.*, **14a**, 848 (1959)†.

La_2

Occurrence. Emission and absorption in a furnace at $2000°C$.

Appearance. Two systems, both in the region 6100–6040 Å, with individual bands degraded to the violet. The B system has a red-degraded head of heads at 6047 and the A system shows a strong sequence at 6073 Å and a weaker group at 6066. The vibrational analyses are not convincing.

Reference. P. Carette and J. M. Blondeau, *C. R. Acad. Sci.* (*Paris*), B, **269**, 16 (1969)†.

LaF

References. E. A. Shenyavskaya, L. V. Gurvich and A. A. Mal'tsev, *Vestnik. Moskov Univ.*, *Ser. II, Khim.*, **20**, 10 (1965); *Opt. Spectrosc.*, **24**, 556 (1968).

R. F. Barrow, N. W. Bastin, D. L. G. Moore and C. J. Pott, *Nature*, **215**, 1072 (1967).

SYSTEMS IN VISIBLE REGION

Occurrence. Emission and absorption in a furnace.

Appearance. A number of systems, all degraded to the red, mostly with clear sequences, have been observed.

The following are the (0, 0) heads and suggested transitions:

λ R head	λ Q head	Trans.	λ R head	Trans.
8570·3		$^1\Sigma - X^1\Sigma$	4768·0	$^1\Pi - X^1\Sigma$
6175·8		$^1\Pi - X^1\Sigma$	4446·1	$^1\Sigma - X^1\Sigma$
5461·2	5464·2	$^3\Phi_2 - {}^3\Delta_1$	4428·6	$^1\Sigma - X^1\Sigma$
5355·6	5398·1	$^3\Phi_3 - {}^3\Delta_2$		
5312·0	5315·0	$^3\Phi_4 - {}^3\Delta_3$		

LaF (*contd.*)

ULTRA-VIOLET

Shenyavskaya *et al.* (1968) reported a violet-degraded system in a discharge. The stronger (0, 0) sequence has a P head at 3307·7 and Q head at 3306·1. Other P heads 3374·6 (0, 1) and 3178·3 (1, 0). Barrow *et al.* have some doubt about the assignment of this system to LaF.

LaO

Strong systems have been observed in the far red, the yellow and the blue, and there are weaker systems in the near red and near ultra-violet.

Occurrence.   In arcs fed with lanthanum salts and in flames.

References.   W. F. Meggers and J. A. Wheeler, *J. Res. Nat. Bur. Stand.*, **6**, 239 (1931)†.
W. Jevons, *Proc. Phys. Soc.*, **41**, 520 (1929).
F. A. Jenkins and A. Harvey, *P.R.*, **39**, 922 (1931).
S. Hautecler and B. Rosen, *Bull. Acad. Roy. Belg.*, **45**, 790 (1959).
D. W. Green, *J. Mol. Spec.*, **38**, 155 (1971)†; *Canad. J. Phys.*, **49**, 2552 (1971)†.
P. Carette and J. Houdart, *C. R. Acad. Sci.* (*Paris*), B, **272**, 595 (1971)†.

FAR RED SYSTEM, 9729–6867 Å, $A^2\Pi - X^2\Sigma$

Appearance.   Degraded to longer wavelengths. Marked sequences.

The following are the strongest heads at the beginning of the main sequences. Measurements are by Meggers and Wheeler; intensities have been reduced to a scale of 10. The two sub-bands due to the doubling of the $^2\Pi$ state are denoted by i and ii.

λ	I	v', v"	λ	I	v', v"
6994·5	1	1, 0 i R	7877·2	6	0, 0 ii R
7011·2	2	1, 0 i Q	7910·5	8	0, 0 ii Q
7023·6	1	2, 1 i R	7912·3	3	1, 1 ii R
7040·8	2	2, 1 i Q	7944·9	6	1, 1 ii Q
7054·8	1	3, 2 i R	7947·9	2	2, 2 ii R
7070·8	2	3, 2 i Q	7979·7	5	2, 2 ii Q
7379·8	8	0, 0 i R	8406·0	1	0, 1 ii R
7403·5	10	0, 0 i Q	8443·3	2	1, 2 ii R
7411·3	5	1, 1 i R	8453·5	3	0, 1 ii Q
7434·3	6	1, 1 i Q	8481·0	1	2, 3 ii R
7442·9	4	2, 2 i R	8489·9	3	1, 2 ii Q
7465·2	5	2, 2 i Q	8526·6	3	2, 3 ii Q

NEAR RED SYSTEM, 6825–6500 Å, $C^2\Pi - A'^2\Delta$

Appearance.   Degraded to the violet. Strong (0, 0) sequences. The system is split into two sub-systems, and each band shows close double P heads and close double Q heads, due to nuclear hyperfine structure. P heads:

v', v"	$C^2\Pi_{\frac{1}{2}} - A'^2\Delta_{\frac{3}{2}}$	$C^2\Pi_{\frac{3}{2}} - A'^2\Delta_{\frac{5}{2}}$
0, 0	6607·8*	6821·5
1, 1	6597·0	6808·9
2, 2		6796·4

* The Q head is at 6598·9.

LaO (*contd.*)

YELLOW SYSTEM, λλ6450–5015, $B^2\Sigma - X^2\Sigma$

Appearance. Degraded to the red. Marked sequences. The bands show close double heads (R_1 and R_2) separation about 2·5 Å.

The following measurements of the first of the close double heads are by Jevons. Intensities, where given, are based on Meggers and Wheeler's reduced to a scale of 10. Only the prominent heads are given.

λ	I	v', v"	λ	I	v', v"	λ	I	v', v"
5178·3		2, 0	5599·9	10	0, 0	5920·7	4	2, 3
5202·7		3, 1	5626·0	5	1, 1	6157·4	0	0, 2
5380·4	1	1, 0	5652·3	3	2, 2	6185·7	1	1, 3
5405·6	2	2, 1	5866·3	3	0, 1	6214·1	1	2, 4
5430·9	2	3, 2	5893·4	4	1, 2	6242·6	2	3, 5

BLUE SYSTEM, 4348–4622 Å, $C^2\Pi - X^2\Sigma$

Appearance. Degraded to the red. Two strong and some weaker sequences of very close (separation 0·1 Å) double-headed bands.

The following are the bands forming the heads of the sequences. Measurements compiled from Jevons, and Meggers and Wheeler. Intensities, where given, are based on those given by Meggers and Wheeler reduced to a scale of 10. The two sub-bands due to the $^2\Pi$ level are denoted by i and ii.

λ	I	v', v"	λ	I	v', v"	λ	I	v', v"
4348·2			4379·6	4	2, 2 i	4433·0	6	3, 3 ii
4352·2			4383·4	3	3, 3 i	4580·7	1	0, 0 ii
4356·3			4418·1	10	0, 0 ii	4585·0	1	1, 2 ii
4371·9	8	0, 0 i	4423·1	8	1, 1 ii	4589·4	1	2, 3 ii
4375·7	6	1, 1 i	4428·0	6	2, 2 ii	4593·8	1	3, 4 ii

ULTRA-VIOLET SYSTEMS, λλ3457–3709

Appearance. Degraded to shorter wavelengths.

Transition.   Analysed by Hautecler and Rosen into two overlapping systems, $D(?^2\Sigma) - X^2\Sigma$ and $F(?^2\Sigma) - X^2\Sigma$.

The following are the stronger bands listed by Jevons, with assignments by Hautecler and Rosen:

System D			*System F*		
λ	I	v', v"	λ	I	v', v"
3709·6	3	?	3666·4	2	1, 2
3614·9	6	3, 3	3660·9	2	2, 3
3611·5	6	2, 2	3566·2	5	0, 0
3608·1	6	1, 1	3560·9	4	1, 1
3604·6	6	0, 0	3556·3	3	2, 2
			3457·1	1	2, 1

LaS

FAR RED SYSTEM, 8500–7000 Å, $B^2\Sigma^+ - X^2\Sigma^+$

Occurrence. Emission and absorption in a furnace.

Appearance. Degraded to the red. Four close (1 Å) heads due to spin splitting and hyperfine structure.

LaS (*contd.*)

Reference.   M. Marcano and R. F. Barrow, *J. Phys.*, B, **3**, L 121 (1970).
First of close R heads:

λ	v', v''	λ	v', v''
7533·2	1, 2	7075·2	2, 1
7507·9	0, 1	7051·0	1, 0
7260·0	0, 0		

Li₂

Occurrence.   Two systems have been observed in the red and blue-green in absorption, magnetic rotation and fluorescence. Two more systems in the ultra-violet have been observed in absorption.

RED SYSTEM, λλ7700–6550, $A^1\Sigma_u^+ - X^1\Sigma_g^+$

Appearance.   Degraded to longer wavelengths.
Reference.   K. Wurm, *Z.P.*, **59**, 35 (1930).
Strong bands:

λ	I	v', v''	λ	I	v', v''
7690·3	8	0, 2	6883·9	10	2, 0
7309·8	8	0, 1	6768·7	8	3, 0
7177·4	8	1, 1	6659·3	8	4, 0
7003·7	8	1, 0			

BLUE-GREEN SYSTEM, λλ5600–4500, $B^1\Pi_u - X^1\Sigma_g^+$

Appearance.   Degraded to the red.
Reference.   A. McKellar, *P.R.*, **44**, 155 (1933).
Strongest bands:

λ	v', v''
4985·6	0, 1
4901·0	0, 0
4838·3	1, 0
4778·6	2, 0

ULTRA-VIOLET SYSTEM, λλ3500–3100, $C^1\Pi_u - X^1\Sigma_g^+$

References.   S. P. Sinha, *Proc. Phys. Soc.*, **60**, 443 (1948).
R. F. Barrow, N. Travis and C. V. Wright, *Nature*, **187**, 141 (1960).
Appearance.   Degraded to the red.
Strongest bands (from Sinha):

λ	I	v', v''	λ	I	v', v''
3431·2	4	0, 4	3277·6	6	0, 0
3404·4	4	1, 4	3253·1	10	1, 0
3392·1	6	0, 3	3206·9	6	3, 0
3353·6	10	0, 2	3184·2	6	4, 0
3315·6	9	0, 1	3162·0	5	5, 0

Li₂ (*contd.*)

ULTRA-VIOLET SYSTEM, λλ3100–2500, $D^1\Pi_u - X^1\Sigma_g^+$

No data available. See Barrow *et al.* above.

LiBr

Occurrence. In absorption in a furnace.

Reference. R. S. Berry and W. Klemperer, *J. Chem. Phys.*, **26**, 724 (1957).

A continuum with regularly spaced diffuse bands on the long wavelength side at λλ3167, 3223, 3281, 3346, 3395.

LiCl

Occurrence. In absorption in a furnace.

Reference. R. S. Berry and W. Klemperer, *J. Chem. Phys.*, **26**, 724 (1957).

A continuum with regularly spaced diffuse bands on the long wavelength side at λλ2805, 2854, 2899.

LiCs

Bands, degraded to the red, have been observed in absorption by a mixture of lithium and cæsium vapours.

References. J. M. Walter and S. Barratt, *P.R.S.*, **119**, 257 (1928).

W. Weizel and M. Kulp, *Ann. Phys. (Lpz.)*, **4**, 971 (1930).

Strongest bands: λλ6255, 6217, 6180, 6146 and 6116.

LiH

References. G. Nakamura, *Z.P.*, **59**, 218 (1930).

F. H. Crawford and T. Jorgensen, *P.R.*, **47**, 932 (1935).

R. Velasco, *Canad. J. Phys.*, **35**, 1204 (1957).

3900 Å SYSTEM, $A\,^1\Sigma^+ - X\,^1\Sigma^+$

This system presents a many-lined appearance which recalls the spectrum of molecular hydrogen. Weak R heads degraded to the red are formed near the origins, but are not obvious on casual inspection. The system is readily obtained in absorption from a mixture of hydrogen and lithium vapour and in emission from a lithium arc in hydrogen or a discharge through a mixture of hydrogen and lithium vapour.

Origins of the strongest bands:

v', v''	λ_0	v', v''	λ_0	v', v''	λ_0
0, 2	4297·5	3, 1	3918·5	6, 0	3574·6
1, 2	4245·2	2, 0	3767·2	7, 0	3526·4
2, 2	4189·5	3, 0	3720·2	8, 0	3478·8
1, 1	4020·7	4, 0	3672·0	9, 0	3432·2
2, 1	3970·7	5, 0	3623·3		

LiH (*contd.*)

3080–2880 SYSTEM, B $^1\Pi - X\,^1\Sigma^+$

A system of red-degraded bands has been observed in absorption by Velasco. The heads are not well developed, because of the strong degradation. Origins:

v', v''	λ_0	v', v''	λ_0
0, 1	3033·8	0, 0	2913·5
1, 1	3021·8	1, 0	2902·5
2, 1	3017·6	2, 0	2898·6

TRIPLET SYSTEM ?

Crawford and Jorgensen note a close grouping of lines in the far red which they suggest belong to another system.

LiI

Occurrence. In absorption in a furnace.

Reference. R. S. Berry and W. Klemperer, *J. Chem. Phys.*, **26**, 724 (1957).

A continuum with regularly spaced diffuse bands on the long wavelength side at λλ3430, 3504, 3573, 3713, 3788, 3863, 3948, 4020, 4079.

LiK

References. See LiCs.

Absorption bands degraded to the red. Strongest bands: λλ5838·7, 5769·3, 5724·0, 5700·0, 5658·0 and 5620·0.

LiRb

References. See LiCs.

Absorption bands degraded to the red. Strongest bands: λλ5815·5, 5778·0, 5743·0 and 5712·5.

LuO

A band at about 7500 Å is reported when lutecium chloride is introduced into a flame.

Reference. L. A. Ovchar, S. A. Mishchenko and N. S. Poluektov, *Zh. Priklad. Spektrosk.*, **3**, 306 (1965).

LuF

Seven systems, each consisting of about 20 red-degraded bands, have been observed in a hollow-cathode discharge.

Reference. J. d'Incan, C. Effantin and R. Bacis, *J. Phys.*, B, **5**, L 189 (1972).

This is a preliminary note. The following are the heads of the (0, 0) bands: λλ6182·8, 5950·3, 5291·9, 4990·0, 4087·1, 3872·8 and 3009·4.

LuH

References. J. d'Incan, C. Effantin and R. Bacis, *J. Phys.*, B, **5**, L 187 (1972).
C. Effantin and J. d'Incan, *Canad. J. Phys.*, **51**, 1395 (1973); **52**, 523 (1974).

Eight systems have been observed in a hollow-cathode discharge. Most show only a (0, 0) sequence of bands which are weakly degraded to the red. The following are the origins, λ_0, which are always close to the Q heads for $\Pi - \Sigma$ transitions, and the R heads, λ_H, where available, for most of the known bands. The systems C and H are strongest, while B and D are very weak.

λ_0	λ_H	v', v''	System	λ_0	λ_H	v', v''	System
7804·2		2, 2	$A^1\Pi - X^1\Sigma$	5863·5	5818·3	0, 0	$D^1\Pi - X^1\Sigma$
7747·2		1, 1	"	5648		1, 1	$E^1\Sigma - X^1\Sigma$
7696·9		0, 0	"	5637·7		0, 0	"
6588·7		1, 1	$B^1\Sigma - X^1\Sigma$	5283·5		0, 0	$F^1\Pi - X^1\Sigma$
6547·0		0, 0	"	5075·7		1, 1	$G^1\Sigma - X^1\Sigma$
6025·7	6007·4	1, 1	$C^1\Pi - X^1\Sigma$	5057·5		0, 0	"
5978·5	5959·3	0, 0	"	4259		1, 1	$H^1\Pi - X^1\Sigma$
5909·0	5871·3	1, 1	$D^1\Pi - X^1\Sigma$	4249·6		0, 0	"

LuO

Occurrence. Lutecium salts in an arc or flame.

Appearance. Marked red-degraded sequences at 4661 and 5171 Å and a number of weaker bands.

References. W. W. Watson and W. F. Meggers, *J. Res. Nat. Bur. Stand.*, **20**, 125 (1938)†.
A. Gatterer and S. G. Krishnamurthy, *Proc. Phys. Soc.*, A, **65**, 151 (1952)†.
C. B. Suaréz, *J. Phys.*, B, **3**, 1389 (1970).
C. Effantin, R. Bacis and J. d'Incan, *C. R. Acad. Sci. (Paris)*, B, **273**, 605 (1971).

The bands have been arranged into three systems α, β and γ. The β may be a $^2\Pi_{\frac{3}{2}} - {}^2\Sigma$ transition. The following are the outstanding sequences or strong bands from the first two references; we have adjusted intensities. Effantin *et al.* give different data and analysis for the γ bands.

λ	I	v', v''	System	λ	I	v', v''	System	λ	I	v', v''	System
5463·9	7	4, 5	γ	5034·6	6	5, 4	γ	4560·9	2	6, 5	β
5448·6	8	3, 4	γ	5019·3	5	4, 3	γ	4252·5	4	1, 2	γ
5217·4	8	3, 3	γ	4695·5	6	3, 3	β	4241·8	6	0, 1	γ
5185·4	7	1, 1	γ	4684·2	8	2, 2	β	4105·8	4	1, 1	γ
5171·4	10	0, 0	γ	4672·3	9	1, 1	β	4094·0	7	0, 0	γ
5050·2	5	6, 5	γ	4661·7	10	0, 0	β	3978·4	2	2, 1	γ

There are also two regions of continuum in the red.

Mg₂

References. H. Hamada, *Phil. Mag.*, **12**, 50 (1931)†.
W. J. Balfour and A. E. Douglas, *Canad. J. Phys.*, **48**, 901 (1970)†.

Balfour and Douglas found a band system, $A^1\Sigma_u^+ - X^1\Sigma_g^+$, of violet-degraded bands with rather indistinct heads, to the red of the Mg line 2852 Å. Of the large number of absorption bands recorded, the following in the region 3600–3400 may be prominent: $\lambda\lambda$3538·7 (12, 0), 3518·7 (13, 0), 3498·9 (14, 0), 3479·8 (15, 0), 3461·0 (16, 0), 3442·8 (17, 0), 3425·0 (18, 0), 3407·5 (19, 0).

Hamada also reported emission in a hollow-cathode discharge.

MgBr

VIOLET SYSTEM, $A^2\Pi - X^2\Sigma^+$

Occurrence. In absorption and in emission from flames, high-frequency discharges and probably arcs.

References. F. Morgan, *P.R.*, **50**, 603, (1936).

M. M. Patel and P. D. Patel, *J. Phys.*, B, **2**, 515 (1969).

Main features:

3936·8 P_1 head of weak (0, 1) band and sequence.

3880·5 P_1 head of strong (0, 0) band and sequence.

3877·8 P_1 head of (1, 1) band.

3864·1 P_2 head of strong (0, 0) band and sequence.

3823·4 P_1 head of rather weak (1, 0) band and sequence.

ULTRA-VIOLET SYSTEM, 2720–2540 Å, $C^2\Sigma^+ - X^2\Sigma^+$

Occurrence. In high-frequency discharge.

Appearance. Degraded to the red. Single-headed.

The following are among the strongest of the 18 bands recorded by Patel and Patel (see above):

λ	I	v', v''	λ	I	v', v''	λ	I	v', v''
2672·3	6	0, 5	2628·6	7	1, 4	2597·0	2	0, 2
2660·9	6	2, 6	2621·8	3	0, 3	2579·5	5	1, 2
2653·6	7	1, 5	2611·4	6	2, 4	2572·4	3	0, 1
2646·9	7	0, 4	2603·9	10	1, 3	2548·1	3	0, 0

MgCl

MAIN VIOLET SYSTEM, $A^2\Pi - X^2\Sigma^+$

Occurrence. In absorption and in emission from arcs, flames and high-frequency discharge.

Appearance. Double double-headed bands degraded to the violet.

References. F. Morgan, *P.R.*, **50**, 603 (1936).

M. M. Patel and P. D. Patel, *Indian J. Phys.*, **42**, 254 (1968)†.

3845·2 P_1 head of (0, 1) band and sequence.

3778·9 P_1 head of strong (0, 0) band and sequence.

3775·3 P_1 head of (1, 1) band.

3770·2 P_2 head of (0, 0) band and sequence.

3711·0 P_1 head of (1, 0) band and sequence.

WEAK BLUE SYSTEM, $B^2\Sigma - A^2\Pi$

Occurrence. High frequency discharge.

Appearance. Double double-headed bands shaded to the violet.

Reference. V. S. Rao and P. T. Rao, *Indian J. Phys.*, **37**, 640 (1963).

Outstanding heads:

λ	I	Head	λ	I	Head
4843·5	4	0, 1 $^0P_{12}$	4717·5	10	0, 0 Q_1
4827·7	5	0, 1 Q_1	4599·4	3	1, 0 P_1
4731·7	5	0, 0 $^0P_{12}$	4597·4	6	1, 0 Q_1
4729·8	10	0, 0 P_2			

MgF

There are three systems due to MgF in the ultra-violet, the strongest $\lambda\lambda 3686-3468$, and weaker systems $\lambda\lambda 2742-2649$ and $\lambda\lambda 2387-2249$.

STRONG SYSTEM, 3686–3468 Å, $A^2\Pi - X^2\Sigma^+$

Occurrence. Carbon arc fed with MgF_2, and also in absorption.
Appearance. Degraded to shorter wavelengths. Well-marked double-headed sequences.
References. S. Datta, *P.R.S.*, **99**, 436 (1921)†.
C. A. Fowler, *P.R.*, **59**, 645 (1941)†.

Outstanding features:

λ	I	
3685·8	5	Head of (0, 1) sequence.
3594·2	10	P_1 head of (0, 0) band and sequence.
3592·8		P_2 head of (0, 0) band.
3588·4		Q_2 head of (0, 0) band.
3503·4	4	P_1 head of (1, 0) band and sequence.

SYSTEM 2742–2649 Å, $B^2\Sigma^+ - X^2\Sigma^+$

Occurrence. Carbon arc fed with MgF_2, and also in absorption.
Appearance. Degraded to shorter wavelengths. Well-marked sequences of single-headed bands.
Reference. W. Jevons, *P.R.S.*, **122**, 211 (1929)†.

Strongest bands:

λ	I	v', v''
2741·6	2	0, 1
2689·3	5	0, 0
2686·5	4	1, 1
2636·5	3	1, 0

SYSTEM 2387–2249 Å, $C^2\Sigma^+ - X^2\Sigma^+$

Occurrence. In absorption.
Appearance. Degraded to shorter wavelengths. Well-marked sequences.
Reference. C. A. Fowler, *P.R.*, **59**, 645 (1941)†.

Outstanding heads:

λ	I	v', v''
2387·9	2	0, 1
2347·8	10	0, 0
2342·1	5	1, 1
2303·6	5	1, 0
2298·6	4	2, 1

Absorption bands around 2275 Å reported by Jenkins and Grinfeld (*P.R.*, **43**, 943 (1933)) are due to AlF.

MgH

Ten systems have been reported for MgH, of which the 5211 Å and 2430 Å systems are the most prominent.

MgH (*contd.*)

5211 Å SYSTEM, $A\,^2\Pi - X\,^2\Sigma^+$

References. A. Guntsch, *Dissertation, Stockholm* (1939).

W. W. Watson and P. Rudnick, *Astrophys. J.*, **63**, 20 (1926).

W. W. Watson and P. Rudnick, *P.R.*, **29**, 413 (1927).

Bands degraded to the violet with P, Q and R branches consisting of narrow doublets. Obtained in emission in the magnesium arc in hydrogen, in discharge tubes containing magnesium vapour and hydrogen, and in absorption in spectra of sun spots. See Plate 5.

	Heads	
v', v''	P	Q
0, 2	6083	
0, 1	5621·4	5609·5
1, 2	5568·3	5549·9
2, 3	5516·4	
0, 0	5211·0	5186·4
1, 1	5182·3	
2, 2	5155·2	
1, 0	4845	

2420 Å SYSTEM, $C\,^2\Pi - X\,^2\Sigma^+$

References. R. W. B. Pearse, *P.R.S.*, **122**, 442 (1929)†.

A. Guntsch, *Z.P.*, **93**, 534 (1935).

L. A. Turner and W. T. Harris, *P.R.*, **52**, 626 (1937).

Bands degraded to the violet similar in appearance to the above bands, except that the doubling is much smaller. They occur under similar circumstances.

	Heads	
v', v''	P	Q
0, 1	2515·3	2510·5
1, 2		2495
0, 0	2429·2	2424·1
1, 1		2413·0
1, 0		2332

4550 Å SYSTEM, $C\,^2\Pi - A\,^2\Pi$

Reference. A. Guntsch, *Z.P.*, **87**, 312 (1934).

Weak bands of headless type with P and R branches composed of narrow doublets.

v', v''	Origins
0, 0	4553
1, 1	4535

4405 Å SYSTEM, $D\,^2\Sigma^- - A\,^2\Pi$

Reference. A. Guntsch, *Z.P.*, **104**, 584 (1937).

Band with two strong heads degraded to the violet.

	Heads			
v', v''	Q_1	Q_2	P_1	P_2
0, 0	4371·9	4373·0	4404·5	4404·6

MgH (*contd.*)

2590 Å SYSTEM, $B^2\Sigma^+ - X^2\Sigma^+$

Reference.   A. Guntsch, *Z.P.*, **104**, 584 (1937).

Band with P and R branches. The R branch shows a peculiar crowding of lines at 2590·4 which gives an intensity maximum without forming a head. Obtained in absorption and emission.

v', v''	Origin
0, 0	2597

2819 Å SYSTEM, $E^2\Sigma - X^2\Sigma^+$

Appearance.   Degraded to shorter wavelengths.

Occurrence.   In absorption, using Xe arc as background to heavy-current discharge through H_2 with Mg in capillary.

Reference.   M. A. Khan, *Proc. Phys. Soc.*, **77**, 1133 (1961)†; **80**, 209 (1962)†.

λ	v', v''
2819·2	0, 0

2702 Å SYSTEM, $F^2\Sigma - X^2\Sigma^+$

Occurrence and reference as above. This is a headless (0, 0) band with origin at 2702.

2172 Å SYSTEM, $G^2\Sigma^+ - X^2\Sigma^+$

Occurrence and reference as for 2819 Å. Degraded to shorter wavelengths. P head of (0, 0) at 2172·0 Å.

2100 Å SYSTEM, $H^2\Sigma^+ - X^2\Sigma^+$

Occurrence and reference as 2819 Å band. Weakly degraded to the red. R head of (0, 0) at 2100·4.

2088 Å SYSTEM, $I^2\Pi - X^2\Sigma$

Occurrence and reference as 2819 Å. Shaded to shorter wavelengths. The (0, 0) band has a P head at 2087·9 and Q at 2082·8.

MgH⁺

References.   A. Guntsch, *Dissertation, Stockholm* (1939).
A. Guntsch, *Z.P.*, **107**, 420 (1937).
R. W. B. Pearse, *P.R.S.*, **125**, 157 (1929).
W. J. Balfour, *Canad. J. Phys.*, **50**, 1082 (1972)†.

2806 Å SYSTEM, $A^1\Sigma - X^1\Sigma$

An extensive system of bands each with a single P and single R branch. Obtained from a magnesium arc in hydrogen at low pressures or from a discharge tube containing magnesium vapour and hydrogen under conditions favouring ionization. See Plate 5.

Origins of Strongest Bands

v', v''	λ	v', v''	λ
0, 4	3388·8	0, 1	2940·6
0, 3	3232·5	1, 1	2847·0
0, 2	3083·0	0, 0	2805·8

MgH^+ (*contd.*)

Origins of Strongest Bands

v', v''	λ	v', v''	λ
1, 0	2720·5	3, 0	2567·9
2, 0	2641·3	4, 0	2499·7

$B^1\Pi - X^1\Sigma$ SYSTEM

Bands degraded to the red, each with single P, Q and R branches.

v', v''	*R Heads*	*Q Heads*	v', v''	*R Heads*	*Q Heads*
0, 5	2357	2358	0, 3	2211	2214
0, 4	2284	2285	0, 2	2142	2144

MgI

Occurrence. In absorption, and in flames.
Appearance. Degraded to violet. Marked sequences.
Reference. F. Morgan, *P.R.*, **50**, 603 (1936).

λ	
4163·7	Head of (0, 1) sequence.
4110·3	Head of strong (0, 0) sequence.
4057·3	Head of (1, 0) sequence.

Walters and Barratt (*P.R.S.*, **118**, 120 (1928)) obtained these bands and other at 3983 (strong, deg. V), 3951 (strong, deg. R), 3927, 3902, 3657, 3628, 3602, etc., in absorption, but the assignment to MgI does not seem quite certain.

MgO

There is a strong system in the green and a weaker system in the red. Complex bands in the violet probably involve three systems of MgO and are also superposed by bands of MgOH when moisture is present. Singh has also reported three more systems in the ultra-violet. The ground state of MgO is assumed to be $X^1\Sigma^+$ but there is also a very low-lying triplet level.

RED SYSTEM, 6900–4700 Å, $B^1\Sigma - A^1\Pi$

Occurrence. Core of magnesium arc in air, and burning magnesium ribbon.
Appearance. Single-headed bands, degraded to violet. Sequences not very obvious.
References. P. C. Mahanti, *P.R.*, **42**, 609 (1932)†.
P. C. Mahanti, *Indian J. Phys.*, **9**, 455 (1935).
A. Lagerqvist and U. Uhler, *Nature*, **164**, 665 (1949).

The following are a few of the strongest bands:

λ	*I*	v', v''	λ	*I*	v', v''	λ	*I*	v', v''
6581·0	3	0, 2	6060·3	6	0, 0	5518·8	4	2, 0
6311·7	4	0, 1	5775·5	5	1, 0	5475·9	3	3, 1
6246·4	3	1, 2	5726·3	3	2, 1	5285·7	3	3, 0

MgO (*contd.*)

GREEN SYSTEM, 5210–4760 Å, $B^1\Sigma - X^1\Sigma$

Occurrence. Mg arc in air, burning Mg ribbon, and in flames, shock tubes and King furnace. Also in sun-spots.

Appearance. Degraded to violet. Very marked (0, 0) sequence.

References. P. C. Mahanti, *P.R.*, **42**, 609 (1932).

A. Lagerqvist, *Ark. Mat. Astr. Fys.*, **A29**, No. 25 (1943)†.

A few weak bands are omitted.

λ	I	v', v''	λ	I	v', v''	λ	I	v', v''
5206·0	4	0, 1	4996·7	9	1, 1	4935·3	4	6, 6
5192·0	4	1, 2	4985·9	8	2, 2	4923·9	3	7, 7
5177·4	3	2, 3	4974·5	7	3, 3	4818·5	3	1, 0
5162·5	3	3, 4	4962·1	6	4, 4	4810·1	3	2, 1
5146·8	3	4, 5	4949·5	5	5, 5	4801·5	3	3, 2
5007·3	10	0, 0						

VIOLET SYSTEMS

References. L. Brewer and R. F. Porter, *J. Chem. Phys.*, **22**, 1867 (1954).

D. Pešić and A. G. Gaydon, *Proc. Phys. Soc.*, **73**, 244 (1959)†.

D. Pešić, *Proc. Phys. Soc.*, **76**, 844 (1960)†.

L. Brewer, S. Trajmar and R. A. Berg, *Astrophys. J.*, **135**, 955 (1962)†.

S. Trajmar and G. E. Ewing, *Astrophys. J.*, **142**, 77 (1965).

The many complex bands 3960–3640 Å which occur readily in arcs, in the flame of Mg ribbon and in furnaces, have proved very difficult to sort out. Overlapping MgOH bands are discussed below. A polyatomic oxide has also been suggested, but now appears less likely. Two systems of red-degraded bands $C^1\Sigma^- - A^1\Pi$ and $D^1\Delta - A^1\Pi$ have now been found and also (L. Brewer, personal communication) a violet-degraded $^3\Delta - ^3\Pi$ system. Wavelengths and relative intensities are based on Pesic and Gaydon, but analyses from the other papers.

$D^1\Delta - A^1\Pi$. Degraded to the red.

λ	I	v', v''	λ	I	v', v''
3821·5	7	3, 3 Q	3805·3	8	0, 0 Q
3815·7	9	2, 2 Q	3804·1	6	1, 1 R
3810·3	8	1, 1 Q	3798·2	7	0, 0 R

$C^1\Sigma^- - A^1\Pi$. Degraded to the red.

λ	I	v', v''	λ	I	v', v''
3788·5	5	3, 3 Q	3777·4	7	1, 1 Q
3784·2	6	3, 3 R	3772·3	8	0, 0 Q
3782·6	5	2, 2 Q	3771·8	6	1, 1 R
3777·8	8	2, 2 R	3766·1	7	0, 0 R

$^3\Delta - ^3\Pi$. Degraded to shorter wavelengths.

Heads and intensities: 3720·7 (10), 3721·0 (9), 3721·4 (9), 3724·9 (6).

SINGH'S ULTRA-VIOLET SYSTEMS

Three red-degraded systems have been found in a direct-current arc.

Reference. M. Singh, *J. Phys.*, B, **4**, 565 (1971)†; B, **6**, 1339 (1973)†; B, **6**, 1917 (1973)†.

MgO (*contd.*)

2652 Å, $E^1\Sigma - X^1\Sigma$. No conspicuous head. (0, 0) origin 2652·9.

2637 Å, $F^1\Pi - X^1\Sigma$. (0, 0) R head 2637·5.

2556 Å, $G^1\Pi - X^1\Sigma$ R heads 2607·6 (0, 1), 2556·7 (0, 0), 2507·1 (1, 0).

There is also a violet-degraded band, $G^1\Pi - A^1\Pi$ at 2750·1.

MgOH

Occurrence.   Mg salts in flame, and arc between Mg poles in water vapour.

References.   D. Pešić and A. G. Gaydon, *Proc. Phys. Soc.*, **73**, 244, (1959)†.

L. Brewer and S. Trajmar, *J. Chem. Phys.*, **36**, 1585 (1962).

In the flame there are two groups of diffuse bands around 3700 and 3830 Å. In the arc the structure is more complex and somewhat overlaid by oxide bands. The following are the strongest heads in the arc source, from Pešić and Gaydon; the letters V or M denote a violet-degraded head or maximum of a headless feature.

λ	I	λ	I	λ	I
3919·1	5 V	3846·9	6 M	3719·6	10 V
3901·2	5 V	3834·8	6*M	3709·4	6 M
3882·5	6 V	3819·2	6 V	3708·8	8 V
3880·2	6 V	3810·2	8*V	3707·9	8*V
3876·8	6 V	3808·8	8*V	3704·1	8 V
3859·7	6 V	3802·4	8 M	3703·2	6*V
3854·9	7 V	3783·4	8 V	3695·1	7*V
3849·7	8 V	3770·6	7 V	3684·4	8 V
3848·6	8 M	3731·8	8 V		

* Relatively strong in the flame.

MgS

4600–4000 Å SYSTEM, $B^1\Sigma - X^1\Sigma$

Occurrence.   In absorption in a furnace at 2000° C. Also reported in emission.

Appearance.   Single-headed bands, shaded to the red.

Reference.   M. Marcano and R. F. Barrow, *Trans. Faraday Soc.*, **66**, 2936 (1970)†.

R heads of all bands. The (0, 0) is the strongest.

λ	v′, v″	λ	v′, v″
4444·3	1, 2	4338·6	0, 0
4439·3	0, 1	4254·3	2, 1
4344·4	1, 1	4247·9	1, 0

MnBr

There are several systems of bands, all rather complex because of the high multiplicity.

Occurrence.   Manganous bromide in high-frequency discharge, or in hollow-cathode discharge. Also in absorption by strongly heated $MnBr_2$.

References.   W. Müller, *Helv. Phys. Acta*, **16**, 1 (1943).

P. Mesnage, *Thesis, Paris* (1938).

J. Bacher, *Helv. Phys. Acta*, **21**, 379 (1948)†.

MnBr (*contd.*)

W. Hayes, *Proc. Phys. Soc.*, A, **68**, 670 and 1097 (1955)†.
W. Hayes and T. E. Nevin, *Proc. Phys. Soc.*, A, **68**, 665 (1955)†.

ULTRA-VIOLET SYSTEM, 3650–4050 Å

This is due to a $^7\Pi - {}^7\Sigma$ transition. Most of the strong heads degrade to shorter wavelengths. The (0, 0) sequence in the region 3834 to 3771 Å has heads at $\lambda\lambda 3823\cdot5$, $3816\cdot5$, $3809\cdot6$, $3808\cdot1$, $3800\cdot9$, $3793\cdot9$, $3783\cdot1$ and $3781\cdot9$. Mesnage reports other sequence heads at $3866\cdot0$, $3772\cdot1$ and $3720\cdot5$.

BLUE-GREEN SYSTEMS, 4900–5100 Å

Degraded to the red. These are analysed into two systems by Hayes. There are three groups, 4915, 4986 and 5068 Å. The following are prominent individual heads, with intensities, $\lambda\lambda 4986\cdot3$ (5), $4990\cdot8$ (8), $4993\cdot7$ (10), $4998\cdot0$ (5), $5072\cdot5$ (7), $5081\cdot2$ (7), $5082\cdot5$ (7), $5083\cdot8$ (6), $5085\cdot1$ (6), $5086\cdot3$ (6).

RED SYSTEM, 6246–6285 Å

Degraded to violet. Two sequences of double-headed bands, possibly due to a $0 - {}^5\Sigma$ transition. Strongest heads:

λ	I	v', v''		λ	I	v', v''
6277·2	8	0, 0 P		6252·4	10	0, 0 Q
6276·3	6	1, 1 P		6251·4	7	1, 1 Q
6257·8	6	0, 0 O		6250·6	6	2, 2 Q

Note. These wavelengths are close to those of CaBr.

INFRA-RED SYSTEMS, 8850–9700 Å

Degraded to larger wavelengths. Hayes has analysed these into two systems one attributed to a $\Pi - \Sigma$ transition, with double heads, and the other to $\Sigma - \Sigma$ with single heads.

About 70 heads of the $\Pi - \Sigma$ are listed; there are sequence heads at $9348\cdot6$ and $9336\cdot0$ Å, with individual strongest heads at $\lambda\lambda 9360\cdot8$, $9362\cdot0$, $9363\cdot6$ and $9365\cdot3$.

The $\Sigma - \Sigma$ system has (2, 0), (1, 0) and (0, 0) sequence heads at $\lambda\lambda 8886\cdot0$, $9124\cdot0$ and $9337\cdot4$.

MnCl

Occurrence. Manganous chloride in high-frequency discharge, or hollow-cathode. Also in absorption by strongly heated $MnCl_2$.

References. W. Müller, *Helv. Phys. Acta*, **16**, 1 (1943).
P. Mesnage, *Thesis, Paris* (1938).
J. Bacher, *Helv. Phys. Acta*, **21**, 379 (1948)†.
W. Hayes, *Proc. Phys. Soc.*, A, **68**, 1097 (1955)†.
W. Hayes and T. E. Nevin, *Nuovo Cimento*, **2**, 734 (1955)†; *Proc. Roy. Irish Acad.*, **57**, 15 (1955).

ULTRA-VIOLET SYSTEM, 3500–4000 Å

Appearance. Degraded to shorter wavelengths. A very complex system with numerous heads (Hayes and Nevin list over 300).

Transition. Probably $^7\Pi - {}^7\Sigma$.

The following are the main heads of the three strongest sequences:

MnCl (*contd.*)

λ	I	Sequence
3770·4	8	(0, 1)
3716·8	10	(0, 0)
3661·3	8	(1, 0)

The strongest heads of the (0, 0) band are as follows: λλ3716·9, 3712·7, 3710·9, 3702·4, 3695·9, 3689·6 and 3686·0; in the (1, 0) sequence: λλ3655·5, 3655·3, 3652·5 and 3651·3.

BLUE VIOLET

Mesnage listed diffuse bands with maxima at λλ4517·3, 4409·5, 4341·2 and 4331·1. They may be due to either MnCl or $MnCl_2$.

BLUE-GREEN SYSTEMS, 4800–5100 Å

Hayes has analysed these into two systems, a $^5\Pi - {}^5\Sigma$ system with (0, 0) sequence 4964 to 5009 Å and another with (0, 0) sequence 5013 to 5030 Å. Most of the heads are degraded to the red. About 150 heads are listed. The (0, 0) band of the first system has R and Q heads at 4966·5 and 4970·8 Å, and there is a strong group of heads 4976–4981 Å. The (0, 1) sequence has a head at 5061·8. The other system has the (0, 0) head at 5012·8.

INFRA-RED SYSTEM, 8180–9150 Å

Degraded to longer wavelengths. The (0, 0) sequence commences at 8750 Å and consists of close double-headed bands. Over 100 heads are listed; the following are the strongest, λλ8752·5, 8758·7, 8761·7, 8762·2, 8765·7, 8767·9, 8771·2. In the (1, 0) sequence, there are fairly strong heads at 8473·4, 8481·0 and in the (0, 1) at 9068·2, 9090·6.

MnF

Several systems have been reported. In a discharge through flowing MnF_2 vapour the ultra-violet system near 3500 Å is strongest, but systems in the blue-green, red and infra-red also occur; these are obtained best in a hollow-cathode discharge in argon and MnF_2. The ultra-violet systems have also been obtained in absorption.

References. G. D. Rochester and E. Olsson, *Z.P.*, **114**, 495 (1939).
J. Bacher, *Helv. Phys. Acta*, **21**, 379 (1948)†.
W. Hayes and T. E. Nevin, *Proc. Phys. Soc.*, A, **68**, 665 (1955)†.
W. Hayes and T. E. Nevin, *Nuovo Cimento*, **2**, 734 (1955)†.
W. Hayes, *Proc. Phys. Soc.*, A, **68**, 1097 (1955).
S. V. K. Rao, S. P. Reddy and P. T. Rao, *Proc. Phys. Soc.*, **79**, 741 (1962)†.

INFRA-RED SYSTEM, 8180–8495 Å

A single complex (0, 0) sequence, with heads degraded in both directions. The following are the strongest of the 36 heads recorded; R or V indicate direction of degradation.

λ	I	λ	I
8225·7 V	7	8282·0 R	3
8238·0 V	7	8364·3 R	6
8254·3 V	5	8377·4 R	5
8266·7 V	4	8390·9 R	3

MnF (*contd.*)

RED SYSTEM, 7300–6230 Å

Five sequences of complex structure, attributed by Hayes and Nevin to a $0 - {}^5\Sigma$ transition. Degraded to longer wavelengths. The (1, 0), (0, 0) and (0, 1) sequence heads at $\lambda\lambda 6621 \cdot 9, 6890 \cdot 7, 7206 \cdot 8$ are most prominent. The following are the strongest individual heads of the T, S, R or Q branches.

λ	v', v''	λ	v', v''	λ	v', v''
6624·6	1, 0 R	6890·7	0, 0 T	7211·2	0, 1 R
6625·1	1, 0 Q	6892·9	0, 0 S	7211·8	0, 1 Q
6649·1	2, 1 R	6894·3	0, 0 R*	7233·4	1, 2 R
6649·5	2, 1 Q	6894·6	0, 0 Q*	7234·0	1, 2 Q

* Strongest.

BLUE-GREEN SYSTEMS

Degraded to the red, with complex structure. Analysed into two systems by Hayes. A $\Pi - \Sigma$ system has its (0, 0) sequence in the region 4920–4989 Å, and its weaker (1, 0) and (0, 1) sequences at 4776 and 5100 Å. The other system shows a sequence of strong heads from 4993 Å.

Strong individual heads:

λ	I	λ	I	λ	I
4931·1	5	4946·2	5	5043·1	9
4934·3	6	5005·7	6	5084·9	5
4936·3	10	5031·5	10	5085·9	8

NEAR ULTRA-VIOLET, 3500 Å

Appearance. Well marked sequences, of which the (0, 0) is the strongest. Each sequence contains numerous heads. Degraded to shorter wavelengths.

Transition. ${}^7\Pi - {}^7\Sigma$.

The following are the sequence heads: 3671 (0, 2), 3594·5 (0, 1), 3517·8 (0, 0), 3439·1 (1, 0) and 3360 (2, 0).

The strongest heads in the (0, 0) sequence, with assignments from Rao *et al.* are:

λ	I	v', v''	λ	I	v', v''
3516·9	7	0, 0 Q_1	3508·9	10	0, 0 Q_4
3514·5	8	0, 0 Q_2	3506·6	9	0, 0 P_5
3513·2	6	1, 1 O_1	3504·3	6	2, 2 Q_1
3510·7	7	1, 1 Q_1	3499·1	7	0, 0 Q_7

FAR ULTRA-VIOLET

Observed in absorption by Rochester and Olsson. Well marked sequences of bands degraded to shorter wavelengths. The (0, 1), (0, 0) and (1, 0) are strongest. Prominent heads:

λ	v', v''	λ	v', v''
2499·1	0, 2	2423·2	1, 1
2460·6	0, 1	2387·7	1, 0
2424·1	0, 0	2352·9	2, 0

MnH

Complex systems of high multiplicity exist near 5677 Å, 4800 Å, 4500 Å and in the infra-red.

References. R. W. B. Pearse and A. G. Gaydon, *Proc. Phys. Soc.*, **50**, 201 (1938)†.

T. E. Nevin, *Proc. Roy. Irish Acad.*, A, **48**, 1 (1942)†; **50**, 123 (1945).

T. E. Nevin and P. J. Doyle, *Proc. Roy. Irish Acad.*, A, **52**, 35 (1948)†.

T. E. Nevin and D. V. Stephens, *Proc. Roy. Irish Acad.*, A, **55**, 109 (1953)†.

Occurrence. The spectrum of MnH has been observed in a high-tension arc between manganese electrodes in a hydrogen flame and in a discharge tube containing hydrogen and manganese vapour. A. Heimer has observed the spectrum in emission and absorption using a furnace. R. E. Smith found all except the infra-red system in absorption in shock-heated Mn powder in H_2 + Ar.

5677 Å SYSTEM, A $^7\Pi - X\,^7\Sigma$

Degraded to the violet. The head at 5677 Å is the most prominent feature, other heads being largely lost in the complexity of the structure due to the high multiplicity. The $^0P_{12}$ heads are weak. P_1 and P_2 heads are stronger, but not always so outstanding. See Plate 5.

v', v''	$^0P_{12}$	P_1	P_2	
0, 2	6849	(6814)		1
0, 1	6244·5	6209	6199	3
0, 0	5724	5683·5	5677·5	10
1, 0	(5212)	5205·8	5203	4

4800 Å SYSTEM

A strong red-degraded band of complex quintet structure. See Plate 5.

	Heads						
v', v''	λ	I	λ	I	λ	I
0, 0 ?	4724	1	4741	4	4794*	5

        * This is not a true head when seen under large dispersion.

4500 Å SYSTEM

Rather weaker open quintet structure, slightly degraded to the red. Heads are partly masked by Mn lines. See Plate 5.

	Heads						
v', v''	λ	I	λ	I	λ	I
0, 0 ?	4475	2	4467·1	1	4428	1

INFRA-RED, 9000–6900 Å

Nevin and Stephens have obtained complex structure, especially strong in the regions 8680–8480 and 8420–8230 Å, in a discharge tube. A photograph is shown, but there are no obvious heads and measurements are not available.

MnI

Occurrence. In absorption by heated MnI_2 vapour.

Appearance. Degraded to shorter wavelengths. A complex system around 4050 Å.

Transition. Probably $^7\Pi - {}^7\Sigma$.

Reference. J. Bacher, *Helv. Phys. Acta*, **21**, 379 (1948)†.

MnI (*contd.*)

The following are amongst the strongest heads, $\lambda\lambda 4108\cdot3, 4086\cdot7, 4079\cdot8, 4068\cdot8, 4047\cdot8,$ $4027\cdot7, 4009\cdot2, 3990\cdot9.$

MnO

VISIBLE SYSTEM

Occurrence. In flames and arcs. Has been observed in steel furnaces. Also in absorption following flash-initiated explosions of organic fuels containing cyclopentadienyl manganese tricarbonyl.

Appearance. Degraded to the red. Well marked sequences. Bands with $v' = 0$ are diffuse without sharp heads.

References. J. M. Das Sarma, *Z.P.*, **157**, 98 (1959)†.

K. C. Joshi, *Spectrochim. Acta*, **18**, 625 (1962).

B. Pinchemel and J. Schamps, *Canad. J. Phys.*, **53**, 421 (1975)†.

Strong bands only, from Das Sarma; measurements by Joshi do not agree at all well.

λ	I	v', v''	λ	I	v', v''	λ	I	v', v''
6556·6	3	3, 6	5943·7	4	3, 4	5582·2	10	0, 0
6524·0	3	2, 5	5910·1	5	2, 3	5390·0	7	2, 1
6497·6	3	1, 4	5879·5	9	1, 2	5359·6	8	1, 0
6236·5	6	3, 5	5853·1	6	0, 1	5227·7	5	4, 2
6203·6	7	2, 4	5672·1	6	3, 3	5192·0	5	3, 1
6174·7	7	1, 3	5639·7	7	2, 2	5158·9	4	2, 0
6148·3	2	0, 2	5608·4	8	1, 1	5050·3	3	5, 2

ULTRA-VIOLET BANDS

Reference. A. B. Callear and R. G. W. Norrish, *P.R.S.*, **259**, 304 (1960).

In absorption, following flash photolysis of cyclopentadienyl manganese tricarbonyl, diffuse bands 2500–2600 Å were observed. The strongest features lie between 2556 and 2578 Å.

MnOH

Occurrence. In oxy-hydrogen flame containing manganese salts.

Reference. P. J. Padley and T. M. Sugden, *Trans. Faraday Soc.*, **55**, 2054 (1959)†.

A system of diffuse bands 3500–4000 Å, centred at 3750, is attributed to MnOH because of the variation with H_2 and O_2 concentrations in the flame. No measurements are available. They also occur in a shock tube. Bands 2550–2650 Å are possibly also due to MnOH.

MnS

Occurrence. Emission in a furnace.

Appearance. Two systems of red-degraded bands with marked sequences.

References. A. Monjazeb and H. Mohan, *Spectrosc. Lett.*, **6**, 143 (1973).

M. Biron, H. Boulet and J. Ruamps, *C. R. Acad. Sci.* (*Paris*), B, **278**, 835 (1974).

System I. 5890–4900 Å

Sequence heads $\lambda\lambda 5739\cdot9$ (0, 3), $5586\cdot5$ (0, 2), $5440\cdot9$ (0, 1), $5300\cdot8$ (0, 0), $5199\cdot8$ (1, 0).

System II. 4750–4200 Å

Sequence heads $\lambda\lambda 4679\cdot9$ (0, 2), $4576\cdot6$ (0, 1), 4479 (0, 0), $4788\cdot4$ (1, 0), $4301\cdot5$ (2, 0), $4217\cdot5$ (3, 0).

MoN

These bands, which occur in an arc in air between molybdenum poles and also in a plasma jet, were originally attributed to MoO but have now been shown by isotopic displacements to be due to MoN or Mo_xN. They are degraded to the red.

Reference. J. C. Howard and J. G. Conway, *J. Chem. Phys.*, **43**, 3055 (1965)†.

The following are all known heads. No intensities are available, but bands most clearly visible in the small reproduction are marked with an asterisk: λλ6311·6, 6284·0, 6276·7, 6272·3, 6269·5, 6264·8, 6256·2, 6248·7*, 6244·7*, 6228·5*, 6148·9, 6135·2, 6110·7*, 5990·7.

MoO, MoO_2

Bands in the orange previously assigned to MoO are now known to be due to MoN (see above). Two red-degraded bands with open structure at 6445·4 and 6509·4 are believed to be due to MoO.

Infra-red bands reported in an arc by Gatterer *et al.* (8656·3, 8613·9, 8599·0 shaded to longer wavelengths, and 8595·1 degraded to the violet) are attributed to MoO_2 by Howard and Conway (see MoN).

N_2

There are a large number of band systems attributed to the neutral nitrogen molecule. In emission the First and Second Positive systems are the most readily developed, the Second Positive being the more easily obtained in a discharge through air (such as a leak in a discharge tube), the presence of oxygen appearing to favour this system relatively to the First Positive bands. Other systems, which appear in emission are the Fourth Positive bands, the Vegard-Kaplan bands, a part of the Lyman-Birge-Hopfield system, the Fifth Positive system and a number of weaker systems reported by Kaplan, Gaydon, Herman and Lofthus. There are also a number of systems observed by Birge, Hopfield, Worley, Watson and Koontz and others in the vacuum ultra-violet. The spectrum has been fully summarised in a review by A. Lofthus (*The Molecular Spectrum of Nitrogen*, Spectroscopic Report No. 2, Oslo, 1960). The energy level scheme and main systems are summarised in the accompanying figure, in which we have used Lofthus's notation. Some levels of high energy are omitted and the unplaced levels of Gaydon's Green and Herman's infra-red systems are not shown. The singlet upper levels of the Gaydon-Herman systems are mostly $^1\Sigma_u^+$ or $^1\Pi_u$ (see later).

FIRST POSITIVE SYSTEM, $B\,^3\Pi_g - A\,^3\Sigma_u^+$

Occurrence. In the positive column of discharge tubes containing nitrogen or air. The bands appear very readily. The bands are the main feature of the spectrum of active nitrogen.

Appearance. Degraded to the violet. Under small dispersion the appearance is of waves of regularly spaced triple-headed bands strongest in the orange, the red and the yellow-green. Under large dispersion the rotational structure is seen to be complex, and there are several heads to each band. See Plate 3.

References. A. Fowler and R. J. Strutt, *P.R.S.*, **85**, 377 (1911)†.

A. H. Poetker, *P.R.*, **30**, 812 (1928)†.

H. Birkenbeil, *Z.P.*, **88**, 1 (1934).

P. K. Carroll and N. D. Sayers, *Proc. Phys. Soc.*, A, **66**, 1138 (1953)†.

Y. Tanaka and A. S. Jursa, *J. Opt. Soc. Amer.*, **51**, 1239 (1961)†.

The following measurements are taken partly from the above references and partly from Lofthus's report. Intensities I_d and I_a are for a discharge and for the nitrogen afterglow; they are reduced to an uncorrected scale of 10. Some weak bands are omitted.

Tanaka and Jursa have observed bands to $v' = 26$ and $v'' = 20$ in an auroral-type afterglow.

N₂ *(contd.)*

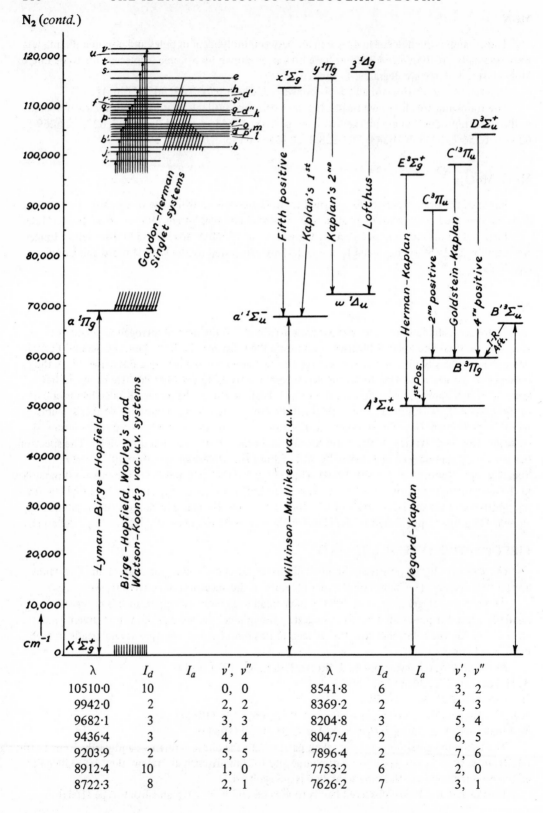

λ	I_d	I_a	v′, v″	λ	I_d	I_a	v′, v″
10510·0	10		0, 0	8541·8	6		3, 2
9942·0	2		2, 2	8369·2	2		4, 3
9682·1	3		3, 3	8204·8	3		5, 4
9436·4	3		4, 4	8047·4	2		6, 5
9203·9	2		5, 5	7896·4	2		7, 6
8912·4	10		1, 0	7753·2	6		2, 0
8722·3	8		2, 1	7626·2	7		3, 1

N₂ (*contd.*)

λ	I_d	I_a	v', v"		λ	I_d	I_a	v', v"
7503·9	7		4, 2		6069·7	7		6, 2
7386·6	5		5, 3		6013·6	7		7, 3
7273·3	3		6, 4		5959·0	8		8, 4
7164·8	2		7, 5		5906·0	8		9, 5
7059·0	2		8, 6		5854·4	8	4	10, 6
6967·8	1		9, 7		5804·3	7	10	11, 7
6875·0	2		3, 0		5755·2	7	8	12, 8
6788·6	6		4, 1		5632·7	1		5, 0
6704·8	8		5, 2		5592·9	1		6, 1
6623·6	9		6, 3		5553·7	1		7, 2
6544·8	10		7, 4		5515·6	2		8, 3
6468·5	10		8, 5		5478·5	2		9, 4
6394·7	9		9, 6		5442·3	3	1	10, 5
6322·9	7	2	10, 7		5407·1	3	5	11, 6
6252·8	3	5	11, 8		5372·8	3	5	12, 7
6185·2	3	1	12, 9 4, 0		5053·6		2	11, 5
6127·4	3		5, 1		5030·8		2	12, 6

SECOND POSITIVE SYSTEM, $C^3\Pi_u - B^3\Pi_g$

Occurrence. In the positive column of discharge tubes containing nitrogen or air and in arcs at low pressure. The bands appear very readily and are of frequent occurrence as an impurity.

Appearance. Degraded to shorter wavelengths. Close triple-headed bands forming fairly obvious sequences. See Plate 3.

References. R. Mecke and P. Lindau, *Phys. Zeit.*, **25**, 277 (1924).

D. Coster, H. Brons and A. van der Ziel, *Z.P.*, **84**, 304 (1933).

D. C. Tyte and R. W. Nicholls, *Identification Atlas of Molecular Spectra*, 2, Univ. Western Ontario (1964)†.

The following measurements are by R. C. Pankhurst and A. G. Gaydon from spectrograms taken on a Hilger E.1 and on a 20-ft. concave grating spectrograph:

λ	I	v', v"		λ	I	v', v"		λ	I	v', v"
4976·4	0	4, 11		4059·4	8	0, 3		3371·3	10	0, 0
4916·8	0	1, 7		3998·4	9	1, 4		3339	2*	1, 1
4814·7	1	2, 8		3943·0	8	2, 5		3309	2*	2, 2
4723·5	1	3, 9		3894·6	7	3, 6		3285·3	3	3, 3
4667·3	0	0, 5		3857·9	5	4, 7		3268·1	4	4, 4
4649·4	1	4, 10		3804·9	10	0, 2		3159·3	9	1, 0
4574·3	2	1, 6		3755·4	10	1, 3		3136·0	8	2, 1
4490·2	3	2, 7		3710·5	8	2, 4		3116·7	6	3, 2
4416·7	3	3, 8		3671·9	6	3, 5		3104·0	3	4, 3
4355·0	3	4, 9		3641·7	3	4, 6		2976·8	6	2, 0
4343·6	4	0, 4		3576·9	10	0, 1		2962·0	6	3, 1
4269·7	5	1, 5		3536·7	8	1, 2		2953·2	6	4, 2
4200·5	6	2, 6		3500·5	4	2, 3		2819·8	1	3, 0
4141·8	5	3, 7		3469	0	3, 4		2814·3	1	4, 1
4094·8	4	4, 8		3446	0	4, 5				

For footnote, please see over.

N₂ (*contd.*)

    * The intensities of the (1, 1) and (2, 2) bands relative to the (0, 0) band seem to vary considerably in different sources. In some sources using pure nitrogen they are often weak, but on spectrograms of a discharge through air they may be quite strong.

    On very heavy exposures, weak heads of the isotope molecule, $N^{14}N^{15}$, may be observed in front of the main heads of the (1, 0) and (2, 0) sequences.

    Bands usually terminate at $v' > 4$ because of predissociation, but in an auroral-type afterglow Y. Tanaka and A. S. Jursa (*J. Opt. Soc. Amer.*, **51**, 1239 (1961)†) have obtained triple-headed red-degraded bands provisionally assigned to $v' = 5$.

λ	v', v''	λ	v', v''
3246·5		2940·3	
3244·2	5, 5	2938·4	5, 3
3241·3		2936·1	
3087·8		—	
3085·1	5, 4	2804·6	5, 2
3083·5		2802·7	

FOURTH POSITIVE SYSTEM, $D^3\Sigma_u^+ - B^3\Pi_g$

    Occurrence. In mildly condensed discharge through nitrogen.

    Appearance. Degraded to shorter wavelengths. A single progression of bands each with five close heads. See Plate 6.

    References. A. Fowler and R. J. Strutt, *P.R.S.*, **85**, 377 (1911)†.
L. Gerö and R. Schmid, *Z.P.*, **116**, 598 (1940).

    Bands as listed by Fowler and Strutt:

λ	I	v', v''	λ	I	v', v''
2903·9	1	0, 6	2448·0	10	0, 2
2902·0			2447·0		
2900·3			2445·6		
2898·1			2444·0		
2896·6			2442·8		
2777·9	2	0, 5	2351·4	6	0, 1
2776·5			2350·3		
2775·1			2349·0		
2772·8			2347·5		
2771·4			2346·4		
2660·5	5	0, 4	2260·8	2	0, 0
2659·3			2259·6		
2657·9			2258·4		
2655·8			2257·1		
2654·5			2256·0		
2550·7	8	0, 3			
2549·7					
2548·4					
2546·6					
2545·3					

N$_2$ (*contd.*)

VEGARD-KAPLAN BANDS, $A^3\Sigma_u^+ - X^1\Sigma_g^+$

Occurrence. This is a relatively weak 'intercombination' system, and is only observed strongly under rather special conditions of excitation. The system was first reported by Vegard in the spectrum of the luminescence of solid nitrogen. Kaplan produced the bands in a special discharge tube consisting of a short length of 1 mm. capillary and a 500 c.c. bulb; an uncondensed discharge was used and the bands also appeared in the afterglow. Bernard has succeeded in observing another part of the system by exciting a mixture of nitrogen and argon with an electron beam. Some of the bands occur in the aurora and the night sky. Bands 1800–1500 Å are also observed in absorption with 12 m.atm. of nitrogen.

Appearance. Degraded to the red. Kaplan and Bernard report the bands as single headed, but in the luminescence of solid nitrogen Vegard observed a weak and a strong head to each.

References. J. Kaplan, *P.R.*, **45**, 675 (1934)[†], and numerous short letters and abstracts, mostly in *Phys. Rev.*

R. Bernard, *C. R. Acad. Sci. (Paris)*, **200**, 2074 (1935)

Y. Tanaka and A. S. Jursa, *J. Opt. Soc. Amer.*, **51**, 1239 (1961)[†].

The following are the bands as observed by Kaplan; intensities are our own estimates from his small published photograph:

λ	I	v', v''	λ	I	v', v''	λ	I	v', v''
3424·6	4	1, 10	2710·1	0	2, 7	2461·6	9	0, 4
3197·5	5	1, 9	2655·5	1	1, 6	2424·2	0	2, 5
2997·0	2	1, 8	2603·8	10	0, 5	2377·5	7	1, 4
2935·7	9	0, 7	2560·1	2	2, 6	2332·8	4	0, 3
2760·6	9	0, 6	2509·8	6	1, 5			

The bands observed by Vegard from solid nitrogen have their stronger heads 125 cm.$^{-1}$ (around 10 Å) to the red of the heads listed by Kaplan.

The following are the strong bands as observed by Bernard:

λ	I	v', v''	λ	I	v', v''	λ	I	v', v''
5060·1	4	0, 14	4319·8	2	1, 13	3889·2	4	0, 11
4837·1	8	2, 15	4171·2	4	3, 14	3854·7	3	3, 13
4649·7	4	4, 16	4072·5	3	2, 13	3603·0	4	0, 10
4616·5	3	7, 18	3979·1	3	1, 12	3502·7	3	2, 11
4535·5	5	3, 15	3940·3	4	7, 16			

LYMAN-BIRGE-HOPFIELD SYSTEM, $a^1\Pi_g - X^1\Sigma_g^+$

Occurrence. The main part of the system, which lies in the vacuum ultra-violet, is readily observed in absorption and emission. With a high current-density discharge the system can be extended to longer wavelengths, and weak bands may be observed as far as 2600 Å.

Appearance. Degraded to the red. Apparently single-headed bands.

References. R. T. Birge and J. J. Hopfield, *Astrophys. J.*, **68**, 257 (1928)[†].

E. T. S. Appleyard, *P.R.*, **41**, 254 (1934).

R. Herman, *Thesis, Paris* (1945)[†].

A. Lofthus, *Canad. J. Phys.*, **34**, 780 (1956).

M. Ogawa and Y. Tanaka, *J. Chem. Phys.*, **32**, 754 (1960)[†].

The following bands have been recorded beyond 2000 Å:

N₂ (*contd.*)

λ	v', v''	λ	v', v''	λ	v', v''	λ	v', v''
2601·5	15, 27	2425·1	11, 22	2278·3	3, 14	2125·9	5, 14
2565·2	13, 25	2406·3	10, 21	*2271·7	8, 18	**2108·1	4, 13
2528·3	11, 23	2387·6	9, 20	2253·4	7, 17	2089·7	3, 12
2509·7	10, 22	2369·0	8, 19	2234·8	6, 16	**2059·0	6, 14
2491·3	9, 21	2366·7	13, 23	2216·6	5, 15	**2041·2	5, 13
2481·7	14, 25	2328·3	11, 21	2198·7	4, 14	2023·5	4, 12
2472·6	8, 20	2309·4	10, 20	2181·1	3, 13	*2006·0	3, 11
2462·9	13, 24	2290·5	9, 19	*2144·0	6, 15		

* Moderately strong.
** Relatively strong on the plate shown by Ogawa and Tanaka.

GAYDON-HERMAN SINGLET SYSTEMS

Some 14 progressions involving transitions from various singlet levels of u symmetry to the $a\,{}^1\Pi_g$ state have been observed. Gaydon and Herman independently observed eight of these systems, and Janin and Lofthus have found others.

Occurrence. These are all weak systems. Gaydon used a mildly condensed discharge through N_2. Herman used an ordinary discharge at low pressure through a tube 25 m. long. Janin obtained the l, o and d'' to $a\,{}^1\Pi$ progressions in a high voltage discharge through NH_3.

Transitions. All to $a\,{}^1\Pi_g$. Lofthus's notation has been used here. The upper states are as follows,

b	${}^1\Pi_u$	o	${}^1\Pi_u$	d'	?
l	${}^1\Pi_u$	r'	${}^1\Sigma^+_u$	s'	${}^1\Sigma^+_u$
p'	${}^1\Sigma^+_u$	k	${}^1\Sigma^+_u$	h	${}^1\Sigma^+_u$
d	?	d''	${}^1\Pi_u$	e	${}^1\Sigma^+_u$
m	${}^1\Pi_u$	g	${}^1\Sigma^+_u$		

References.   A. G. Gaydon, *P.R.S.*, **182**, 286 (1944)†.
R. Herman, *Ann. Phys. (Paris)*, **20**, 241 (1945).
A. G. Gaydon and R. Herman, *Proc. Phys. Soc.*, **58**, 292 (1946).
J. Janin, *Cahiers de Physique*, **No. 16**, 16 (1943).
J. Janin and A. Crozet, *C. R. Acad. Sci. (Paris)*, **223**, 1141 (1946).
A. Lofthus, *Canad. J. Phys.*, **35**, 216 (1957).

The following table is based on Lofthus's review, but with intensity estimates added; these, when given, are taken from Gaydon's plate. The letters R or V indicate a head degraded to longer or shorter wavelengths, and RQ or VQ indicate a narrow piled up Q branch, slightly degraded to longer or shorter wavelengths.

λ	I	Deg.	Upper state	v', v''	λ	I	Deg.	Upper state	v', v''
3661·1		V	p'	0, 5	3075·1		R	b	0, 0
3463·3		V	p'	0, 4	3020·3	4	R	m	0, 2
3416·5		R	d	0, 4	3008·1		V	d''	0, 4
3283·3		V	p'	0, 3	2980·1		R	l	0, 1
3241·3		R	b	0, 1	2967·0	6	V	p'	0, 1
3240·8		R	d	0, 3	2932·0		R	d	0, 1
3175·0		R	m	0, 3	2878·0	7	R	m	0, 1
3118·6	5	V	p'	0, 2	2871·3		V	d''	0, 3
3079·9		R	d	0, 2	2853·3	4	R	o	0, 1

N₂ (*contd.*)

λ	I	Deg.	Upper state	v', v''	λ	I	Deg.	Upper state	v', v''
2839·4	5	R	*l*	0, 0	2569·6	7	RQ	*h*	0, 3
2827·1	9	V	*p'*	0, 0	2558	2		*d'*	0, 2
2796·0	6	VQ	*r'*	0, 1	2524·9	5	R	*k*	0, 0
2795·6		RQ	*h*	0, 5	2516·0	5	V	*d''*	0, 0
2795·4		R	*d*	0, 0	2498·6	6	R	*g*	0, 0
2753·8	3	R	*k*	0, 2	2496·8	8	RQ	*s'*	0, 1
2746·2	5	R	*m*	0, 0	2492·4		V	*e*	0, 4
2744·3	2	V	*d''*	0, 2	2455·1	2		*d'*	0, 1
2723·6	5	R	*o*	0, 0	2397·8		V	*e*	0, 3
2699·9		V	*e*	0, 6	2397·1	9	RQ	*s'*	0, 0
2678·5	2	RQ	*h*	0, 4	2371·6	8	RQ	*h*	0, 1
2671·7	10	VQ	*r'*	0, 0	2358	2		*d'*	0, 0
2626·2	3	V	*d''*	0, 1	2308·6	3	V	*e*	0, 2
2603·3	6	RQ	*s'*	0, 2	2281·5	6	RQ	*h*	0, 0
2592·8	1	V	*e*	0, 5	2224·4	2	V	*e*	0, 1

FIFTH POSITIVE (VAN DER ZIEL) SYSTEM, $x^1\Sigma_g^- - a'^1\Sigma_u^-$

Occurrence. High current-density discharge or mildly condensed discharge through nitrogen.

Appearance. Degraded to shorter wavelengths. Single-headed.

References. A. van der Ziel, *Physica*, **1**, 513 (1934).

A. G. Gaydon, *P.R.S.*, **182**, 286 (1944)†.

Most of the following measurements and intensities are by Gaydon, who has modified van der Ziel's vibrational analysis.

λ	I	v', v''	λ	I	v', v''	λ	I	v', v''
2781·6	3	1, 8	2525·6	2	0, 4	2235·8	3	2, 3
2743·0	1	2, 9	2496·9	3	1, 5	2198·9	4	0, 0
2681·2	5	1, 7	2469·7	4	2, 6	2181·5	4	1, 1
2647·0	2	2, 8	2411·7	7	1, 4	2165·1	5	2, 2
2619·3	4	0, 5	2353·6	4	0, 2	2112·1	5	1, 0
2586·5	7	1, 6	2331·0	2	1, 3	2098	2	2, 1
2556·0	1	2, 7	2274·2	6	0, 1	2033·6	5	2, 0

Some bands of this system are masked by other bands of N_2.

KAPLAN'S FIRST SYSTEM, $y^1\Pi_g - a'^1\Sigma_u^-$

Occurrence. A weak system obtained in a mildly condensed discharge (Gaydon), in a long tube (Herman) or in a special discharge tube for studying afterglow effects (Kaplan).

Appearance. Single-headed bands. Degraded to shorter wavelengths.

References. J. Kaplan, *P.R.*, **46**, 631 (1934); **47**, 259 (1935).

A. G. Gaydon, *P.R.S.*, **182**, 286 (1944)†.

R. Herman, *Thesis, Paris* (1944).

A. G. Gaydon and R. Herman, *Proc. Phys. Soc.*, **58**, 292 (1946).

N_2 (*contd.*)

λ	I	v', v''	λ	I	v', v''
2466·0	2	0, 4	2288·5	1	1, 3
2381·7	3	0, 3	2225·8	5	0, 1
2366·4	2	1, 4	2153·6	4	0, 0
2301·9	4	0, 2	2077·3	–	1, 0

KAPLAN'S SECOND SYSTEM (HERMAN'S α) $y\,^1\Pi_g - w\,^1\Delta_u$

Appearance. Strong single heads. Degraded to shorter wavelengths.
Occurrence and References as for Kaplan's First System.

λ	I	v', v''	λ	I	v', v''
2854·9	?	0, 5	2536·6	5	0, 2
2741·9	3	0, 4	2522·3	3	1, 3
2722·0	3	1, 5	2431·0	?	1, 2
2636·2	5	0, 3	2354·5	4	0, 0
2619·3	? 5	1, 4	2263·4	4	1, 0

LOFTHUS'S $z\,^1\Delta_g - w\,^1\Delta_u$ SYSTEM

Two weak red-degraded bands obtained in a transformer discharge at low pressure.
Reference. A. Lofthus, *Canad. J. Phys.*, **35**, 216 (1957).

λ	v', v''
2477·3	$n+3$, 4
2368·8	n , 2

HERMAN-KAPLAN SYSTEM, $E\,^3\Sigma_g^+ - A\,^3\Sigma_u^+$

Occurrence. A weak system originally reported by Kaplan (Third System) in his tube for examining the afterglow, and extended and analysed by Herman using a very long discharge tube.
Appearance. Degraded to shorter wavelengths.
References. J. Kaplan, *P.R.*, **47**, 259 (1935).
R. Herman, *Thesis, Paris* (1944).

λ	v', v''	λ	v', v''	λ	v', v''
2733·2	0, 7	*2471·4	0, 4	2272·9	1, 3
2642·1	0, 6	2419·8	1, 5	*2242·3	0, 1
2554·9	0, 5	*2391·6	0, 3	2203·8	1, 2
2497·8	1, 6	*2315·3	0, 2	2137·6	1, 1

* These bands may be relatively strong.

GOLDSTEIN-KAPLAN BANDS, $C'\,^3\Pi_u - B\,^3\Pi_g$

Occurrence. Silent (ozonizer type) discharge or Tesla discharge through N_2 at relatively high gas pressure. Also in afterglow under some conditions.
Appearance. Degraded to the red. Complex structure with several heads.
References. H. Hamada, *Phil. Mag.*, **23**, 25 (1937).
A. G. Gaydon, *Proc. Phys. Soc.*, **56**, 85 (1944)†.
Y. Tanaka and A. S. Jursa, *J. Opt. Soc. Amer.*, **51**, 1239 (1961).

Most bands are triple-headed with an overall separation varying from 8 Å for ultra-violet bands to 15 Å for those in the visible. The following are the first (short-wave) heads; data mainly from Tanaka and Jursa.

N$_2$ (*contd.*)

λ	I	v', v''	λ	I	v', v''	λ	I	v', v''
5059·4	1	0, 12	3706·2	5	0, 7	3161·9	6	0, 4
4729·1	2	0, 11	3506·1	6	0, 6	3025·8	–	1, 4
4432·6	4	0, 10	3325·2	5	0, 5	3004·8	5	0, 3
4166·0	4	0, 9	3178·4	–	1, 5	2863·1	2	0, 2
3925·0	4	0, 8						

GAYDON'S GREEN SYSTEM, H $^3\Phi_u$ − G $^3\Delta_g$

Occurrence. Silent (ozonizer type) and Tesla discharge through N$_2$ at relatively high gas pressure. The bands are not obtained in discharges through air. They were obtained by Herman (κ system) in a very long tube, and by Carroll *et al.* in a Geissler tube cooled in liquid air.

Appearance. Degraded to the violet. Under small dispersion the bands appear to show five heads, of which the fifth (shortest wavelength) is the strongest. Under large dispersion Grün observed only three heads, the first of which is very weak, the second corresponds to Gaydon's first head, and the last is the strong (fifth) head. The appearance varies somewhat from band to band.

References. A. G. Gaydon, *Proc. Phys. Soc.*, **56**, 85 (1944)†.
A. E. Grün, *Z. Naturforsch.*, **9a**, 1017 (1954).
P. K. Carroll, C. C. Collins and J. T. Murnaghan, *J. Phys.*, B, **5**, 1634 (1972)†.
The following are the strongest (shortest λ) heads, as observed by Grün.

λ	I	v', v''	λ	I	v', v''	λ	I	v', v''
6336·3	5	0, 3	5815·1	10	0, 1	5435·0	3	3, 3
6246·3	5	1, 4	5755·1	3	1, 2	5308·6	8	1, 0
6160·5	3	2, 5	5640	1	3, 4	5272·0	5	2, 1
6068·6	8	0, 2	5574·4	9	0, 0	5073·4	4	2, 0
5994·5	6	1, 3	5527·1	2	1, 1	5047·0	2	3, 1
5924	1	2, 4	5479·6	6	2, 2			

The following are the heads of the (0, 0) and (1, 0) bands under small dispersion.

(0, 0)			(1, 0)		
λ	I	Head	λ	I	Head
5602·1	0	$^0P_{12}$	5333·8	1	$^0P_{12}$
5595·0	6	P_1	5326·9	5	P_1
5587·6	6	P_2	5320·7	5	P_2
5582·9	6	Q_2	5316·7	5	Q_2
5579·3	6	P_3	5312·9	5	P_3
5574·8	8	Q_3	5309·5	10	Q_3

HERMAN'S INFRA-RED SYSTEM

Occurrence. In low-current discharge and Tesla discharge.

Appearance. Degraded to the violet. Each band has about six close heads, of which the last (shortest λ) is probably the strongest.

Reference. P. K. Carroll and N. D. Sayers, *Proc. Phys. Soc.*, A, **66**, 1138 (1953)†.
The (0, 0) band has heads at λλ8101·1, 8094·1, 8084·7, 8077·6, 8070·9 and 8057·6.
The following are the shortest λ heads of all bands:

N$_2$ (*contd.*)

λ	I	v', v"	λ	I	v', v"
8549	2	0, 1	7521·0	4	1, 0
8397	1	1, 2	7435·0	5	2, 1
8057·6	10	0, 0	7061·7	6	2, 0
7828·5	8	2, 2	7001·2	4	3, 1

INFRA-RED AFTERGLOW SYSTEM, B' $^3\Sigma_u^-$ — B $^3\Pi_g$

Occurrence. Discovered by Carroll and Sayers in transformer discharge through very pure N$_2$. Obtained best from low-temperature afterglow of nitrogen in argon. These weak bands are usually masked by the First Positive system.

Appearance. Degraded to longer wavelengths. They resemble First Positive bands.

References. P. K. Carroll and H. E. Rubaclava, *Proc. Phys. Soc.*, **76**, 337 (1960)†.
F. LeBlanc, Y. Tanaka and A. Jursa, *J. Chem. Phys.*, **28**, 979 (1958)†.

The following are the first heads of the strong bands, from LeBlanc *et al.*

λ	I	v', v"	λ	I	v', v"
6735	6	7, 1	7780	10	8, 3
6896	8	8, 2	8247	5	5, 1
7408	6	6, 1	8458	8	6, 2
7591	10	7, 2	8675	10	7, 3

The slightly stronger last heads lie from 9 to 16 Å to longer wavelengths.

B' $^3\Sigma_u^-$ — X $^1\Sigma_g^+$ SYSTEM

Weak double-headed bands observed in discharge through N$_2$ and xenon. Degraded to red.

Reference. M. Ogawa and Y. Tanaka, *J. Chem. Phys.*, **30**, 1353 (1959); **32**, 754 (1960).

λ	v', v"	λ	v', v"
2232·2	2, 11	2135·1	2, 10
2204·0	1, 10	2081·2	0, 8
2176·5	0, 9	2018·6	1, 8

The system extends into the far ultra-violet.

LEDBETTER'S RYDBERG SYSTEMS

Occurrence. Absorption following flash discharges.

Reference. J. W. Ledbetter, *J. Mol. Spec.*, **42**, 100 (1972)†.

Three violet-degraded bands are attributed to the (0, 0) bands of separate systems of Rydberg series.

λ	
5977·3	C$_4$$^1\Pi_u$ — a" $^1\Sigma_g^+$
4783·5	C$_5$$^1\Pi_u$ — a" $^1\Sigma_g^+$
4358·9	C$_6$$^1\Pi_u$ — a" $^1\Sigma_g^+$

N$_2^+$

FIRST NEGATIVE SYSTEM, B $^2\Sigma_u^+$ — X$^2\Sigma_g^+$

Occurrence. This is the main system of N$_2^+$ and occurs readily in discharge tubes at very low pressure or at moderate pressure in the presence of excess helium. Strongly in negative glow and in

N_2^+ *(contd.)*

hollow cathode. Observed in shock tubes. Also in aurorae.

Appearance. The strong bands form fairly well marked sequences and are degraded to shorter wavelengths. A number of weaker 'tail' bands are degraded to the red. Single headed. See Plate 3.

References.   G. Herzberg, *Ann. Phys. (Lpz.)*, **86**, 189 (1928).

J. Janin and I. Eyraud, *J. Phys. Radium*, **15**, 888 (1954).

A. E. Douglas, *Astrophys. J.*, **117**, 380 (1953)†.

D. C. Tyte and R. W. Nicholls, *Identification Atlas of Molecular Spectra*, **3**, Univ. Western Ontario (1965)†.

The following table gives all the more prominent bands; Tyte and Nicholls list many more weak ones. Intensities are our own estimates for a hollow-cathode discharge, based on our own plates and on published photographs.

λ	I	v', v''	λ	I	v', v''	λ	I	v', v''
5864·7	1	0, 4	4599·7	4	2, 4	3857·9	4	2, 2
5754·4	1	1, 5	4554·1	3	3, 5	3582·1	9	1, 0
5653·1	1	2, 6	4515·9	2	4, 6	3563·9	9	2, 1
5228·3	2	0, 3	4278·1	9	0, 1	3548·9	7	3, 2
5148·8	3	1, 4	4236·5	8	1, 2	3538·3	7	4, 3
5076·6	3	2, 5	4199·1	6	2, 3	3532·6	5	5, 4
5012·7	2	3, 6	4166·8	3	3, 4	3308·0	5	2, 0
4709·2	7	0, 2	3914·4	10	0, 0	3298·7	5	3, 1
4651·8	5	1, 3	3884·3	3	1, 1	3293·4	5	4, 2

The following are the strongest of the weak red-degraded 'tail' bands, from Janin and Eyraud.

λ	I	v', v''	λ	I	v', v''	λ	I	v', v''
5292·9	7	20, 21	3782·8	8	18, 15	3655·7	7	19, 15
5240·2	5	17, 19	3761·6	10	20, 16	3646·1	5	21, 16
5136·4	8	19, 20	3756·1	5	22, 17	3381·5	–	10, 8
4988·2	5	18, 19	3733·1	6	12, 11	3349·6	8	18, 13
4850·7	8	14, 16	3730·3	5	15, 13	3345·7	8	22, 15
4743·1	5	13, 15	3691·5	7	27, 19	3345·7	8	13, 10
3891·8	6	19, 16	3682·1	7	17, 14	3341·7	8	20, 14
3875·1	6	10, 10	3668·1	6	11, 10	3263·0	8	17, 12
						3250·1	8	19, 13

SECOND NEGATIVE SYSTEM, $C\,^2\Sigma_u^+ - X\,^2\Sigma_g^+$

Occurrence.   As first negative.

Appearance.   Degraded to red; single headed. Appearance of apparent well-marked sequences; actually bands with $v' \geqslant 3$ are usually much enhanced, probably because of an inverse predissociation, so that those bands with $v' = 3$ form the apparent heads.

References.   W. W. Watson and P. G. Koontz, *P.R.*, **46**, 32 (1934)†.

Y. Tanaka, *J. Chem. Phys.*, **21**, 1402 (1953).

P. K. Carroll, *Canad. J. Phys.*, **37**, 880 (1959)†.

Y. Tanaka, T. Namioka and A. S. Jursa, *Canad. J. Phys.*, **39**, 1138 (1961)†.

The following are the bands in the quartz region; the system extends to 1377 Å.

N$_2^+$ (*contd.*)

λ_{air}	I	v', v''	λ_{vac}	I	v', v''
2059·0	3	3, 11	1984·6	7	3, 10
2051·6	2	4, 12	1979·2	6	4, 11
2044·0	1	5, 13	1973·5	5	5, 12
2036·4	0	6, 14	1967·7	3	6, 13
2028·6	0	7, 15	1913·5	6	3, 9

In discharges through He + N$_2$ those bands with $v' = 3$ (and to a lesser extent $v' = 4, 5$ and 6) are strong. With neon, however, bands with $v' = 0$ and 1 are strong; see Tanaka *et al.*

MEINEL'S AURORAL SYSTEM, A $^2\Pi_u - X^2\Sigma_g^+$

Occurrence. This is a weak system in the extreme red and near infra-red and was first observed in aurorae. It has been obtained in low-pressure and hollow-cathode discharges through N$_2$ or through N$_2$ and He.

Appearance. Triple-headed bands. Degraded to longer wavelengths.

References. A. B. Meinel, *Astrophys. J.*, **113**, 583 (1951)†; **114**, 431 (1951).

A. E. Douglas, *Astrophys. J.*, **117**, 380 (1953)†.

I. D. Liu, *Astrophys. J.*, **129**, 516 (1959).

The following are the first (R$_2$) and third (R$_1$) heads, from Douglas; intensities where given are reduced to a scale of 10 based on observations in the aurora.

λ	I	v', v''	λ	I	v', v''	λ	I	v', v''
9502·3	–	2, 1	8105·3	8	3, 1	6890·5	3	3, 0
9431·2			8053·6			6853·0		
9212·1	10	1, 0	7874·6	10	2, 0	6298·9	1	5, 1
9145·3			7825·7			6267·6		
8603·9	1	5, 3	7281·8	3	5, 2	6136·2	1	4, 0
8545·7			7239·9			6106·4		
8348·3	5	4, 2	7081·4	4	4, 1			
8293·4			7036·8					

JANIN-D'INCAN SYSTEM, 3070–2050 Å, D $^2\Pi_g - A^2\Pi_u$

Occurrence. A weak system observed in a microwave discharge through N$_2$ + Ne and in the pink afterglow of nitrogen.

Appearance. Degraded to the red. Double-headed, with separation between heads about 3 Å.

References. J. Janin and J. d'Incan, *C. R. Acad. Sci. (Paris)*, **246**, 3436 (1958).

Y. Tanaka, T. Namioka and H. S. Yursa, *Canad. J. Phys.*, **39**, 1138 (1961).

H. H. Brömer, K. Fette and J. Hirsch, *Z. Naturforsch.*, **20a**, 643 (1965)†.

The following measurements of the first, R$_1$, heads of most of the bands are from Tanaka *et al.*

λ	I	v', v''	λ	I	v', v''	λ	I	v', v''
2920·7	b	7, 8	2713·3	6	6, 6	2578·6	3*	4, 4
2860·3	4	8, 8	2703·2	2*	4, 5	2543·4	7	7, 5
2804·3	b	9, 8	2659·7	6	7, 6	2527·5	2	5, 4
2764·1	2*	3, 5	2634·4	3*	3, 4	2518	*	3, 3

N_2^+ *(contd.)*

λ	I	v', v''	λ	I	v', v''	λ	I	v', v''
2465·6	2*	4, 3	2316·7	4	5, 2	2169·6	6	9, 2
2418·5	5	5, 3	2276·4	7	6, 2	2149·7	6	7, 1
2377·2	6	11, 5	2238·4	10	7, 2	2116·9	5	8, 1
2374·2	5	6, 3	2224·3	4	10, 3	2086·1	4	9, 1
2332·2	5	7, 3	2202·9	7	8, 2	2057·0	3	10, 1

* Strong in pink afterglow (Brömer *et al.*).
b Recorded as strong ($I = 10$) but probably a blend.

N_3

In absorption after the flash photolysis of hydrogen azide, HN_3, a diffuse band 2730–2670 Å is observed and a much weaker absorption near 2600. The strongest narrow maxima or violet-degraded heads are at $\lambda\lambda 2723\cdot0$, 2719·5, 2713·6, 2708·3, 2700·7, 2700·1.

References. B. A. Thrush, *P.R.S.*, **235**, 143 (1956)†.
A. E. Douglas and W. J. Jones, *Canad. J. Phys.*, **43**, 2216 (1965)†.

NBr

Occurrence. In afterglow of nitrogen containing bromine.
Appearance. Three sequences of bands degraded to the violet. The system is not unlike the first positive bands of nitrogen, but on a smaller scale.
Transition. $^1\Sigma^+ - X\,^3\Sigma^-$.
References. A. Elliott, *P.R.S.*, **169**, 469 (1939)†.
E. R. V. Milton, H. B. Dunford and A. E. Douglas, *J. Chem. Phys.*, **35**, 1207 (1961)†.

The following are the strong bands; the intensities are our estimates from the published photograph:

λ	I	v', v''	λ	I	v', v''	λ	I	v', v''
6404·8	5	1, 0	6075·9	4	3, 1	5905·0	10	9, 7
6370·1	5	2, 1	6047·9	7	4, 2	5876·2	5	10, 8
6335·4	7	3, 2	6019·5	9	5, 3	5741·9	6	7, 4
6300·2	7	4, 3	5990·7	10	6, 4	5718·6	8	8, 5
6265·9	8	5, 6	5962·4	10	7, 5	5695·1	8	9, 6
6231·1	8	6, 5	5933·8	10	8, 6	5671·5	4	10, 7
6196·0	6	7, 6						

In addition, there are a few unassigned bands, but these mostly appear to be weak.

NCO

Two systems, one near 4400 Å and the other between 3200 and 2650 Å have been observed in absorption following flash photolysis of organic cyanates and isocyanates. The first system has also been obtained in emission in short-wave fluorescence.

References. R. N. Dixon, *Phil. Trans. Roy. Soc.*, A, **252**, 21 (1960)†; *Canad. J. Phys.*, **38**, 10 (1960)†.
H. Okabe, *J. Chem. Phys.*, **53**, 3507 (1970).
P. S. H. Bolman *et al.*, *P.R.S.*, **343**, 17 (1975)†.

NCO (*contd.*)

4400 Å SYSTEM

The bands are degraded to the violet, and a full analysis has been made, showing that the molecule is linear and the transition is $A^2\Sigma^+ - X^2\Pi$. The following are the outstanding heads, with intensities reduced to a scale of 10, in flash photolysis:

λ	I	λ	I	λ	I
4403·6	10	4356·6	3	4133·6	3
4389·0	3	4347·9	3	3992·6	4
4384·8	10	4166·9	3	3977·0	4
4375·0	3	4150·2	3	3641·8	2

D. W. G. Style has also observed heads at 4664 and 4645 Å in fluorescence.

Okabe has found emission following photodissociation of HNCO; bands at 4258 and 4167 are strongest.

3200–2650 Å SYSTEM

Degraded to the red, and attributed to a $^2\Pi - {}^2\Pi$ transition. The following heads are indicates as being strong, or appear prominently on the photograph:
λλ3149·9, 3148·4, 3051·3, 3045·9, 2964·3, 2959·5, 2953·9, 2866·9.

NCS

Absorption bands following flash photolysis of methyl cyanate and other compounds are due to overlapping $A^2\Pi - X^2\Pi$ and $B^2\Sigma - X^2\Pi$ systems of linear NCS.

Strongest red-degraded heads:

λ	λ	λ	λ
3853·8	3750·6	3712·7*	3559·2*
3821·5	3746·3	3701·9	3526·8
3819·8**	3730·3	3611·9*	

* Strong. ** Strongest.

These bands also occur in emission in high-frequency discharges and in fluorescence excited by far ultra-violet light. The strongest bands in emission have maxima around λλ4295, 4170, 4079, 3965, 3857 and 3755.

References. R. Holland, D. W. G. Style, R. N. Dixon and D. A. Ramsay, *Nature*, **182**, 336 (1958)†.

R. N. Dixon and D. A. Ramsay, *Canad. J. Phys.*, **46**, 2619 (1968)†.

NCl

RED SYSTEM, $b^1\Sigma^+ - X^3\Sigma^-$

Occurrence. Microwave discharge through $N_2 + Cl_2$.

Appearance. Double-headed bands, shaded to the violet, in well-marked sequences.

Reference. R. Colin and W. E. Jones, *Canad. J. Phys.*, **45**, 301 (1967)†.

The following wavelengths and intensities are based on our study of the published photograph:

NCl (*contd.*)

λ	I	v', v''	λ	I	v', v''
7050	5	0, 1 ^{0}P	6599·4	8	1, 1 ^{Q}P
7028·4	7	0, 1 ^{Q}P	6552·8	3	2, 2 ^{Q}P
6990	3	1, 2 ^{0}P	6283	2	1, 0 ^{0}P
6970·4	5	1, 2 ^{Q}P	6261·7	6	1, 0 ^{Q}P
6668·3	9	0, 0 ^{0}P	6223·9	4	2, 1 ^{Q}P
6646·7	10	0, 0 ^{Q}P	6186·7	3	3, 2 ^{Q}P

Note. The above values of λ_{air} are derived from the published wave-numbers. The few wavelengths quoted in the paper appear to be λ_{vac}, unless in deriving wave-numbers the correction to vacuum has not been made.

ULTRA-VIOLET BANDS

In the flash photolysis of NCl_3 diffuse bands at 2401·3, 2397·8, 2392·5 and 2391·0 and continua 2459–2439, 2380–2355 and 2344–2328 have been attributed to NCl by Briggs and Norrish, but Colin and Jones appear to double this assignment.

Reference. A. G. Briggs and R. G. W. Norrish, *P.R.S.*, **278**, 27 (1964)†.

NCl₂

In the flash-photolysis of NCl_3 diffuse bands with centres at 3141, 3085, 3037·5, 2987·5, 2890, 2835, 2794 and 2755 are attributed to NCl_2 by Briggs and Norrish (see above).

NF

Occurrence. Green afterglow of a discharge through argon + NF_3.

References. W. E. Jones, *Canad. J. Phys.*, **45**, 21 (1967)†.

A. E. Douglas and W. E. Jones, *Canad. J. Phys.*, **44**, 2251 (1966)†.

8742 Å SYSTEM, $a^1\Delta - X^3\Sigma^-$

A single band, the (0, 0), with violet-shaded Q head at 8742·4. The weaker P branch extends in front of this but does not close up to form a proper head.

5288 Å SYSTEM, $b^1\Sigma^+ - X^3\Sigma^-$

Line-like Q branches, weakly degraded to the violet with P and R-type structure on each side. The (0, 0) band with head at 5287·8 Å is relatively very strong. There are weaker bands at 5621·7 (0, 1) and 5271·8 (1, 1), and very weak bands at 5598 (1, 2) and 5255 (2, 2).

NF₂

Occurrence. In absorption by N_2F_4 heated to around 100°C.

Reference. L. A. Kuznetsova, Y. Y. Kuzyakov and V. M. Tatevskii, *Opt. Spectrosc.*, **16**, 295 (1964).

16 diffuse bands 2777–2377 Å with strongest maxima λλ2651, 2620, 2593, 2568, 2542 and 2516.

NH

There is a very strong triplet system at 3360 Å. Three weaker singlet systems and a very weak intercombination band at 4709 Å are also known.

3360 Å SYSTEM, $A^3\Pi - X^3\Sigma^-$

Occurrence. Obtained very readily from many sources, including ammonia/oxygen or hydrogen/nitrous oxide flames, discharge tubes, low-pressure arcs, active nitrogen and chemiluminescence. Also absorption, especially after flash photolysis.

Appearance. The very strong (0, 0) band has an almost line-like Q branch at 3360 Å with triple P and R branches spreading out from it, the R branch forming a weak head in high-temperature sources. See Plate 4. Bands of the (0, 1) and (1, 0) sequences are very much weaker, some heads being degraded to the red and others to shorter wavelengths.

References. A. Fowler and C. C. L. Gregory, *Phil. Trans. Roy. Soc.*, A, **218**, 251 (1919)†.
G. Pannetier, H. Guenebaut and A. G. Gaydon, *C. R. Acad. Sci. (Paris)*, **246**, 958 (1955)†.
R. N. Dixon, *Canad. J. Phys.*, **37**, 1171 (1959)†.
J. Malicet, J. Brion and H. Guenebaut, *J. Chim. Phys.*, **67**, 25 (1970).

The following are the more certainly established heads, with the direction of degradation (R = to the red, V = to shorter wavelengths and M = maximum). Intensities, where given, are our estimates:

λ	I	deg.	v', v''		λ	I	deg.	v', v''
3803·7	–	V	P 0, 1		3360·1	10	M	Q 0, 0
3795	–	V	P 1, 2		3317·0	3	R	R 1, 1
3755·1	–	V	Q_1 0, 1		3302·5	4	R	R 0, 0
3751·8	–	V	Q_2 0, 1		3076	–	R	Q 2, 1
3749·9	–	V	Q_3 0, 1		3053·6	–	R	R_3 2, 1
3746	–	V	Q_1 1, 2		3050·3	1	R	Q_3 1, 0
3742·7	–	V	Q_2 1, 2		3047·3	0	R	Q_2 1, 0
3740·2	–	V	Q_3 1, 2		3046·6	–	R	Q_1 1, 0
3383	3	M	Q 2, 2		3022·1	0	R	R 1, 0
3370	8	M	Q 1, 1					

3240 Å SYSTEM, $c^1\Pi - a^1\Delta$

References. R. W. B. Pearse, *P.R.S.*, **143**, 112 (1933)†.
G. Nakamura and T. Shidei, *Japan J. Phys.*, **10**, 5 (1934).

Bands degraded to the red. Branches consist of very narrow doublets whose spacing increases with rotation; the R branch is much weaker than the P and Q. The system occurs in discharge tubes; in cheminluminescence and in active nitrogen where hydrogen and nitrogen are present together. See Plate 4.

	Heads	
v', v''	R	Q
0, 1	3609·6	3627·2
0, 0	3240·1	3253·4
1, 0	3035·2	3042·6

4502 Å SYSTEM, $c^1\Pi - b^1\Sigma$

References. R. W. Lunt, R. W. B. Pearse and E. C. W. Smith, *P.R.S.*, **151**, 602 (1935)†.
F. L. Whittaker, *J. Phys.*, B, **1**, 977 (1968)†.

NH (*contd.*)

Bands degraded to the red. Single P, Q and R branches. Observed in controlled electron discharges in ammonia and in hollow-cathode discharges with rapidly streaming ammonia. The (0, 0) is strongest:

v', v''	R heads	Q heads
0, 1	5254·0	5290·3
0, 0	4502·0	4523·2

2530 Å SYSTEM, $d^1\Sigma - c^1\Pi$

Occurrence. In mildly condensed discharge through $N_2 + H_2$ or NH_3, and in a hollow cathode.

Appearance. Degraded to shorter wavelengths with strong Q heads and weaker P heads. The (0, 0) band is strongest.

References. R. W. Lunt, R. W. B. Pearse and E. C. W. Smith, *P.R.S.*, **155**, 173 (1936)†.
N. A. Narasimhan and G. Krishnamurthy, *Proc. Indian Acad. Sci.*, A, **64**, 97 (1966)†.

v', v''	R heads	Q heads
0, 1	2683·0	2673·4
0, 0	2557·3	2530·2
1, 1	2516·3	2504

4709 Å INTERCOMBINATION SYSTEM, $b^1\Sigma^+ - X^3\Sigma^-$

Occurrence. A very weak band observed in emission when ammonia is photolysed by far ultraviolet light.

Appearance. A line-like Q branch at 4710 Å with weak ^sR and ^oP branches extending on each side of it.

Reference. A. Gilles, G. Masanet and C. Vermeil, *Chem. Phys. Lett.*, **25**, 346 (1974).

NH⁺

Occurrence. Hollow-cathode discharge through flowing ammonia.

Appearance. All bands have open rotational structure and are shaded to the red.

4629 Å SYSTEM, $A^2\Sigma^- - X^2\Pi$

Reference. R. Colin and A. E. Douglas, *Canad. J. Phys.*, **46**, 61 (1968)†.
The (0, 0) band is strongest. R heads:

λ	v', v''
5349·4	0, 1
4628·9	0, 0
4312·7	1, 0

4348 Å SYSTEM, $B^2\Delta - X^2\Pi$

Reference. R. Colin and A. E. Douglas, *Canad. J. Phys.*, **46**, 61 (1968)†.
There is a single (0, 0) band with R head at 4348·5 Å; the Q branch starts at about 4372.

2885 Å SYSTEM, $C^2\Sigma - X^2\Pi$

References. R. W. Lunt, R. W. B. Pearse and E. C. W. Smith, *Nature*, **136**, 32 (1935).
M. W. Feast, *Astrophys. J.*, **114**, 344 (1951)†.

NH$^+$ (*contd.*)

The following are the R heads. Lunt *et al.* also report a band at 2980 Å.

λ	I	v', v''
2614	2	2, 0
2725	6	1, 0
2835	2	2, 1
2885	10	0, 0

NH$_2$, Ammonia α band

Occurrence. In flames such as NH_3/O_2 and H_2/N_2O, in decomposition of hydrazoic acid, weakly in discharge through flowing ammonia, and in absorption following flash photolysis of ammonia.

Appearance. This is a many-line system extending throughout the visible and near infra-red; it is strongest in the yellow and green.

References. W. B. Rimmer, *P.R.S.*, **103**, 696 (1923)†.

A. Fowler and J. S. Badami, *P.R.S.*, **133**, 325 (1931)†.

A. G. Gaydon, *P.R.S.*, **181**, 197 (1942)†.

G. Herzberg and D. A. Ramsay, *Disc. Faraday Soc.*, **14**, 11 (1953).

K. Dressler and D. A. Ramsay, *Phil. Trans. Roy. Soc.*, A, **251**, 553 (1959)†.

Rimmer lists about 300 lines in emission, and Dressler and Ramsay give a very large number in absorption. Fowler and Badami report apparent bands (maxima) under small dispersion at 6652*, 6470, 6302*, 6042*, 5870, 5713*, 5575, 5436*, 5384 and 5265; stronger bands are marked with an asterisk. Gaydon found a group of lines from 7099 leading to a head at 7350. In absorption the strongest features, which tend to degrade to the violet, are at 7354, 6619, 6302*, 5977*, 5708*, 5385, 5166*, 4925, 4718 and 4524 Å.

NH$_3$, Ammonia

SCHUSTER'S EMISSION BAND

Occurrence. First observed in an uncondensed discharge through flowing ammonia, they have also been studied in a high-frequency discharge.

References. W. B. Rimmer, *P.R.S.*, **103**, 696 (1923)†.

H. Schüler, A. Michel and A. E. Grün, *Z. Naturforsch*, **10a**, 1, (1955)†.

There is a strong continuum in the yellow green with strong intensity maxima at 5670 and 5635 Å and related weaker bands at 5282, 6497 and 7666.

Another system with bands at 6635, 7350 and 7413 occurs in an electron beam. The emitter of both systems is uncertain.

ULTRA-VIOLET ABSORPTION

Occurrence. Absorption by gaseous ammonia.

Appearance. Rather diffuse double- (or triple-) headed bands, shaded to the red, occur in the region below 2300 Å, the bands getting stronger in the shorter wavelength region. In the vacuum ultra-violet the bands are strong and well defined.

References. M. Ferrieres, *C. R. Acad. Sci.* (*Paris*), **178**, 202 (1924).

A. B. F. Duncan, *P.R.*, **47**, 822 (1935).

A. D. Walsh and P. A. Warsop, *Trans. Faraday Soc.*, **57**, 345 (1961)†.

A. E. Douglas, *Disc. Faraday Soc.*, **35**, 158 (1963).

NH₃, Ammonia (*contd.*)

The following are the maxima of the principal bands as compiled from the above references:

λ	I	λ	I	λ	I
2245	1	2167·3	4	2086·4	5
2239		2163·7		2084·1	
2211	2	2126·5	4	2048·4	6
2206		2123·9		2045·7	
				2010·9	7

VIBRATION-ROTATION SPECTRUM

Occurrence. In absorption by gaseous ammonia.

Appearance. Bands composed of complex line-structure. The strongest band probably shows a maximum of intensity (the Q branch) at 7919 Å and perhaps a head degraded to the red at 7874 Å.

Reference. R. M. Badger and R. Mecke, *Z. phys. Chem.*, **5**, 333 (1929).

λ		
6450	$5\nu_1$	weak
7919	$4\nu_1$	strong
8800	$3\nu_1 + \nu_2$	weak

N₂H₂, Diimide

In absorption by long paths (72 m at 50 torr) there are about 30 diffuse bands, slightly shaded to shorter wavelengths 4300–3000 Å. Strongest, λλ3807, 3642, 3597, 3492; the last is strongest.

Reference. R. A. Back, C. Willis and D. A. Ramsay, *Canad. J. Chem.*, **52**, 1006 (1974)†.

N₂H₄, Hydrazine

Hydrazine vapour shows roughly equidistant absorption bands λλ2326, 2320, 2276, 2250 and 2225 Å; there is continuous absorption below 2200 Å.

Reference. S. Imanishi, *Nature*, **127**, 782 (1931).

N₃H

Herzberg notes absorption bands 3150–2950 Å.

Reference. G. Herzberg, *Electronic Spectra of Polyatomic Molecules*, Van Nostrand (1966).

NO

Four band systems of nitric oxide are well known in both emission and absorption. They are known as the β, γ, δ and ε systems. The γ system is the most readily obtained with quartz instruments; the δ and ε systems are stronger, but lie at the limit of the quartz region. Two weaker ultraviolet systems have been reported by Migeotte, there is a fairly strong band at 6000 Å and there are other weak systems in the visible and infra-red.

NO (*contd.*)

β SYSTEM, $B^2\Pi - X^2\Pi$

Occurrence. In discharge tubes containing oxygen and nitrogen and in the nitrogen afterglow, and especially strongly when a trace of oxygen is introduced into active nitrogen. Weakly in absorption.

Appearance. Degraded to red. Double-headed. See Plate 3.

References. R. C. Johnson and H. G. Jenkins, *Phil. Mag.*, **2**, 621 (1926).

F. A. Jenkins, H. A. Barton and R. S. Mulliken, *P.R.*, **30**, 150 (1927)†.

A. G. Gaydon, *Proc. Phys. Soc.*, **56**, 160 (1944)†.

M. Ogawa, *Science of Light*, **3**, 90 (1955)†.

M. Brook and J. Kaplan, *P.R.*, **96**, 1540 (1954)†.

The wavelengths of the R heads of both sub-bands are given. The strong bands are from Jenkins, Barton and Mulliken's measurements, others by Johnson and Jenkins, and Gaydon. Ogawa, and Brook and Kaplan, have measured a number of weaker bands at longer wavelengths.

λ	I	v',	v''	λ	I	v',	v''	λ	I	v',	v''
5270·1	0	3,	18	3962·7	2	2,	13	3206·9	10	0,	8
5252·7				3949·8				3198·0			
4912·1	2	3,	17	3880·7	5	1,	12	3168·3	3	2,	9
4892·1				3868·3				3159·8			
4810·0	1	2,	16	3800·9	10	0,	11	3131·1	2	4,	10
4791·4				3788·5				3125·0			
4590·8	5	3,	16	3658·5	4	1,	11	3043·0	10	0,	7
4574·0				3647·2				3034·9			
4496·2	3	2,	15	3583·5	10	0,	10	3010·7	1	2,	8
4479·8				3572·4				3002·8			
4401·5	2	1,	14	3456·9	2	1,	10	2950·8	3	1,	7
4385·7				3446·0				2943·2			
4309·7	3	0,	13	3408·5	1	3,	11	2923·1	4	3,	8
4293·7				3398·9				2915·9			
4303·1	3	3,	15	3386·4	10	0,	9	2892·6	10	0,	6
4288·2				3376·4				2885·2			
4215·2	4	2,	14	3364·3	0			2809·4	4	1,	6
4200·7				3355·2				2802·6			
4127·9	4	1,	13	3340·3	3	2,	10	2786·0	3		
4113·6				3330·7				2779·5			
4041·8	6	0,	12	3298·6	0	4,	11	2754·3	9	0,	5
4027·8				3289·7				2747·6			

NO (*contd.*)

λ	I	v', v''	λ	I	v', v''	λ	I	v', v''
2731·7	1	2, 6	2493·4	7	2, 4	2229·8	2	5, 3
2725·4			2487·8			—		
2678·7	7	1, 5	2433·0	7	3, 4	2183·2	0	6, 3
2672·2			2427·8			2179·2		
2626·6	6	0, 4	2386·4	1	2, 3	—	3	3, 1
2620·5			2381·5			2143·5		
2608·3	6	2, 5	2376·2	1	4, 4	2142·8	3	5, 2
2602·1			2371·6			2139·1		
2557·9	5	1, 4	2331·4	3	3, 3	2103·6	2	4, 1
2551·8			2326·6			2099·8		
2542·3	3	3, 5	2287·6	2	2, 2	—	1	6, 2
2536·3			—			2096·4		
2528·7	1	5, 6	2236·1	6	3, 2	2021·1	0	6, 1
2523·6			2232·0			2018·1		

γ SYSTEM (NITROGEN THIRD POSITIVE), $A^2\Sigma^+ - X^2\Pi$

Occurrence. In discharge tubes containing nitrogen and oxygen, in active nitrogen, in flames containing nitric oxide, and in shock-heated air. In absorption, bands with low v'' appear with moderate intensity.

Appearance. Degraded to shorter wavelengths. Double double-headed bands, the four heads being $^0P_{12}$, P_2, P_1 and Q_1. The strong bands in emission form a single v'' progression. See Plate 3.

References. R. J. Strutt, *P.R.S.*, **93**, 254 (1917)†.

W. H. Bair, *Astrophys. J.*, **52**, 301 (1920).

A. G. Gaydon, *Proc. Phys. Soc.*, **56**, 95 and 160 (1944)†.

For the first nine (long wavelength) bands, measurements are for the $^0P_{12}$ heads from Bair; the wavelengths have been raised slightly to bring them into agreement with other measurements, but are probably not of high accuracy. The bulk of the measurements are by Gaydon and are for the $^0P_{12}$ and P_2 heads of the bands. Intensities are for emission in a discharge tube.

λ	I	v', v''	λ	I	v', v''	λ	I	v', v''
3458·5	0	1, 10	3008·8	4	0, 6	2859·5	7	0, 5
3375·5	0	0, 8	2997·6			2849·8		
3303·0	0	3, 12						
3278·5	0	1, 9	2952·0	3	1, 7	2810·4	5	1, 6
3201·1	0	2, 10	2941·9			2800·8		
3170·7	1	0, 7						
3120·6	0	3, 11	2898·3	2	2, 8	2763·7	3	2, 7
3112·4	0	1, 8	2882·2			2755·2		
3044·3	1	2, 9						

NO (*contd.*)

λ	I	v', v"	λ	I	v', v"	λ	I	v', v"
2722·2	8	0, 4	2447·0	3	1, 3	2199·6	1	3, 3
2713·2			2440·0			2194·0		
2680·0	5	1, 5	2370·2	10	0, 1	2154·9	7	1, 0
2671·4			2363·3			2149·1		
2639·1	3	2, 6	2316·3	2	2, 3	2135·0	1	2, 1
2630·7			2309·5			2129·6		
2595·7	9	0, 3	2289·8	2	3, 4	2115·0	0	3, 2
2587·5			2284·1			2109·5		
2559·0	4	1, 4	2269·4	8	0, 0	2052·8	4	2, 0
2550·0			2262·8			2047·5		
2523·6	1	2, 5	2245·4	3	1, 1	2035·7	2	3, 1
2516·4			2239·4			2030·7		
2478·7	10	0, 2	2222·4	3	2, 2	1961·1	2	3, 0
2471·1			2216·3			1956·1		

δ SYSTEM, $C^2\Pi - X^2\Pi$

Occurrence. In active nitrogen and in condensed discharge through air. Also strongly in absorption.

Appearance. A single progression of double double-headed bands, degraded to shorter wavelengths.

References. H. P. Knauss, *P.R.*, **32**, 417 (1928).

R. Schmid, *Z.P.*, **64**, 279 (1930)†.

A. G. Gaydon, *Proc. Phys. Soc.*, **56**, 160 (1944)†.

G. Herzberg, A. Lagerqvist and E. Miescher, *Canad. J. Phys.*, **34**, 622 (1956).

The following are the $^0P_{12}$ and P_2 heads, some from unpublished data by Gaydon:

λ	I	v', v"	λ	I	v', v"
2414·6	1	0, 6	2060·9	5	0, 2
2407·6			2055·9		
2317·7	3	0, 5	1985·4	6	0, 1
2311·4					
2226·8	4	0, 4	1914·2	8	0, 0
2220·8					
2141·3	5	0, 3	1828	10	1, 0
2135·8					

NO (*contd.*)

ϵ SYSTEM, $D^2\Sigma^+ - X^2\Pi$

The bands of this system have at times been confused with those of the γ system, but this is now recognised as a separate system.

Occurrence. In discharge tubes containing nitrogen and air. Strongly in absorption. The strongest bands of this system lie in the vacuum ultra-violet, and only a few of the weaker bands are recorded with quartz instruments.

Appearance. Double double-headed bands degraded to shorter wavelengths.

References. M. Guillery, *Z.P.*, **42**, 121 (1927).

M. Hellermann, *Z.P.*, **104**, 417 (1936).

A. G. Gaydon, *Proc. Phys. Soc.*, **56**, 95 and 160 (1944)†.

The following are the $^0P_{12}$ and P_2 heads of bands in the quartz region:

λ	I	v', v''	λ	I	v', v''
2181·8	3	0, 4	2022·3	3	0, 2
2176·1			2017·5		
2157·5	0	1, 5	2003·6	1	1, 3
—			1998·6		
2099·8	3	0, 3	1949·7	4	0, 1
2094·5			1945·0		
2078·2	3	1, 4			
2073·0					

6000 Å SYSTEM, $E^2\Sigma^+ - A^2\Sigma^+$

A fairly strong band at 6000 Å was at first attributed to NO^+ but has now been assigned to a transition between two Rydberg states of NO. It occurs in discharges through flowing NO + Ar or NO_2 and in emission following photolysis of NO by far ultra-violet light (1470 Å). The (0, 0) band is hardly degraded at all, showing an open origin at 6001 Å with apparently single P and R branches extending for about 50 Å each side of the origin. Lines of weak (1, 1) and (2, 2) bands have been observed.

References. M. W. Feast, *Canad. J. Res.*, A, **28**, 488 (1950)†.

R. Suter, *Canad. J. Phys.*, **47**, 881 (1969)†.

R. A. Young, G. Black and T. G. Slanger, *J. Chem. Phys.*, **48**, 2067 (1965).

VISIBLE SYSTEMS (OGAWA'S b AND b')

These bands, first reported by Tanaka and Ogawa, occur weakly in ordinary discharges through flowing NO, and have been analysed into two systems $b - B^2\Pi$ and $b' - B^2\Pi$. Suter refers to the first system as $B'^2\Delta - B^2\Pi$. They consist of double double-headed bands degraded to the violet; the b bands form sequences but the b' form a single progression.

References. M. Ogawa, *Science of Light*, **2**, 87 (1953)†.

R. Suter, *Canad. J. Phys.*, **47**, 881 (1969)†.

We list the first heads only of all bands, except the b (0, 0) for which all four heads are given. No intensities are available.

NO (*contd.*)

b system

λ	v', v''	λ	v', v''	λ	v', v''	λ	v', v''
8020·9*	0, 2	6898·4* ⎫		6811·8	1, 1	6324·1	2, 1
7420·2*	0, 1	6889·3 ⎪		6754·3*	2, 2	5940·6	2, 0
7313·9	1, 2	6880·5 ⎬ 0, 0		6700·2	3, 3	5907·4	3, 1
7170·1	3, 4	6872·4 ⎭		6377·8*	1, 0	5571·4	3, 0

b' system

λ	v', v''	λ	v', v''
6213·2	0, 4	5224·8	0, 1
5856·2	0, 3	(4970·5)	0, 0
5334·3*	0, 2		

* These bands appear fairly strong on the plate.

WEAK RYDBERG TRANSITIONS

References. M. Huber, *Helv. Phys. Acta*, **37**, 329 (1964)†.

R. A. Young, G. Black and T. G. Slanger, *J. Chem. Phys.*, **48**, 2067 (1968).

R. Suter, *Canad. J. Phys.*, **47**, 881 (1969)†.

A number of weak systems involving transitions from high Rydberg levels to the states $A^2\Sigma^+$, $C^2\Pi$ and $D^2\Sigma^+$ have been reported in emission in discharges through flowing NO + Ar or in emission following photolysis of NO by far ultra-violet light. They are rather difficult to sort out, but the (0, 0) bands (usually the only ones observed) of some of the better established of these systems are given below. The direction of degradation, to the violet (V), red (R) or origin of headless band (0), is indicated.

λ	*deg.*	*System*
9690	V	$H^2\Sigma^+ - C^2\Pi$ and $H'^2\Pi - C^2\Pi$
6800	V	$O'^2\Pi - D^2\Sigma^+$
6569	R ⎫	$N^2\Delta - C^2\Pi$
6535·7	V ⎭	
6400	O	$O'^2\Pi - C^2\Pi$
5402	V	$H^2\Sigma^+ - A^2\Sigma^+$ and $H'^2\Pi - A^2\Sigma^+$

INFRA-RED QUARTET SYSTEM

Occurrence. Trace of O_2 in active nitrogen. Discharge through flowing NO at 10 to 18 torr.

Appearance. Degraded to violet. Six heads of about equal strength. Regularly spaced heads, 8733 to 8425 and 7939 to 7787 Å.

Transition. Perhaps $^4\Pi - {}^4\Pi$ or $^4\Sigma - {}^4\Pi$.

References. M. Brook and J. Kaplan, *P.R.*, **96**, 1540 (1954)†.

M. Ogawa, *Science of Light*, **3**, 37 (1954)†.

Analyses and wavelengths given in above references differ somewhat. The following are the 1st and 6th heads from Brook and Kaplan. Own estimates of intensity from published photographs.

NO (*contd.*)

λ	I	v', v''	λ	I	v', v''	λ	I	v', v''
9732·5	?	0, 0	8619·8	10	2, 1	7939·1	5	2, 0
9617·3			8529·9			7861·9		
9563·1	?	1, 1	8514·2	3	3, 2	7861·9	5	3, 1
9453·5			8425·6			7787·6		
8733·3	8	1, 0						
8640·3								

NO⁺

See NO 6000 Å system.

NO₂

The spectrum of nitrogen dioxide is extremely complex. The main 'Visible' system, $A^2B_1 - X^2A_1$ extends from 11 000 to 3000 Å; it has been studied mainly in absorption but also occurs in emission. The ultra-violet system, $B^2B_2 - X^2A_1$, with origin at 2491 Å, shows better defined band heads in absorption.

VISIBLE ABSORPTION SYSTEM, $A^2B_1 - X^2A_1$

Occurrence. Readily obtained in absorption from NO_2 vapour. To obtain the spectrum in a pure state the temperature should be chosen so that N_2O_4 is decomposed but NO_2 is still stable (about 120° C.).

Appearance. With small amounts of gas a number of bands are obtained in the violet and ultra-violet regions, showing sharp rotational structure at the violet end but becoming more and more diffuse toward the ultra-violet and finally merging into a continuum. As the quantity of gas is increased the absorption spreads step by step to the red and sharp bands may be followed down to about 9000 Å. It is possible that more than one electronic system is involved. See Plate 8.

References. L. Harris and R. W. B. Pearse, unpublished data.
T. C. Hall and F. E. Blacet, *J. Chem. Phys.*, **20**, 1745 (1952).
A. E. Douglas and K. P. Huber, *Canad. J. Phys.*, **43**, 74 (1965)†.

Individual bands vary greatly in appearance. The following are some of the more outstanding edges and maxima:

λ		I	λ		I	λ		I	λ		I
8361	V	–	4795	M	3	4480	M	10	4270	R	5
5095	M	3	4740	R	5	4448	R	8	4133	M	5
5048	M	3	4630	R	6	4390	R	8	4102	M	4
5027	M	3	4605	M	3	4366·9	R	5	4081	R	3
4945	M	3	4580	M	4	3450	M	6	3909·8	R	–
4880	R	3	4544·7	R	4	4304	R	5	3778·5	R	–

These are followed by many bands growing wider and more diffuse to about 3200 Å.

NO$_2$ (*contd.*)

ULTRA-VIOLET SYSTEM, 2491 Å, $B^2B_2 - X^2A_1$

Occurrence. Obtained readily in absorption under the same conditions as the system in the visible.

Appearance. With a small quantity of gas two bands have sharp heads degraded to the red and show a very open rotational structure resembling a single Q branch. Other bands have been observed as far as 2000 Å but are all diffuse. With a greater quantity of gas other sharp bands to the red are observed.

References. As for Visible System.

L. Harris, G. W. King, W. S. Benedict and R. W. B. Pearse, *J. Chem. Phys.*, **8**, 765 (1940).

J. B. Coon, F. A. Cesani and F. P. Huberman, *J. Chem. Phys.*, **52**, 1647 (1970)†.

The strongest bands, degraded to the red, are as follows:

λ	I	$(v_1\ v_2\ v_3)'$			$(v_1\ v_2\ v_3)''$			λ	I	$(v_1\ v_2\ v_3)'$			$(v_1\ v_2\ v_3)''$		
2665·6	h	0	0	0	2	0	0	2459·0	5	0	1	0	0	0	0
2628·8	h	0	1	0	2	0	0	2447·6	5d	0	0	1	0	0	0
2626·1	h	0	0	0	1	1	0	2430·5	2d	0	2	0	0	0	0
2575·8	h	0	0	0	1	0	0	2421	10d	1	0	0	0	0	0
2541·4	h	0	1	0	1	0	0	2389	8d	1	1	0	0	0	0
2538·5	h	0	0	0	0	1	0	2373	8d	1	0	1	0	0	0
2529·4	h	0	0	1	1	0	0	2363	4dm	1	2	0	0	0	0
2491·1	6	0	0	0	0	0	0	2351	9dm	2	0	0	0	0	0

h = observed only in heated gas. d = diffuse. dm = diffuse maximum.

EMISSION BANDS (AIR AFTERGLOW), $A^2B_1 - X^2A_1$

Occurrence. In the yellow-green air afterglow, the reactions $NO + O_3$ or $NO + O$ and thermal emission from NO_2 at 1200° C.

Appearance. Narrow diffuse bands superposed on a continuum from 3975 Å to well down in the infra-red. This appears to be the same system as the 'Visible' absorption system.

References. V. Kondratiev, *Phys. Z. Sowjet.*, **11**, 320 (1937).

J. C. Greaves and D. Garvin, *J. Chem. Phys.*, **30**, 348 (1959).

H. P. Broida, H. I. Schiff and T. M. Sugden, *Trans. Faraday Soc.*, **57**, 259 (1961)†.

M. A. A. Clyne, B. A. Thrush and R. P. Wayne, *Trans. Faraday Soc.*, **60**, 359 (1964)†.

The following are averaged wavelengths of maxima of strong bands observed by Clyne *et al.*, and Greaves and Garvin: λλ8515, 8400, 8024, 7990, 7919, 7533, 6745, 6513, 6305, 6088. Kondratiev records a further 35 bands, spaced at about 50 Å, from 5900–4070.

NO$_3$

Absorption bands have been observed during the reaction of NO_2 with ozone and in pyrolysis of N_2O_5 in a shock tube.

References. G. Schott and N. Davidson, *J. Amer. Chem. Soc.*, **80**, 1841 (1958).

E. J. Jones and O. R. Wulf, *J. Chem. Phys.*, **5**, 873 (1937)†.

D. A. Ramsay, *Proc. Xth Colloq. Spectroscopium Internat.*, 583 (1962)†.

The following maxima of diffuse bands are from Ramsay's photograph, with our estimates of intensity: λλ6625 (10), 6233 (9), 6045 (4), 5893 (7), 5589 (5), 5474 (4).

N₂O

Nitrous oxide shows strong continuous absorption with maximum intensity in the far ultraviolet, reaching about 2200 Å at room temperature but spreading to longer wavelengths at higher temperature, e.g. to about 2800 Å at 680° C. With long paths (150 m.atm.) there is weak continuous absorption with maximum near 2900 Å.

Reference. M. G. Holliday and B. G. Reuben, *Trans. Faraday Soc.*, **64**, 1735 (1968).

N₂O⁺

Occurrence. Hollow-cathode discharge and electron beam through nitrous oxide.

Appearance. Degraded to shorter wavelengths. Double-headed bands due to a $^2\Sigma - {}^2\Pi$ transition in the linear molecule.

Reference. J. H. Callomon, *Proc. Chem. Soc.*, 313 (1959)†.

Strongest bands:

λ	I	v'_1, v''_1	λ	I	v'_1, v''_1
3864·2	3	0, 2	3558·4	9	0, 0
3845·4			3541·6		
3707·0	10	0, 1	3396·7	4	1, 0
3688·8			3380·4		

N₂O₃

Reference. E. H. Melvin and O. R. Wulf, *J. Chem. Phys.*, **3**, 755 (1935)†.

Continuous absorption by mixtures of NO and NO₂, starting at 3000 Å and becoming stronger below 2700 Å.

N₂O₄

Reference. T. C. Hall and F. E. Blacet, *J. Chem. Phys.*, **20**, 1745 (1952).

The equilibrium between N₂O₄ and N₂O has been studied. N₂O₄ contributes a continuum starting at 3900 Å with maxima around 3400, 3600 and below 2400 Å.

N₂O₅

Reference. E. J. Jones and O. R. Wulf, *J. Chem. Phys.*, **5**, 873 (1937).

Weak continuous absorption by nitrogen pentoxide starts at 3800 Å and becomes stronger to shorter wavelengths.

NS

The two main systems of NS are usually called the β and γ systems to maintain the analogy with NO, although the transitions are actually different. Four more weaker systems are now known; they are made difficult to sort out by a change in naming of states by Jenouvrier and Pascat; we have retained the older nomenclature.

References. A. Fowler and C. J. Bakker, *P.R.S.*, **136**, 28 (1932)†.

R. F. Barrow, G. Drummond and P. B. Zeeman, *Proc. Phys. Soc.*, A, **67**, 365 (1954).

NS (*contd.*)

N. A. Narasimham and T. K. B. Subramanian, *J. Mol. Spec.*, **29**, 294 (1969)†; **40**, 511 (1971)†.
J. J. Smith and B. Meyer, *J. Mol. Spec.*, **14**, 160 (1964).
A. Jenouvrier and B. Pascat, *Canad. J. Phys.*, **51**, 2143 (1973)†.

β SYSTEM, $A^2\Delta - X^2\Pi$

Occurrence. In discharge through nitrogen mixed with sulphur vapour.

Appearance. Pairs of bands degraded to the red.

Main heads from Fowler and Bakker; the two sub-bands are denoted i and ii:

λ	I	v', v''	λ	I	v', v''	λ	I	v', v''
2697·4	5	1, 3 i	2597·1	6	0, 1 i	2477·8	3	2, 1 i
2683·3	4	1, 3 ii	2584·8	5	0, 1 ii	2465·6	3	2, 1 ii
2680·0	4	0, 2 i	2553·1	3	2, 2 i	2460·7	7	1, 0 i
2667·0	3	0, 2 ii	2540·1	2	2, 2 ii	2448·7	6	1, 0 ii
2614·6	4	1, 2 i	2518·5	8	0, 0 i	2406·0	4	2, 0 i
2601·2	3	1, 2 ii	2506·7	8	0, 0 ii	2394·4	4	2, 0 ii

γ SYSTEM, $C^2\Sigma - X^2\Pi$

Occurrence. In discharge through nitrogen mixed with sulphur vapour.

Appearance. Degraded to shorter wavelengths. A single progression of double double-headed bands.

The two sub-bands are indicated by i and ii. Only the P heads are listed. The Q heads lie about 1·2 Å to the violet of these; data from Fowler and Bakker.

λ	I	v', v''	λ	I	v', v''	λ	I	v', v''
2587·0	1	0, 4 ii	2453·0	5	0, 2 i	2371·1	10	0, 1 ii
2525·7	3	0, 3 i	2439·7	4	0, 2 ii	2317·2	10	0, 0 i
2511·7	3	0, 3 ii	2383·6	10	0, 1 i	2305·2	8	0, 0 ii

$B^2\Pi - X^2\Sigma$ SYSTEM

Occurrence. In microwave discharge through neon $+ N_2$ and a trace of sulphur, and when sulphur compounds react with active nitrogen. Two very different vibrational intensity distributions seem to occur; in the microwave discharge (Narasimhan and Subramanian) there is an ultra-violet progression with $v'' = 0$ and high v'; with pulsed active nitrogen (Smith and Meyer) the $v' = 0$ progression in the blue and violet is prominent.

Appearance. Double-headed bands, shaded to the red.

The following are probably the strongest heads in the pulsed active nitrogen:

λ	v', v''	λ	v', v''	λ	v', v''
4355·3	0, 6 ii	4152·2	0, 5 ii	3842·4	1, 4 ii
4330·2	0, 6 i	4129·0	0, 5 i	3823·0	1, 4 i
4214·7	1, 6 ii	3963·4	0, 4 ii	3789·2	0, 3 ii
4191·7	1, 6 i	3942·4	0, 4 i	3770·0	0, 3 i

and in the microwave discharge:

λ	v', v''	λ	v', v''
2890	6, 0 ii	2770·3	8, 0 ii
2879·8	6, 0 i	2761·8	8, 0 i
2828·3	7, 0 ii	2714·9	9, 0 ii
2819·3	7, 0 i	2707·1	9, 0 i

NS (*contd.*)

F $^2\Pi - X^2\Pi$ SYSTEM

Occurrence. Discharge through $Ne + N_2$ + trace S, and in microwave discharge.

Appearance. Double-headed. Degraded to the red.

The following heads are from Narasimhan and Subramanian (1971):

λ	v', v''		λ	v', v''	
2661·8	0,	5 ii	2583·0	0,	4 ii
2634·1	0,	5 i	2556·9	0,	4 i

G $^2\Sigma^- - X^2\Pi$ SYSTEM

Observed by Jenouvrier and Pascat in a microwave discharge. Double double-headed and degraded to the red.

Band			λ origin	λ heads	
$G - X^2\Pi_{\frac{1}{2}}$	0,	1	2381·6	R 2380·6	
$G - X^2\Pi_{\frac{3}{2}}$	0,	0	2327·1	R 2326·4	Q 2327·0
$G - X^2\Pi_{\frac{1}{2}}$	1,	0	2268·9		

I $^2\Sigma - X^2\Pi$ SYSTEM

Observed in microwave discharges. Shaded to the red, double double-headed. Data from Narasimhan and Subramanian (1969). First and third heads:

v', v''		λ	
0,	2	2386·1	2398·9
0,	1	2320·4	2332·4

BANDS 2703 AND 2687 Å

Reference. N. A. Narasimham, K. Raghuveer and T. K. Balasubramanian, *J. Mol. Spec.*, **54**, 160 (1975)†.

Two headless bands with origins at 2703 and 2687 are attributed to a $^2\Sigma - {}^2\Pi$ transition.

NS$^+$

Four bands obtained by Dressler on addition of helium to a discharge giving NS bands.

Appearance. Double-headed bands degraded to the red.

Transition. A $^1\Pi - X^1\Sigma$.

Reference. K. Dressler, *Helv. Phys. Acta*, **28**, 563 (1955)†.

Dressler gives the following data. Actual values of v' and v'' have not been assigned.

λ	I	v', v''	Head
2326·5	10	$v' + 1, v''$	R
2327·2	10	$v' + 1, v''$	Q
2380·8	10	v', v''	R
2381·7	10	v', v''	Q

NSF

References. T. Barrow *et al.*, *Trans. Faraday Soc.*, **65**, 2295 (1969)†.

T. Barrow and R. N. Dixon, *Chem. Phys. Lett.*, **4**, 547 (1970).

D. P. Craig and G. Fischer, *Chem. Phys. Lett.*, **4**, 227 (1970).

NSF (*contd.*)

Thiazyl fluoride shows three banded absorption systems, the strongest with origin at 3923, another at 2362 and a weak triplet-singlet system at 5360 Å.

3923 Å SYSTEM, 4050–2500 Å, A – X

Some heads are degraded to the red. Barrow *et al.* record about 55 bands in their main array. The following are strongest, with our estimates of intensity:

λ	I	$(v_1\ v_2\ v_3)'$	$(v_1\ v_2\ v_3)''$	λ	I	$(v_1\ v_2\ v_3)'$	$(v_1\ v_2\ v_3)''$
3930·5	5	0 0 1	0 0 1	3874·0	8	0 0 1	0 0 0
3922·2	5	0 0 0	0 0 0	3827·1	10	0 0 2	0 0 0
3882·1	6	0 0 2	0 0 1	3753·0	10	0 0 3	0 0 0

2362 Å SYSTEM, 2400–2150 Å, B – X

Strongest bands 2361·8 (000, 000), 2302·6 (100, 000), 2247·4 (200, 000), 2195·6 (300, 000).

NSe

References. K. V. Subbaram and D. R. Rao, *J. Mol. Spec.*, **36**, 163 (1970)†.
L. Harding *et al.*, *Canad. J. Phys.*, **49**, 2033 (1971).

Occurrence. Microwave discharge and selenium compounds in active nitrogen.

VISIBLE SYSTEM, 5675–3950 Å, $A^2\Pi - X^2\Pi$

Degraded to the red. Wide-spaced doublets. Data from Subbaram and Rao, with our estimates of intensity (N.B. these values are taken from the plate, but differ from the listed values of wavenumbers by about the correction from air to vacuum).

$A^2\Pi_{\frac{1}{2}} - X^2\Pi_{\frac{1}{2}}$			$A^2\Pi_{\frac{3}{2}} - X^2\Pi_{\frac{3}{2}}$		
λ	I	$v',\ v''$	λ	I	$v',\ v''$
5674·6	5	1, 8	5552·5	4	1, 7
5607·5	4	0, 7	5472·8	5	0, 6
5408·9	7	1, 7	5294·9	4	1, 6
5345·2	8	0, 6	5219·2	6	0, 5
5102·6	7	0, 5	4984·9	8	0, 4
4878·5	10	0, 4	4768·3	10	0, 3

$B^2\Sigma - X^2\Pi$ SYSTEM

Appearance. Shaded to the red with close double-headed bands and wide $^2\Pi_{\frac{1}{2}} - {}^2\Pi_{\frac{3}{2}}$ spacing. R heads from Harding *et al.*

$v',\ v''$	$^2\Sigma - {}^2\Pi_{\frac{1}{2}}$	$^2\Sigma - {}^2\Pi_{\frac{3}{2}}$
0, 2	3167	3081
0, 1	3076·4	2994·5
0, 0	2989·6	2912·1

$C^2\Delta_{\frac{5}{2}} - X^2\Pi_{\frac{3}{2}}$ SYSTEM

Degraded to the red. R heads from Harding *et al.*

NSe (*contd.*)

λ	v', v''	λ	v', v''
3070·4	1, 2	2903·1	1, 0
3002·4	2, 2	2842·3	2, 0
2984·9	1, 1		

Na$_2$

Band systems occur in the red and the blue-green and several are reported in the near ultra-violet.

RED SYSTEM, 8000–6000 Å, $A^1\Sigma_u^+ - X^1\Sigma_g^+$

Occurrence. In absorption, in fluorescence, in magnetic rotation and in emission in a discharge tube.

Appearance. Single-headed bands degraded to the red. About 80 bands have been recorded.

References. W. R. Fredrickson and W. W. Watson, *P.R.*, **30**, 429 (1927).

W. R. Fredrickson and C. R. Stannard, *P.R.*, **44**, 632 (1933).

The following are the strongest bands (mostly $I = 5$) in absorption:

λ	v', v''	λ	v', v''	λ	v', v''
6751·2	4, 2	6630·1	5, 1	6465·8	7, 0
6732·0	3, 1	6561·5	5, 0	6418·4	8, 0
6679·0	4, 1	6513·2	6, 0	6374·2	9, 0

BLUE-GREEN SYSTEM, 5040–4560 Å, $B^1\Pi_u - X^1\Sigma_g^+$

Occurrence. Similar to red system and also in flame of metal burning in air.

Appearance. Degraded to the red.

References. W. R. Fredrickson and W. W. Watson, *P.R.*, **30**, 429 (1927).

F. W. Loomis and R. W. Wood, *P.R.*, **32**, 223 (1928).

The intensities as listed in the above references do not agree very well, but the following include the strong bands as observed in absorption:

λ	v', v''	λ	v', v''
5040·4	0, 3	4894·5	1, 0
5001·4	0, 2	4865·5	2, 0
4962·8	0, 1	4837·2	3, 0
4932·6	1, 1	4819·5	5, 1
4924·2	0, 0	4809·8	4, 0

ULTRA-VIOLET SYSTEMS

References. S. P. Sinha, *Proc. Phys. Soc.*, **59**, 610 (1947).

S. P. Sinha, *Proc. Phys. Soc.*, **62**, 124 (1949).

C. S. Chang, *Chinese J. Phys.*, **7**, 377 (1950).

R. F. Barrow, N. Travis and C. V. Wright, *Nature*, **187**, 141 (1960).

Sodium vapour gives numerous bands in absorption in the region $\lambda\lambda$3640–2400. Some of the bands have also been obtained in emission. The bands are degraded to the red and originate in transitions involving the $^1\Sigma$ ground state. Different arrangements of the bands into systems have been given by various authors. Sinha (1947) finds seven different systems of which only three are

Na$_2$ (*contd.*)

well developed. In the later paper Sinha investigates the strongest system in the region $\lambda\lambda 3600$–3200 with high dispersion. The data for the strongest bands are given below.

$C^1\Pi_u - X^1\Sigma_g^+$, 3600–3200 Å

λ	I	v', v''	λ	I	v', v''	λ	I	v', v''
3486·3	4	2, 6	3403·8	5	0, 0	3338·8	10	5, 0
3468·1	5	2, 5	3400·4	8	3, 2	3326·3	10	6, 0
3450·0	7	2, 4	3382·4	6	3, 1	3314·0	10	7, 0
3432·2	7	2, 3	3369·2	10	4, 1	3290·0	9	9, 0
3414·0	5	2, 2	3356·5	8	5, 1	3278·4	8	10, 0
3409·4	4	1, 1	3351·5	7	4, 0	3266·8	7	11, 0

The $\lambda 3303$ line of Na overlaps the system at the position of the (8, 0) band.

$D^1\Pi_u - X^1\Sigma_g^+$, 3325–3030 Å

About 70 bands. Strongest ($I = 2$) from Chang: $\lambda\lambda 3151\cdot 6$ (1, 2), $3145\cdot 2$ (3, 3), $3140\cdot 0$ (2, 2), $3135\cdot 7$ (1, 1), $3131\cdot 2$ (0, 0), $3125\cdot 1$ (2, 1), $3120\cdot 5$ (1, 0), $3109\cdot 5$ (2, 0).

$E - X^1\Sigma_g^+$, 3120–2880 Å

Strongest bands from Sinha: $\lambda\lambda 2945\cdot 5$ (7, 0), $2936\cdot 2$ (8, 0), $2927\cdot 6$ (9, 0).

NaAr

A diffuse band at 5565 Å accompanies the D lines in emission from rapidly compressed sodium in argon.

    Reference.   G. T. Lalos and G. L. Hammond, *Astrophys. J.*, **135**, 616 (1962)†.

NaCd

Diffuse bands degraded to the violet, observed in absorption by Barratt (see CsCd). Heads $\lambda\lambda 4400$ and 3974.

NaCs

    References.   See LiCs.

    Absorption bands observed by Walter and Barratt, and arranged into three systems by Weizel and Kulp. Degraded to the red. Strongest heads (intensity 10): $\lambda\lambda 5631$, 5602, 5571, 5542, 5492, 5463, 5444, 5425, 4136, 4098.

NaF

    Occurrence.   In absorption in a furnace.

    Reference.   R. F. Barrow and A. D. Caunt, *P.R.S.*, **219**, 120 (1953).

    Continuous absorption with maximum intensity below 2200 Å. Diffuse bands $\lambda\lambda 2540$, 2595, 2635, 2685 and 2732.

NaH

References. T. Hori, *Z.P.*, **62**, 352 (1930); **71**, 478 (1931).
E. Olsson, *Z.P.*, **93**, 206 (1934).
R. C. Pankhurst, *Proc. Phys. Soc.*, **62**, 191 (1949).

4500 Å SYSTEM, $A^1\Sigma - X^1\Sigma$

A many lined system with weak R heads degraded to the red. The system is readily obtained in absorption from a mixture of hydrogen and sodium vapour and in emission from a sodium arc in hydrogen or a discharge through a mixture of hydrogen and sodium vapour, and in flames. See Plate 5.

Origins of the Strongest Bands

v', v''	λ	I	v', v''	λ	I	v', v''	λ	I
5, 1	4376·5	4	5, 0	4169·7	2	10, 0	3879·9	10
6, 1	4309·7	5	6, 0	4109·0	5	11, 0	3826·5	9
7, 1	4244·3	5	7, 0	4049·5	7	12, 0	3774·8	9
4, 0	4231·4	0	8, 0	3991·5	9	13, 0	3724·9	9
8, 1	4180·6	4	9, 0	3934·9	10			

NaHg

Diffuse bands, degraded to the violet, observed in absorption by Barratt (see CsCd). Heads $\lambda\lambda$6499, 6454, 4649, 4425 and 4406.

NaI

The heated vapour shows strong continuous absorption in the near ultra-violet, merging into diffuse banded structure in the violet.

Reference. K. Sommermeyer, *Z.P.*, **56**, 548 (1929).

NaK

Four systems, in the near infra-red, the yellow, the green and the violet respectively, have been attributed to NaK.

Occurrence. All four systems have been observed in absorption by a mixture of sodium and potassium vapours. The yellow and green systems, have also been observed in magnetic rotation, and the latter has been obtained in fluorescence.

Appearance. All the systems are degraded to longer wavelengths.

INFRA-RED SYSTEM, 9100–7200 Å, $A^1\Sigma - X^1\Sigma$

Reference. F. W. Loomis and M. J. Arvin, *P.R.*, **46**, 286 (1934).
About 120 bands, degraded to longer wavelengths. Strongest head 8338 Å (3, 3).

YELLOW SYSTEM, 5970–5650 Å, $B^1\Pi - X^1\Sigma$

References. F. W. Loomis and M. J. Arvin, *P.R.*, **46**, 286 (1934).
S. Barratt, *P.R.S.*, **105**, 221 (1923–4)†.
S. P. Sinha, *Proc. Phys. Soc.*, **60**, 447 (1948)†.
Measurements, intensities in absorption and vibrational assignments given by Sinha:

NaK (*contd.*)

λ	I	v', v''	λ	I	v', v''	λ	I	v', v''
6066·2	3	0, 4	5869·5	5	1, 0	5711·7	9	9, 0
6040·5	2	1, 4	5845·2	7	2, 0	5696·2	8	10, 0
6022·1	2	0, 3	5823·5	6	3, 0	5680·8	8	11, 0
5997·0	3	1, 3	5802·6	7	4, 0	5676·0	3	13, 2
5973·0	2	2, 3	5783·0	6	5, 0	5665·3	6	12, 0
5954·2	3	1, 2	5763·4	7	6, 0	5650·4	5	13, 0
5935·1	3	0, 1	5745·4	9	7, 0	5640·8	4	14, 0
5912·0	4	1, 1	5728·2	10	8, 0			

GREEN SYSTEM, 5260–4730 Å, $C^1\Pi - X^1\Sigma$

References. F. W. Loomis and M. J. Arvin, *P.R.*, **46**, 286 (1934).
S. P. Sinha, *Proc. Phys. Soc.*, **60**, 447 (1948)†.
Bands observed by Sinha:

λ	I	v', v''	λ	I	v', v''	λ	I	v', v''
5261·8	1	0, 9	5011·9	5	3, 3	4836·5	8	11, 2
5230·5	1	0, 8	5001·8	6	2, 2	4830·3	10	8, 0
5198·9	1	0, 7	4982·2	7	3, 2	4819·3	7	12, 2
5167·5	2	0, 6	4962·4	7	4, 2	4813·0	8	9, 0
5136·3	3	0, 5	4943·2	5	5, 2	4796·2	7	10, 0
5115·0	2	1, 5	4932·2	8	4, 1	4779·7	6	11, 0
5105·4	3	0, 4	4913·5	8	5, 1	4764·1	5	12, 0
5083·8	4	1, 4	4894·5	8	6, 1	4748·5	4	13, 0
5063·5	3	2, 4	4876·8	10	7, 1	4733·1	4	14, 0
5052·6	4	1, 3	4866·0	8	6, 0	4717·8	4	15, 0
5032·4	4	2, 3	4847·4	10	7, 0	4702·7	4	16, 0

ULTRA-VIOLET BANDS, $D^1\Sigma - X^1\Sigma$

References. J. M. Walter and S. Barratt, *P.R.S.*, **119**, 257 (1928).
Y. Uchida, *Japan. J. Phys.*, **5**, 145 (1929).
W. Weizel and M. Kulp, *Ann. Phys.* (*Lpz.*), **4**, 971 (1930).
S. P. Sinha, *Proc. Phys. Soc.*, **60**, 447 (1948).
Sinha has measured eighty-nine bands in the region λλ4080–3820. Strongest bands:

λ	I	v', v''	λ	I	v', v''	λ	I	v', v''
3959·9	6	3, 2	3898·4	6	11, 4	3885·8	8	12, 4
3954·9	6	2, 1	3894·3	8	8, 2	3880·1	6	16, 6
3907·9	10	4, 0	3888·6	8	7, 1	3876·2	8	8, 1

Walter and Barratt also obtained bands in the region λλ3650–3550. These were not obtained by Sinha, who suggests that they belong to Na_2.

NaMg

An absorption band, degraded to the violet, is reported at 5290·8 Å.
Reference. S. Barratt, *P.R.S.*, **109**, 194 (1925).

NaRb

Occurrence. Absorption by mixed vapours and in magnetic rotation.

Appearance. Two rather extensive systems of red-degraded bands.

References. J. M. Walter and S. Barratt, *P.R.S.*, **119**, 257 (1928).
P. Kusch, *P.R.*, **49**, 218 (1936).

$A^1\Pi - X^1\Sigma$, 6336–5990 Å

Strongest, from Kusch, $\lambda\lambda 6255 \cdot 2$ (0, 4), $6216 \cdot 0$ (0, 3), $6192 \cdot 7$ (1, 3), $6152 \cdot 8$ (1, 2), $6113 \cdot 9$ (1, 1), $6091 \cdot 9$ (2, 1).

$B - X^1\Sigma$

Strongest bands from Walter and Barratt, 5432, 5397, 5368, 5339 and 5331 Å.

NbN see NbO⁺

NbO

There are three systems, usually referred to as A, B and C. There is some doubt about the reality of system B, which is not included in the tabulation by Rosen *et al.*, but is included again here.

Occurrence. In emission from an arc containing niobium (columbium) salts and in a discharge through niobium pentachloride.

References. V. R. Rao, *Indian J. Phys.*, **24**, 35 (1950)†.
V. R. Rao and D. Premaswarup, *Indian J. Phys.*, **27**, 399 (1953)†.
U. Uhler, *Ark. Fys.*, **8**, 265 (1954).

BLUE SYSTEM, A, 5100–4200 Å, $A^2\Delta - X^2\Delta$

Degraded to the red, with marked sequences. Strongest, from Rao:

λ	I	v', v''	λ	I	v', v''	λ	I	v', v''
5161·8	4	0, 2	4915·1	6	0, 1	4510·7	8	1, 0
4976·8	5	2, 3	4689·0	10	0, 0	4346·7	3	2, 0
4946·4	4	1, 2	4540·3	5	2, 1			

B SYSTEM, 6300–5000 Å

The bands are weakly degraded to the red. Rao has made a provisional analysis in which each band has four heads, with the (0, 0) at 5470 Å. Strongest heads, with intensities, $\lambda\lambda 5200 \cdot 9$ (8), $5238 \cdot 8$ (4), $5470 \cdot 3$ (4), $5591 \cdot 2$ (4), $5599 \cdot 9$ (4), $5780 \cdot 2$ (4), $5826 \cdot 5$ (4), $5894 \cdot 3$ (4), $6144 \cdot 5$ (4), $6184 \cdot 9$ (5), $6251 \cdot 6$ (4), $6268 \cdot 4$ (6).

EXTREME RED SYSTEM, C

A large number of bands, shaded to longer wavelengths. Rao has given an analysis in which each band has four heads with the (0, 0) at 6474 Å. The following are the strongest heads, with intensities: $\lambda\lambda 7027 \cdot 3$ (4), $6969 \cdot 4$ (4), $6950 \cdot 9$ (4), $6756 \cdot 8$ (4), $6737 \cdot 1$ (5), $6495 \cdot 6$ (5), $6494 \cdot 8$ (5), $6119 \cdot 7$ (4).

NbO⁺

Reference. T. M. Dunn and K. M. Rao, *Nature*, **222**, 266 (1969).
Three sub-bands of a (0, 0) sequence observed 'in emission'.

NbO$^+$ *(contd.)*

Sub-band	R head	Q head
$^3\Phi_2 - {}^3\Delta_1$	6174·4	6192·3
$^3\Phi_3 - {}^3\Delta_2$	6028·6	6043·2
$^3\Phi_4 - {}^3\Delta_3$	5916·0	5929·6

These bands are attributed to ionized niobium (colombium) oxide, but may possible be due to NbN.

NdO

Occurrence. Neodymium salts in oxy-hydrogen flame.

References. G. Piccardi, *Rend. Accad. Linc.*, **21**, 584 (1935).

R. Herrmann and C. T. J. Alkemade, *Flame Photometry* (1963).

The following are among the strongest of over 250 heads listed:

λ	I	λ	I	λ	I	λ	I
6629·6	8	6585·2	15	5999	8	5250·2	6
6624·6	8	6370·9	8	5996	8	4768·7	4
6620·2	7	6368·6	8	5990·7	10	4620·0	8
6606·7	10	6353·5	9	5975	10	4510·8	3
6602·4	10	6249·9	10	5971·1	8	4413·5	3
6598·3	10	6002	8	5313·5	7		

Herrmann and Alkemade also give a very long list of bands from 8661 to 4414 Å, obtained in flames. They include strong ones ($I = 2000$) at 7115, 7092, 7033 and 7011 Å; many are shaded to longer wavelengths.

Ne$_2$

Occurrence. In a transformer discharge and in recombination of Ne$^+$ ions on a carbon-film surface.

References. D. G. Dhavale, *Nature*, **125**, 276 (1930).

V. V. Gritsyna *et al.*, *Opt. Spectrosc.*, **29**, 343 (1970)†.

Headless bands with rather open rotational structure λλ7393, 7208, 7003, 6963 and 6847. Gritsyna *et al.* also record weak bands λλ5528, 5385, 5338 and 5253 Å.

NiBr

References. P. Mesnage, *Thesis, Paris* (1938).

V. G. Krishnamurty, *Indian J. Phys.*, **26**, 429 (1952)†.

S. P. Reddy and P. T. Rao, *Proc. Phys. Soc.*, **75**, 275 (1960)†.

N. Sundarachary, *Proc. Nat. Acad. Sci. India*, A, **32**, 311 (1962).

Complex red-degraded bands, mostly in sequences, are obtained from high-frequency or heavy direct-current discharges. Krishnamurty lists about 200 heads and shows the best photographs. There is some lack of agreement in both measurements and analyses. The following sequence heads are mainly from Reddy and Rao, with the addition of Sundarachary's E system.

NiBr (*contd.*)

λ	I	v', v''	System	λ	I	v', v''	System
4658·3	2	0, 1	γ	4354·1	10	0, 0	β_1
4588·0	9	0, 0	γ	4300·0	2	1, 0	β_1
4526·2	2	1, 0	γ	4259·1	2	0, 1	α_3
4520·6	1	0, 1	E	4202·9	4	0, 0	α_3
4456·3	4	0, 0	E	4182·1	2	0, 0	α_2
4416·8	7	0, 0	β_2	4163·3	2	0, 1	α_1
4406·3	2	1, 0	β_2	4110·9	7	0, 0	α_1
4399·6	1	1, 0	E	4062·3	2	1, 0	α_1

NiCl

Occurrence. In high-frequency and in heavy-current D.C. discharge.

References. V. G. Krishnamurty, *Indian J. Phys.*, **26**, 207 (1952)†.

S. P. Reddy and P. T. Rao, *Proc. Phys. Soc.*, **75**, 276 (1960)†.

S. V. K. Rao, S. P. Reddy and P. T. Rao, *Z.P.*, **166**, 261 (1961)†.

N. V. K. Rao and P. T. Rao, *Curr. Sci. (India)*, **38**, 589 (1969).

A number of sequences of bands, degraded to the red, have been reported and analysed into 17 systems or sub-systems. Main features:

λ	I	v', v''	Sys.	λ	I	v', v''	Sys.	λ	I	v', v''	Sys.
8685·4	10	0, 0	J	6738·3	7	1, 0	F	4597·0	10	0, 0	E_1
8398·4	4	1, 0	J	5352·1		0, 0	IV	4562·8	10	0, 0	D
8394·6	10	0, 0	I	5006·9		0, 0	III	4396·5	7	0, 0	C
8144·0	8	0, 0	H	4933·5		0, 0	II	4304·9	10	0, 0	B_2
7880·0	8	1, 0	H	4868·5		0, 0	I	4284·5	2	0, 0	B_1
7713·0	8	0, 0	G	4705·4	5	0, 1	E_2	4146·1	6	1, 1	A_3
7486·6	8	1, 0	G	4625·5	10	2, 2	E_2	4143·1	4	0, 0	A_3
7274·8	7	2, 0	G	4622·5	9	1, 1	E_2	4094·2	7	0, 0	A_2
6926·1	8	0, 0	F	4619·9	4	0, 0	E_2	4061·4	10	0, 0	A_1

NiF

Occurrence. Heavy-current discharge tube.

Appearance. Degraded to red. Marked sequences.

Transition. Perhaps $^2\Pi - {}^2\Sigma$.

Reference. V. G. Krishnamurty, *Indian J. Phys.*, **27**, 354 (1953)†.

λ	I	v', v''		λ	I	v', v''	
4514·6	7	0, 0	$^SR_{21}$	4695·1	6	0, 1	R_1
4518·7	9	0, 0	R_2	4700·0	5	0, 1	Q_1
4523·9	5	0, 0	Q_2	4832·7	4	0, 2	$^SR_{21}$
4535·8	6	0, 0	R_1	4843·7	4	0, 2	Q_2
4669·1	6	0, 1	$^SR_{21}$	4859·7	5	0, 2	R_1
4673·0	8	0, 1	R_2	4864·4	3	0, 2	Q_1
4678·8	5	0, 1	Q_2				

NiH

References. A. G. Gaydon and R. W. B. Pearse, *P.R.S.*, **148**, 312 (1935)†.
A. Heimer, *Z.P.*, **105**, 56 (1936)†.
N. Åslund, H. Neuhaus, A. Lagerqvist and E. Andersér, *Ark. Fys.*, **28**, 271 (1965).

6442 Å SYSTEM, $^2\Delta_{2\frac{1}{2}} - {}^2\Delta_{2\frac{1}{2}}$

Band showing widely-spaced R and P branches with narrow doublets. A Q branch is observed near the origin but quickly decreases in intensity with increasing rotation. Occurs in flames fed with nickel carbonyl, in high-tension arcs in hydrogen flame, and in discharges where hydrogen and nickel vapour are present together. Also in absorption in shock-heated Ni in $H_2 + Ar$. Heads:

v', v''	R	Q	I
0, 0	6425·1	6443·3	4

6257 Å SYSTEM, $^2\Delta_{2\frac{1}{2}} - {}^2\Delta_{2\frac{1}{2}}$

Bands similar in appearance to the above system and occurring under similar conditions. See Plate 5.

	Heads		
v', v''	R	Q	I
0, 0	6246·0	6260·1	10
1, 0	5712·5	5724·8	7
2, 0	5290·0	5300·5	2
3, 0	4952·1	4962·0	1

4207 Å SYSTEM, $^2\Delta_{2\frac{1}{2}} - {}^2\Delta_{2\frac{1}{2}}$

Bands similar in appearance and occurrence to those of the other two systems.

	Heads	
v', v''	R	Q
0, 1	4570·5	4579·5
0, 0	4200·6	4208·3

4320 Å SYSTEM, $^2\Delta_{\frac{3}{2}} - {}^2\Delta_{\frac{3}{2}}$

A weak red-degraded band found by Åslund *et al.*, (0, 0) R head 4320·9.

A still weaker $^2\Delta_{\frac{5}{2}} - {}^2\Delta_{\frac{5}{2}}$ band with R head at 4140·3 has also been found.

2934 Å SYSTEM

A diffuse red-degraded band with maxima at 2934 and 2943 Å has been found by R. E. Smith in absorption in a shock tube.

There is also much weak unsystematised band-structure in the orange and red; there are weak heads (deg. R) at λλ6391·8, 6171·5 and 5834·3.

NiO

Three strong sequences in the infra-red and numerous bands in the visible have been observed in a nickel arc in air, in a flame containing nickel carbonyl, and in exploding wires. They have mostly been arranged by Malet and Rosen into six systems.

References. L. Malet and B. Rosen, *Bull. Soc. Roy. Sci. Liège*, 382 (1945).
B. Rosen, *Nature*, **156**, 570 (1945).

NiO (*contd.*)

In the following tables wavelengths given to 0·1 Å are by Gaydon; others by Malet and Rosen, whose values often differ from Gaydon's by several Å. All bands are degraded to greater wavelengths.

SYSTEM I. INFRA-RED

7900 (0, 0), 7983 (1, 1), 8094 (2, 2), 8187 (3, 3), 8300 (0, 1), 8392 (1, 2), 8502 (2, 3), 8600 (3, 4), 8739 (0, 2), 8843 (1, 3), 8950 (2, 4), 9070 (3, 5), 9195 (4, 6).

SYSTEM II. RED

7611 (0, 1), 7404 (1, 1), 7330 (0, 0), 7122 (2, 1), 7090 (1, 0).

SYSTEM III

λ	I	v', v''	λ	I	v', v''	λ	I	v', v''
6604·5	1	1, 3	6325	1	0, 1	5935·5	2	3, 2
6581	1	0, 2	6152·1	5	3, 3	5914·1	5	2, 1
6385·8	5	3, 4	6133·3	8	2, 2	5703	3	2, 0
6368	1	2, 3	6111	1	1, 1	5529·5	8	3, 0
6342·2	8	1, 2						

SYSTEM IV

λ	I	v', v''
5407·7	5	0, 1
5174·5	10	0, 0
5024	4 ?	1, 0
4889·5	5	2, 0

SYSTEM V

λ	I	v', v''
5323·4	2	0, 1
5098·3	3	0, 0
4951	0	1, 0

SYSTEM VI

λ	I	v', v''	λ	I	v', v''	λ	I	v', v''
4849·2	2	2, 2	4606·5	2	1, 0	4382	2	3, 0
4790·1	1	1, 1	4545·6	2	3, 1	4334	1	5, 1
4730	2 ?	0, 0	4494·3	1	2, 0	4238	0	6, 1
4721·4	1	3, 2	4436	2	4, 1	4145	0	7, 1
4665·7	5	2, 1						

Unclassified bands, $\lambda\lambda$6704, 6537, 6460, 6264, 6038, 5454·9, 5006·6 (5), 4938·8, 4876·7 (5), 4751·0 (8), 4714·6, 4630, 4405.

O$_2$

The neutral oxygen molecule does not readily show an emission spectrum in discharge tubes at low pressure. The main Schumann-Runge absorption system is strong in the far ultra-violet and extends to longer wavelengths in hot gas; this system also occurs in emission in an arc at atmospheric

O$_2$ (*contd.*)

pressure and some other sources, the emission lying in the nearer ultra-violet. Weak absorption bands due to forbidden transitions are observed with long paths; these include the Atmospheric and Herzberg systems. These also occur in emission in special sources such as afterglows and the night sky. The absorption spectrum of liquid oxygen is also discussed briefly.

SCHUMANN-RUNGE SYSTEM, $B^3\Sigma_u^- - X^3\Sigma_g^-$

Occurrence. The main part of this system lies in the vacuum ultra-violet below 1900 Å. Using heated gas and greater thicknesses, band structure may be extended well into the quartz region. The bands may sometimes be observed as an impurity in absorption using furnaces in which a considerable path of air becomes heated. In emission from a high tension arc at atmospheric or higher pressure, from shock-heated oxygen or air and in H_2—O_2 flames. Also CO—O_2 explosion flames. Durie obtained the bands with very clear heads in a glow of fluorine reacting at the surface of cold water. Although this is the main resonance system, the bands do not occur in low-pressure sources such as discharge tubes.

Appearance. Degraded to longer wavelengths. Relatively open rotational structure. In high temperature sources the heads are not well developed. See Plates 7 and 8.

References. S. W. Leifson, *Astrophys. J.*, **63**, 73 (1926)†.

H. P. Knauss and S. S. Ballard, *P.R.*, **48**, 796 (1935)†.

W. Lochte-Holtgreven and G. H. Dieke, *Ann. Phys. (Lpz.)*, **3**, 937 (1929).

M. W. Feast, *Proc. Phys. Soc.*, A, **62**, 114 (1949)†; A, **63**, 549 (1950).

R. A. Durie, *P.R.S.*, **211**, 110 (1952)†.

A. Herczog and K. Wieland, *Helv. Phys. Acta*, **23**, 432 (1950)†.

K. L. Wray and S. S. Fried, *J. Quant. Spec. Rad. Transfer*, **11**, 1171 (1973)†.

V. N. Snapko, *Opt. Spectrosc.*, **29**, 445 (1970).

G. R. Hébert, S. H. Innanen and R. W. Nicholls, *Identification Atlas of Molecular Spectra*, **4**, York Univ., (1967)†.

Absorption. Absorption by oxygen sets the limit to work with quartz spectrographs in air, and the bands are often seen at the end of spectra taken with small quartz instruments. The strong bands lie below 1900 Å, and merge into continuous absorption beyond 1759 Å. Intensities are rough estimates from Leifson's and Knauss and Ballard's photographs and published descriptions for absorption at atmospheric temperature. The bands with $v'' = 1$ are stronger in heated oxygen, and bands with higher v'' are only observed with the heated gas. With strongly heated oxygen absorption may be observed as far as 2500 Å, but in this region it consists only of complex line structure and definite heads are not observed.

$\lambda_{air.}$	I	v', v''	$\lambda_{vac.}$	I	v', v''	$\lambda_{vac.}$	I	v', v''
2221·0	—	5, 5	1997·5	0	1, 0	1863·0	10	7, 0
2193·4	—	6, 5	1971·4	1	2, 0	1845·8	10	8, 0
2167·9	—	7, 5	1959·5	0	5, 1	1830·1	10	9, 0
2150·8	—	5, 4	1946·7	2	3, 0	1815·6	10	10, 0
2125·0	—	6, 4	1938·0	0	6, 1	1803·1	10	11, 0
2101·1	—	7, 4	1923·5	4	4, 0	1792·0	9	12, 0
2084·2	—	5, 3	1918·3	1	7, 1	1782·3	7	13, 0
2060·0	—	6, 3	1901·9	6	5, 0	1774·3	6	14, 0
2037·6	—	7, 3	1900·1	1	8, 1	1767·7	4	15, 0
2020·7	—	5, 2	1887·7	8	6, 0			

O₂ (*contd.*)

Emission. The following measurements are from Durie using the fluorine/water flame. In the arc bands with $v' > 0$ are slightly stronger.

λ	I	v', v"	λ	I	v', v"	λ	I	v', v"
2440	2	3, 7	2923	7	1, 11	3671	9	0, 16
2480	2	2, 7	2984	8	0, 11	3743	5	1, 17
2528	4	3, 8	3039	5	1, 12	3841	8	0, 17
2570	5	2, 8	(3104)	8	0, 12	3914	8	1, 18
2613	5	1, 8	3145	7	?	4021	7	0, 18
2663	6	2, 9	3232	9	0, 13	4096	5	1, 19
2710	6	1, 9	3370	10	0, 14	4179	6	2, 20
2762	6	0, 9	3434	2	1, 15	4294	2	1, 20
2813	6 ?	1, 10	3517	10	0, 15	4375	3	2, 21
2870	7	0, 10	3582	3	1, 16			

Feast has photographed the emission spectrum over the range 5000–2000 Å and has extended the vibrational and rotational analysis. He finds the intensity of the system to be very pressure dependent.

ATMOSPHERIC ABSORPTION SYSTEM, $b\,^1\Sigma_g^+ - X\,^3\Sigma_g^-$

Occurrence. In atmospheric absorption (especially solar spectrum) and absorption by long path lengths of O_2. Also weakly in emission from oxygen afterglow, the night sky, $CO-O_2$ explosion flames and in discharge through O_2 and neon.

Appearance. The bands are degraded to longer wavelengths, but the heads are weak in some cases.

References. G. H. Dieke and H. D. Babcock, *Proc. Nat. Acad. Sci., U.S.A.*, **13**, 670 (1927). W. Ossenbruggen, *Z.P.*, **49**, 167 (1928). R. Mecke and W. Baumann, *Z.P.*, **73**, 139 (1932). R. C. Herman, G. A. Hornbeck and S. Silverman, *J. Chem. Phys.*, **17**, 220 (1949)†. G. A. Hornbeck and H. S. Hopfield, *J. Chem. Phys.*, **17**, 982 (1949). H. D. Babcock and L. Herzberg, *Astrophys. J.*, **108**, 167 (1948)†. F. D. Findlay, *Canad. J. Phys.*, **47**, 687 (1969).

Intensities I_a and I_e are for absorption and emission and are our estimates based on published data; they are not very reliable.

λ_{Head}	I_e	v', v"	λ_{Head}	λ_{origin}	I_a	I_e	v', v"
9970	1	0, 2	7684·3	7708	2	7	1, 1
8803	0	2, 3	7593·7	6969	10	10	0, 0
8697·8	1	1, 2	6954·0	7620	1	0	2, 1
8597·8	4	0, 1	6867·2	6895	8	2 ?	1, 0
7879·2	2	3, 3	6360·0	–	0	0	3, 1
7779·0	2	2, 2	6276·6	6287	3	0	2, 0
			5788·1	5796	1	0	3, 0
			5380	–	0	0	4, 0

INFRA-RED ATMOSPHERIC SYSTEM, $a\,^1\Delta_g - X\,^3\Sigma_g^-$

This is another strongly forbidden system which gives an atmospheric absorption band at 12683 Å (0, 0) and a weaker one at 10674 Å (1, 0). It has also been observed in emission in the twilight airglow and some afterglows.

O$_2$ (*contd.*)

HERZBERG SYSTEM I AND NIGHT-SKY BANDS, $A^3\Sigma_u^+ - X^3\Sigma_g^-$

Occurrence. In absorption by long (25 m.) paths of O$_2$. In emission from oxygen afterglow in a discharge tube. This is the main feature of the night sky emission spectrum. These bands appear to be strengthened at high pressure, perhaps due to 'induced absorption'; see Finkelnberg and Steiner, and Plate 8.

References. G. Herzberg, *Canad. J. Phys.*, **30**, 185 (1952)†.

H. P. Broida and A. G. Gaydon, *P.R.S.*, **222**, 181 (1954)†.

W. Finkelnberg and W. Steiner, *Z.P.*, **79**, 69 (1932)†.

C. A. Barth and J. Kaplan, *J. Mol. Spec.*, **3**, 583 (1959)†.

V. Degan *et al.*, *Identification Atlas of Molecular Spectra*, 6, York Univ. (1968)†.

The following are the origins of the absorption bands, from Herzberg, with revised v' and our estimates of intensity.

λ_0	I	v', v''	λ_0	I	v', v''	λ_0	I	v', v''
2794·0	1	1, 0	2593·5	7	5, 0	2463·2	10	9, 0
2737·1	2	2, 0	2554·3	9	6, 0	2443·0	6	10, 0
2684·9	4	3, 0	2519·3	10	7, 0	2429·2	2	11, 0
2637·0	6	4, 0	2488·9	10	8, 0			

The following are the heads of emission bands in the afterglow, from Broida and Gaydon.

λ	I	v', v''	λ	I	v', v''	λ	I	v', v''
2563	1	9, 1	2945	5	5, 3	3633	8	2, 6
2588	2	8, 1	3002	5	4, 3	3734	8	1, 6
2622	3	7, 1	3026	3	6, 4	3829	8	2, 7
2644	1	10, 2	3066	? 5	3, 3	3840	? 5	0, 6
2667	2	9, 2	3080	?	5, 4	3938	7	1, 7
2696	4	8, 2	3142	7	4, 4	4044	?	2, 8
2734	5	7, 2	3211	10	3, 4	4064	5	0, 7
2775	3	6, 2	3285	7	2, 4	4170	6	1, 8
2820	? 3	5, 2	3292	? 4	4, 5	4309	7	0, 8
2850	5	7, 3	3370	10	3, 5	4577	5	0, 9
2873	2	4, 2	3453	8	2, 5	4880	2	0, 10
2895	6	6, 3	3459	?	4, 6			
2931	1	3, 2	3542	8	3, 6			

HERZBERG SYSTEM II, $c^1\Sigma_u^- - X^3\Sigma_g^-$

Occurrence. Very weakly in absorption and en emission from an O$_2$ + Ar afterglow and in the night sky glow.

Appearance. Degraded to the red.

References. G. Herzberg, *Canad. J. Phys.*, **31**, 657 (1953)†.

V. Degan, *Canad. J. Phys.*, **46**, 783 (1968).

Absorption bands $\lambda\lambda 2714\cdot5$ (6, 0), 2672·3 (7, 0), 2634·0 (8, 0), 2599·2 (9, 0), 2568·0 (10, 0), 2540·0 (11, 0).

Emission $\lambda\lambda 4791\cdot5$ (0, 8), 4491·3 (0, 7).

HERZBERG'S SYSTEM III, $C^3\Delta_u - X^3\Sigma_g^-$

Reference. G. Herzberg, *Canad. J. Phys.*, **31**, 657 (1953)†.

O_2 *(contd.)*

Very weakly in absorption. Degraded to the red. Heads $\lambda\lambda 2619\cdot8$, $2588\cdot1$ and $2578\cdot4$ Å.

CHAMBERLAINS AIRGLOW SYSTEM, $C\,^3\Delta_u - a\,^1\Delta_g$

Reference. J. W. Chamberlain, *Astrophys. J.*, **128**, 713 (1958)†.

Very weak emission bands in the blue (night-sky) airglow are attributed to this forbidden system. They occur under the same conditions as the stronger Herzberg I system. Bands are triple-headed and shaded to the red.

The following are all the heads provisionally assigned to this system. The three sub-bands are due to transitions from $^3\Delta_1$, $^3\Delta_2$ and $^3\Delta_3$, and are indicated i, ii and iii:

λ	v', v''	λ	v', v''	λ	v', v''
4378	5, 5 i	4114	3, 3 i	3887	3, 2 i
4326	5, 5 iii	4107	5, 4 ii; 1, 2 ii	3866	3, 2 ii
4317	3, 4 iii	4090	3, 3 ii	3861	5, 3 iii
4244	4, 4 i	4086	5, 4 iii; 1, 2 iii	3844	3, 2 iii
4240	6, 5 ii	4071	3, 3 iii	3813	6, 3 i
4221	4, 4 ii	4031	6, 4 i	3792	6, 3 ii
4215	6, 5 iii	4009	6, 4 ii	3771	6, 3 iii
4135	1, 2 i	3985	6, 4 iii	3698	5, 2 i
4127	5, 4 i				

BROIDA AND GAYDON'S SYSTEM, $A\,^3\Sigma_u^+ - b\,^1\Sigma_g^+$

Reference. H. P. Broida and A. G. Gaydon, *P.R.S.*, **222**, 181 (1954).

Broida and Gaydon considered the possibility that some of the emission bands of the afterglow and night sky might be due to this other forbidden system, but, although probable, Barth and Kaplan failed to find definite evidence of these bands.

DOUBLE ELECTRONIC TRANSITIONS (O_4)

Weak diffuse bands involving a collision complex between two oxygen molecules occur in both emission and absorption. Metastable molecules, e.g. $O_2(a\,^1\Delta_g)$, are involved; the processes are basically:

$$O_2(a\,^1\Delta_g) + O_2(a\,^1\Delta_g) \rightleftharpoons O_4 \rightleftharpoons O_2(X\,^3\Sigma_g^-) + O_2(X\,^3\Sigma_g^-) + h\nu \quad \text{and}$$

$$O_2(a\,^1\Delta_g) + O_2(b\,^1\Sigma_g^+) \rightleftharpoons O_4 \rightleftharpoons O_2(X\,^3\Sigma_g^-) + O_2(X\,^3\Sigma_g^-) + h\nu$$

Occurrence. In emission from an oxygen afterglow, especially when produced by a microwave discharge with mercury or mercuric oxide present to remove free oxygen atoms. Also recorded in chemiluminescent emission from hydrogen peroxide reacting with sodium hypochlorite. In absorption the strength of these bands varies roughly as the square of the oxygen pressure.

Appearance. All bands are quite broad and diffuse. Wavelengths quoted by various authors differ appreciably.

References. L. W. Bader and E. A. Ogryzlo, *Disc. Faraday Soc.*, **37**, 46 (1964).
V. I. Dianov-Klokov, *Opt. Spectrosc.*, **16**, 224 (1964).
S. H. Whitlow and F. D. Findlay, *Canad. J. Chem.*, **45**, 2087 (1967).
R. P. Blickensderfer and G. E. Ewing, *J. Chem. Phys.*, **51**, 5284 (1969).
E. W. Gray and E. A. Ogryzlo, *Chem. Phys. Lett.*, **3**, 658 (1969).
F. D. Findlay, *Canad. J. Phys.*, **48**, 2107 (1970).

The following are the more certainly established bands, from the last two references, with our guesses at intensities I_e and I_a for emission and absorption:

O$_2$ (*contd.*)

$$({}^1\Delta_g)_{v'_1} + ({}^1\Delta_g)_{v'_2} \to ({}^3\Sigma_g^-)_{v''_1} + ({}^3\Sigma_g^-)_{v''_2}$$

λ	I_e	I_a	v'_1	v'_2	v''_1	v''_2
7030	4	–	0	0	0	1
6320	10	8	0	0	0	0
5788	3	10	1	0	0	0

$$({}^1\Sigma_g^+)_{v'_1} + ({}^1\Delta_g)_{v'_2} \to ({}^3\Sigma_g^-)_{v''_1} + ({}^3\Sigma_g^-)_{v''_2}$$

λ	I_e	I_a	v'_1	v'_2	v''_1	v''_2
5165	2	–	0	1	1	0
4784	10	5	0	0	0	0

$$({}^1\Sigma_g^+)_{v'_1} + ({}^1\Sigma_g^+)_{v'_2} \to ({}^3\Sigma_g^-)_{v''_1} + ({}^3\Sigma_g^+)_{v''_2}$$

λ	I_e	v'_1	v'_2	v''_1	v''_2
4000	1	0	0	0	1
3800	1	0	0	0	0

These double transitions involving collision complexes are also involved in some of the absorption by liquid oxygen (see below). They are also probably involved in some chemiluminescent processes in solution.

LIQUID ABSORPTION

Occurrence. Absorption by liquid oxygen.

Appearance. Degraded to longer wavelengths.

References. J. C. McLennan, H. D. Smith and J. O. Wilhelm, *Trans. Roy. Soc. Canada*, iii, **24**, 65 (1930)†.

V. I. Dianov-Klokov, *Opt. i. Spek.*, **6**, 457 (1959).

The following measurements represent the limits of the bands; most of the bands have fairly sharp heads and the longer wavelength edge is the head. Intensities are our estimates from the published photographs. The bands have been analysed into several systems.

λ *limits*	I	λ *limits*	I	λ
10420–10220	?	4802–4710	3	2863
9300–9100	?	4628–4605	0	2800
8300–8200	?	4481–4456	2	2763
7665–7590	3	4208–4180	1	2735
6916–6875	0	3930–3920	0	2686
6368–6160	10	3825–3785	3	2639
5826–5640	9	3628–3585	3	continuous
5364–5290	4	3455–3423	?	absorption
4982–4916	1	3305–3274	?	below 2609

The bands in the visible region have also been observed in absorption by solid oxygen, with slight changes of wavelength and intensity.

O$_2^+$

Three systems are now known. The First Negative System extends from the red to the green. The extensive Second Negative System has its strongest bands in the near ultra-violet. The Hopfield System, previously attributed to neutral oxygen, lies near the limit of the quartz ultra-violet.

O₂⁺ (*contd.*)

FIRST NEGATIVE SYSTEM, $b^4\Sigma_g^- - a^4\Pi_u$

Occurrence. In discharges through oxygen at low pressure, and in hollow-cathode and high frequency discharges. Also excited by electron or proton bombardment and occurring in auroral spectra.

Appearance. Degraded to the violet. Bands of complex structure. With small dispersion the appearance is of five strong bands somewhat similar in spacing and region to the Ångstrom bands of CO. See Plate 7.

References. R. Frerichs, *Z.P.*, **35**, 683 (1926).
L. Bozóky and R. Schmid, *P.R.*, **48**, 465 (1935).
T. E. Nevin, *Phil. Trans. Roy. Soc.*, A, **237**, 471 (1938)†.
N. L. Singh and L. Lal, *Sci. Cult.*, **9**, 89 (1943).
S. Weniger, *J. Phys. Radium*, **23**, 225 (1962)†.

Wavelengths of first heads from the above references with intensities mainly from Singh and Lal:

λ	I	v', v''	λ	I	v', v''	λ	I	v', v''
8527·1		0, 5	5973·4	10	1, 1	5295·7	9	2, 0
8347·3		1, 6	5925·6	9	2, 2	5274·7	10	3, 1
7899·7	(1)	0, 4	5883·4	8	3, 3	5259	6	4, 2
7757·8		1, 5	5847·3	2	4, 4	5251	10	5, 3
7334·7	(10)	0, 3	5814	1	5, 5	5241	8	6, 4
7235·0		1, 4	5631·9	10	1, 0	5234	9	7, 5
6856·3	(9)	0, 2	5597·5	10	2, 1	5005·6	2	3, 0
6418·7	10	0, 1	5566·6	6	3, 2	4998	2	4, 1
6351·0	10	1, 2	5540·7	2	4, 3	4992	2	5, 2
6026·4	10	0, 0	5521	2	5, 4			

SECOND NEGATIVE SYSTEM, $A^2\Pi_u - X^2\Pi_g$

Occurrence. In discharge tubes containing oxygen at low pressure, especially in the negative glow, or in the presence of excess of helium. Also in high-frequency discharge, and electron-beam excitation.

Appearance. Degraded to the red. An extensive system of double-headed bands, separation about 200 cm.⁻¹. See Plate 7.

References. R. C. Johnson, *P.R.S.*, **105**, 683 (1923–24)†.
V. M. Ellsworth and J. J. Hopfield, *P.R.*, **29**, 79 (1927)†.
R. S. Mulliken and D. S. Stevens, *P.R.*, **44**, 720 (1933).
M. W. Feast, *Proc. Phys. Soc.*, A, **63**, 557 (1950).
J. Byrne, *Proc. Phys. Soc.*, **78**, 1074 (1961).
G. L. Bhale and P. R. Rao, *Proc. Indian Acad. Sci.*, A, **67**, 350 (1968)†.

In the following table the bands with intensities are from Johnson's measurements; bands without recorded intensities are from the other references quoted above; they are relatively weak bands. The R_2 and R_1 heads are denoted by ii and i following the vibrational quantum numbers. Values of v'' have been raised by one unit (Bhale and Rao).

O_2^+ (contd.)

λ	I	v', v''	λ	I	v', v''	λ	I	v', v''
6102·9		1, 15 ii	3210·8	8	2, 6 i	2458·6	1	8, 3 ii
5678·3		1, 14 i	3141·0	5	1, 5 ii	2446·9	1	8, 3 i
5498·4		0, 13 ii	3123·1	5	1, 5 i	2433·5	3	6, 2 ii
5443·0		0, 13 i	3062·8	8	2, 5 ii	2421·8	2	6, 2 i
5086·3		0, 12 ii	3043·6	8	2, 5 i	2392·6	3	7, 2 ii
5035·1		0, 12 i	2987·5	8	3, 5 ii	2381·0	3	7, 2 i
4877·6		1, 12 ii	2970·0	7	3, 5 i	2354·3	1	8, 2 ii
4820·3		1, 12 i	2919·8	8	4, 5 ii	2343·3	1	8, 2 i
4720·7		0, 11 ii	2907·1	5	2, 4 ii	2328·7		6, 1 ii
4678·5		0, 11 i	2901·9	7	4, 5 i	2317·9		6, 1 i
4399·4	(8)	0, 10 ii	2890·3	7	2, 4 i	2307·2		9, 2 i
4363·1	(8)	0, 10 i	2839·7	9	3, 4 ii	2291·8		7, 1 ii
4115·8	8	0, 9 ii	2823·7	8	3, 4 i	2285·8		10, 2 ii
4082·4	8	0, 9 i	2776·7	7	4, 4 ii	2281·3		7, 1 i
3859·5	8	0, 8 ii	2761·9	7	4, 4 i	2275·3		10, 2 i
3830·5	8	0, 8 i	2720·0	7	5, 4 ii	2252·8		11, 2 ii
3733·9	8	1, 8 ii	2705·3	7	5, 4 i	2246·9		8, 1 i
3706·6	8	1, 8 i	2688·5	2	3, 3 i	2243·5		11, 2 i
3629·8	8	0, 7 ii	2666·5	4	6, 4 ii	2224·3		9, 1 ii
3620·1	2	2, 8 ii	2652·3	4	6, 4 i	2213·8		9, 1 i
3603·7	7	0, 7 i	2646·7	6	4, 3 ii	2183·9		10, 1 i
3594·5	2	2, 8 i	2632·7	6	4, 3 i	2164·0		11, 1 ii
3517·7	8	1, 7 ii	2594·3	8	5, 3 ii	2155·3		11, 1 i
3494·2	7	1, 7 i	2581·0	8	5, 3 i	2138·6		12, 1 ii
3421·2	8	0, 6 ii	2545·5	7	6, 3 ii	2128·4		12, 1 i
3416·2	2	2, 7 ii	2532·8	6	6, 3 i	2112·2		13, 1 ii
3397·8	8	0, 6 i	2512·9	3	4, 2 i	2103·7		13, 1 i
3393·1	4	2, 7 i	2500·6	6	7, 3 ii	2090·3		14, 1 ii
3322·6	6	1, 6 ii	2488·3	6	7, 3 i	2080·8		14, 1 i
3300·3	6	1, 6 i	2478·0	4	5, 2 ii	2068·3		15, 1 ii
3231·2	8	2, 6 ii	2465·8	2	5, 2 i	2059·7		15, 1 i

Many additional weak bands have been found by Feast, by Byrne and by Bhale and Rao.

HOPFIELD SYSTEM, $c\,^4\Sigma_u^- - b\,^4\Sigma_g^-$

Occurrence. Condensed discharge through oxygen and helium. Bands have not been observed with argon.

Appearance. A single progression of single-headed bands, shaded to shorter wavelengths.

References. Y. Tanaka, A. S. Jursa and F. J. Le Blanc, *J. Chem. Phys.*, **24**, 915 (1956).
F. J. Le Blanc, *J. Chem. Phys.*, **38**, 487 (1963).

The middle bands of the progression are strongest.

λ_{air}	v', v''	λ_{air}	v', v''	$\lambda_{vac.}$	v', v''
2362·4	0, 9	2169·9	0, 5	1985·0	0, 1
2314·1	0, 8	2122·7	0, 4	1940·3	0, 0
2265·7	0, 7	2075·9	0, 3		
2217·6	0, 6	2029·8	0, 2		

O_3

ULTRA-VIOLET SYSTEM (HUGGINS BANDS), $^1B_2 - X\,^1A_1$

Occurrence. These bands occur strongly in absorption by ozone; this absorption is responsible for the ultra-violet limit of solar radiation reaching the earth.

Appearance. The usual appearance of ozone absorption is a fairly sharp cut-off in the ultra-violet, between 3000 Å and 3600 Å, according to the thickness of ozone, with a few narrow bands showing up clearly in the neighbourhood of this cut-off. The bands are headless, but in some cases appear slightly degraded to the red. The bands are clearest in the region 3200–3400 Å; they extend to below 2500 Å, but are only faintly visible above the continuous absorption which is strongest at 2550 Å.

References. A. Fowler and R. J. Strutt, *P.R.S.*, **93**, 577 (1917)†.
O. R. Wulf and E. H. Melvin, *P.R.*, **38**, 330 (1931)†.
A. Jakowlewa and V. Kondratiev, *Phys. Zeit. Sowjetunion*, **1**, 471 (1932).
J. W. Simons *et al.*, *J. Chem. Phys.*, **59**, 1203 (1973).

The following measurements are by Fowler and Strutt; these agree well with those of Jakowlewa and Kondratiev, whose measures are however lower by about 1 Å. The bands have been partly analysed by Wulf and Melvin, and by Jakowlewa and Kondratiev.

λ	I	λ	I	λ	I	λ	I
3432·2	1	3311·5	5d	3227·2	10	3171·6	4
3421·4	1	3304·1	3	3221·5	10	3162·6	2d
3402·6	1	3284·0	2	3206·8	2	3156·1	8
3377·7	1	3279·8	8d	3201·0	8d	3137·4	10d
3374·1	3	3272·0	3	3194·8	6	3114·3	8d
3365·2	1	3255·5	5	3188·8	1	3105·0	5
3346·0	1	3249·7	8	3181·5	1	3096·5	4
3338·5	4	3243·0	1d	3177·0	8d	3089·5	8d
3331·2	1	3232·8	1				

d = diffuse.

Emission. A number of emission bands have been reported by J. Janin, *C. R. Acad. Sci.* (*Paris*), **207**, 145 (1938) in the violet region.

Feast (see O_2^+) has identified bands attributed by Johnson to O_3 with bands of O_2^+ and lines of O^+.

VISIBLE ABSORPTION BANDS (CHAPPUIS BANDS)

Wulf has observed absorption bands of ozone in the orange and in the infra-red, using long columns of the gas. There are two strong diffuse bands at about 5730 and 6020 Å. The cut-off in the infra-red varies from around 9000 Å for short paths of ozone, up to the limit of the visible, and finally the absorption merges with the orange bands for very long paths.

Reference. O. R. Wulf, *Proc. Nat. Acad. Sci., U.S.A.*, **16**, 507 (1930)†.

O_4

See O_2, Double Electronic Transitions.

OH

The 3064 Å system of OH occurs very readily and is one of the commonest and best known band systems. A weak system has been found in the visible, and another in the ultra-violet, 2530–2250 Å. The rotation-vibration bands in the red and infra-red are known as the Meinel bands.

3064 Å SYSTEM, $A^2\Sigma^+ - X^2\Pi$

Occurrence. In emission from almost all sources where water vapour is present, such as flames, arcs and discharge tubes. Also in absorption in flames, flash photolysis, etc.

Appearance. Degraded to the red. Four main heads R_1, R_2, Q_1 and Q_2; the Q_2 is rather overlapped. There is also a weak $^SR_{21}$ head. See Plate 4.

References. W. W. Watson, *Astrophys. J.*, **60**, 145 (1924)†.

D. Jack, *P.R.S.*, **115**, 373 (1927)†; **118**, 647 (1928); **120**, 222 (1928).

T. Tanaka and Z. Koana, *Proc. Phys. Math. Soc. Japan*, **15**, 272 (1933).

C. E. Moore and H. P. Broida, *Nat. Bur. Stand. J. Res.*, A, **63**, 279 (1959).

G. H. Dieke and H. M. Crosswhite, *J. Quant. Spectrosc. Rad. Transfer.*, **2**, 97 (1962).

A. Stoebner and R. Delbourgo, *J. Chim. Phys.*, **64**, 1115 (1967).

v', v''	$^SR_{21}$	R_1	R_2	Q_1	Q_2	I
3, 0	—	2444	—	—	—	0
2, 0	—	2608·5	2613·4	2613·4	2622·1	3
3, 1	—	2677·3	2683·1	2681·8	2691·1	2
1, 0	—	2811·3	2816·0	2819·1	2829·0	6
2, 1	—	2875·3	2880·6	2882·3	2892·7	3
3, 2	—	2945·2	2951·2	2951·2	2962·4	1
0, 0	3021·2	3063·6	3067·2	3078	3089	10
1, 1	—	3121·7	3126·4	—	—	1
2, 2	—	3184·8	3190·2	3195·9	3208·7	1
0, 1	—	3428·1	3432·1	3458·5	3472·1	2
0, 2	—	—	3847	—	3934·7	0
1, 3	—	—	3902	—	3975·5	1
2, 4	—	—	—	—	4021·8	0

Under some conditions (e.g. low-pressure hydrogen flame) bands with $v' = 2$ are enhanced by an inverse predissociation process.

VISIBLE SYSTEM (SCHULER AND WOELDIKE) 5800–3700 Å, $B^2\Sigma^+ - A^2\Sigma^+$

Occurrence. This is a very weak system occurring in discharges through water vapour. It probably occurs in sun-spots.

Appearance. Open P and R branches; strongly degraded to the red.

References. H. Schüler and L. Reinebeck, *Z. Naturforsch.*, A, **4**, 560 (1949).

R. F. Barrow and A. R. Downie, *Proc. Phys. Soc.*, A, **69**, 178 (1956).

A. Michel, *Z. Naturforsch.*, **129**, 887 (1957).

S. Benoist, *Ann. Phys. (Paris)*, **10**, 363 (1955)†.

P. Felenbok, *Ann. Astrophys.*, **26**, 393 (1963)†.

No intensities are available but the (0, 7) and (0, 8) are relatively strong. R heads:

λ	v', v''	λ	v', v''	λ	v', v''
5534·1	1, 9	4957	1, 7	4336·6	0, 5
5480·3	0, 8	4729·7	0, 6	4216·4	1, 5
5123·5	0, 7	4587·3	1, 6	3864	0, 4

OH (*contd.*)

ULTRA-VIOLET SYSTEM (SCHÜLER-MICHEL-BENOIST) 2550–2250 Å, $C^2\Sigma^+ - A^2\Sigma^+$

Occurrence. A very weak system observed in high frequency and other discharges through water vapour. Found in sun spots.

Appearance. Probably open P and R branches only; degraded to red.

References. S. Benoist, *Ann. Phys. (Paris)*, **10**, 363 (1955).

A. Michel, *Z. Naturforsch.*, A, **12**, 887 (1957).

C. Carlone and F. W. Dalby, *Canad. J. Phys.*, **47**, 1945 (1969)†.

R heads from Michel and from Carlone and Dalby, with the latter authors values of v':

λ	v', v''	λ	v', v''
2682·9*	0, 9	2464·7	1, 7
2599·6*	1, 9	2333·9	3, 7
2544·7*	1, 8	2248·6	3, 6

* These may be relatively strong.

MEINEL (VIBRATION-ROTATION) BANDS $(X^2\Pi)$

Occurrence. In spectrum of the night sky, and in reactions of ozone with atomic hydrogen or with acetaldehyde. Under these conditions bands with $v' = 8$ and 9 are strong. Also in oxyhydrogen flames, when bands with lower v' are excited.

Appearance. Degraded to longer wavelengths. In cool sources, such as the night sky glow, the Q branch is strongest, with a few R lines closing up towards a weak head 20 to 50 Å on the short-wave side, and resolved P lines on the long-wave side. In hotter laboratory sources the R heads are better developed.

References. A. B. Meinel, *Astrophys. J.*, **111**, 555 (1950)†.

J. D. McKinley, D. Garvin and M. J. Boudart, *J. Chem. Phys.*, **23**, 784 (1955)†.

G. Dejardin, J. Janin and M. Peyron, *Cahiers de Phys.*, **46**, 1 (1953).

A. M. Bass and D. Garvin, *J. Mol. Spec.*, **9**, 114 (1962)†.

B. J. Finlayson, S. S. Gaffney and J. N. Pitts, *Chem. Phys. Lett.*, **17**, 22 (1972).

R heads from Bass and Garvin in $H + O_3$ reaction; Q heads from older references. Intensities are our estimates from Broida and Garvin's plate:

λ_R	λ_Q	I	v', v''	λ_R	λ_Q	I	v', v''
9946	10000	3	9, 5	6665·0		3	10, 4
9652·0		2	3, 0	6463·8		5	6, 1
9310·1	9400	8	8, 4	6235·4	6258	10	9, 3
8761·3	8829·4	7	7, 3	6137·0		3	5, 0
8281·0	8344	5	6, 2	5666·3		7	8, 2
7850·0	7918·5	6	5, 1	5541·1		5	7, 1
7714·7	7755·8	10	9, 4	5530·6		3	10, 3
7462·2	7529·1	3	4, 0	5188·5		5	9, 2
7238·0	7284·2	6	8, 3	4629·4		3	7, 0
6828·9	6879	8	7, 2				

OH⁺

3565 Å SYSTEM, $A^3\Pi - X^3\Sigma^-$

Occurrence. Electrodeless and high-frequency discharges through water vapour at low pressure.

OH⁺ *(contd.)*

   Appearance. Degraded to the red. Rather complex bands with nine main branches.
   References.   F. W. Loomis and W. H. Brandt, *P.R.*, **49**, 55 (1936)†.
W. H. Rodebush and M. H. Wahl, *J. Chem. Phys.*, **1**, 696 (1933)†.
S. Weniger and R. Herman, *J. Phys. Radium*, **19**, 582 (1958).
D. Rakotoaryijimy, *Physica*, **49**, 360 (1970)†.
   The following are the R heads. No intensities are available but the (0, 0) is clearly the strongest:

λ	v', v''	λ	v', v''
3983	0, 1	3695	1, 1
3958	3, 3	3565	0, 0
3830	2, 2	3332	1, 0

   The (0, 0) band shows the following heads:
$3573 \cdot 5$ R_{11}, $3568 \cdot 5$ R_{22}, $3568 \cdot 4$ $^{R}Q_{21}$, $3566 \cdot 8$ $^{S}R_{21}$, $3565 \cdot 1$ $^{R}P_{31}$, $3564 \cdot 8$ R_{33}, $3564 \cdot 7$ $^{R}Q_{32}$, $3541 \cdot 3$ $^{S}R_{22}$, $3541 \cdot 4$ $^{S}Q_{31}$, $3505 \cdot 3$ $^{T}R_{31}$.

OsO

   Reference.   V. Raziunas, G. Macur and S. Katz, *J. Chem. Phys.*, **43**, 1010 (1965).
   The following band heads have been observed in a DC arc; the direction of degradation is not stated but is probably to the red: $\lambda\lambda 7435 \cdot 1$, $7134 \cdot 9$, $6936 \cdot 5$, $6727 \cdot 5$, $6118 \cdot 0$.

P₂

   A strong extensive system covers the ultra-violet below 3500 Å. Four weaker systems are known, in the near infra-red, the visible (orange), the violet and ultra-violet. Many strong systems have also been found in absorption and emission below 2000 Å, but these are not discussed here.

MAIN SYSTEM, 3500–1800 Å, $C^1\Sigma_u^+ - X^1\Sigma_g^+$

   Occurrence. In emission, especially from discharge tubes with hydrogen or helium as a carrier gas containing a small amount of phosphorus. Also in fluorescence and absorption.
   Appearance. Degraded to the red. This is a very extensive system of single-headed bands. See Plate 6.
   References.   A. Jakowlewa, *Z.P.*, **69**, 548 (1931).
G. Herzberg, *Ann. Phys. (Lpz.)*, **15**, 677 (1932).
M. F. Ashley, *P.R.*, **44**, 919 (1933)†.
G. Herzberg, L. Herzberg and G. Milne, *Canad. J. Res.*, A, **18**, 139 (1940).
E. J. Marias, *P.R.*, **70**, 499 (1946)†.
   Emission.   The following measurements are from Herzberg:

λ	I	v', v''	λ	I	v', v''	λ	I	v', v''
3222·9	4	11, 30	3082·0	5	10, 27	2845·7	5	3, 18
3201·9	4	10, 29	3069·5	5	6, 24	2830	—*	5, 19
3184·1	4	9, 28	3064·2	4	9, 26	2774·5	4*	5, 18
3166·2	5	8, 27	3046·4	4	8, 25	2757·1	6*	4, 17
3141·1	4	10, 28	3028·7	4	7, 24	2707·4	4*	7, 18
3123·3	5	9, 27	2953·6	7	6, 22	2690·5	5	6, 17
3105·4	5	8, 26	2898·0	4*	6, 21	2689·3	4*	3, 15

P$_2$ *(contd.)*

λ	I	v', v''	λ	I	v', v''	λ	I	v', v''
2645·2	5	9, 18	2565·6	4	7, 15	2468·6	4	1, 9
2642·1	4	6, 16	2562·1	5	4, 13	2461·0	6	6, 12
2628·1	5	8, 17	2520·8	5	7, 14	2456·9	6	3, 10
2625·5	5*	5, 15	2516·6	6	4, 12	2450·6	5	8, 13
2586·6	5	11, 18	2513·6	5	1, 10	2413·7	4	3, 9
2582·0	5	8, 16	2509·6	4	9, 15	2403·3	5	5, 10
2578·0	4	5, 14	2472·5	4	4, 11	2393·3	4	7, 11

* These bands appear prominently on the reproduction by Ashley.

Only bands listed by Herzberg as intensity 4 or greater are given above; this only represents a small fraction of the whole system.

Absorption. Strongest bands listed by Jakowlewa:

λ	I	v', v''	λ	I	v', v''	λ	I	v', v''
2261·6	4	1, 4	2128·6	4	2, 1	2074·9	5	3, 0
2186·4	4	1, 2	2122·6	5	4, 5	2069·0	5	5, 1
2164·3	5	2, 3	2108·1	6	3, 1	2055·0	5	4, 0
2150·0	5	1, 1	2094·3	4	2, 0	2050·0	4	6, 1
2143·0	5	3, 4	2088·3	4	4, 1	2036·0	5	5, 0

$A^1\Pi_g - X^1\Sigma_u^+$ SYSTEM (DOUGLAS AND RAO), 3140–2758 Å

Occurrence. This system, overlapping the main system, was observed by Douglas and Rao in a discharge through helium containing a trace of P vapour. It has been found, free from overlapping by other P$_2$ bands, in the glow of phosphorous reacting with atomic oxygen.

Appearance. Degraded to the red. Clear sequences, with the (0, 1) band strongest.

References. A. E. Douglas and K. S. Rao, *Canad. J. Phys.*, **36**, 565 (1958)†.

R. D. Verma and H. P. Broida, *Canad. J. Phys.*, **48**, 2991 (1970).

R heads of all known bands:

λ	v', v''	λ	v', v''	λ	v', v''	λ	v', v''
3140·2	2, 5	3053·7	1, 3	2969·8	0, 1	2853·2	1, 0
3125·8	1, 4	3039·3	0, 2	2903·0	0, 0	2818·4	3, 1
3112·4	0, 3	2983·5	1, 2	2866·0	2, 1	2803·7	2, 0
						2758·2	3, 0

NEAR INFRA-RED SYSTEM, 10 000–7500 Å, $c(?\,^3\Pi) - b(?\,^3\Pi)$

Occurrence. High-frequency discharge in He and P vapour. Also in PH$_3$/atomic hydrogen flame.

Appearance. Waves of triple-headed bands, slightly degraded to the violet, rather similar in appearance to the N$_2$ First Positive system. The waves, or sequences, have maximum intensity around 9200, 8782, 8226, 7820 and 7427 Å. Best identified from photographs.

References. H. Guenebaut, B. Pascat and J. Brion, *C. R. Acad. Sci. (Paris)*, **259**, 3545 (1964)†.

J. Brion, J. Malicet and H. Guenebaut, *C. R. Acad. Sci. (Paris)*, B, **264**, 622 (1967).

S. Mrozowski and C. Santaram, *J. Opt. Soc. Amer.*, **57**, 522 (1967)†.

This system appears to have been later attributed to P$_2^+$ by Brion, Malicet and Guenebaut (see later under P$_2^+$), but although there is considerable but not complete agreement in head measurements in the papers by Brion *et al.* with those by Mrozowski and Santaram, analyses are quite

P$_2$(*contd.*)

different. The occurrence in the PH$_3$ + H reaction seems to indicate that the neutral molecule is the more likely emitter.

VISIBLE SYSTEM, 6900–4900 Å

A number of weak violet-degraded bands have been observed in a high-frequency discharge through He + P. Bands have been analysed around 6025, 5985, 5940, 5808, 5770 and 5700 Å. They are probably due to a $^3\Pi - {}^3\Pi$ transition.

References. J. Brion *et al.*, *C. R. Acad. Sci.* (*Paris*), B, **272**, 127 and 999 (1971).

VIOLET SYSTEM, 4380–3520 Å, $b'\,{}^3\Sigma_u^- - X\,{}^1\Sigma_g^+$

Occurrence. High-frequency discharge through He + P.

Appearance. Degraded to the red. The $v' = 0$ progression is prominent.

References. S. Mrozowski and C. Santaram, *J. Opt. Soc. Amer.*, **57**, 522 (1967)†.

J. Brion, J. Malicet and H. Guenebaut, *Canad. J. Phys.*, **52**, 2143 (1974)†.

Strong heads:

λ	I	v', v''	λ	I	v', v''	λ	I	v', v''
4215·7	3	1, 7	3943·1	5	0, 4	3638·9	5	1, 2
4190·7	3	0, 6	3828·8	6	0, 3	3616·5	4	0, 1
4063·7	3	0, 5	3720·0	6	0, 2	3539·5	3	1, 1

P$_2^+$

Three, possibly four, systems of P$_2^+$ occur in emission, together with bands of P$_2$, in a hollow-cathode discharge through helium containing a little phosphorus vapour; the He pressure was 2–3 torr and the P temperature 40–44°C.

References. N. A. Narasimham, *Canad. J. Phys.*, **35**, 1242 (1957)†.

J. Brion, J. Malicet and H. Guenebaut, *C. R. Acad. Sci.* (*Paris*), B, **266**, 82 (1968); C, **276**, 471 (1973); C, **276**, 551 (1973).

3850–3400 Å SYSTEM, $C\,{}^2\Pi - X\,{}^2\Pi$

Appearance. Degraded to the red. Overlapping due to the $^2\Pi$ spin splitting and one of the vibrational frequencies being nearly equal complicates both the appearance and the analysis.

The following band origins are very close to the heads. Data are from Brion *et al.*, whose values of v' are one unit greater than Narasimham's. The 3505, 3453 and 3403 bands are probably strongest. The $^2\Pi_{\frac{3}{2}} - {}^2\Pi_{\frac{3}{2}}$ and $^2\Pi_{\frac{1}{2}} - {}^2\Pi_{\frac{1}{2}}$ sub-bands are denoted ii and i:

λ_0	v', v''	λ_0	v', v''	λ_0	v', v''
3797·1	0, 2 i	3674·8	1, 2 ii	3505·7	2, 0 i
3796·4	3, 4 i	3645·6	1, 1 i	3505·1	1, 0 ii
3764·8	1, 3 ii	3588·9	2, 1 i	3453·8	3, 0 i
3734·9	1, 2 i	3587·5	1, 1 ii	3453·1	2, 0 ii
3734·4	0, 2 ii	3559·7	1, 0 i	3403·3	3, 0 ii
3705·7	3, 3 i	3533·9	2, 1 ii		

4400–3900 SYSTEM, $B\,{}^2\Sigma^+ - A\,{}^2\Sigma^+$

Appearance. Double P and R branches, shaded to the red. Eight bands forming two progressions. The following band origins are from Narasimham:

P_2^+ (*contd.*)

λ_0	v', v''	λ_0	v', v''	λ_0	v', v''
4304·6	0, 0	4094·3	3, 0	3969·7	7, 1
4230·9	1, 0	4030·9	4, 0	3913·8	8, 1
4160·9	2, 0	3970·4	5, 0	3912·6	6, 0

6250–6520 Å SYSTEM, $D\,^2\Pi_{\frac{3}{2}} - X\,^2\Pi_{\frac{3}{2}}$

Four red-degraded bands found by Brion *et al.*, $\lambda\lambda 6247\cdot6$ (0, 3), 6003·6 (0, 2), 5776·1 (0, 1) and 5627·4 (1, 1).

BAND AT 8905 Å

Brion *et al.* have identified a $^2\Pi - {}^2\Pi$ band with P_1 and P_2 heads at 8905 and 8900 Å and have assigned this as the (2, 3) band of the Near Infra-red System listed under P_2. It is possible the whole system is due to P_2^+, but it seems more likely that it is P_2, although there may be an overlapping weak system of P_2^+ of which this is one band.

PCl

Occurrence. Absorption following flash photolysis of PCl_3.

Appearance. Rather diffuse bands, shaded to shorter wavelengths. The (0, 0) band at 2418 is strongest.

Reference. N. Basco and K. K. Yee, *Chem. Comm.*, 1146 (1967).

λ	v', v''	λ	v', v''
2486·9	0, 2	2418·6	0, 0
2475·8	1, 3	2370·9	1, 0
2452·5	0, 1	2338·8	2, 0

PF

One triplet, four singlets and an intercombination system have been observed, and more recently an additional system in the visible (Skolnik and Goodfriend).

Occurrence. In high frequency discharge through PF_3 + He and especially in afterglow of this discharge.

References. A. E. Douglas and M. Frackowiak, *Canad. J. Phys.*, **40**, 832 (1962)†.
R. Colin, J. Devillers and F. Prevot, *J. Mol. Spec.*, **44**, 230 (1972)†.
E. G. Skolnik and P. L. Goodfriend, *J. Mol. Spec.*, **50**, 202 (1974).

$A^3\Pi - X^3\Sigma^-$, 4000–3000 Å

Multiple-headed bands, degraded to the red. R_{11} heads 3683·0 (0, 3), 3506·0 (0, 1), 3356·1 (1, 0).

$b^1\Sigma^+ - X^3\Sigma^-$, 7483 Å

There is a single weak sequence, found only in the afterglow. Degraded to the violet. Q heads 7483·1 (0, 0), relatively strong, 7472·2 (1, 1), 7461·8 (2, 2), 7451·5 (3, 3), 7441·3 (4, 4).

$d^1\Pi - b^1\Sigma^+$, 4453 Å

Close double-headed bands, shaded to the red. R heads:

PF (*contd.*)

λ	v', v''	λ	v', v''
4819·5	0, 2	4373·0*	1, 0
4630·3*	0, 1	4294·8	2, 0
4543·3	1, 1	4218·9	3, 0
4453·5*	0, 0		

* A rotational analysis has been made for these.

$d\,^1\Pi - a\,^1\Delta$

Close double-headed bands, shaded to the red. R heads:

λ	v', v''	λ	v', v''
4138·0	1, 6	3587·6	0, 1
3940·9	0, 4	3481·5*	0, 0
3877·7*	1, 4		

* A rotational analysis has been made.

$g\,^1\Pi - b\,^1\Sigma^+$, 2612 Å

Degraded to shorter wavelengths. Heads 2672·0 (0, 1), 2612·2 (0, 0).

$g\,^1\Pi - a\,^1\Delta$, 2245 Å

Degraded to shorter wavelengths; heads 2333·1 (0, 2), 2288·2 (0, 1), 2245·3 (0, 0).

VISIBLE BANDS

Skolnik and Goodfriend have observed triple-headed bands, shaded to the red, in the PF_3 + Ar afterglow. First heads 6398·5 (0, 2), 5964·3 (0, 1), 5586·2 (0, 0). The (0, 0) is strongest and its other heads are at 5590·2 and 5595·0.

Some unassigned bands also occur in this source.

PF⁺

Occurrence. High-frequency discharge through PF_3 + He.

Appearance. Double-headed bands degraded to the red. They are due to a $^2\Sigma - {}^2\Pi$ transition and the (0, 0) has its R_{21} head at 2824·8 Å. There are no details of other bands.

Reference. A. E. Douglas and M. Frackowiak, *Canad. J. Phys.*, **40**, 832 (1962)†.

PH

References. R. W. B. Pearse, *P.R.S.*, **129**, 328 (1930)†.
M. Ishaque and R. W. B. Pearse, *P.R.S.*, **173**, 265 (1939)†.
F. Legay, *Canad. J. Phys.*, **38**, 797 (1960)†.

3400 Å SYSTEM, $A^3\Pi - X^3\Sigma$

Complex band obtained in emission from a discharge tube containing phosphorus vapour and hydrogen. See Plate 5. Legay observed the (1, 0) band, with strongest (R_3) head at 3196·2 Å.

PH (*contd.*)

Heads $\lambda(I)$

v', v''					
0, 0	3390·1 (5)	3394·9 (7)	3409·2 (4)	3419·6 (10)	3426·8 (5)

PH⁺

Three bands at 4228, 3854 and 3567 Å have been observed, the strongest at 3854 Å.

Reference. N. A. Narasimham, *Canad. J. Phys.*, **35**, 901 (1957)†.

Occurrence. In emission from a hollow-cathode discharge through helium containing a little phosphorus vapour and hydrogen.

Appearance. Complex bands degraded to red, with 24 branches.

Transition. $A^2\Delta - X^2\Pi$.

Wavelengths corresponding to band origins and heads given by Narasimham are as follows:

v', v''	λ_0	$^sR_{21}$	R_1	Q_1	R_2	Q_2
0, 1	4259·4	4208·9	4228·0	4235·6	4264·9	4281·5
0, 0	3879·3	3840·2	3853·7	3860·0	3885·6	3896·7
1, 0	3679·6	3648·7	3657·4	3662·8	3688·2	3697·2

PH₂

Occurrence. In absorption following flash photolysis of PH_3 and also in emission in a discharge.

Appearance. Complex bands of a $^2A_1 - {}^2B_1$ transition, with fairly prominent violet-shaded Q heads. In absorption there is a progression of 11 bands 5500–3600 Å.

References. D. A. Ramsay, *Nature,* **178**, 374 (1956)†.

R. N. Dixon, J. Duxbury and D. A. Ramsay, *P.R.S.*, **296**, 137 (1967)†.

H. Guenebaut and B. Pascat, *C. R. Acad. Sci. (Paris)*, **259**, 2412 (1964).

B. Pascat *et al.*, *J. Chim. Phys.*, **68**, 2022 (1968)†.

The origin of the system is at 5471·4 Å. In absorption the Q heads are at 5471·4, 5200·6, 4958·8, 4740·4*, 4543·2*, 4364·3*, 4208·4, etc. (* indicates the strongest).

In emission bands occur at 8448·1, 7763·1, 7173·1, 6660 and 6215 Å.

PH₃

References. C. M. Humphries, A. D. Walsh and P. A. Warsop, *Disc. Faraday Soc.*, **35**, 148 (1963)†.

L. Mayor, A. D. Walsh and P. A. Warsop, *J. Mol. Spec.*, **10**, 320 (1963).

In absorption there is a continuum centred at 1800 Å and a band system around 1400 Å. The continuum extends to 2500 Å with long paths (200 cm.atm.). Bands 2400–2200 reported by earlier workers are apparently due to impurities.

PHO

Reference. M. Lam Thanh and M. Peyron, *J. Chim. Phys.*, **60**, 1289 (1963)†; **61**, 1531 (1964)†.

Emitted by flames containing phosphorus compounds; a sensitive test for P. In hot flames the appearance is of a strong pseudocontinuum at 5280 Å, with weaker continua at 5100 and 5600 Å.

PHO (*contd.*)

In the cooler flame of phosphorus reacting with moist atomic hydrogen bands, given below, are obtained.

Band maxima:

λ	I	$(v_1\ v_2\ v_3)'$	$(v_1\ v_2\ v_3)''$	λ	I	$(v_1\ v_2\ v_3)'$	$(v_1\ v_2\ v_3)''$
6436	3	0 0 0	0 2 0	5597	8	0 0 0	0 0 0
6282	3	0 0 0	0 0 2	5535	6	0 1 0	0 0 0
6101	4	0 0 1	0 2 0	5425	5	0 0 2	0 1 0
5991	6	0 0 0	0 1 0	5248·7	10	0 1 0	0 0 0
5923	5	0 0 0	0 0 1	5097	5	0 1 2	0 0 0
5849	4	0 1 0	0 0 1	5020	4	0 1 1	0 0 0
5698	4	0 0 1	0 1 0	4879		0 1 3	0 1 0

PN

Occurrence. In heavy-current discharge tubes containing phosphorus and nitrogen.

Appearance. Degraded to the red. Close double-headed bands (separation about 0·6 Å). Well-marked sequences.

Transition. $A^1\Pi - X^1\Sigma$.

Reference. J. Curry, L. Herzberg and G. Herzberg, *Z.P.*, **86**, 348 (1933)†.

The following are the strong bands. The intensities have been reduced to a scale of 10. The R heads only are given here.

λ	I	$v',\ v''$	λ	I	$v',\ v''$	λ	I	$v',\ v''$
2742·9	3	3, 5	2620·1	8	1, 2	2466·2	7	2, 1
2727·5	4	2, 4	2605·0	9	0, 1	2451·1	8	1, 0
2712·1	4	1, 3	2533·0	3	1, 1	2418·7	3	4, 2
2696·9	3	0, 2	2518·2	10	0, 0	2403·3	3	3, 1
2635·2	4	2, 3	2481·4	3	3, 2	2388·2	2	2, 0

PO

Two strong systems $A^2\Sigma^+ - X^2\Pi$ and $B^2\Sigma^+ - X^2\Pi$ occur in the ultra-violet. These are known as the β and γ systems, although the intended analogy with NO is not correct. Several weaker and partly overlapping systems also occur in the ultra-violet, and probably nine weak systems involving transitions from higher states to $B^2\Sigma^+$ and $A^2\Sigma^+$ have been observed in the visible region. The whole spectrum is very complicated to analyse and sort out because most systems are barely degraded, showing heads degraded in each direction and some headless features. Strong perturbations add to the complexity. Most analyses have needed the use of O^{18} isotopic shifts.

Over 30 papers have been published 1964–74. We retain the original notation of $B^2\Sigma - X^2\Pi$ for the β and $A^2\Sigma - X^2\Pi$ for the γ bands; this is used in most recent literature although Rosen *et al.* reversed the A and B. Energy levels in order from the ground state are believed to be $X^2\Pi$, $B^2\Sigma^+$, $B'(\equiv D')^2\Pi$, $A^2\Sigma^+$, $C'^2\Delta$, $C^2\Sigma^-$, $D^2\Pi$, $F^2\Sigma^+$, $G^2\Sigma^+$, $E^2\Delta$, $I^2\Sigma^+$ and $H^2\Sigma^+$.

β SYSTEM, 3587–3198 Å, $B^2\Sigma^+ - X^2\Pi$

Occurrence. In arcs and discharge tubes containing phosphorus and oxygen and in flames containing phosphorus oxychloride. Also from high-frequency discharges through helium containing a trace of oxygen and phosphorus vapour.

PO (*contd.*)

Appearance. There is a strong sequence of bands commencing at around 3240 Å and getting weaker to longer wavelengths; this sequence has no marked head and the bands of which it is formed show heads degraded in both directions. There are two similar weaker sequences further to the red.

References. A. Petrikaln, *Z.P.*, **51**, 395 (1928)†.

J. Curry, L. Herzberg and G. Herzberg, *Z.P.*, **86**, 364 (1933).

R. Ramanadham, G. V. S. R. Rao and C. Ramasastry, *Indian J. Phys.*, **20**, 161 (1946)†.

K. Dressler, *Helv. Phys. Acta*, **28**, 563 (1955)†.

N. L. Singh, *Canad. J. Phys.*, **37**, 136 (1959)†.

H. Meinel and L. Krauss, *Z. Naturforsch.*, A, **21**, 1878 (1966).

M. N. Dixit and N. A. Narasimham, *Proc. Indian Acad. Sci.*, A, **68**, 1 (1968)†.

Strongest heads and vibrational assignments according to Dressler:

λ	I	v', v''	λ	I	v', v''
3460·1 R	1	5, 6 $^Q R_{12}$	3362·0 R	3	8, 8 R_1
3429·6 V	1	2, 3 P_2	3346·2 R	7	6, 6 $^Q R_{12}$
3414·1 V	5	1, 2 P_2	3340·6 R	7	7, 7 R_1
3409·8 V	2	3, 4 $^Q P_{21}$	3328·2 R	7	5, 5 $^Q R_{12}$
3405·7 V	6	0, 1 P_2	3320·9 R	7	6, 6 R_1
3397·8 V	4	2, 3 $^Q P_{21}$	3311·8 R	5	4, 4 $^Q R_{12}$
3387·9 V	6	1, 2 $^Q P_{21}$	3302·8 R	5	5, 5 R_1
3387·6 R	3	8, 8 $^Q R_{12}$	3270·5 V	9	0, 0 P_2
3385·5 R	2	9, 9 R_1	3255·3 V	8	1, 1 $^Q P_{21}$
3379·8 V	6	0, 1 $^Q P_{21}$	3246·2 V	10	0, 0 $^Q P_{21}$
3365·9 R	7	7, 7 $^Q R_{12}$	3218·1 R	3	7, 6 R_1

γ SYSTEM, 2750–2280 Å, $A^2\Sigma^+ - X^2\Pi$

Occurrence. With phosphorus compounds in carbon arc, in discharge tubes containing phosphorus and oxygen and in flames. With high-frequency discharge through helium containing a trace of oxygen and phosphorus vapour.

Appearance. Degraded to shorter wavelengths. The bands are double double-headed and the sequences are fairly well marked. See Plate 6.

References. A. Petrikaln, *Z.P.*, **51**, 295 (1928)†.

P. N. Ghosh and G. N. Ball, *Z.P.*, **71**, 362 (1931)†.

A. K. Sen Gupta, *Proc. Phys. Soc.*, **47**, 247 (1935).

K. S. Rao, *Canad. J. Phys.*, **36**, 1526 (1958)†.

B. Coquart *et al.*, *J. Chim. Phys.*, **64**, 1197 (1967).

M. N. Dixit and N. A. Narasimham, *Proc. Indian Acad. Sci.*, A, **68**, 1 (1968)†.

The following measurements of the strong bands are by Ghosh and Ball. The intensities are our own estimates from the published photographs by Petrikaln and by Ghosh and Ball. The bands appear double-headed, the separation between the P and Q heads being about 1·3 Å; only the P heads are given. Later rotational analyses use different notation for the branches, P_1 being referred to as $^0P_{12}$ and P_2 as P_1.

λ	I	v', v''	λ	I	v', v''	λ	I	v', v''
2721·5	1	0, 3 P_1	2692·4	4	2, 5 P_1	2662·9	3	3, 6 P_2
2706·8	3	1, 4 P_1	2690·8	4	1, 4 P_2	2636·3	7	0, 2 P_1
2705·1	1	0, 3 P_2	2676·7	4	2, 5 P_2	2623·4	7	1, 3 P_1

PO (*contd.*)

λ	I	v', v''	λ	I	v', v''	λ	I	v', v''
2620·5	8	0, 2 P_2	2477·9	10	0, 0 P_1	2375·2	7	2, 1 P_2
2610·7	2	2, 4 P_1	2468·3	6	1, 1 P_1	2367·3	2	3, 2 P_2
2608·0	7	1, 3 P_2	2464·2	10	0, 0 P_2	2320·6	3	2, 0 P_1
2595·7	2	2, 4 P_2	2459·0	2	2, 2 P_1	2313·7	4	3, 1 P_1
2555·0	10	0, 1 P_1	2454·6	6	1, 1 P_2	2306·9	4	4, 2 P_1
2543·9	7	1, 2 P_1	2396·3	8	1, 0 P_1	2301·7	4	3, 1 P_2
2540·4	10	0, 1 P_2	2387·9	7	2, 1 P_1	2300·4	3	5, 3 P_1
2529·4	7	1, 2 P_2	2383·5	8	1, 0 P_2	2294·9	3	4, 2 P_2
2518·7	5	2, 3 P_2	2379·9	2	3, 2 P_1	2288·2	2	5, 3 P_2

WEAK SYSTEMS INVOLVING $X^2\Pi$

These systems in the violet and ultra-violet, although apparently all allowed transitions to the ground state, are relatively weak.

Occurrence. Microwave or radio-frequency discharges, usually through $POCl_3$ or $POCl_3$ in argon. $B' - X$ also occurs in the phosphorus/atomic oxygen glow.

References. K. Dressler, *Helv. Phys. Acta*, **28**, 563 (1955).
M. N. Dixit and N. A. Narasimham, *Proc. Indian Acad. Sci.*, A, **68**, 1 (1968)†.
R. D. Verma, *Canad. J. Phys.*, **48**, 2391 (1970).
B. Coquart *et al.*, *C. R. Acad. Sci.* (*Paris*), B, **273**, 384 (1971)†.
B. Coquart *et al.*, *Canad. J. Phys.*, **50**, 1014 (1972).
C. Couet, M. Larzillière and H. Guenebaut, *C. R. Acad. Sci.* (*Paris*), **272**, 425 (1971).
J. C. Prudhomme, M. Larzillière and C. Couet, *Canad. J. Phys.*, **51**, 2464 (1973)†.

$B'\,^2\Pi - X^2\Pi$, 4800–3800 Å

Degraded to the red. Double-headed. The vibrational analysis is tentative.

λ	v', v''	λ	v', v''	λ	v', v''
4802·1	1, 11 $^QR_{12}$	4277·1	0, 8 $^QR_{12}$	4040·8	0, 7 $^SR_{21}$
4725·4	0, 10 $^QR_{12}$	4233·9	0, 8 $^SR_{21}$	3898·5	0, 6 $^QR_{21}$
4563·4	1, 10 $^QR_{12}$	4080·1	0, 7 $^QR_{12}$	3862·6	0, 6 $^SR_{21}$
4491·4	0, 9 $^QR_{12}$				

$C'\,^2\Delta - X^2\Pi$, 3000–2200 Å

Strongly degraded to the red; double-headed bands, overall separation 20 Å at longer wavelengths and 10 at shorter. This is an extensive and relatively strong system. The following data for the first, $^SR_{21}$, heads are from Dixit and Narasimham who confirmed the vibrational analysis but referred to the system as $C^2\Sigma^- - X^2\Pi$.

$\lambda\lambda$2958·2 (5, 12), 2930·5 (4, 11), 2903·3 (3, 10), 2839·9 (4, 10), 2813·2 (3, 9), 2702·3 (2, 7), 2677·5 (1, 6), 2621·5 (2, 6), 2597·2 (1, 5), 2520·7 (1, 4), 2497·5 (0, 3), 2447·8 (1, 3), 2425·2 (0, 2), 2356·2 (0, 1), 2290·3 (0, 0).

$C^2\Sigma^- - X^2\Pi$, 2320–1960 Å

Double-headed and degraded to the red. A strong $v'' = 0$ progression. Main heads from Coquart *et al.*

PO (*contd.*)

λ	v', v''	λ	v', v''	λ	v', v''
2208·7	1, 0	2137·5	3, 0	2042·1	6, 0
2197·8		2127·4		2032·9	
2194·7	3, 1	2104·3	4, 0	2013·0	7, 0
2184·0		2094·5		2004·0	
2172·3	2, 0	2072·5	5, 0		
2161·8		2063·0			

$D^2\Pi - X^2\Pi$, 2170–2153 Å

Double-headed; shaded to the red. All heads from Dressler.

λ	I	v', v''	λ	I	v', v''	λ	I	v', v''
2169·9	6	0, 2	2114·5	8	0, 1	2061·3	10	0, 0
2160·9			2105·9			2053·1		

WEAK VISIBLE SYSTEMS INVOLVING $B^2\Sigma^+$

These are all weak systems in the visible and near infra-red, best observed in microwave or radio-frequency discharges through $POCl_3$ + Ar.

References. R. D. Verma and M. N. Dixit, *Canad. J. Phys.*, **46**, 2079 (1968)†.
R. D. Verma *et al.*, *Canad. J. Phys.*, **49**, 3180 (1971)†.
R. D. Verma and S. S. Jois, *Canad. J. Phys.*, **51**, 322 (1973)†.
T. A. Ngo *et al.*, *Canad. J. Phys.*, **52**, 154 (1974).

$A^2\Sigma^+ - B^2\Sigma^+$, 11230–3800 Å

A weak but extensive system. Apparently single-headed; mostly shaded to the violet. Known bands below 10 000 Å from Verma *et al.* and Verma and Jois:

λ	v', v''	λ	v', v''	λ	v', v''	λ	v', v''
10227	0, 0	8602·2	3, 2	7855·1	3, 1	4835·8*	9, 1
9982·3	1, 1	8423·4	4, 3	7719·0	4, 2	4586·5*	9, 0
9736·9	2, 2	8239·0	5, 4	7579·4	5, 3	4107·5*	12, 1
8964·5	1, 0	7991·4	2, 0	7436·2	6, 4	3918·4*	12, 0
8786·8	2, 1						

* Degraded to the red.

B' (formerly D') $^2\Pi - X^2\Sigma^+$

Double double-headed, shaded to the red. The (0, 0) has its first head at 5506·3 Å (Verma and Dixit).

$D^2\Pi - B^2\Sigma^+$, 6900–5500 Å

Double double-headed bands, shaded to the red. The (0, 0) is strongest. R_2 heads from Verma and Dixit:

λ	v', v''	λ	v', v''
6860	0, 3	5962·5	0, 1
6384·7	0, 2	5584·7	0, 0

PO (*contd.*)

$F^2\Sigma^+ - B^2\Sigma^+$, 7324–3964 Å

A weak system of single-headed bands, shaded to the red. Data mainly from Verma *et al.* (1971). λλ7322·8 (0, 5), 4982·6 (4, 2), 4934·2 (3, 1), 4831·2 (2, 0), 4721·4 (4, 1), 4672·1 (3, 0), 4615·8 (5, 2), 4571·4 (5, 1), 4480·8 (4, 0), 4151·3 (8, 1), 3964·1 (8, 0).

$G^2\Sigma^+ - B^2\Sigma^+$, 5023–4248 Å

Single-headed. Most are weakly degraded to the violet. The (0, 0) band, degraded to the red, is strongest. Data from Verma *et al.* λλ5023·5 (1, 3), 4856·8 (0, 1), 4766·4 (1, 2), 4605·8 (0, 0), 4529·2 (1, 1), 4248·7 (2, 1), 4295·0 (1, 0).

$H^2\Sigma^+ - B^2\Sigma^+$

Data from Verma *et al.* λλ4122·8 (1, 1) deg. V; 4075·2 (2, 2) deg. V; 3929·0 (1, 0) deg. R; 3898·8 (2, 1) deg. V.

$I^2\Sigma^+ - B^2\Sigma^+$

Single-headed. Degraded to the violet. Data from Verma *et al.* λλ4219·8 (0, 1), 4027·0 (0, 0), 3987·8 (1, 1).

WEAK TRANSITIONS TO $A^2\Sigma^+$

Verma and Jois (see above) list red-degraded bands:

λ		
8613·1	$F^2\Sigma^+ - A^2\Sigma^+$ (3, 0)	
8412·4	$G^2\Sigma^+ - A^2\Sigma^+$ (0, 0)	
8260·7	″	(1, 1)
8245·8	″	(2, 2)

PO$^+$

FAR ULTRA-VIOLET SYSTEM, $A - X^1\Sigma$

Occurrence. In a discharge tube. A weak system.
Appearance. Degraded to the red.
Reference. K. Dressler, *Helv. Phys. Acta*, **28**, 253 (1955)†.
R heads:

λ_{air}	v', v''	λ_{air}	v', v''	λ_{vac}	v', v''
2192·9	0, 3	2026·1	1, 1	1971·1	1, 0
2129·0	0, 2	2009·9	0, 0	1933·5	2, 0
2067·9	0, 1			1898·0	3, 0

POBr or POBr$_2$

Reference. D. W. Naegeli and H. B. Palmer, *J. Mol. Spec.*, **26**, 277 (1968)†.

In emission from low-pressure flames of POBr$_3$ reacting with potassium vapour. Numerous closely-spaced narrow headless bands from 4515 to about 4900 Å. Judging from the photograph the following are strong: λλ4789, 4732, 4692, 4654, 4638, 4621, 4600, 4584, 4567, 4549, 4515.

POCl or POCl$_2$

Reference. D. W. Naegeli and H. B. Palmer, *J. Mol. Spec.*, **26**, 277 (1968)†.

In emission from low-pressure flame of POCl$_3$ reacting with potassium vapour. Narrow headless bands between 4700 and 4220 Å. Maxima of strong bands, with our estimates of intensity (I) based on the photograph: λλ4484 (7), 4462 (7), 4423 (8), 4363 (10), 4305 (8), 4267 (10), 4230 (8).

PS

Occurrence. In discharge through helium containing small amounts of P and S vapour, and in microwave discharge.

References. K. Dressler, *Helv. Phys. Acta*, **28**, 563 (1955).

N. A. Narasimham and T. K. B. Subramanian, *J. Mol. Spec.*, **29**, 294 (1969)†; **37**, 371 (1971)†.

ULTRA-VIOLET SYSTEM, 3300–2700 Å, C$^2\Sigma$ − X$^2\Pi$

Appearance. Double double-headed bands; separation between $^SR_{21}$ and R_1 and between R_2 and Q_2 only $\frac{1}{2}$ Å; main doubling 321 cm^{-1}.

Data, from Dressler, for $^SR_{21}$ and R_2 heads:

λ	I	v', v''		λ	I	v', v''		λ	I	v', v''	
3084·7	8	0, 3	$^SR_{21}$	2907·4	8	1, 1	$^SR_{21}$	2789·8	8	3, 0	R_2
3047·5	10	0, 2	R_2	2889·8	8	0, 0	$^SR_{21}$	2782·3	6	4, 1	$^SR_{21}$
3017·5	10	0, 2	$^SR_{21}$	2872·9	10	1, 0	R_2	2768·9	4	5, 1	R_2
2981·4	10	0, 1	R_2	2846·6	10	1, 0	$^SR_{21}$	2765·0	8	3, 0	$^SR_{21}$
2952·6	10	0, 1	$^SR_{21}$	2830·5	10	2, 0	R_2	2750·3	6	4, 0	R_2
2934·8	8	1, 1	R_2	2822·5	4	3, 1	$^SR_{21}$	2744·1	4	5, 1	$^SR_{21}$
2917·2	8	0, 0	R_2	2804·9	10	2, 0	$^SR_{21}$	2726·7	6	4, 0	$^SR_{21}$

VISIBLE SYSTEM, 6000–4200 Å, B$^2\Pi$ − X$^2\Pi$

Appearance. Double-headed bands, degraded to the red.

Measurements from Narasimham and Subramanian, with our intensity estimates from the plate:

λ	I	v', v''	λ	I	v', v''	λ	I	v', v''
5698·2	6	0, 7	5085·8	6	0, 4	4628·2	6	1, 2
5625·1			5027·7			4579·9		
5480·0	6	0, 6	4957·8	8	1, 4	4522·5	8	2, 2
5412·5			4902·7			4476·7		
5276·4	6	0, 5	4788·0	10	1, 3	4422·7	8	3, 2
5213·9			4736·4			4378·8		
5138·6	6	1, 5	4675·2	4	2, 3	−	−	3, 1
5079·9			4625·7			4243·6		

PS$^+$

2750–2330 Å SYSTEM, A$^1\Sigma$ − X$^1\Sigma$

Occurrence. Discharge through He containing a little P and S vapour.

PS⁺ *(contd.)*

> *Appearance.* Single-headed bands, degraded to the red.
> *Reference.*   K. Dressler, *Helv. Phys. Acta*, **28**, 563 (1955).

λ	I	v', v"	λ	I	v', v"	λ	I	v', v"
2748·3	2	0, 5	2574·6	10	0, 2	2398·2	6	2, 0
2704·1	2	1, 5	2520·6	10	0, 1	2380·4	2	4, 1
2688·5	6	0, 4	2483·2	6	1, 1	2365·2	4	3, 0
2645·9	2	1, 4	2468·5	6	0, 0	2349·0	2	5, 1
2630·5	8	0, 3	2432·5	6	1, 0	2333·8	2	4, 0

PSb see SbP

Pb₂

> *Occurrence.* Absorption in a furnace or by flash heating. Some systems also occur in thermal emission and in resonance fluorescence using a laser.
> *Appearance.* Degraded to the red.Probably seven systems.
> *References.* E. N. Shawhan, *P.R.*, **48**, 343 (1935)†.

J. G. Kay, N. A. Kuebler and L. S. Nelson, *Nature*, **194**, 671 (1962).
S. Weniger, *J. Physique*, **28**, 595 (1967)†.
S. E. Johnson *et al.*, *J. Chem. Phys.*, **56**, 5723 (1972).

RED SYSTEM, 6950–6400 Å

Clear sequences. Main features from Weniger:

λ	I	v', v"	λ	I	v', v"
6625·9	9	0, 3	6440·3	10	1, 1
6553·8	10	0, 2	6417·5	10	0, 0
6485·8	10	0, 1	6370·4	10	1, 0

BLUE-GREEN SYSTEM

A strong v" = 0 progression in absorption. Wavelengths from Shawhan; intensities based on Weniger.

λ	I	v', v"	λ	I	v', v"
4963·1	7	4, 0	4855·0	8	7, 0
4926·2	10	5, 0	4821·0	7	8, 0
4890·2	9	6, 0	4787·7	7	9, 0

ULTRA-VIOLET SYSTEMS

Weniger's B. 22 bands, strongest:

λ	I	v', v"	λ	I	v', v"
2929·3	7	0, 0	2901·9	7	3, 0
2920·3	10	1, 0	2892·8	6	4, 0
2911·2	8	2, 0	2885·9	8	5, 0

Weniger's E system, 13 bands. The following are probably strongest, 2398·2 (0, 1), 2385·3 (0, 0), 2378·8 (1, 0), 2373·4 (2, 0).

There are also diffuse bands 3193–3006 and a progression 2718–2627 Å. Kay *et al.* report bands 2167, 2161, 2143 and 2136 Å.

PbBr

VISIBLE SYSTEM, $A\frac{1}{2} - X\frac{1}{2}$ (case c)

Occurrence. In absorption and in a microwave discharge.

Appearance. Degraded to the red. At least 250 bands. In absorption the $v'' = 0$ progression is strong.

References. F. Morgan, *P.R.*, **49**, 47 (1936).

K. M. Lal and B. N. Khanna, *Canad. J. Phys.*, **45**, 3663 (1967); **46**, 1991 (1968)†.

P. Deschamps, *Thesis (Paris)*, (1967)†.

The following are the strong absorption bands from Morgan:
λλ5092·1 (0, 6), 5040·4 (0, 5), 5002·0 (1, 5), 4902·5 (1, 3), 4866·4 (2, 3), 4818·5 (2, 2), 4597·1 (6, 0), 4566·9 (7, 0), 4537·5 (8, 0), 4509·5 (9, 0). Lal and Khanna have studied the rotational structure in emission of the bands 4818·5 (2, 2), 4784·3 (3, 2), 4738·0 (3, 1), 4704·7 (4, 1).

ULTRA-VIOLET SYSTEM, 3012–2705 Å, $B - X\frac{1}{2}$

Occurrence. Absorption by heated $PbBr_2$ at low pressure.

Appearance. Degraded to shorter wavelengths.

Reference. K. Wieland and R. Newburgh, *Helv. Phys. Acta*, **25**, 87 (1952)†.

Strongest bands:

λ	I	v', v''	λ	I	v', v''	λ	I	v', v''
2964·0	7	0, 4	2906·6	9	1, 2	2868·3	9	2, 1
2946·1	8	0, 3	2893·8	7	0, 0	2851·9	10	2, 0
2928·6	9	0, 2	2889·5	10	1, 1	2831·2	10	3, 0
2911·0	8	0, 1	2872·6	9	1, 0	2811·4	10	4, 0
						2791·7	8	5, 0

PbCl

VISIBLE SYSTEM, 5730–4100 Å, $A\frac{1}{2} - X^2\Pi_{\frac{1}{2}}$

Occurrence. High-frequency discharge through $PbCl_2$ vapour, and in absorption. Also fluorescent emission.

Appearance. Degraded to red.

References. G. D. Rochester, *P.R.S.*, **153**, 407 (1935)†.

F. Morgan, *P.R.*, **49**, 47 (1936).

O. N. Singh and I. S. Singh, *Curr. Sci. (India)*, **37**, 282 (1968)†.

S. P. Singh, *Indian J. Pure Appl. Phys.*, **8**, 114 (1970).

Strongest bands (intensities listed by Rochester as 9 or 10):

λ	v', v''	λ	v', v''	λ	v', v''
5358·3	4, 14	5062·0	0, 7	4596·0	1, 1
5342·1	3, 13	4988·2	0, 6	4548·9	2, 1
5263·1	3, 12	4916·5	0, 5	4399·0	4, 0
5169·5	2, 10	4846·0	0, 4	4356·5	5, 0
5153·4	1, 9	4660·3	1, 2	4315·3	6, 0

FAR RED SYSTEM, 8815–6208 Å, $A\frac{1}{2} - {}^2\Pi_{\frac{3}{2}}$

Occurrence. In Schüler-type discharge tube.

Appearance. Shaded to longer wavelengths. Strong $v' = 0$ and $v'' = 0$ progressions. 33 bands recorded.

PbCl (*contd.*)

Reference.   P. Deschamps, *Thesis (Paris)*, (1967)†.

λ	I	v', v"	λ	I	v', v"	λ	I	v', v"
8360	8	0, 5	7359	7	3, 2	6921·9	8	4, 0
8146	7	0, 4	7191·3	10	3, 1	6818·7	7	5, 0
7611	8	1, 2	7029·6	9	3, 0	6718·8	7	6, 0

ULTRA-VIOLET SYSTEM, 2960–2590 Å, B − X$^2\Pi_{\frac{1}{2}}$

Occurrence.   Absorption by heated $PbCl_3$ at low pressure.

Appearance.   Degraded to shorter wavelengths. Marked sequences.

References.   K. Wieland and R. Newburgh, *Helv. Phys. Acta*, **25**, 87 (1952)†.
H. Cordes and F. Gehrke, *Z. Phys. Chem.*, **51**, 281 (1966).

Analyses differ. The following are the strongest, from Cordes and Gehrke:

λ	I	v', v"	λ	I	v', v"
2882·7	7	0, 1	2826·9	6	1, 0
2858·0	10	0, 0	2796·7	5	2, 0

PbF

The ground state of PbF is a widely separated $^2\Pi$. Six systems, A to F, involve transitions to $X_1{}^2\Pi_{\frac{1}{2}}$ and in some cases (A, B, E) transitions to $X_2{}^2\Pi_{\frac{3}{2}}$ are known. The main systems are all obtained in absorption in a carbon-tube furnace containing PbF_2. Some systems, including the sub-systems to X_2, occur in emission in high frequency discharges.

References.   G. D. Rochester, *P.R.S.*, **153**, 407 (1936)†; **167**, 567 (1938)†.
F. Morgan, *P.R.*, **49**, 47 (1936).
R. F. Barrow *et al.*, *Proc. Phys. Soc.*, **73**, 317 (1959).
S. P. Singh, *Indian J. Pure Appl. Phys.*, **5**, 292 (1967)†.
O. N. Singh, M. P. Srivastava and I. S. Singh, *Canad. J. Phys.*, **47**, 1639 (1969)†.
I. S. Singh and O. N. Singh, *Canad. J. Phys.*, **50**, 2206 (1972)†.

Wavelengths and intensity estimates vary somewhat. Most of the data are from Rochester.

SYSTEM A, 5277–4110 Å, A$\frac{1}{2}$ − X$_1{}^2\Pi_{\frac{1}{2}}$; 8340–7300 Å, A$\frac{1}{2}$ − X$_2{}^2\Pi_{\frac{3}{2}}$

Appearance.   System of double-headed bands degraded to the red. Well marked sequences.

Strong bands and intensities in emission according to Rochester. Absorption was observed by Morgan.

λ	I	v', v"	λ	I	v', v"	λ	I	v', v"
4912·7	6	2, 6	4690·6	8	2, 4	4364·8	10	1, 0
4891·3	6	1, 5	4669·1	9	1, 3	4291·6	5	2, 0
4799·6	7	2, 5	4647·6	9	0, 2	4221·3*		3, 0
4778·1	7	1, 4	4542·5	10	0, 1	4153·7*		4, 0
4756·6	7	0, 3	4441·2	8	0, 0			

* These bands were observed in absorption.

For the A − X_2 component, Rosen *et al.* note bands at 8337·6 (0, 4), 7897·2 (0, 3), 7588·6 (0, 2) and 7299·5 (0, 1).

PbF (*contd.*)

SYSTEM B, 2925–2665 Å, $B - X_1{}^2\Pi_{\frac{1}{2}}$; 3795–3565 Å, $B - X_2{}^2\Pi_{\frac{3}{2}}$

Appearance. Rochester divides the system into two groups of bands; B_1 (2925–2665) fifteen double-headed bands degraded to shorter wavelengths and B_2 (3795–3565) a further set of six double-headed bands degraded to shorter wavelengths but observed only in emission.

P heads of strongest bands according to Rochester.

Group B_2, $B - X_2$			Group B_1, $B - X_1$		
λ	I	v', v''	λ	I	v', v''
3792·8	2	0, 2	2841·2	8	0, 1
3718·9	5	0, 1	2801·2	9	0, 0
3707·7	3	1, 2	2754·5	10	1, 0
3647·2	10	0, 0	2747·2	5	2, 1
3568·7	3	1, 0	2709·8	7	2, 0

SYSTEM C, 2690–2570 Å, $C - X_1{}^2\Pi_{\frac{1}{2}}$

Appearance. Weak absorption bands grouped in sequences.

Q heads according to Rochester.

λ	I	v', v''	λ	I	v', v''
2653·3	1	1, 2	2612·8	2	2, 2
2646·9	2	2, 3	2606·9	2	3, 3
2618·6	0	1, 1			

SYSTEM D, 2305–2245 Å, $D - X_1{}^2\Pi_{\frac{1}{2}}$

Appearance. Three bands with single sharp heads.

λ	I	v', v''
2305·6	1	0, 1
2279·1	2	0, 0
2248·5	2	1, 0

SYSTEM E, 2250–2120, $E - X_1{}^2\Pi_{\frac{1}{2}}$; 2684–2763 Å, $E - X_2{}^2\Pi_{\frac{3}{2}}$

Appearance. Group of intense bands degraded to shorter wavelengths.

Q heads of strongest bands in absorption.

λ	I	v', v''	λ	I	v', v''
2224·8	3	0, 1	2173·6	10	1, 0
2199·9	8	0, 0	2147·4	5	2, 0

Singh (1967) records $E - X_2$ bands at 2762·9 (0, 2), 2723·6 (0, 1) and 2685·0 (0, 0).

SYSTEM F, 2130–2060, $F - X_1{}^2\Pi_{\frac{1}{2}}$

Appearance. Single sharp heads.

λ	I	v', v''	λ	I	v', v''
2130·5	0	0, 2	2085·8	2	0, 0
2108·1	2	0, 1	2058·8	1	1, 0

In addition, Rochester records three unclassified bands, λ (I), 2095·5 (1), 2092·3 (1), 2048·4 (1) and two broad regions of continuum, one in absorption having a maximum at 2440 Å and one in emission with a maximum at 3050 Å.

PbH

VISIBLE SYSTEM, 8450–5180 Å

Occurrence. In absorption in a King furnace and in emission from a 1000 V arc in H_2 at 4 to 5 atm. pressure.

Appearance. A many-line type of spectrum, each band possessing four branches. Strongly degraded to the red.

Transition. Bands show P and R branches and were originally assigned to a $^2\Sigma - ^2\Sigma$ transition, but Rosen *et al.* favour $^4\Sigma$ (case c) to $X^2\Pi_{\frac{1}{2}}$.

References. W. W. Watson, *P.R.*, **54**, 1068 (1938).
W. W. Watson and R. Simon, *P.R.*, **57**, 708 (1940).

Origins:

λ_0	v', v''	λ_0	v', v''	λ_0	v', v''
8446	0, 4	6459·0	2, 2	5561·0	1, 0
7595	0, 3	6068·4	1, 1	5425·8	2, 0
7329	1, 3	5907·7	2, 1	5299·5	3, 0
6870·4	0, 2	5758·5	3, 1	5180·7	4, 0
6651·5	1, 2				

A further faint band at 3815 Å was also reported.

PbI

Reference. K. Wieland and R. Newburgh, *Helv. Phys. Acta*, **25**, 87 (1952)†.

VISIBLE SYSTEM, 6510–4416 Å, $A - X^2\Pi_{\frac{1}{2}}$

Occurrence. In high-frequency discharge and in absorption.

Appearance. Degraded to the red. In emission the strongest bands are 6500–5400, and in absorption 4800–4500 Å.

The following are strongest in emission:

λ	I	v', v''	λ	I	v', v''	λ	I	v', v''
6218·7	8	1, 30	5948·6	10	1, 25	5746·5	9	2, 22
6162·8	9	1, 29	5845·7	10	1, 23	5699·0	8	2, 21
6105·9	10	0, 27	5795·5	9	1, 22	5650·7	8	2, 20

ULTRA-VIOLET SYSTEM, 3086–2680 Å, $B - X^2\Pi_{\frac{1}{2}}$

Occurrence. In absorption through heated PbI_2 at low pressure.

Appearance. Degraded to violet.

Reference. K. Wieland and R. Newburgh, *Helv. Phys. Acta*, **25**, 87 (1952)†.

Strongest bands:

λ	I	v', v''	λ	I	v', v''	λ	I	v', v''
2994·2	8	1, 2	2931·8	10	3, 0	2863·4	9	8, 1
2980·1	7	1, 1	2915·0	10	4, 0	2861·0	8	9, 2
2963·1	8	2, 1	2898·7	10	5, 0	2848·0	8	9, 1
2949·0	8	2, 0	2882·5	7	6, 0	2827·4	7	12, 3

PbO

Six band systems known as A, B, C, C', D, and E, are attributed to lead oxide.

Occurrence. All six systems have been observed in absorption in a carbon arc furnace. Systems A, B, and D are emitted by an arc (between carbon or copper electrodes) containing lead or lead salts and in flame sources. System C also occurs in emission in a heavy-current discharge.

References. S. Bloomenthal, *P.R.*, **35**, 34 (1930).

A. Christy and S. Bloomenthal, *P.R.*, **35**, 46 (1930).

H. G. Howell, *P.R.S.*, **153**, 683 (1935–6).

L. Withrow and G. M. Rassweiler, *Ind. Eng. Chem.*, **23**, 769 (1931)†.

R. F. Barrow, J. L. Deutsch and D. N. Travis, *Nature*, **191**, 374 (1961).

D. N. Travis, *Thesis (Oxford)*, (1963).

R. S. Ram, J. Singh and K. N. Upadhya, *Spectrosc. Lett.*, **6**, 515 (1973).

SYSTEM A, 6720–4748 Å, A 0^+–$X^1\Sigma^+$

Appearance. Degraded to red.

Strong bands. Intensities (for emission) on scale of 6.

λ	I	v', v''	λ	I	v', v''	λ	I	v', v''
6433·6	3	3, 8	6160·5	4	0, 5	5459·4	6	0, 2
6427·7	3	0, 6	5910·7	6	0, 4	5331·1	3	1, 2
6342·0	3	2, 7	5677·8	6	0, 3	5138·2	3	1, 1
6250·7	5	1, 6	5617·6	3	2, 4	5068·8	1	0, 0

SYSTEM B, 5770–4146 Å, B $1 - X^1\Sigma^+$

Appearance. Degraded to red. A strong progression on each side of the weak (0, 0) band.

Strong bands. Intensities (for emission) on scale of 6.

λ	I	v', v''	λ	I	v', v''
5353·8	3	0, 5	4553·7	6	1, 1
5162·3	6	0, 4	4509·2	1	0, 0
4983·8	6	0, 3	4410·4	5	1, 0
4816·9	6	0, 2	4317·1	4	2, 0
4658·0	5	0, 1	4229·0	4	3, 0

SYSTEM C, 4156–3607 Å, C 0^+–$X^1\Sigma^+$

Degraded to red. Intensities for absorption. Strong bands:

λ	I	v', v''	λ	I	v', v''
4156·2	3	2, 1	3955·0	7	3, 0
4037·6	3	2, 0	3877·8	8	4, 0
3987·7	4	4, 1	3804·9	6	5, 0

SYSTEM C', C' $- X^1\Sigma^+$

A single progression of absorption bands observed by Howell. λλ3673·7 (5, 0), 3612·9 (6, 0), 3554·8 (7, 0), 3499·3 (8, 0), 3446·2 (9, 0), 3395·3 (10, 0). Degraded to the red.

SYSTEM D, 3594–3209 Å, D $1 - X^1\Sigma^+$

Appearance. Degraded to red.

PbO (*contd.*)

Bands as observed by Bloomenthal in emission. Many more bands have been observed in absorption.

λ	I	v', v''	λ	I	v', v''
3594·2	1	1, 4	3341·8	2	1, 1
3485·7	6	0, 2	3320·7	1	0, 0
3442·8	1	2, 3	3264·4	2	1, 0
3401·9	5	0, 1	3209·2	2	2, 0

SYSTEM E, 3062–2780 Å, E 0^+–$X^1\Sigma^+$

Reference. E. E. Vago and R. F. Barrow, *Proc. Phys. Soc.*, **59**, 449 (1947).

This is Howell's system F, but with modified analysis. The following are the strongest bands as recorded by Vago and Barrow:

λ	I	v', v''	λ	I	v', v''
3062·7	4	1, 3	2925·6	3	3, 2
2998·5	4	1, 2	2900·2	4	2, 1
2960·7	3	2, 2	2866·2	5	3, 1

PbS

Numerous bands, from the near infra-red to the ultra-violet, all degraded to the red, have been obtained in absorption. They have been analysed into seven systems, denoted Far Red and A to F.

References. G. D. Rochester and H. G. Howell, *P.R.S.*, **148**, 157 (1935)†.

E. E. Vago and R. F. Barrow, *Proc. Phys. Soc.*, **59**, 449 (1947)†.

R. F. Barrow, P. W. Fry and R. C. Le Bargy, *Proc. Phys. Soc.*, **81**, 697 (1963).

FAR RED SYSTEM, $a\,1 - X0^+$

Three bands observed by Barrow *et al.*, 7081·8 (2, 3), 6876·4 (2, 2), 6745·4 (3, 2). Degraded to the red.

The following are the strongest bands of systems A to E, from Rochester and Howell, nearly all listed as intensity 10.

SYSTEM A, 7670–4545 Å, A 0^+–$X^1\Sigma^+$

λ	v', v''	λ	v', v''	λ	v', v''
5999·1	0, 5	5228·5	3, 1	4919·9	6, 0
5630·2	1, 3	5159·0	4, 1	4858·5	7, 0
5549·1	2, 3	5047·5	4, 0	4798·5	8, 0
5499·8	1, 2	4982·7	5, 0		

SYSTEM B, 5080–3950 Å, B $1 - X^1\Sigma^+$

λ	v', v''	λ	v', v''
4563·5	2, 1	4316·5	5, 0
4421·3	3, 0	4266·4	6, 0
4368·1	4, 0		

PbS (*contd.*)

SYSTEM C, 4400–3675 Å, C – X$^1\Sigma^+$

λ	I	v', v''
3923·5	8	8, 0
3881·4	9	9, 0
3840·4	8	10, 0

SYSTEM D, 4080–3500 Å, C' – X$^1\Sigma^+$

λ	I	v', v''
3796·0	8	5, 0
3757·5	8	6, 0
3720·0	8	7, 0

SYSTEM E, 3660–3130 Å, D 1 – X$^1\Sigma^+$

λ	v', v''	λ	v', v''
3530·6	0, 3	3428·3	0, 1
3478·7	0, 2	3393·9	1, 1
3443·4	1, 2	3313·2	2, 0

SYSTEM F, 2175–2060 Å, F – X$^1\Sigma^+$

The following heads are prominent on Vago and Barrow's plate: 2151·9 (0, 3), 2132·5 (0, 2), 2115·9 (1, 2), 2097·2 (1, 1).

There are also unanalysed bands, probably E – X$^1\Sigma^+$ in the region 3200–2750 Å.

PbSe

Five systems, denoted A, B, C, D and F, have been obtained in absorption. They are all degraded to the red.

References. J. W. Walker, J. W. Straley and A. W. Smith, *P.R.*, **53**, 140 (1938).

E. E. Vago and R. F. Barrow, *Proc. Phys. Soc.*, **59**, 449 (1947)†.

Strongest bands only:

SYSTEM A, 6100–4800 Å, A – X$^1\Sigma^+$

λ	I	v', v''	λ	I	v', v''
5979·2	8	1, 7	5278·7	10	5, 2
5674·3	9	2, 5	5203·2	10	10, 4
5372·6	10	3, 2	5202·6	10	5, 1
5325·1	10	4, 2	5158·6	10	6, 1

SYSTEM B, 5410–4160 Å, B – X$^1\Sigma^+$

4921·2 (1, 3), 4855·9 (1, 2), 4813·1 (2, 2), 4791·3 (1, 1), 4749·0 (2, 1), 4708·4 (3, 1), 4570·9 (5, 0).

SYSTEM C, 4500–4050 Å, C – X$^1\Sigma^+$

4165·9 (4, 0), 4134·9 (5, 0), 4104·9 (6, 0).

PbSe (*contd.*)

SYSTEM D, 3850–3250 Å, D – $X^1\Sigma^+$

λ	I	v′, v″	λ	I	v′, v″
3713·4	7	6, 1	3557·8	10	1, 0
3676·2	7	5, 1	3534·1	8	1, 1
3628·8	7	3, 0	3500·0	7	0, 1
3593·2	10	2, 0	3477·1	8	0, 2

SYSTEM F, 2298–2158 Å, F – $X^1\Sigma^+$

Strongest bands on Vago and Barrow's photograph: λλ2267·1 (0, 4), 2253·1 (0, 3), 2239·3 (0, 2), 2228·1 (1, 2), 2225·6 (0, 1), 2214·5 (1, 1), 2190·0 (2, 0), 2179·4 (3, 0), 2169·0 (4, 0).

PbTe

References.   J. W. Walker, J. W. Straley and A. W. Smith, *P.R.*, **53**, 140 (1938).
E. E. Vago and R. F. Barrow, *Proc. Phys. Soc.*, **59**, 449 (1947)†.
R. C. Le Bargy and R. F. Barrow, *Proc. Phys. Soc.*, **82**, 332 (1963).
    All degraded to the red and observed in absorption. Strongest heads:

FAR RED, a – X

λλ6839·2 (1, 2), 6771·5 (2, 2), 6676·3 (2, 1), 6610·1 (3, 1), 6521·7 (3, 0), 6460·5 (4, 0).

SYSTEM A, 6430–5620 Å, A – X

λλ5777·3, 5731·7, 5686·8.

SYSTEM B, 5550–4650 Å, B – X

λλ5054·4, 5018·0, 4965·2, 4930·8, 4896·4, 4863·3.

SYSTEM D, 3864–3579 Å, D – X

λλ3771·2 (0, 3), 3741·6 (0, 2), 3722·4 (1, 2), 3703·4 (2, 2), 3693·3 (1, 1), 3674·6 (2, 1).

SYSTEM F, 2490–2309 Å, F – X

λλ2437·9 (0, 3), 2425·4 (0, 2), 2415·1 (1, 2), 2402·8 (1, 1), 2380·8 (2, 0), 2371·3 (3, 0), 2361·7 (4, 0), 2352·4 (5, 0), 2342·9 (6, 0).

SYSTEM G, 2240–2120 Å, G – X

λλ2209·3 (0, 6), 2199·1 (0, 5), 2188·9 (0, 4), 2178·9 (0, 3), 2161·5 (1, 2), 2151·9 (1, 1).

PdH

Reference.   A. Lagerqvist, H. Neuhaus and R. Scullman, *Proc. Phys. Soc.*, **83**, 498 (1964).
    Complex structure 4700–4100 has been observed in absorption in a King furnace containing palladium and hydrogen. No heads are prominent. The corresponding system of PdH shows 12 red-degraded bands of a $^2\Sigma - {}^2\Sigma$ transition.

Po$_2$

In electrodeless discharges some 450 band heads degraded to longer wavelengths have been obtained with Po^{210} and more than 500 with mixtures of Po^{208}–Po^{209}. About 350 have been assigned to one band system which is taken to be the main system.

Reference. G. W. Charles, D. L. Timma, D. J. Hunt and G. Pish, *J. Opt. Soc. Amer.*, **47**, 291 (1957)†.

MAIN SYSTEM, λλ5130–3600

Strong heads for the isotope combination Po$_2^{210}$.

λ	I	v', v''	λ	I	v', v''
4288·7	6	0, 12	3960·2	10	4, 2
4261·8	9	0, 11	3944·4	10	5, 2
4235·0	7	0, 10	3943·9	7	6, 2
4208·5	10	0, 9	3904·8	6	6, 1
4182·1	8	0, 8	3873·9	6	8, 1
4156·0	10	0, 7	3835·8	7	9, 0
4130·0	6	0, 6	3821·3	7	10, 0

PrO

Occurrence. In carbon arc containing praseodymium metal or Pr$_6$O$_{11}$ and when Pr salts are introduced into a flame.

Appearance. A number of systems of red-degraded bands.

References. W. W. Watson, *P.R.*, **53**, 639 (1938).

R. Herrmann and C. T. J. Alkemade, *Flame Photometry*, (1963).

T. V. Venkitachalam, G. Krishnamurty and N. A. Narasimham, *Proc. Indian Acad. Sci.*, A, **76**, 113 (1972)†.

A large number of bands have been listed. The following are likely to be prominent, being mostly (0, 0) bands of systems or sub-systems. Intensities are tentative estimates, partly based on Herrmann and Alkemade, but also influenced by observations of other authors.

λ	I	System	λ	I	System
8488·9	10	A$^2\Delta_{\frac{5}{2}}$ – X$^2\Pi_{\frac{3}{2}}$ (0, 0)	6589	6	VII
7986·4	9	” ” (1, 0)	6474	9	VIII
7866	9	II	6019	9	IX
7662·8	10	B$^2\Delta_{\frac{5}{2}}$ – X$^2\Pi_{\frac{3}{2}}$ (0, 0)	5763·4	8	X
7320·2	9	IV (0, 0)	5691·1	9	XI
7095	10	–	5352·0	7	XII
6923·9	8	V	5137	5	XIII
6822	6	VI			

PtC

Occurrence. Emission and absorption in a King furnace at high temperature.

Appearance. Four systems of red-degraded bands, probably with clear sequences.

Reference. O. Appelblad, C. Nilsson and R. Scullman, *Phys. Scripta*, **7**, 65 (1973)†.

PtC (*contd.*)

R heads of main sequences:

$A'' \, ^1\Sigma - X^1\Sigma$

16 bands. 8605·9 (0, 1), 7899·3 (0, 0), 7358·6 (1, 0).

$A' \, ^1\Pi - X^1\Sigma$

8 bands. 8217·0 (0, 1), 7570·1 (0, 0), 7084·3 (1, 0).

$A \, ^1\Pi - X^1\Sigma$

7 bands. 6079·8 (0, 2), 5721·1 (0, 1), 5399·5 (0, 0), 5173·8 (1, 0), 4968·9 (2, 0).

$B \, ^1\Sigma - X^1\Sigma$

10 bands. 3379·5 (0, 3), 3266·6 (0, 2), 3160·1 (0, 1), 3059·4 (0, 0), 2982·4 (1, 0), 2912·1 (2, 0).

PtH

Occurrence. Platinum hollow-cathode discharge in H_2. The 4547 and 3752 bands have also been obtained in absorption.

Appearance. 14 red-degraded bands with open rotational structure have been analysed into four sub-systems.

References. R. Scullman, *Ark. Fys.*, **28**, 255 (1965)†.
B. Kaving and R. Scullman, *Canad. J. Phys.*, **49**, 2264 (1971)†.

The following are the heads listed by Kaving and Scullman. No intensities are available.

λ	System	v', v''	λ	System	v', v''
5720·9	$^2\Delta_{\frac{3}{2}} - X^2\Delta_{\frac{3}{2}}$	0, 1	4194·7	$^2\Phi_{\frac{7}{2}} - X^2\Delta_{\frac{5}{2}}$	0, 0
5088·6	" "	0, 0	4104·4	$B^2\Delta_{\frac{5}{2}} - X^2\Delta_{\frac{5}{2}}$	0, 1
5076·3	$A^2\Delta_{\frac{5}{2}} - X^2\Delta_{\frac{5}{2}}$	0, 1	4004·4	$A^2\Delta_{\frac{5}{2}} - X^2\Delta_{\frac{5}{2}}$	2, 0
4729·4	$^2\Delta_{\frac{3}{2}} - X^2\Delta_{\frac{3}{2}}$	1, 0	3958	$^2\Phi_{\frac{7}{2}} - X^2\Delta_{\frac{5}{2}}$	1, 0
4640·9	$^2\Phi_{\frac{7}{2}} - X^2\Delta_{\frac{5}{2}}$	0, 1	3814·0	$A^2\Delta_{\frac{5}{2}} - X^2\Delta_{\frac{5}{2}}$	3, 0
4547·2	$A^2\Delta_{\frac{5}{2}} - X^2\Delta_{\frac{5}{2}}$	0, 0	3751·7	$B^2\Delta_{\frac{5}{2}} - X^2\Delta_{\frac{5}{2}}$	0, 0
4245·4	" "	1, 0	3546·6	" "	1, 0

PtO

$A^1\Sigma - X^1\Sigma$, 7000–4000 Å

Occurrence. In arcs and hollow-cathode discharge.

Appearance. Single-headed bands, shaded to the red.

Reference. C. Nilsson, R. Scullman and N. Mekendale, *J. Mol. Spec.*, **35**, 177 (1970)†.

No intensities available. Bands marked * have been analysed and may be strong.

λ	v', v''	λ	v', v''	λ	v', v''
6594	1, 3	*5902·3	0, 0	5489	3, 1
*6548·5	0, 2	5707	2, 1	*5445·6	2, 0
6255	1, 2	*5663·0	1, 0	5248	3, 0
*6210·6	0, 1				

RaCl

Two intense red-degraded sequences beginning at λ6763 and λ6498 and a weak sequence beginning at λ6607 have been obtained with radium chloride in an atmosphere of helium in a hollow-cathode discharge.

Transition. $C^2\Pi - X^2\Sigma$.

Reference.   A. Lagerqvist, *Ark. Fys.*, **6**, 141 (1953).

Initial bands of sequences have heads as follows:

v', v''		λ			v', v''	λ
0, 0	6763·3 Q_1	6498·1 Q_2	6489·2 R_2		0, 1	6607·6 Q_2
1, 1	6764·7 Q_1	6499·6 Q_2	6490·6 R_2		1, 2	6608·4 Q_2
2, 2	6765·8 Q_1	6500·9 Q_2	6492·1 R_2		2, 3	6609·2 Q_2

RaOH

Reference.   R. Herrmann and C. T. J. Alkemade, *Flame Photometry*, (1963).

Bands at 6653, 6349, 6329, 6285, 6269, 6247, 6210, 6020 and 5200 Å have been provisionally assigned to radium hydroxide. The strongest are at 6653 and 6269.

Rb$_2$

Four, possibly five, systems are known. That in the red is strongest.

References.   J. M. Walter and S. Barratt, *P.R.S.*, **119**, 257 (1928).

E. Matuyama, *Nature*, **133**, 567 (1934).

P. Kusch, *P.R.*, **49**, 218 (1936).

Ny Tsi-Zé and Tsien San-Tsiang, *P.R.*, **52**, 91 (1937).

INFRA-RED SYSTEM, $A - X^1\Sigma$

Bands around 8800 Å observed by Matuyama in absorption. No data available.

RED SYSTEM, $B^1\Pi - X^1\Sigma$

Observed in absorption, fluorescence and magnetic rotation. Strongest bands, degraded to the red, from Kusch (all $I = 10$):

λλ6926·9 (0, 4), 6899·7 (0, 3), 6872·9 (0, 2), 6824·2 (1, 1), 6797·8 (1, 0), 6775·7 (2, 0), 6754·5 (3, 0), 6733·6 (4, 0).

BLUE SYSTEM, 5017–4695 Å, $C - X^1\Sigma$

About 160 bands, degraded to the red, observed in absorption. Strongest from Ny and Tsien:

λ	I	v', v''	λ	I	v', v''	λ	I	v', v''
4818·5	8	1, 2	4775·8	8	3, 0	4749·0	9	8, 1
4805·4	8	1, 1	4764·6	8	6, 1	4746·5	10	10, 2
4797·1	9	2, 1	4754·1	8	9, 2	4739·1	8	11, 2

VIOLET SYSTEM, 4540–4220 Å, $D - X^1\Sigma$

118 red-degraded bands observed in absorption. Strongest from Ny and Tsien:

Rb₂ (*contd.*)

λ	I	v′, v″	λ	I	v′, v″	λ	I	v′, v″
4341·1	9	8, 1	4319·7	9	11, 1	4281·3	9	15, 0
4338·9	9	9, 1	4295·2	9	13, 0	4274·4	8	16, 0
4326·8	10	10, 11	4288·2	10	14, 0			

Walter and Barratt also recorded orange bands 6054, 6033 and 5992 but these have not been confirmed by later workers.

RbAr

Reference.   G. T. Lalos and G. L. Hammond, *Astrophys. J.*, **135**, 616 (1962)†.
Diffuse band at 7630 Å accompanies the Rb resonance line in emission from rapidly compressed rubidium vapour in argon.

RbBr

Occurrence. In absorption in a furnace.
Reference.   R. F. Barrow and A. D. Caunt, *P.R.S.*, **219**, 120 (1953).
A continuum and regularly spaced diffuse bands which have a maximum at about 2800 Å and extend to longer wavelengths with increasing temperature, the longest wavelength recorded being 3750 Å.

RbCd

Diffuse bands, degraded to the violet, observed in absorption by Barratt (see CsCd). Heads λλ4487 and 4423. Also bands 4400–4200 Å.

RbCl

Occurrence. In absorption in a furnace.
Reference.   R. F. Barrow and A. D. Caunt, *P.R.S.*, **219**, 120 (1953).
A continuum and regularly spaced diffuse bands which have a maximum at about 2480 Å and extend to longer wavelengths with increasing temperature, the longest wavelength recorded being 3336 Å.

RbCs

J. M. Walter and S. Barratt (*P.R.S.*, **119**, 257 (1928)) attribute a diffuse band at 5640 Å to RbCs, but Kusch (*P.R.*, **49**, 218 (1936)) only found red-degraded absorption bands 7400–7230 Å; λλ7409 (0, 5), 7381 (0, 4), 7355 (0, 3), 7328 (0, 2), 7307 (1, 2), 7301 (0, 1), 7281 (1, 1), 7266 (3, 2), 7255 (1, 0), 7225 (2, 0).

RbF

Occurrence. In absorption in a furnace.
Reference.   R. F. Barrow and A. D. Caunt, *P.R.S.*, **219**, 120 (1953).

RbF (*contd.*)

A continuum and regularly spaced diffuse bands which have a maximum at about 2150 Å and extend to longer wavelengths with increasing temperature, the longest wavelength recorded being 2982 Å.

RbH

5300 Å SYSTEM, $A^1\Sigma - X^1\Sigma$

Occurrence. In discharge through rubidium vapour and H_2, and in absorption in a furnace at 650°C.

Appearance. A very extensive many-line system, with weak R heads, degraded to the red.

References. A. G. Gaydon and R. W. B. Pearse, *P.R.S.*, **173**, 28 (1939)†.
I. R. Bartky, *J. Mol. Spec.*, **21**, 1 (1966).

Band origins (close to R heads):

λ_0	I	v', v''	λ_0	I	v', v''	λ_0	I	v', v''
5871·4	8	4, 2	5345·8	9	7, 1	4959·3	5	9, 0
5783·4	10	5, 2	5268·9	9	8, 1	4892·1	7	10, 0
5696·2	10	6, 2	5193·5	8	9, 1	4826·3	5	11, 0
5610·1	8	7, 2	5098·1	5	7, 0	4762·2	5	12, 0
5502·9	7	5, 1	5028·1	5	8, 0	4699·9	4	13, 0
5423·9	8	6, 1						

RbHe

Reference. G. T. Lalos and G. L. Hammond, *Astrophys. J.*, **135**, 616 (1962)†.

A diffuse band at 7250 Å accompanies the Rb resonance lines in emission from rapidly compressed rubidium vapour in helium.

RbHg

Diffuse bands, degraded to the violet, observed in absorption by Barratt (see CsCd). Heads λλ6364, 6335 and 4879. Also mention of bands λλ4400–4200.

RbI

Occurrence. In absorption in a furnace.

Reference. R. F. Barrow and A. D. Caunt, *P.R.S.*, **219**, 120 (1953).

A continuum and regularly spaced diffuse bands which have a maximum at about 3240 Å and extend to longer wavelengths with increasing temperature, the longest wavelength recorded being 4318 Å.

RbMg

S. Barratt (*P.R.S.*, **109**, 194 (1925)) observed a violet-degraded absorption band at 4728 Å.

RbZn

Mention of absorption bands λλ4400–4200. See CsCd.

ReO

Reference.   V. Raziunas, G. Macur and S. Katz, *J. Chem. Phys.*, **43**, 1010 (1965).

Red-degraded bands observed in low-current DC arc containing rhenium, at λλ7119·1, 7108·7, 6443·8, 6092·4, 6088·5, 6077·5, 6061·0 and 5763·4.

RhC

Occurrence.   Emission and absorption in a King furnace containing rhodium at high temperature.

Appearance.   Four systems of bands degraded to the red. The $A^2\Pi - X^2\Sigma$ bands are double-headed.

References.   A. Lagerqvist and R. Scullman, *Ark. Fys.*, **32**, 479 (1966)†.
B. Kaving and R. Scullman, *J. Mol. Spec.*, **32**, 475 (1969)†.

R heads of bands likely to be prominent:

$A^2\Pi_{\frac{3}{2}} - X^2\Sigma$

10591·1 (0, 0), 9636·5 (1, 0), 9047 (4, 2), 8947 (3, 1), 8369 (5, 2).

$A^2\Pi_{\frac{1}{2}} - X^2\Sigma$

9789·1 (0, 0), 9076·4 (2, 1), 8976·1 (1, 0), 8387·7 (3, 1), 8294·7 (2, 0).

$B^2\Sigma - X^2\Sigma$

4936·5 (0, 1), 4695·5 (0, 0).

$C^2\Sigma - X^2\Sigma$

4895·8 (0, 1), 4658·7 (0, 0), 4520·7 (2, 1), 4484·9 (1, 0), 4317·7 (2, 0).

RhO

Reference.   V. Raziunas, G. Macur and S. Katz, *J. Chem. Phys.*, **43**, 1010, (1965).

Observed in low-current DC arc containing rhodium; degraded to the red. λλ6592·0, 6563·5, 6373·3, 6268·1, 6228·9, 6173·3, 6165·6 (? 0, 0).

RuC

Occurrence.   Emission from a King furnace containing ruthenium.

Appearance.   Eleven sub-systems, probably involving triplets. Degraded to the red.

References.   R. Scullman and B. Thelin, *Phys. Scripta*, **3**, 19 (1971)†; **5**, 201 (1972)†.

The following are believed to be the (0, 0) bands of the various sub-systems; those marked * appear prominent in the plate:

λλ7909, 7884*, 7754, 7623*, 7514, 7499, 7224*, 6509, 4383, (4337), 4317.

RuO

References.   V. Raziunas, G. Macur and S. Katz, *J. Chem. Phys.*, **43**, 1010 (1965)†.
R. Scullman and B. Thelin, *J. Mol. Spec.*, **56**, 64 (1975)†.

Occurrence.   Arc between ruthenium electrodes in argon + oxygen.

RuO (*contd.*)

Appearance. Degraded to the red. Prominent sequences, with the (0, 0) relatively strong.

λ	v', v''	λ	v', v''	λ	v', v''
6744·5	0, 4	5799·8	0, 1	5525·8	0, 0
6705·3	1, 5	5544·6	2, 2	5324·5	3, 2
6406·3	0, 3	5532·0	1, 1	5301·8	2, 1

Scullman and Thelin also found systems with (0, 0) bands at 5532 and 5544 Å.

S_2

The main $B^3\Sigma - X^3\Sigma$ system, which is very extensive, occurs readily and has sometimes been obtained as impurity in other spectra. Three weak systems in the far red to near infra-red, one in the ultra-violet and others in the far ultra-violet have been reported in recent years.

MAIN SYSTEM, 7110–2400 Å, $B^3\Sigma_u^- - X^3\Sigma_g^-$

Occurrence. In sources containing sulphur vapour, or sulphur compounds, including discharges, arcs and flames. Readily as impurity in flames containing H_2. Also in absorption.

Appearance. Degraded to the red. This is a very extensive system of roughly equally spaced bands. In flames containing H_2, bands of the $v' = 0$ progression are usually outstanding, but in low-pressure flames and atomic flames bands with $v' \geqslant 9$ occur strongly because of a reaction $H + OH + S_2 = H_2O + S_2^*$. See Plate 9.

References. A. Christy and S. M. Naudé, *P.R.*, **37**, 903 (1931).
W. E. Curtis and S. Tolansky, *Durham Phil. Soc.*, 323 (1931)†.
A. Fowler and W. M. Vaidya, *P.R.S.*, **132**, 310 (1931)†.
T. M. Sugden and A. Demerdache, *Nature*, **195**, 596 (1962).

Most of the strong heads of the very extensive main system of bands are listed below. The v', v'' values are from Fowler and Vaidya. Wavelengths greater than 4800 are from Curtis and Tolansky, and other wavelengths are those given by Fowler and Vaidya reduced by 0·2 Å. Intensities are for a CS_2 flame, on a scale of 1 to 6.

λ	I	v', v''	λ	I	v', v''	λ	I	v', v''
6165·8	1	9, 30	5309·2	3	4, 22	4523·3	2	0, 14
6102·3	2	8, 29	5249·7	5	6, 23	4478·6	4	2, 15
6039·1	2	7, 28	5194·2	5	5, 22	4433·4	6	1, 14
5980·8	1	6, 27	5145·4	4	4, 21	4394·8	2	0, 13
5962·2	3	9, 29	5090·2	5	6, 22	4354·8	2	2, 14
5900·7	4	8, 28	5036·2	6	5, 21	4310·8	6	1, 13
5840·6	4	7, 27	4989·5	6	4, 20	4274·2	3	0, 12
5783	2	6, 26	4937	5	3, 19	4193·6	6	1, 12
5769·5	3	9, 28	4893·6	5	2, 18	*4157·0	5	0, 11
5710·1	4	8, 27	4842·1	6	4, 19	4080·8	3	1, 11
5651·6	4	7, 26	4790·6	6	3, 18	*4045·6	6	0, 10
5596·4	3	6, 25	4747·4	5	2, 17	*3938·9	6	0, 9
5530·0	3	8, 26	4698·8	3	1, 16	3909·4	1	2, 10
5472·8	4	7, 25	4651·1	4	3, 17	*3837·1	6	0, 8
5418·8	4	6, 24	4609·8	5	2, 16	*3739·8	6	0, 7
5359·0	3	5, 23	4563·0	4	1, 15	3677·4	3	1, 7

S_2 (contd.)

λ	I	v', v"	λ	I	v', v"	λ	I	v', v"
*3645·0	5	0, 6	3244·5	3	3, 3	3017·8	2	7, 2
3587·2	5	1, 6	3215·9	2	2, 2	2996·8	1	4, 0
3555·6	3	0, 5	3203·0	2	4, 3	*2989·5	4	6, 1
3500·3	5	1, 5	3171·3	2	3, 2	2959·9	2	5, 0
3469·4	2	0, 4	3160·9	1	5, 3	*2954·0	4	7, 1
3450·8	2	2, 5	3143·5	1	2, 1	2926·4	2	6, 0
3416·8	4	1, 4	3132·2	3	4, 2	*2920·2	4	8, 1
3386·8	1	0, 3	3101·3	1	3, 1	2892·3	4	7, 0
3369·4	4	2, 4	3091·5	5	5, 2	2887·9	4	9, 1
3336·5	2	1, 3	3063·4	3	4, 1	*2860·0	3	8, 0
3321·0	1	3, 4	3054·7	3	6, 2	*2829·1	3	9, 0
3290·5	3	2, 3	3032·9	1	3, 0	2798·8	3	10, 0
3259·7	2	1, 2	3024·6	4	5, 1	2769·4	3	11, 0

* Fainter head to violet.

INFRA-RED AND FAR RED SYSTEMS

Occurrence. In discharge tubes, usually with helium carrier or microwave excitation.

Appearance. Three weak overlapping systems, all shaded to the violet.

References. J. E. Meakin and R. F. Barrow, *Canad. J. Phys.*, **40**, 377 (1962).

N. A. Narasimham, *Curr. Sci. (India)*, **33**, 261 (1964)†.

N. A. Narasimham and K. V. S. R. Apparao, *Nature*, **210**, 1034 (1966)†.

R. F. Barrow and R. P. du Parcq, *J. Phys.*, B, **1**, 283 (1968).

$B'^3\Pi_g - A^3\Sigma_u^+$, 8083–7434 Å

There appear, in each case, to be sub-bands due to transitions $B'^3\Pi_2$ and $^3\Pi_1$. Heads λλ8083 (0, 2 i), 7996 (0, 2 ii), 7785·6 (0, 1 i), 7707·4 (0, 1 ii), 7506·8 (0, 0 i), 7433·6 (0, 0 ii).

$B'^3\Pi_g - A'^3\Delta_u$, 7761–6984 Å

Sub-bands from $^3\Pi_2$ and $^3\Pi_1$. Heads λλ7760·6 (0, 3 i), 7584 (0, 2 ii), 7485 (0, 2 i), 7319·5 (0, 1 ii), 7227·8 (0, 1 i), 7096·6 (0, 0 ii), 6983·8 (0, 0 i).

$e^1\Pi_g - c^1\Pi_u$, 7430–7152 Å

Heads 7430 (0, 0), 7152 (1, 0).

ULTRA-VIOLET SYSTEM, $f^1\Delta_u - a^1\Delta_g$

Occurrence. In emission from microwave discharge and in absorption following flash photolysis or flash discharges.

Appearance. Degraded to the red. 29 bands have been recorded 3347–2424 Å.

References. N. A. Narasimham and J. K. Reddy, *Proc. Indian Acad. Sci.*, A, **59**, 345 (1964)†.

R. F. Barrow and R. P. du Parcq, *J. Phys.*, B, **1**, 283 (1968).

M. Carleer and R. Colin, *J. Phys.*, B, **3**, 1715 (1970).

The following are probably strong:

S$_2$ (*contd.*)

λ	v', v''	λ	v', v''	λ	v', v''
3188·9	0, 8	2813·2**	2, 3	2668·0*	5, 2
3123·9	0, 7	2760·1*	2, 2	2619·8*	5, 1
3060·8	0, 6	2728·3*	3, 2	2495·8	8, 0
2847·5**	1, 3	2697·6	4, 2	2471·8	9, 0

* Relatively strong in emission.   ** The two strongest emission bands.

There are three other systems; $g\,^1\Delta_u - a\,^1\Delta_g$, $h\,^1\Sigma^+ - b\,^1\Sigma^+$ and $i - b\,^1\Sigma^+$ lie mainly just below the quartz limit, with weak bands reaching 2100 Å, however.

S$_3$

Reference.   B. Meyer, T. Stroyer-Hansen and T. V. Oommen, *J. Mol. Spec.*, **42**, 335 (1972).

Absorption by sulphur vapour, best at 500° C and 50 torr, shows 34 red-degraded bands 4400–3600, maximum intensity around 3950 Å.

S$_4$

A weak absorption continuum at 5300 Å, best at 400° C and 10 torr, is attributed to S$_4$. See Meyer *et al.* as S$_3$, above.

SF

4000–3300 Å SYSTEM, $A\,^2\Pi - X\,^2\Pi$

Occurrence.   Absorption following flash photolysis and in the reaction between F atoms and COS.

Appearance.   Degraded to the red. The strong bands form a $v'' = 0$ progression.

Reference.   G. DiLonardo and A. Trombetti, *Trans. Faraday Soc.*, **66**, 2694 (1970)†.

About 40 heads are recorded. Strongest $^2\Pi_{\frac{3}{2}} - {}^2\Pi_{\frac{3}{2}}$:

λ	I	v', v''	λ	I	v', v''
3783·8	6	3, 0	3597·0	10	6, 0
3718·6	8	4, 0	3540·5	8	7, 0
3654·4	10	5, 0	3487·1	4	8, 0

The strongest $^2\Pi_{\frac{1}{2}} - {}^2\Pi_{\frac{1}{2}}$ head is 3571·0 (6, 0).

SH

References.   M. N. Lewis and J. V. White, *P.R.*, **55**, 894 (1939).
J. W. C. Johns and D. A. Ramsay, *Canad. J. Phys.*, **39**, 210 (1961)†.

3237 Å SYSTEM, $A\,^2\Sigma^+ - X\,^2\Pi$

Analogous to the OH bands and degraded to the red. Obtained in absorption by passing repeated flashes from a source of continuum through a discharge tube in which HS radicals were formed by pulses of radio-frequency current synchronised to precede the flashes by a very short interval. The band does not appear at all readily in emission, but has been observed by Gaydon in flames of hydrocarbons containing SO_2 or H_2S. Also in flash photolysis.

SH (*contd.*)

Heads λ (*I*)

v', v''	R_1	Q_1	Q_2
0, 0	3236·6 (8)	3240·7 (7)	3279·1 (4)
1, 0	3060·6		3098·3
2, 0	2918·9		2953·1

SH⁺

Occurrence. Electron beam excitation.

Reference. M. Horani, S. Leach and J. Rostas, *J. Mol. Spec.*, **23**, 115 (1967).

The (0, 0) band is in the region 3470–3310 Å. It is due to a $^3\Pi - {}^3\Sigma^-$ transition and is degraded to the red with sub-origins at 3369, 3342 and 3316 Å. Details are not readily available.

SH₂, SH₂⁺, S₂H see H₂S etc.

SO

The main system lies in the near ultra-violet. Weaker systems have recently been reported in the near infra-red and further down in the ultra-violet.

MAIN SYSTEM $B\,{}^3\Sigma^- - X\,{}^3\Sigma^-$

Occurrence. In almost all sources containing sulphur and oxygen. Strong in discharges through SO_2 and in flames containing sulphur compounds. Also in absorption following flash photolysis and flash discharges through SO_2.

Appearance. Degraded to the red. Single-headed bands with rather extended rotational structure. See Plate 10.

References. V. Henri and F. Wolff, *J. Phys. Radium*, **10**, 81 (1929)†.

E. V. Martin, *P.R.*, **41**, 167 (1932)†.

R. G. W. Norrish and G. A. Oldershaw, *P.R.S.*, **249**, 498 (1958)†.

R. Colin, *Canad. J. Phys.*, **47**, 979 (1969)†.

D. Abadie and L. Herman, *J. Quant. Spectrosc. Rad. Transfer*, **4**, 195 (1964)†.

Measurements of strong bands by Martin, weaker bands by Henri and Wolff. Values of v'' have been adjusted. Additional weak bands have been listed by Abadie and Herman.

λ	I	v', v''	λ	I	v', v''	λ	I	v', v''
3941·6	0	1, 16	3271·0	10	0, 10	2699·1	4	2, 5
3903·6	1	0, 15	3247·5	1	2, 11	2664·8	5	1, 4
3862·7	3	2, 16	3164·8	10	0, 9	2655·6	1	3, 5
3811·8	5	1, 15	3064·1	10	0, 8	2630·1	1	0, 3
3761·6	2	0, 14	3007·9	2	1, 8	2622·2	3	2, 4
3724·7	3	2, 15	2968·5	5	0, 7	2589·0	2	1, 3
3676·2	7	1, 14	2915·4	4	1, 7	2581·1	1	3, 4
3628·2	3	0, 13	2877·7	4	0, 6	2555·5	0	0, 2
3548·7	6	1, 13	2827·4	8	1, 6	2548·6	2	2, 3
3502·1	4	0, 12	2791·3	2	0, 5	2516·4	1	1, 2
3428·1	3	1, 12	2779·8	1	2, 6	2510·4	1	3, 3
3383·1	6	0, 11	2744·0	6	1, 5	2477·7	1	2, 2
3314·8	1	1, 11	2708·6	1	0, 4	2442·0	1	3, 2

SO (*contd.*)

INFRA-RED SYSTEM, $b\,^1\Sigma^+ - X\,^3\Sigma^-$

Occurrence. In flame of $H_2S + O_2$ and afterglow of discharge through $COS + O_2$.

Appearance. Degraded to longer wavelengths. Strong ^{S}R heads and weaker Q heads.

References. A. M. Bouchoux, J. Marchand and J. Janin, *Spectrochim. Acta*, A, **27**, 1909 (1971)†; A, **28**, 1771 (1972).

R. Colin, *Canad. J. Phys.*, **46**, 1539 (1968).

The following are probably strong heads; the ^{S}R heads of the (1, 0) and (2, 0) sequences are most prominent:

λ	v', v''		λ	v', v''		λ	v', v''	
10708	0,	1 Q	8689·1	2,	1 ^{S}R	7987·5	3,	1 ^{S}R
9626·1	1,	1 Q	9675·5	1,	0 Q	7958	2,	0 Q
9548·1	0,	0 Q	8613·3	1,	0 ^{S}R	7915·2	2,	0 ^{S}R
8750	2,	1 Q	8030	3,	1 Q			

ULTRA-VIOLET SYSTEM, 2500 Å, $A\,^3\Pi - X\,^3\Sigma^-$

Occurrence. In absorption following flash photolysis, and in emission in a discharge through flowing SO_2 (B. N. Bhaduri, *Thesis, London*).

Appearance. Degraded to the red. A strong $v'' = 0$ progression of bands with fairly widely spaced triple heads, each head being closely double. See Plate 10.

Reference. R. Colin, *Canad. J. Phys.*, **47**, 979 (1969)†.

13 bands are known. The following are the outstanding heads, with our estimates of intensities in absorption from Colin's plate:

v', v''	R_{11}	I	R_{21}	I	R_{32}	I
0, 0	2634·7	3	2623·3	5	2613·4	4
1, 0	2605·5	7	2595·1	8	2585·4	8
2, 0	2578·1	9	2568·0	10	2558·5	9
3, 0	2551·4	8	2541·7	9	2532·5	8
4, 0	2525·5	3	2516·1	5	2507·2	5

SO_2

Sulphur dioxide possesses a very full spectrum both in absorption and emission. The main system $^1B_1 - {}^1A_1$ gives strong absorption in the near ultra-violet and also occurs readily in emission. There is a weaker $^3B_1 - {}^1A_1$ giving absorption 3880–3400 and emission in the afterglow 4500–3880 Å. Bands further in the ultra-violet, below 2350 Å, are attributed to another system $^1B_2 - {}^1A_1$.

MAIN SYSTEM, $^1B_1 - {}^1A_1$

References. T. Chow, *P.R.*, **44**, 638 (1933).

J. H. Clements, *P.R.*, **47**, 224 (1935).

W. Lotmar, *Z.P.*, **83**, 765 (1933).

J. C. D. Brand and R. Nanes, *J. Mol. Spec.*, **46**, 194 (1973).

Y. Hamada and A. J. Merer, *Canad. J. Phys.*, **52**, 1443 (1974)†.

Occurrence. Strongly in absorption, and in emission from discharge through flowing SO_2 and in fluorescence.

Appearance. Degraded to the red. In absorption the system extends from 3400–2500 Å with fairly closely spaced bands 3200–2750. Below 2750 bands become diffuse. See Plate 10. The system

SO$_2$ (*contd.*)

origin is probably at 3540 Å and in emission there is a very complex mass of red-degraded bands strongest 4000–3000 Å.

The following are maxima of the strongest absorption bands; these maxima are usually within 1 Å of the heads. Intensities are based on quantitative measurements by Clements:

λ	I	λ	I	λ	I	λ	I
3190·9	1	3065·9	5	2906·5	8	2780·0	6
3181·1	1	3043·3	7	2900·9	5	2772·0	4
3173·0	1	3022·1	9	2887·7	9	2765·2	4
3167·0	1	3001·0	10	2868·9	8	2754·6	4
3159·0	2	2980·0	9	2852·0	9	2751·2	4
3151·8	2	2961·2	10	2832·3	8	2738·1	4
3131·3	3	2943·8	9	2818·1	7	2734·6	5
3129·5	?	2937·7	8	2815·5	7	2727·5	3
3108·4	3	2924·8	8	2797·0	8	2685·0	4
3087·7	4	2923·1	8	2789·4	7	2646·6	3

In emission the following heads may be strong, λλ3691·8, 3668·7, 3557·7, 3359·2, 3343·4, 3228·7, 3158·1, 3122·0.

Lotmar has studied the fluorescence.

3880 Å SYSTEM, $^3B_1 - {}^1A_1$

References. A. G. Gaydon, *P.R.S.*, **146**, 901 (1934)†.
A. J. Merer, *Disc. Faraday Soc.*, **35**, 127 (1963)†.
J. C. D. Brand, V. T. Jones and C. Di Lauro, *J. Mol. Spec.*, **45**, 404 (1973).

Occurrence. These bands are the main feature of the afterglow of SO$_2$ in a discharge (Gaydon). In absorption they occur weakly.

Appearance. Narrow bands. Under large dispersion the K structure shows red-degraded heads.

In emission (SO$_2$ afterglow) the strongest bands are at λλ4461·0, 4361·1, 4265·3, 4244·6, 4152·9, 4066·5, 4048·3, 3963·7 and 3883·0 (0 0 0, 0 0 0).

In absorption the strongest bands are λλ3880 (0 0 0, 0 0 0), 3826 (0 1 0, 0 0 0), 3748 (1 0 0, 0 0 0) and 3699 (1 1 0, 0 0 0).

ULTRA-VIOLET SYSTEM, $^1B_2 - {}^1A_1$

References. B. N. Bhaduri, *Ph. D. Thesis, London* (1935).
W. C. Price and D. M. Simpson, *P.R.S.*, **165**, 272 (1938)†.
T. Chow, *P.R.*, **44**, 638 (1933).
J. C. D. Brand and K. Srikameswaram, *Chem. Phys. Lett.*, **15**, 130 (1972).

This system gives strong absorption bands in the far ultra-violet. The origin is probably at 2350 Å. Strongest red-degraded heads from Bhaduri λλ2348·0, 2326·5, 2323·9, 2303·5, 2295·5, 2276·2, 2257·5, 2241·5, 2238·6, 2222·9, 2204·5, 2186·6, 2168·3.

In emission, bands at similar wavelengths have been recorded and Chow lists about 150 heads. Emission bands 2640–2490 Å are now known to be due to the $A^3\Pi - X^3\Sigma$ system of SO.

S$_2$O

Absorption bands found in the products of a discharge through SO$_2$ were initially attributed to a monoxide S$_2$O$_2$ but are now, on microwave evidence, attributed to S$_2$O (Meschi and Myers).

S₂O (*contd.*)

Jones reports 11 red-degraded bands 3352–3193 Å of which the strongest are 3351·7, 3321·3, 3307·0, 3234·4 (probably the strongest), 3222·5, 3206·8 and 3193·1.

References. A. V. Jones, *J. Chem. Phys.*, **18**, 1263 (1950)†.
D. J. Meschi and R. J. Meyers, *J. Mol. Spec.*, **3**, 405 (1959).

SO₃

Reference. E. Fajans and C. F. Goodeve, *Trans. Faraday Soc.*, **32**, 511 (1936).

Absorption by sulphur trioxide vapour is usually masked by the much stronger dioxide bands, but with careful purification and avoidance of photochemical decomposition, weak bands were reported at λλ2483, 2455, 2432, 2405, 2382, 2360 and 2337, all superposed on a continuum.

Sb₂

EMISSION BANDS

References. S. Mrozowski and C. Santaram, *J. Opt. Soc. Amer.*, **57**, 522 (1967)†.
J. Sfeila *et al.*, *J. Mol. Spec.*, **42**, 239 (1972).

Mrozowski and Santaram report near infra-red system of 24 bands 8400–7300 Å (strongest 8316 and 7788) and other systems 4200–3600 and 3410–3100 in a discharge tube. Sfeila *et al.* report a single extensive system of 75 red-degraded bands 6700–4725 Å in a high-frequency discharge.

ULTRA-VIOLET ABSORPTION

References. S. M. Naudé, *P.R.*, **45**, 280 (1934).
G. Nakamura and T. Shidei, *Japan. J. Phys.*, **10**, 11 (1935).

There are two systems of red-degraded bands. A third, near 2900 Å, is now attributed to SO₂.

A system 2333–2178 has strongest heads λλ2272·2 (0, 3), 2258·5 (0, 2), 2244·9 (0, 1), 2222·8 (2, 1) the strongest, and 2209·4 (2, 0).

A system 2168–2048 Å has its strongest heads at 2138·6, 2126·8, 2115·0 and 2104·3.

SbBi

A system of bands λλ2528–2399 have been observed in absorption by a mixture of antimony and bismuth vapours. They are degraded to the red.

Reference. G. Nakamura and T. Shidei, *Japan. J. Phys.*, **10**, 11 (1935)†.

The following are the strongest bands:

λ	I	v', v″	λ	I	v', v″
2514·0	2	0, 4	2462·2	2	1, 1
2500·6	4	0, 3	2460·4	3	0, 0
2487·4	4	0, 2	2449·0	2	1, 0
2473·7	4	0, 1			

SbBr

Three systems, in the visible and near ultra-violet, have been reported by Avasthi in a microwave discharge, and three sub-systems in the region 2425–2260 Å occur in flash-photolysis of SbBr₃ (Danon *et al.*).

SbBr (*contd.*)

References. M. N. Avasthi, *Z. Naturforsch.*, **26a**, 250 (1971)†.
N. Danon *et al.*, *C. R. Acad. Sci.* (*Paris*), C, **269**, 1249 (1969).

5340–4905 Å. Degraded to the red. 5069 (0, 0).
3340–3050 Å. 70 red-degraded bands. 3217·6 (3, 8), 3136·4 (0, 2), 3143·9 (2, 4).
3020–2950 Å. 16 red-degraded bands. 3011·1 (1, 3), 3007·5 (0, 2), 2999·4 (4, 5), 2986·1 (0, 1).
2425–2260 Å. Degraded to shorter wavelengths. 2380·3 (0, 1 i), 2365·3 (0, 0 i), 2323·9 (0, 1 ii),
                2310·4 (0, 0 ii), 2260·7 (0, 0 iii).

SbCl

VISIBLE SYSTEM

Occurrence. $SbCl_3$ in active nitrogen.
Appearance. Degraded to the red. Two sub-systems of bands between 4200 and 5600 Å. The strongest bands form two long $v' = 0$ progressions.
Reference. W. F. C. Ferguson and I. Hudes, *P.R.*, **57**, 705 (1940)†.
The following are the strongest heads:

λ	I	v', v''	λ	I	v', v''	λ	I	v', v''
5333·8	5	0, 10	4902·0	4	1, 6	4575·0	4	0, 11
5236·0	4	0, 9	4872·9	3	0, 5	4503·6	4	0, 10
5141·0	4	0, 8	4816·1	3	1, 5	4434·0	4	0, 9
5048·6	4	0, 7	4680·8	3	2, 4	4365·0	4	0, 8
4959·7	4	0, 6				4298·4	4	0, 7
						4233·0	3	0, 6

ULTRA-VIOLET SYSTEMS, 2530–2010 Å

Occurrence. Absorption following flash photolysis of $SbCl_5$.
Appearance. Degraded to shorter wavelengths. Three sub-systems, perhaps due to a $^3\Pi - {}^3\Sigma$ transition.
Reference. N. Danon, A. Chatalic and G. Pannetier, *C. R. Acad. Sci.* (*Paris*), C, **272**, 1411 (1971).
Strongest heads λλ2339·2 (0, 1 i), 2319·3 (0, 0 i), 2296·0 (1, 0 i), 2188·2 (1, 0 ii), 2109·3 (0, 0 iii).

SbF

Occurrence. In discharges of various types, including high-frequency, and also in active nitrogen. nitrogen.
References. G. D. Rochester, *P.R.*, **51**, 486 (1937)†.
T. A. P. Rao and P. T. Rao, *Indian J. Phys.*, **36**, 85 (1962)†.
M. M. Patel and K. C. Abraham, *Indian J. Pure Appl. Phys.*, **7**, 641 (1969)†.
K. C. Abraham and M. M. Patel, *J. Phys.*, B, **3**, 882 (1970)†; B, **4**, 1398 (1971)†.
R. S. Shanker, S. C. Srivastava and I. S. Singh, *Indian J. Pure Appl. Phys.*, **10**, 395, 541 (1972)†.
M. Chakravorty, K. C. Abraham and M. M. Patel, *J. Phys.*, B, **6**, 757 (1973)†.
K. K. W. Wang *et al.*, *J. Mol. Spec.*, **49**, 377 (1974).

SbF (*contd.*)

There appear to be five overlapping red-degraded systems in the visible and other violet-degraded systems, one in the near infra-red, one around 3200 Å and several more below 2750 Å. Chakravorty *et al.* summarize probable electronic transitions.

Infra-red.	7776·3 (0, 0).
A″, 5770–4730 Å.	17 bands. 5343·9 (0, 2), 5179·3 (0, 1), 5021·4 (0, 0), 4919·3 (1, 0).
A′, 5450–4050 Å.	23 bands. 4444·4 (0, 0), 4364·2 (1, 0), 4287·7 (2, 0), 4214·8 (3, 0).
$A_1, A_2, A_3.$	Strongest 4406·6, 4292·6, 4183·6, 4111·9, 4043·2.
B, 3274–3140 Å.	3274·0 (0, 1), 3209·8 (0, 0), 3139·8 (1, 0).
$C_1, C_2, C_3, 2750–2050$ Å.	2632·3 (0, 0 C_1), 2295·1 (0, 0 C_2), 2262·0 (0, 1 C_3), 2259·0 (1, 0 C_2), 2231·6 (0, 0 C_3).

SbH

Reference. P. Bollmark and B. Lindgren, *Chem. Phys. Lett.*, **1**, 480 (1967).

Complex structure 3450–3300 Å observed in flash photolysis of SbH_3, perhaps due to a $^3\Pi - {}^3\Sigma$ transition. Strongest band 3392 Å, degraded to the red. Weaker bands 3433 and 3318 Å.

SbI

Reference. N. Danon *et al.*, *C. R. Acad. Sci. (Paris)*, C, **269**, 1249 (1969).

Bands 2510–2350 Å observed in flash photolysis of SbI_3. Three overlapping systems of bands degraded to shorter wavelengths. Prominent heads 2457·7 (0, 1 A), 2446·1 (0, 0 A) the strongest, 2431·6 (1, 0 A), 2414·1 (0, 0 B), 2399·8 (1, 0 B), 2383·5 (0, 0 C).

SbN

Occurrence. Discharge through mixture of antimony vapour and N_2, and in flash photolysis.

Appearance. Degraded to red. Close double-headed bands forming fairly well-marked sequences.

References. N. H. Coy and H. Sponer, *P.R.*, **58**, 709 (1940)†.

S. Ciach and P. J. Thistlewaite, *J. Chem. Phys.*, **53**, 5381 (1970).

The following are the R heads. Intensities are our estimates from published photograph and description. Strong bands only.

λ	I	v', v''	λ	I	v', v''
3166·1	5	1, 4	2995·3	7	1, 2
3062·3	5	8, 9	2985·7	9	0, 1
3052·8	5	7, 8	2924·1	4	2, 2
3015·5	5	3, 4	2912·0	7	1, 1
3004·5	7	2, 3	2904·3	10	0, 0

SbO

Six or seven systems are known, all involving the ground state.

Occurrence. Metallic antimony in a carbon arc in air and in discharge tubes containing heated antimony oxide. Systems C, D, E and F have been found in absorption as well.

SbO (*contd.*)

References. B. C. Mukherji, *Z.P.*, **70**, 552 (1931)†.

A. K. SenGupta, *Indian J. Phys.*, **13**, 145 (1939); **17**, 216 (1943).

S. V. J. Lakshman, *Z.P.*, **158**, 367 and 386 (1960)†.

M. Shimauchi, *Science of Light*, **9**, 109 (1960)†.

B. Rai, K. N. Upadhya and D. K. Rai, *J. Phys.*, B, **3**, 1374 (1970).

The following classification is mainly due to Lakshman:

SYSTEM A, 6800–4500 Å, $A^2\Pi - X^2\Pi$

Appearance. Single-headed bands degraded to the red, falling into two sub-systems.

$^2\Pi_{\frac{3}{2}} - {}^2\Pi_{\frac{3}{2}}$		$^2\Pi_{\frac{1}{2}} - {}^2\Pi_{\frac{1}{2}}$			
λ	v', v''	λ	v', v''	λ	v', v''
6559·1	0, 4	5750·7	0, 4	5065·0	0, 1
6540·2	0, 3	5505·6	0, 3	4926·2	1, 1
5949·4	0, 2	5341·2	1, 3	4795·8	2, 1
5757·9	1, 2	5277·7	0, 2	4675·0	3, 1
		5189·9	2, 3	4617·2	2, 0
		5126·3	1, 2	4504·7	3, 0

SYSTEM B, 4500–3400 Å, $B^2\Sigma - X^2\Pi$

Degraded to the red. Two sub-systems.

$^2\Sigma - {}^2\Pi_{\frac{3}{2}}$		$^2\Sigma - {}^2\Pi_{\frac{1}{2}}$			
λ	v', v''	λ	v', v''	λ	v', v''
4272·7	0, 1	3894·8	0, 1	3656·8	3, 1
4130·2	0, 0	3774·0	0, 0	3621·7	2, 0
4075·0	2, 1	3695·0	1, 0	3460·6	6, 1
3987·9	3, 1	3672·2	9, 5	3402·4	7, 1

SYSTEM C, 3600–2800 Å, $C^2\Delta - X^2\Pi$

About 80 bands observed. Degraded to the red.

$^2\Delta_{\frac{3}{2}} - {}^2\Pi_{\frac{1}{2}}$			$^2\Delta_{\frac{5}{2}} - {}^2\Pi_{\frac{3}{2}}$		
λ	I	v', v''	λ	I	v', v''
3568·3	5	0, 2	3687·3	5	0, 1
3469·1	5	0, 1	3580·7	5	0, 0
3374·8	6	0, 0	3510·6	4	1, 0
3311·4	4	1, 0	3443·7	3	2, 0
3251·1	4	2, 0	3260·4	4	5, 0
3194·1	4	3, 0	3205·1	3	6, 0

SYSTEM D, 3077–2960 Å, $D^2\Pi_{\frac{1}{2}} - X^2\Pi_{\frac{1}{2}}$

A weak system of six red-degraded bands.

λ	I	v', v''	λ	I	v', v''
3077·0	5	1, 3	2960·0	4	2, 2
3030·9	3	2, 3	2933·2	4	1, 1
3003·9	5	1, 2	2891·2	4	2, 2

SbO (*contd.*)

SYSTEM E, 2910–2450 Å, $E^2\Sigma - X^2\Pi$

Degraded to shorter wavelengths. P heads from Sen Gupta.

$^2\Sigma - ^2\Pi_{\frac{3}{2}}$		$^2\Sigma - ^2\Pi_{\frac{1}{2}}$	
λ	v', v''	λ	v', v''
2724·0	0, 1	2565·2	0, 1
2665·9	0, 0	2513·2	0, 0
2661·0	1, 1	2509·5	1, 1
2605·4	1, 0	2459·6	1, 0

SYSTEM F, $F(?\,^2\Delta) - X^2\Pi$

Shimauchi found red-degraded (0, 0) sub-bands at 2733·4 and 2573·6 Å.

He also found another system of diffuse bands $G - X^2\Pi$ in this region.

SbP

References. K. K. Yee and W. E. Jones, *J. Mol. Spec.*, **33**, 119 (1970).
W. E. Jones, C. G. Flinn and K. K. Yee, *J. Mol. Spec.*, **52**, 344 (1974)†.

Red-degraded $^1\Pi - ^1\Sigma$ bands 3970–3460 Å obtained in a microwave discharge through flowing $PCl_3 + SbCl_5$. Strongest:

λ	I	v', v''	λ	I	v', v''
3689·7	8	0, 2	3510·4	8	1, 0
3623·6	10	0, 1	3462·2	8	2, 0
3559·6	8	0, 0			

SbS

Occurrence. In hollow-cathode discharge and in absorption using Sb_2S_3.

Appearance. Seven systems totalling 100 bands, mostly shaded to the red and involving transitions to a $^2\Pi$ ground state with large (2000 cm.$^{-1}$) spin splitting.

Reference. M. Shimauchi and Y. Nishiyama, *Science of Light*, **17**, 76 (1968)†.

Strongest bands of the various systems:

System A, *5427–4420* Å. (0, 0) heads 5134·0, 4487·3.

System B$_2$, *5057–4439* Å. 4732·0 (0, 3), 4631·5 (0, 2).

System C, *4574–3845* Å. 4068·2 (0, 3), 3991·8 (0, 2).

System D$_2$, *4637–4047* Å. 4450·4 (0, 4), 4362·1 (0, 3), 4276·4 (0, 2).

System E, *3953–3680* Å. 3881·5, 3878·4, 3812·4, 3808·7, 3745·4; the analysis is complicated
      due to overlapping.

System F, *3914–3317* Å. 3777·0 (0, 2 ii), 3670·2 (0, 3 i), 3607·7 (0, 2 i), 3546·8 (0, 1 i).

System G, *4294–3648* Å. 4293·9, 4215·1, 4136·8, 3775·6, 3711·4, 3647·8.

SbSe

Four band systems have been reported in the furnace absorption spectrum of a mixture of antimony and selenium vapours.

Reference. C. B. Sharma, *Proc. Phys. Soc.*, A, **66**, 1109 (1953).

SbSe (*contd.*)

SYSTEM I, 3685–3289 Å

Bands degraded to the red. The strongest of the 32 bands given are as follows:

λ	I	v′, v″	λ	I	v′, v″
3659·3	9	0, 5	3496·7	8	0, 1
3644·3	10	2, 6	3483·5	8	2, 2
3630·0	10	1, 5	3470·1	8	1, 1
3588·3	10	1, 4	3457·6	10	0, 0
3577·7	6	0, 3	3431·6	10	1, 0
3537·2	8	0, 2	3406·3	8	2, 0

SYSTEM II, 2827–2620 Å

Bands degraded to the violet. Of 30 bands the most intense heads are as follows:

λ	I	v′, v″	λ	I	v′, v″
2770·3	10	0, 0	2732·0	6	2, 1
2763·0	4	1, 1	2708·2	5	2, 0
2739·0	7	1, 0	2704·0	4	6, 5

SYSTEM III, λλ2450–2351

Bands degraded to the violet.

λ	I	λ	I	λ	I
2455·8	5	2396·0	10	2363·6	1
2436·4	8	2377·5	3	2361·3	4
2417·4	9	2374·6	4	2351·5	2

SYSTEM IV, λλ2335–2222

Bands degraded to the violet. Strongest heads of the 27 given are as follows:

λ	I	v′, v″	λ	I	v′, v″	λ	I	v′, v″
2300·8	7	0, 1	2264·8	10	1, 0	2246·4	8	2, 0
2283·7	8	0, 0	2262·8	6	2, 1	2240·4	4	5, 3
2281·5	7	1, 1	2261·0	6	3, 2	2222·6	6	6, 3
2279·3	7	2, 2						

SbTe

A band system, degraded to the violet, has been reported in the absorption spectrum from a furnace containing antimony and tellurium.

Reference.   C. B. Sharma, *Proc. Phys. Soc.*, A, **66**, 1109 (1953).

In the region λλ2384–2260, 31 bands have been obtained. The strongest are as follows:

λ	I	v′, v″	λ	I	v′, v″	λ	I	v′, v″
2368·2	7	2, 7	2309·6	10	0, 1	2294·6	8	0, 0
2354·1	7	1, 5	2307·9	10	1, 2	2293·0	8	1, 1
2340·3	6	0, 3	2306·4	8	2, 3	2278·2	8	1, 0
2324·8	10	0, 2	2304·7	8	3, 4	2276·8	8	2, 1

ScCl

Occurrence. Schuler-type discharge tube.

Reference. E. A. Shenyavskaya *et al.*, *Opt. Spectrosc.*, **26**, 509 (1969).

108 heads of nine systems are listed. No photographs or intensities are available, but it is likely that most systems show prominent sequences, and main sequence heads are given here.

I. $A^1\Sigma - X^1\Sigma$.	Degraded to longer wavelengths. 8685·5 (0, 2), 8365·6 (0, 1), 8066·0 (0, 0).
II. $b^3\Phi - a^3\Delta$.	Degraded to longer wavelengths; triple-headed. 8194·0 (0, 1), 7936·7 (0, 0).
III. $c^3\Delta - a^3\Delta$.	Degraded to the red. Close triple-headed. 7871·8 (0, 1), 7633·7 (0, 0), 7434·8 (1, 0), 7247·3 (2, 0).
IV. $B^1\Pi - X^1\Sigma$.	Degraded to the red. 5835·4 (0, 1), 5687·8 (0, 0), 5570·6 (1, 0), 5459·5 (2, 0).
V. $D^1\Pi - X^1\Sigma$.	Degraded to the red. 4751·6 (0, 1), 4653·4 (0, 0), 4574·6 (1, 0), 4499·1 (2, 0).
VI. $e^3\Pi - d^3\Sigma$.	Degraded to the violet. Multi-headed. 4495·0 (0, 0 N_1), 4470·7 (0, 0 P_2), 4467·7 (1, 1 P_2), 4428·2 (1, 0 P_2).
VII. $E^1\Pi - X^1\Sigma$.	Degraded to the red. 3696·3 (0, 0), 3692·6 (1, 1), 3633·2 (1, 0).
VIII. 3680–3600 Å.	Degraded to shorter wavelengths. 3675·5 (0, 0 P).
IX. $F^1\Pi - X^1\Sigma$.	Degraded to the red. 3296·8 (0, 2), 3249·6 (0, 1), 3202·3 (0, 0).

ScF

Occurrence. Twelve systems are reported, mostly in absorption, but some (e.g. systems V and VI) occur in a hollow-cathode discharge or thermal emission.

References. L. V. Gurvich and E. A. Shenyavskaya, *Opt. Spectrosc.*, **14**, 161 (1963).

R. F. Barrow *et al.*, *Nature*, **215**, 1072 (1967); *Proc. Phys. Soc.*, **83**, 889 (1964); **84**, 615 (1964).

The following data, mostly R heads of sequences, are also based on Barrow's summary in Rosen's *Spectroscopic Data Related to Diatomic Molecules.*

I. $B^1\Pi - X^1\Sigma$. 9500–8400 Å.	Degraded to longer wavelengths. 9374·4 (0, 0), 8889·8 (1, 0).
II. $C^1\Sigma^+ - X^1\Sigma$.	Degraded to the red. 6506·9 (0, 1), 6211·5 (0, 0), 5792·8 (2, 0).
III. $E^1\Pi - X^1\Sigma$. 5300–4600.	5099·7 (0, 1), 4917·2 (0, 0), 4773·0 (1, 0). Deg. R.
IV. $F^1\Pi - X^1\Sigma$. 3860–3650.	Deg. R. 3832·6 (0, 1), 3728·5 (0, 0), 3651·6 (1, 0).
V. $G^1\Pi - X^1\Sigma$. 2960–2800.	Deg. R. 2922·3 (0, 1), 2861·5 (0, 0), 2815·5 (1, 0).
VI. $H^1\Pi - X^1\Sigma$. 2640–2575.	Deg. R. 2625·0 (0, 1), 2575·8 (0, 0).
VII. $^3\Phi - a^3\Delta$. 7100–6300.	Deg. R. Triple-headed. 6808·2 (0, 1 first head), 6557·7, 6539·9, 6523·2 (0, 0).
VIII. 5450 Å.	Probably $^3\Pi - a^3\Delta$. Deg. R. Q heads 5452·2, 5444·7.
IX. 4560 Å.	Probably $^3\Delta - a^3\Delta$. Deg. V. 4559·2 (0, 0), 4543·8 (1, 1).
X. $^3\Phi - a^3\Delta$.	Weak triple-headed bands, deg. R., 3582·9 3579·5 3574·4 (0, 0).
XI. 3800 Å.	Deg. R. Complex band.
XII. 2780 Å.	Deg. V. P head 2781·7, Q 2781·4.

ScH

Reference. R. E. Smith, *P.R.S.*, **332**, 113 (1973)†.

In absorption in shock-heated scandium powder in H_2 + Ar there is a complex headless band 5650–5450 Å, maximum 5570 Å.

ScO

A strong band system in the orange and a weaker system in the blue-green have been attributed to scandium oxide.

Occurrence. Scandium salts in an arc.

References. W. F. Meggers and J. A. Wheeler, *J. Res. Nat. Bur. Stand.*, **6**, 239 (1931)†.

A. Adams, W. Klemperer and T. M. Dunn, *Canad. J. Phys.*, **46**, 2213 (1968)†.

R. Stringat, C. Athénour and J. L. Féménias, *Canad. J. Phys.*, **50**, 395 (1972)†.

ORANGE SYSTEM, 7300–5740 Å, $A^2\Pi - X^2\Sigma$

Appearance. Degraded to longer wavelengths. Rather widely-spaced double double-headed bands. Long sequences.

The following are the strong heads at the beginning of the three principal sequences. Intensities are based on Meggers and Wheeler's, but reduced to a scale of 10.

λ	I	v', v''	λ	I	v', v''	λ	I	v', v''
6525·6	2	3, 4 R_2	6079·1	8	0, 0 Q_1	5849·1	3	3, 2 Q_1
6495·9	2	1, 2 Q_1	6072·5	8	1, 1 R_2	5847·7	3	4, 3 R_2
6446·2	5	1, 2 R_2	6064·2	7	0, 0 R_1	5811·6	2	2, 1 Q_1
6437·1	1	0, 1 R_1	6036·1	10	0, 0 R_2	5809·8	3	3, 2 R_2
6116·0	6	1, 1 Q_1	6017·0	6	0, 0 $^SR_{12}$	5772·7	2	2, 1 R_2
6109·8	6	2, 2 R_2	5928·1	2	5, 4 Q_1	5764·4	1	1, 0 R_1
6101·6	5	1, 1 R_1	5887·4	3	4, 3 Q_1	5736·8	1	1, 0 R_2

BLUE-GREEN SYSTEM, 5330–4500 Å, $B^2\Sigma - X^2\Sigma$

Appearance. Close double-headed bands (separation 0·3 Å) degraded to the red. Close doubling due to hyperfine structure.

The following are the first (R) heads of the strong bands:

λ	I	v', v''	λ	I	v', v''
5096·7	4	0, 1	4571·8	2	4, 2
4857·8	5	0, 0	4536·6	1	3, 1
4707·0	2	2, 1	4502·8	1	2, 0
4672·6	2	1, 0			

Se$_2$

Selenium possess a very extensive band system 6700–3250 Å. Other systems in the red have been reported but Barrow believes these are part of the same main system.

MAIN SYSTEM, 6700–3250 Å, $B^3\Sigma_u^- - X^3\Sigma_g^-$

Occurrence. Readily in absorption, fluorescence and emission in discharges, including high-frequency and microwave. Also in flames by Miyanisi.

Appearance. A very extensive system of red-degraded bands; Barrow *et al.* list about 120 heads. In absorption the $v'' = 0$ progression in the near ultra-violet is readily observed, but in emission, bands at longer wavelengths are stronger and there is much overlapping and complexity due to isotope splitting and perturbations. The rotational structure shows apparently single P and R branches like a $^1\Sigma - {}^1\Sigma$ transition, but Barrow *et al.* have shown that the coupling is between Hund's cases *b* and *c*, so that there are really two overlapping sub-systems, a strong B $0_u^+ - X0_g^+$ and a weaker B $1_u - X1_g$.

Se$_2$ (*contd.*)

References. B. Rosen, *Physica*, **6**, 205 (1939).

T. E. Nevin, *Phil. Mag.*, **20**, 347 (1935).

R. K. Asundi and Y. P. Parti, *Proc. Indian Acad. Sci.*, A, **6**, 207 (1937)†.

M. Miyanisi, *Inst. Phys. Chem. Res. Tokyo, Res. Pap.*, 955 (1940).

V. Leelavathi and P. T. Rao, *Indian J. Phys.*, **29**, 1 (1955)†.

R. F. Barrow, G. G. Chandler and C. B. Meyer, *Phil. Trans. Roy. Soc.*, A, **260**, 395 (1966)†.

In absorption the following bands, from Nevin with corrected v', are strong:

λ	I	v', v''	λ	I	v', v''	λ	I	v', v''
3545·4	7	15, 3	3432·0	9	14, 0	3337·1	9	18, 0
3530·4	7	12, 1	3407·1	9	15, 0	3315·3	9	19, 0
3483·5	9	12, 0	3383·2	9	16, 0	3293·9	9	20, 0
3457·3	9	13, 0	3359·7	9	17, 0	3274·0	8	21, 0

Analysis of emission bands is difficult. The wavelengths recorded are probably intensity maxima associated with overlapping features, and there is little agreement between observations in various sources by Asundi and Parti, by Rosen, by Miyanisi and by Leelavathi and Rao. It is possible overlapping bands of SeO$_2$ also contribute to the emission.

SeBr

Occurrence. Absorption following flash photolysis of Se$_2$Br$_2$.

Appearance. Degraded to shorter wavelengths. Probably due to a transition $B - X^2\Pi$, forming two sequences.

Reference. G. A. Oldershaw and K. Robinson, *Trans. Faraday Soc.*, **67**, 907 (1971).

Bands 2164–2064. The two main (0, 0) sequence heads are at 2164·1 and 2115·0. Weak (1, 0) sequences at 2145·9 and 2097·8.

SeBr$_2$

Diffuse bands have been observed in absorption. Maxima at $\lambda\lambda$5402, 4334, 5267, 5205, 5138*, 5075, 5017 and 4963.

                 * Strongest band.

Reference. M. Wehrli, *Helv. Phys. Acta*, **9**, 329 (1936).

SeCl

Occurrence. In absorption following flash photolysis of SeCl$_2$.

Reference. S. Ciach, G. Power and P. Thistlewaite, *Chem. Phys. Lett.*, **9**, 349 (1971).

A progression of red-degraded bands 3687–3267 Å, strongest 3531·9, 3460·7, 3392·0, 3327·9.

SeCl$_2$

Diffuse bands have been observed in absorption. Maxima at $\lambda\lambda$5974, 5854, 5742, 5634, 5531, 5439, 5346, 5257, 5179, 5084, 5031, 4960, 4885, 4806, 4735 and 4659.

Reference. M. Wehrli, *Helv. Phys. Acta*, **9**, 637 (1936).

SeH

Occurrence.   In absorption following flash photolysis of H_2Se.
Reference.   B. Lindgren, *J. Mol. Spec.*, **28**, 536 (1968).
A single progression of diffuse bands, degraded to the red. $\lambda\lambda 3219 \cdot 9$ (0, 0) strong, 3097 (1, 0) moderate, 3000 (2, 0) weak. The transition is probably $A^2\Sigma - X^2\Pi_{\frac{1}{2}}$.

SeO

The main $^3\Sigma - {}^3\Sigma$ system lies in the near ultra-violet. Weaker systems are reported in the infra-red (two systems), the visible and the further ultra-violet (4 systems).

MAIN SYSTEM, $A^3\Sigma^- - X^3\Sigma^-$

Occurrence.   In flames and discharge tubes of various types.
Appearance.   Degraded to the red. The strongest emission bands form a $v' = 0$ progression. They are double-headed because of the large spin splitting between the F_1 and the $F_2 + F_3$ components of the $^3\Sigma^-$ states.
References.   Choong Shin-Piaw, *Ann. Phys. (Paris)*, **10**, 173 (1938)†.
R. F. Barrow and E. W. Deutsch, *Proc. Phys. Soc.*, **82**, 548 (1963).
P. B. V. Haranath, *Indian J. Pure Appl. Phys.*, **3**, 75 (1965)†.
The following measurements of $F_2 + F_3$ and F_1 heads respectively are from Choong Shin Piaw, with Barrow and Deutsch's revision of the analysis. The system has been extended to 5100 Å by Haranath.

λ	I	v', v''	λ	I	v', v''	λ	I	v', v''
3942·9	10	0, 10	3481·8	9	0, 6	3229·0	6	1, 4
3930·7	8		3473·1	6		3222·0	5	
3818·5	10	0, 9	3380·5	6	0, 5	3139·8	5	1, 3
3807·3	8		3372·1	5		3133·1	2	
3700·3	9	0, 8	3322·1	5	1, 5	3088·8	4	2, 3
3690·2	7		3314·7	5		3084·2	3	
3588·3	9	0, 7						
3578·9	7							

INFRA-RED SYSTEMS

Reference.   M. Azam and S. P. Reddy, *Canad. J. Phys.*, **51**, 2166 (1973)†.
Two weak systems, shaded to longer wavelengths, observed in a microwave discharge.
$a^1\Delta - X^3\Sigma^-$, 10780–10500 Å.   The (0, 0) band at 10503 Å is strongest.
$b^1\Sigma - X^3\Sigma^-$, 10750–9490 Å.   45 heads have been observed. The (0, 0) sequence at 10260 Å is strongest. The (1, 0) head is at 9494 Å.

VISIBLE SYSTEM, 6150–5250 Å

Reference.   V. S. Kushawaha and C. M. Pathak, *Spectrosc. Lett.*, **5**, 393 (1972).
A system of 24 red-degraded bands is reported. However Azam and Reddy (see above) doubt the analysis as $b^1\Sigma - X^3\Sigma$.

SeO (*contd.*)

FAR ULTRA-VIOLET SYSTEMS, 2400–1800 Å

Reference. P. B. V. Haranath, *J. Mol. Spec.*, **13**, 168 (1964)†.

Four fairly strong systems, degraded to shorter wavelengths, have been found in a radio-frequency discharge.

$D - x$, 2435–2165 Å.   17 bands. Strongest 2386·4 (1, 6), 2340·3 (1, 5), 2295·3 (1, 4), 2212·5 (0, 1).

$F - x$, 2170–2080 Å.   7 bands. 2127·8 (0, 1), 2089·1 (0, 0).

$C - X^3\Sigma^-$, 2110–1915 Å.   13 bands. 2072·5 (0, 3), 2035·1 (0, 2).

$E - X^3\Sigma^-$, 2045–1840 Å.   7 double-headed bands. (0, 4) at λ_{vac} 1979·2, 1973·3.

SeO₂

ULTRA-VIOLET SYSTEM, λλ3300–2300

Occurrence. In absorption. Some of the bands also occur in emission in a discharge tube.

Appearance. Complex system of red-degraded bands, somewhat similar to the corresponding system of SO_2.

References. R. K. Asundi, M. Jan-Khan and R. Samuel, *P.R.S.*, **157**, 28 (1936)†.
Choong Shin Piaw, *Ann. Phys. (Paris)*, **10**, 173 (1938)†.
L. Herman, H. Grenat and J. Akriche, *Nature*, **194**, 468 (1962).

The photographs given in the two references are similar, but the recorded wavelengths differ by up to 4 Å, the intensity estimates show little correlation, and the analyses are quite different. The following are averaged wavelengths of the heads which appear prominently in the photographs: λλ2985, 2950, 2928, 2893, 2872, 2851, 2818, 2799, 2767, 2748, 2730, 2719, 2699, 2681, 2669, 2653, 2636, 2609, 2600, 2592.

Duchesne and Rosen (*Physica*, **8**, 540 (1941)) have re-examined this system; their wavelengths agree better with those of Choong Shin Piaw, but their analysis differs from both the earlier ones.

VISIBLE SYSTEM, λλ4774–4005

Reference. J. Duchesne and B. Rosen, *J. Chem. Phys.*, **15**, 631 (1947).

About 70 weak absorption bands are listed, but no intensities are recorded. The bands are degraded to the red and form very long progressions; bands in the centre of the region are diffuse.

SeS

Reference. F. Ahmed and R. F. Barrow, *J. Phys.*, B, **7**, 2256 (1974).

Absorption bands, largely overlapped by S_2 and Se_2, best 4300–3700 Å, are attributed to two systems, $A\,0^+ - X\,0^+$ and $B\,0^+ - X\,0^+$, with system origins at 3673 and 3552 Å; data for individual bands not available.

Si₂

Occurrence. Systems in the visible and near ultra-violet were first studied in discharges through silane with xenon as carrier. Later these systems and four more in the further ultra-violet were obtained in absorption following flash photolysis or flash discharges through silane or other volatile silicon compounds.

Si$_2$ (*contd.*)

   References.   A. E. Douglas, *Canad. J. Phys.*, **33**, 801 (1955)†.
R. D. Verma and P. A. Warsop, *Canad. J. Phys.*, **41**, 152 (1963)†.
A. Lagerqvist and C. Malmberg, *Phys. Scripta*, **2**, 45 (1970)†.
I. Dubois and H. Leclercq, *Canad. J. Phys.*, **49**, 3053 (1971).

NEAR ULTRA-VIOLET SYSTEM, 3785–3489 Å, $L^3\Pi_2 - D^3\Pi_0$

Appearance. Double-headed bands, shaded to the red.

Data from Douglas and from Dubois and Leclercq; no intensities are available, but the (0, 3), (0, 2) and (0, 1) are relatively strong.

λ	v', v''	λ	v', v''	λ	v', v''
3784·6	0, 4 R$_3$	3644·2	1, 3 R$_3$	3562·0	1, 2 R$_1$
3771·2	0, 4 R$_1$	3638·4	0, 2 R$_3$	3556·5	0, 1 R$_1$
3710·4	0, 3 R$_3$	3631·1	1, 3 R$_1$	3507·0	1, 1 R$_3$
3697·5	0, 3 R$_1$	3625·9	0, 2 R$_1$	3495·0	1, 1 R$_1$
		3568·7	0, 1 R$_3$	3489·1	0, 0 R$_1$

VISIBLE SYSTEM, 4630–3863 Å, $H^3\Sigma_u^- - X^3\Sigma_g^-$

Appearance. Strongly degraded to the red. Heads not well developed.

Data for band origins from above references. The values of v' are not quite certain. No intensities available.

λ_0	v', v''	λ_0	v', v''	λ_0	v', v''
4686·9	0, 6	4427·6	1, 4	4060·9	5, 2
4628·1	1, 6	4414·4	5, 6	3979·6	5, 1
4582·3	0, 5	4375·8	2, 4	3942·1	4, 0
4526·0	1, 5	4326·0	3, 4	3900·8	5, 0
4471·9	2, 5	4283·1	2, 3	3863·4	6, 0

3200 Å SYSTEM, $K^3\Sigma_u^- - X^3\Sigma_g^-$

Degraded to the red. Band heads from Verma and Warsop. 3248·9 (0, 0), 3202·0 (1, 0), 3157·8 (2, 0).

2100 Å SYSTEM, $N^3\Sigma_u^- - X^3\Sigma_g^-$

Degraded to the red. The (0, 0) band with head 2138·4 Å is probably strongest. Band origins from Verma and Warsop and from Lagerqvist and Malmberg,

λ_0	v', v''	λ_0	v', v''	λ_0	v', v''
2161·8	0, 1	2101·9	3, 1	2083·5	4, 1
2138·3	0, 0	2098·5	2, 0	2079·7	3, 0
2117·9	1, 0				

Lagerqvist and Malmberg have also observed two systems below 2000 Å. $O^3\Sigma_u^- - X^3\Sigma_g^-$ (1892 (0, 1), 1874 (0, 0)) and $P^3\Pi - D^3\Pi$.

SiBr

There appear to be seven systems definitely assigned to SiBr, all involving the ground state. Another system 3700–3400 is listed under SiBr$^+$, although the emitter is not quite certain. Bands 3320–3170 and 2863–2710 Å listed by Asundi and Karim appear to be due to SiS.

SiBr (*contd.*)

VISIBLE SYSTEM, 6570–4330 Å, $A^2\Sigma^+ - X^2\Pi$

Reference. K. B. Rao and P. B. V. Haranath, *J. Phys.*, B, **2**, 1381 (1969)†.

An extensive system of about 65 heads, degraded to the red, obtained in a high-frequency discharge. There are two sub-systems, but all the strong bands belong to sub-system I. The following, of about equal intensity, are prominent in the published photograph:

λ	v', v''	λ	v', v''	λ	v', v''	λ	v', v''
5959·9	0, 10	5564·8	0, 7	5144·7	1, 4	4667·7	4, 1
5823·2	0, 9	5442·7	0, 6	5037·8	1, 3	4614·8	5, 1
5691·5	0, 8	5255·2	1, 5	4721·9	3, 1	4563·3	6, 1

$B'\,^2\Delta - X^2\Pi$ SYSTEM

Reference. L. A. Kuznetsova and Y. Y. Kuzyakov, *Zh. Priklad. Spektrosk.*, **10**, 413 (1969)†.
Bands at 4350 and 4150 Å observed in a glow discharge through SiBr₄.

$B^2\Sigma^+ - X^2\Pi$, 3233–2750 Å

Reference. W. Jevons and L. A. Bashford, *Proc. Phys. Soc.*, **49**, 554 (1937).
Violet-degraded bands observed in a discharge through SiBr₄.
Two sub-systems. Strongest bands:

λ	I	v', v''	λ	I	v', v''	λ	I	v', v''
3086·8	7	0, 2 i	3009·2	8	0, 1 ii	2958·2	7	1, 0 i
3047·7	9	0, 2 ii	3008·8	10	0, 0 i	2922·0	7	1, 0 ii
3047·3	9	0, 1 i	2958·7	7	1, 1 ii			

$C^2\Pi - X^2\Pi$, 2440–2340 Å

Reference. G. A. Oldershaw and K. Robinson, *Trans. Faraday Soc.*, **67**, 1870 (1971)†.
Observed in flash photolysis. Two sub-systems, shaded to shorter wavelengths. Values of v' are uncertain. Main heads:

λ	v', v''	λ	v', v''
2432·1	0, 0 i	2395·1	2, 0 ii
2425·5	1, 0 ii	2371·6*	2, 0 i
2401·3*	1, 0 i	2365·6	3, 0 ii
		2342·8	3, 0 i

* Relatively strong.

SYSTEMS D – X, E – X, F – X

Observed by Oldershaw and Robinson (see above) in flash photolysis. Degraded to shorter wavelengths.
D – X. 2311·8 (0, 2), 2289·5 (0, 1), 2267·5 (0, 0), 2238·8 (1, 0).
E – X. 2263·9 (0, 1 i), 2242·2 (0, 1 ii), 2234·0 (1, 0 ii), 2214·9 (1, 0 i), 2188·4 (2, 0 i).
F – X. 2220·7 (0, 0), 2196·0 (1, 0).

SiBr⁺

Reference. L. A. Kuznetsova, N. E. Kuz'menko and Y. Y. Kuzyakov, *Opt. Spectrosc.*, **24**, 434 (1968)†.

SiBr⁺ (*contd.*)

Bands of $\Sigma - \Sigma$ type, 3700–3400 Å, observed in a glow discharge through $SiBr_4$ are provisionally assigned to $SiBr^+$. 16 bands degraded to the red. Strongest:

λ	I	v', v"	λ	I	v', v"
3598·6	6	1, 3	3467·2	4	1, 1
3584·2	6	0, 2	3453·2	10	0, 0
3531·9	7	1, 2	3404·4	4	1, 0
3517·7	9	0, 1			

There are also weak bands at 3493·8 and 3430·6.

SiC₂

Occurrence. In emission from a King furnace containing silicon at 2200–2400° C, and in absorption in flash discharges. A strong feature of N-type stars.

Appearance. Degraded to the red. Double-headed.

References. B. Kleman, *Astrophys. J.*, **126**, 162 (1956).
R. D. Verma and S. Nagaraj, *Canad. J. Phys.*, **52**, 1938 (1974)†.

Bands are due to a $^1\Pi - {}^1\Sigma$ transition in linear SiCC. About 20 bands are known. Data for the stronger R heads, mainly from Kleman, with analysis by Verma and Nagaraj. Intensities on scale of 20.

λ	I	$(v_1\ v_2\ v_3)'$	$(v_1\ v_2\ v_3)''$	λ	I	$(v_1\ v_2\ v_3)'$	$(v_1\ v_2\ v_3)''$
5450·0	3	0 0 0	1 0 0	4905·9	3	0 2 0	0 0 0
5198·0	3	0 0 0	0 0 1	4866·9	4	0 0 1	0 0 0
5128·2	3	1 0 0	1 2 0	4639·8	5	1 0 0	0 0 0
4977·4	20	0 0 0	0 0 0				

SiCl

Seven systems involving transitions to the ground state are now known. The $B - X$, $C - X$ and $D - X$ are most readily obtained.

References. W. Jevons, *Proc. Phys. Soc.*, **48**, 563 (1936).
I. E. Ovcharenko and Y. Y. Kuzyakov, *Opt. Spectrosc.*, **13**, 362 (1962); **20**, 14 (1966).
R. D. Verma, *Canad. J. Phys.*, **42**, 2345 (1964)†.
M. Sanii and R. D. Verma, *Canad. J. Phys.*, **43**, 960 (1965)†.
I. E. Ovcharenko, Y. Y. Kuzyakov and V. M. Tatevskii, *Opt. Spectrosc.*, *Mol. Spec. Suppl.*, **2**, 6 (1966).
G. A. Oldershaw and K. Robinson, *J. Mol. Spec.*, **38**, 306 (1971)†.
S. R. Singhal and R. D. Verma, *Canad. J. Phys.*, **49**, 407 (1971)†.
R. K. Pandey, K. N. Upadhaya and K. P. R. Nair, *Indian J. Pure Appl. Phys.*, **9**, 36 (1971).

$A^2\Sigma^+ - X^2\Pi$, 6400–4500 Å

Occurrence. Discharge through flowing $SiCl_4$.

Appearance. Double double-headed. Degraded to the red. Heads split by isotope effect. An extensive system with strong $v' = 0$ progression.

The following are the Q_2 and R_1 heads. Weak $^SR_{21}$ lie in front of the R_1 heads:

SiCl (*contd.*)

λ	I	v', v''	λ	I	v', v''	λ	I	v', v''
5505·1	4	0, 9	5218·3	8	0, 7	4955·5	7	0, 5
5443·6			5162·2			4905·1		
5358·1	6	0, 8	5084·0	8	0, 6	4832·4	6	0, 6
5299·6			5031·1			4784·5		

$B^2\Sigma^+ - X^2\Pi$, 3118–2772 Å

Occurrence. In discharges and flash photolysis.

Appearance. Double double-headed and shaded to shorter wavelengths.

First and third heads from Jevons; the Q heads lie about 0·6 Å to the violet.

λ	I	v', v''	λ	I	v', v''	λ	I	v', v''
3085·6	2	0, 3 i	3017·6	6	0, 2 ii	2955·0	3	1, 2 ii
3068·7	2	1, 4 i	3001·7	4	1, 3 ii	2942·2	8	0, 0 i
3065·8	3	0, 3 ii	2988·8	8	0, 1 i	2924·4	8	0, 0 ii
3049·3	2	1, 4 ii	2973·5	4	1, 2 i	2882·9	8	1, 0 i
3036·6	4	0, 2 i	2970·4	8	0, 1 ii	2865·8	8	1, 0 ii
3020·5	6	1, 3 i						

$B'\,^2\Delta - X^2\Pi$, 2830–2760 Å

Occurrence. In discharges.

Appearance. Two sequences of double double-headed bands, with the rotational structure degraded to shorter wavelengths but the vibrational structure to the red.

The most prominent heads from Ovcharenko *et al.*:

λ	I	v', v''	λ	I	v', v''	λ	I	v', v''
2829·2	6	2, 2 P_2	2823·0	8	0, 0 Q_2	2807·2	10	0, 0 Q_1
2826·1	8	1, 1 P_2	2809·7	9	1, 1 Q_1	2784·5	5	1, 0 P_1
2823·5	10	0, 0 P_2	2809·2	7	0, 0 P_1	2768·5	6	1, 0 Q_1

$C^2\Pi - X^2\Pi$, 2600–2350 Å

Occurrence. In ordinary and glow discharge and in flash photolysis.

Appearance. Double-headed. Shaded to the red.

Heads of sub-bands from Jevons:

λ	I	v', v''	λ	I	v', v''	λ	I	v', v''
2601·5	1	0, 5 i	2521·0	1	0, 3 ii	2424·4	3	0, 0 ii
2588·3	1	0, 5 ii	2500·3	4	0, 2 i	2416·2	2	1, 1 ii
2567·3	1	0, 4 i	2488·2	6	0, 2 ii	2396·7	6	1, 0 i
2557·5	1	1, 5 i	2467·8	6	0, 1 i	2385·6	5	1, 0 ii
2555·5	2	0, 4 ii	2456·0	7	0, 1 ii	2359·1	1	2, 0 i
2544·3	1	1, 5 ii	2435·9		0, 0 i	2348·6	0	2, 0 ii
2533·4	4	0, 3 i	2437·7	2	1, 1 i			

$D^2\Pi - X^2\Pi$, 2350–2090 Å

Occurrence. In discharges and flash photolysis.

Appearance. Degraded to shorter wavelengths. Double-headed.

SiCl (*contd.*)

Strong heads from Oldershaw and Robinson. Own estimates of intensity:

λ	I	v', v''	λ	I	v', v''	λ	I	v', v''
2258·5	5	0, 1	2199·7	10	1, 0	2139·3	7	3, 0
2248·0			2190·0			2129·7		
2231·6	9	0, 0	2168·9	9	2, 0	2125·4	6	4, 1
2221·4			2159·8			2115		

The (0, 1) is strongest in emission.

$E - X^2\Pi$ AND $F - X^2\Pi$

Observed by Oldershaw and Robinson in flash photolysis. Degraded to shorter wavelengths.

E − X		F − X	
λ	v', v''	λ	v', v''
2229·3	0, 1 i	2174·7	0, 0 ii
2213·2	0, 0 ii	2164·8	0, 0 i
2203·2	0, 0 i		

SiF

Silicon monfluoride has a very full spectrum extending from the infra-red to the far ultra-violet, with much overlapping of systems. The region from 6600 to 2500 Å was arranged by Johnson and Jenkins into systems denoted by the Greek letters $\alpha, \beta, \gamma, \delta, \epsilon, \zeta$ and η. These are treated here as a group, apart from the δ bands which were later assigned first to SiF_2 and now to SiF_3. There is also a system 2200–2000 Å found by Dovell and Barrow and a weak system in the infra-red. Systems below 2000 Å are also known but are not discussed here.

BANDS 6600–2500 Å

Occurrence. In discharges of various types through SiF_4.

Appearance. Very complicated overlapping systems, with bands degraded both ways. The β and γ systems have close double heads. The α bands show two strong and two weak heads. See Plate 8.

Transitions. α System, $A^2\Sigma^+ - X^2\Pi$
                β System, $B^2\Sigma^+ - X^2\Pi$
                γ System, $C^2\Delta - X^2\Pi$
                η System, $a^4\Sigma^- - X^2\Pi$

The ϵ system has now been separated into two, $C'\,^2\Pi - A^2\Sigma^+$ with (0, 0) at 4180·7 and $D'\,^2\Pi - A^2\Sigma^+$ with (0, 0) at 5189·5 Å. The ζ system does not appear to have been analysed.

References.   R. C. Johnson and H. G. Jenkins, *P.R.S.*, **116**, 327 (1927)†.
R. K. Asundi and R. Samuel, *Proc. Indian Acad. Sci.*, **3**, 346 (1936).
E. H. Eyster, *P.R.*, **51**, 1078 (1937).
J. W. C. Johns, G. W. Chantry and R. F. Barrow, *Trans. Faraday Soc.*, **54**, 1589 (1958)†.
R. F. Barrow *et al.*, *Proc. Phys. Soc.*, **73**, 317 (1959).
S. Sankaranayanan and P. S. Narayanan, *Proc. Nat. Inst. Sci. India*, A, **32**, 56 (1966)†.
O. Appelblad, R. F. Barrow and R. D. Verma, *J. Phys.*, B, **1**, 274 (1968).

SiF (*contd.*)

Only a selection of the strongest bands are listed below. R, V or M indicate bands degraded to the red, to shorter wavelengths or maxima of headless features. Intensities have been reduced to a common scale of 10.

λ		I	v', v"	System	λ		I	v', v"	System
6594	V	3		ζ	4240·8	R	7	1, 0 R_1	a
6492	V	5		ζ	4229·7	V	3		C' − A
6416	V	5		ζ	4183·3	V	4	0, 0	C' − A
6397	V	5		ζ	4011·8	V	4	1, 0	C' − A
5189·5	V	−		D' − A	3363	M	10	0, 0 i	η
4850·5	R	4	3, 5 $^QR_{12}$	a	3346	M	10	0, 0 ii	η
4569·5	R	5	0, 1 $^QR_{12}$	a	3042·4	V	3	0, 2 P i	β
4535·9	R	5	0, 1 R_1	a	3027·5	V	3	0, 2 P ii	β
4531·6	R	6	4, 4 $^QR_{12}$	a	2967·1	V	6	0, 1 P i	β
4495·8	R	6	3, 3 $^QR_{12}$	a	2952·8	V	6	0, 1 P ii	β
4462·0	R	5	2, 2 $^QR_{12}$	a	2894·4	V	6	0, 0 P i	β
4430·2	R	7	1, 1 $^QR_{12}$	a	2880·8	V	6	0, 0 P ii	β
4429·8	R	6	2, 2 R_1	a	2813·0	V	4	1, 0 P i	β
4400·5	R	7	0, 0 $^QR_{12}$	a	2800·0	V	4	1, 0 P ii	β
4398·3	R	6	1, 1 R_1	a	2652·8	V	3	0, 2 P_2	γ
4388·6	R	2	0, 0 R_2	a	2641·4	V	3	0, 2 Q_1	γ
4368·2	R	10	0, 0 R_1	a	2595·1	V	5	0, 1 P_2	γ
4354·5	R	1	0, 0 $^SR_{21}$	a	2584·3	V	5	0, 1 Q_1	γ
4334·4	R	5	3, 2 $^QR_{12}$	a	2539·2	V	7	0, 0 P_2	γ
4301·3	R	5	2, 1 $^QR_{12}$	a	2528·9	V	7	0, 0 Q_1	γ
4270·2	R	5	1, 0 $^QR_{12}$	a					

DOVELL AND BARROW'S SYSTEM, 2200–2000 Å

A further ultra-violet system of double double-headed bands, degraded to shorter wavelengths, has been obtained by Dovell and Barrow, using a high-frequency electrodeless discharge through SiF$_4$. The doublet separation corresponds closely with the ground-state interval $^2\Pi_{\frac{3}{2}} - {}^2\Pi_{\frac{1}{2}}$.

Reference.   W. H. Dovell and R. F. Barrow, *Proc. Phys. Soc.*, A, **64**, 98 (1951).

Wave-numbers for Q$_1$ and P$_2$ heads of about ten bands are given. The wavelengths and assignments of the main heads are as follows:

λ	v', v"	λ	v', v"
2147·0	0, 1 P_1	2101·5	0, 0 Q_2
2139·6	0, 1 Q_2	2095·2	1, 1 P_1
2108·7	0, 0 P_1	2065·5	1, 0 P_1
2102·2	1, 1 Q_2	2058·7	1, 0 Q_2

Additional violet-degraded bands were obtained with heads at λλ2098·4, 2092·1. 2091·4, 2076·2, 2075·6, 2069·3 and 2068·7. These could not be fitted into the system.

INFRA-RED SYSTEM, $B^2\Sigma^+ - A^2\Sigma^+$

The infra-red region has been observed by Johns and Barrow using powdered silicon and potassium fluorosilicate in a hollow cathode containing helium.

Reference.   J. W. C. Johns and R. F. Barrow, *Proc. Phys. Soc.*, A, **71**, 476 (1958)†.

SiF (*contd.*)

Band origins are given for the system as follows:

λ_0	v', v''	λ_0	v', v''
8964·5	0, 1	7221·2	2, 0
8435·9	0, 0	7076·2	3, 1
7778·5	1, 0	7068·1	4, 2
7605·0	2, 1		

SiF$_2$

Reference. D. R. Rao, *J. Mol. Spec.*, **34**, 284 (1970)†.

Regularly spaced complex but narrow bands, 4183–3645 Å, have been obtained from an electrodeless discharge through flowing SiF$_4$. They form an array in v'_2, v''_2. Principle maxima, with own estimates of intensity:

λ	I	v'_2, v''_2	λ	I	v'_2, v''_2	λ	I	v'_2, v''_2
4064·5	6	0, 5	3901·5	10	0, 2	3800	3	0, 0
4008·5	9	0, 4	3850	7	0, 1	3721	6	2, 0
3954·6	10	0, 3	3809	5	1, 1			

Bands 2700–2100 Å, initially attributed to the δ system of SiF and later to SiF$_2$, are now listed under SiF$_3$ below.

SiF$_3$

References. D. R. Rao and P. Venkateswarlu, *J. Mol. Spec.*, **7**, 287 (1961)†.
V. M. Khanna, G. Besenbruch and J. L. Margrave, *J. Chem. Phys.*, **46**, 2310 (1967).
R. N. Dixon and M. Hallé, *J. Mol. Spec.*, **36**, 192 (1970)†.
J. Ling-Fai Wang, C. N. Krishman and J. L. Margrave, *J. Mol. Spec.*, **48**, 346 (1973).

These bands, obtained in emission from discharges through flowing SiF$_4$ and in absorption by cooled products of reaction of hot silicon with SiF$_4$, were for long attributed to SiF$_2$ but are now believed by Ling-Fai Wang *et al.* to be due to SiF$_3$; this conclusion depends on comparison with frozen matrix studies.

In emission the strongest heads (Rao and Venkateswarlu), degraded to the red, are at $\lambda\lambda(I)$ 2452·5 (7), 2447·3 (9), 2427·4 (10), 2422·0 (7), 2407·3 (9), 2402·2 (10), 2382·7 (7).

In absorption Khanna *et al.* list the following as strong: $\lambda\lambda(I)$ 2240·8 (6), 2228·2 (8), 2215·9 (10), 2203·6 (10), 2191·5 (8), 2179·4 (6).

SiH

References. C. V. Jackson, *P.R.S.*, **126**, 373 (1930)†.
G. D. Rochester, *Z.P.*, **101**, 769 (1936)†.
A. E. Douglas, *Canad. J. Phys.*, **35**, 71 (1957).
R. D. Verma, *Canad. J. Phys.*, **43**, 2136 (1965)†.
L. Klynning and B. Lindgren, *Ark. Fys.*, **33**, 73 (1967).
G. Herzberg, A. Lagerqvist and B. J. McKenzie, *Canad. J. Phys.*, **47**, 1889 (1969)†.
P. Bollmark, L. Klynning and P. Pagès, *Phys. Scripta*, **3**, 219 (1971).

SiH (*contd.*)

4142 Å SYSTEM, $A^2\Delta - X^2\Pi$

Occurrence. In silicon arc in H_2, in discharge tubes (sometimes as impurity from Si in walls), in absorption from a King furnace at $2100°$ C and flash photolysis. Also in sun-spots and weakly in solar disc.

Appearance. Complex red-degraded bands with open rotational structure. The (0, 0) is strongest.

λ	v', v''	λ	v', v''	λ	v', v''
4277·7	2, 2 Q_2	4142·3	0, 0 Q_2	3940·7	2, 1 Q_1
4259·2	2, 2 Q_1	4128·0	0, 0 Q_1	3877·9	1, 0 Q_2
4198·6	1, 1 Q_2	3956·3	2, 1 Q_2	3863·1	1, 0 Q_1
4183·9	1, 1 Q_1				

3250 Å, $B^2\Sigma^+ - X^2\Pi$

Observed by Verma and by Bollmark *et al.* in flash photolysis. Open but irregularly spaced rotational structure without obvious heads.

2058 Å, $D^2\Delta - X^2\Pi$

Band of diffuse lines, obtained in flash photolysis by Verma and by Herzberg *et al.*

SiH⁺

3993 Å SYSTEM, $A^1\Pi - X^1\Sigma$

Occurrence. Hollow-cathode discharge in silane with helium. Also in solar spectrum.
Appearance. Strongly degraded to the red. Open structure with weak R heads.
Reference. A. E. Douglas and B. Lutz, *Canad. J. Phys.*, **48**, 247 (1970)†.
R heads: $\lambda\lambda$4777 (0, 2), 4356·3 (0, 1), 4284·3 (1, 1), 3992·9 (0, 0), 3932·4 (1, 0).

SiH₂

Occurrence. Absorption following flash photolysis of silane.
Appearance. Bands with open but complex rotational structure. The seven strongest form a $(0\ v_2'\ 0, 0\ 0\ 0)$ progression.
Reference. I. Dubois, *Canad. J. Phys.*, **46**, 2485 (1968)†.
The following are the long-wave edges of the Q branches: $\lambda\lambda$6436, 6098*, 5797*, 5527*, 5253, 5056, 4847. Those marked * are strongest.

SiH₄

Reference. J. Bardwell and G. Herzberg, *Astrophys. J.*, **117**, 462 (1953)†.
An absorption band has a complex Q maximum at 9738 Å, with P and R branches on each side. There is another band at 11977 Å.

SiHBr

Reference. G. Herzberg and R. D. Verma, *Canad. J. Phys.*, **42**, 395 (1964)†.
Complex absorption bands 5000–4300 Å in flash photolysis of SiH_3Br. Many bands.

Those with origins at λ_0 5023 (0 0 0, 0 0 0), 4891 (0 1 0, 0 0 0), 4769 (0 2 0, 0 0 0) and 4656 (0 3 0, 0 0 0) may be strong.

SiHCl

Reference. G. Herzberg and R. D. Verma, *Canad. J. Phys.*, **42**, 395 (1964)[†].

Complex absorption bands 4800–4100 Å in flash photolysis of SiH_3Cl. Of the many bands those with origins at 4825 (0 0 0, 0 0 0), 4698 (0 1 0, 0 0 0), 4584 (0 1 1, 0 0 0) and 4578 (0 2 0, 0 0 0) could be strong. Also bands 6000–4800 in fluorescence.

SiHI

Reference. J. Billingsley, *Canad. J. Phys.*, **50**, 531 (1972)[†].

Complex violet-degraded absorption bands 5600–4600 Å, in flash photolysis of SiH_3I. Bands with origins at 5475 (0 0 0, 0 0 0), 5333 (0 1 0, 0 0 0), 5213 (0 2 1, 0 0 1), 5205 (0 2 0, 0 0 0) and 5091 (0 3 0, 0 0 0) could be strong.

SiI

References. G. A. Oldershaw and K. Robinson, *Trans. Faraday Soc.*, **64**, 2256 (1968)[†]. J. Billingsley, *J. Mol. Spec.*, **43**, 128 (1972)[†].

$A^2\Sigma^+ - X^2\Pi_{\frac{1}{2}}$, 4900–4200 Å

Occurrence. Flash photolysis of SiH_3I (Billingsley).

Appearance. A long $v'' = 0$ progression. Of the 15 known bands the following seem strong: $\lambda\lambda$4537·9 (4, 0), 4517·8 (5, 0), 4478·8 (6, 0), 4441·0 (7, 0).

$B^2\Sigma^+ - X^2\Pi_{\frac{1}{2}}$, 3200–2900 Å

Occurrence. Flash photolysis of SiI_4 or SiH_3I.

Appearance. Degraded to shorter wavelengths. The following are prominent on Oldershaw and Robinson's photograph:

λ	v', v''	λ	v', v''
3152·1	0, 2	3038·3	1, 0
3117·1	0, 1	2995·6	2, 0
3082·2	0, 0	2954·3	3, 0

$A' - X^2\Pi_{\frac{3}{2}}$, 5260–4870 Å

Occurrence. In emission in microwave discharge and very weakly in absorption in flash photolysis.

13 red-degraded bands observed by Billingsley. Strongest, with own intensity estimates:

λ	I	v', v''	λ	I	v', v''
5117·4	3	0, 2	4960·3	8	1, 1
5049·1	5	1, 2	4937·6	10	0, 0
5026·3	3	0, 1	4874·0	3	1, 0

SiN

There are two systems attributed to SiN, a strong system $\lambda\lambda$5260–3786 and a weak overlapping

SiN (*contd.*)

system λλ5620–3188.

MAIN SYSTEM, $A^2\Sigma - X^2\Sigma$

Occurrence. In high-frequency discharges and with Si compounds in active nitrogen.
Appearance. Degraded to the red. Single-headed.
References. W. Jevons, *P.R.S.*, **89**, 187 (1913).
F. A. Jenkins and H. de Laszlo, *P.R.S.*, **122**, 105 (1929).
S. Nagaraj and R. D. Verma, *Canad. J. Phys.*, **46**, 1597 (1968)†.
M. Singh *et al.*, *J. Phys.*, B, **6**, 2656 (1973)†.

The following table is compiled from the above references. Wavelengths of R heads. Intensities as given by Jevons, for active nitrogen:

λ	I	v', v"	λ	I	v', v"	λ	I	v', v"
4748·7	3	6, 8	4317·6	5	6, 6	4050·7	8	4, 3
4705·1	4	5, 7	4277·0	5	5, 5	4032·0	4	8, 6
4664·3	5	4, 6	4239·1	9	4, 4	4016·8	6	3, 2
4629·2	3	3, 5	4204·1	10	3, 3	3989·9	4	7, 5
4618·1	2	8, 9	4169·6	6	2, 2	3985·8	5	2, 1
4569·8	4	7, 8	4168·2	4	7, 6	3957·7	2	1, 0
4524·3	5	6, 7	4141·5	3	1, 1	3949·8	4	6, 4
4482·4	6	5, 6	4126·6	8	6, 5	3911·8	4	5, 3
4443·1	8	4, 5	4116·3	1*	0, 0	3814·0	2	2, 0
4406·9	8	3, 4	4087·4	8	5, 4			

* In high-frequency discharges bands with low v' are much stronger and the (0, 0) band is the strongest.

WEAK SYSTEM

Occurrence. Silicon tetrachloride vapour in active nitrogen.
Appearance. Degraded to red. Double-headed, separation 27 cm.$^{-1}$.
Transition. Perhaps $^2\Pi - ^2\Pi$.
Reference. R. S. Mulliken, *P.R.*, **26**, 319 (1925).

In the following table the strongest bands listed by Mulliken are given. Intensities have been increased to a scale of 5. Only the shorter wavelength head is listed here.

λ	I	v', v"	λ	I	v', v"	λ	I	v', v"
5555·2	2	4, 12	4877·6	2	1, 7	3744·1	2	3, 2
5492·1	2	3, 11	4834·7	2	4, 9	3698·4	1	2, 1
5235·0	3	3, 10	4463·6	2	1, 5	3607·8	2	3, 1
5172·6	5	2, 9	3994·7	4	2, 3	3535·0	4	4, 1

At one time the 3832 band of SiO$^+$ was attributed to SiN.

SiO

There is a strong $A^1\Pi - X^1\Sigma$ system in the near ultra-violet and five weaker systems have now been attributed to SiO.

MAIN SYSTEM, $A^1\Pi - X^1\Sigma^+$

Occurrence. In flames into which SiCl$_4$ is introduced, in discharge through SiCl$_4$ + O$_2$, and

SiO (*contd.*)

strongly in arc. Also in absorption.

Appearance. Degraded to the red. Single-headed. See Plate 1.

References. W. Jevons, *P.R.S.*, **106**, 174 (1924).

D. Sharma, *Proc. Nat. Acad. Sci. India*, A, **14**, 37 (1944)†.

A. Lagerqvist and U. Uhler, *Ark. Fys.*, **6**, 95 (1953).

G. Bosser, J. Lebreton and L. Marsigny, *C. R. Acad. Sci. (Paris)*, C, **275**, 531 (1972)†.

The following measurements are by Jevons. Intensities I_a, I_d and I_f are for absorption, discharge tube and flame respectively, the latter being from de Gramont and de Watteville.

λ	I_a	I_d	I_f	v', v''	λ	I_a	I_d	I_f	v', v''		
2925·3		2		4, 10	2486·8	7	6	10	0, 2		
2898·4		3		3, 9	2481·9	3	2	3	3, 4		
2871·6		4		2, 8	2459·0	1	3	4	2, 3		
2845·7		2		1, 7	2436·3	1	3		1, 2		
2832·2		1		4, 9	2413·8	9	7	8	0, 1		
2820·0				0, 6	2410·2				3, 3		
2806·3		8		3, 8	2387·9	6	5	3	2, 2		
2780·5		7	6	2, 7	2365·7	7	6	1	1, 1		
2755·0		6	6	1, 6	2364·5			4	4, 3		
2730·1		2		0, 5	2344·3	8	5	4	0, 0		
2718·8		4		3, 7	2342·4	1	1	4	3, 2		
2693·7		9	7	2, 6	2298·9	10	6	2	1, 0	4, 2
2669·0		9	8	1, 5	2277·2	4	1	1	3, 1		
2644·8		4	4	0, 4	2255·9	7	4		2, 0		
2636·0				3, 6	2236·3	4	2	1	4, 1		
2611·3		4		2, 5	2215·4	6	2	0	3, 0		
2587·1	2	5	8	1, 4	2197·4	4	0		5, 1		
2563·8	3	5	8	0, 3	2176·6	5	1		4, 0		
2509·9	2	4	3	1, 3	2160·3	5			6, 1		

$a\,^3\Pi_1 - X\,^1\Sigma^+$, 3235–2870 Å

Occurrence. A weak system observed in a hollow-cathode discharge using heated SiO_2 in flowing He.

Reference. H. Bredohl *et al.*, *J. Phys.*, B, **7**, L 66 (1974).

Degraded to the red. Heads λλ3227·2 (2, 4), 3205·1 (1, 3), 3107·3 (2, 3), 3086·0 (1, 2), 3065·3 (0, 1), 2953·8 (0, 0), 2889·7 (2, 1), 2869·4 (1, 0).

$c\,^3\Pi - a\,^3\Pi$, 3950–3700 Å

Occurrence. In high-frequency and hollow-cathode discharges.

Appearance. Degraded to the red. Triple-headed.

References. R. Cornet and I. Dubois, *Canad. J. Phys.*, **50**, 630 (1972)†.

H. Bredohl, R. Cornet and I. Dubois, *J. Phys.*, B, **8**, L 16 (1975).

First heads: λλ4139·7 (0, 1), 4060·3 (1, 1), 3975·6 (0, 0), 3902·1 (1, 0), 3831·5 (2, 0), 3763·7 (3, 0) 3699·3 (4, 0).

$d\,^3\Sigma - a\,^3\Pi$, 2970 Å

References. R. Cornet and I. Dubois, *Canad. J. Phys.*, **50**, 630 (1972).

M. Singh *et al.*, *Canad. J. Phys.*, **52**, 569 (1974)†.

SiO (*contd.*)

Cornet and Dubois reported a band, degraded to shorter wavelengths with heads at $\lambda\lambda 2962\cdot 3$, $2955\cdot 6$ and $2948\cdot 8$. Later Singh *et al.* analysed it and gave the following heads: $^OP_{12}$ $2979\cdot 7$, $^PQ_{12}$ $2972\cdot 9$ and Q_1 $2966\cdot 3$.

$^3\Sigma - {}^3\Pi$, 4300–4200 Å

Reference. S. Nagaraj and R. D. Verma, *Canad. J. Phys.*, **48**, 1436 (1970)†.

A single sequence of triple-headed bands, observed in high-frequency discharges. The sequence is degraded to the red, but heads shade both ways.

The (0, 0) has heads at $4228\cdot 6$ R_{11} (deg. R), $4241\cdot 6$ $^QR_{12}$ (deg. R) and $4254\cdot 4$ P_{33} (deg. V). The less strong (1, 1) has heads $4243\cdot 8$ R_{11} (deg. R), $4256\cdot 9$ $^QR_{12}$ (maximum) and $4269\cdot 8$ $^PQ_{23}$ (deg. V). Weaker (2, 2) and (3, 3) bands are known.

$E^1\Sigma^+ - X^1\Sigma^+$, 3650–3200 Å

References. H. Bredohl *et al.*, *Canad. J. Phys.*, **51**, 2332 (1973)†.
R. F. Barrow and T. J. Stone, *J. Phys.*, B, **8**, L 13 (1975).

Fourteen red-degraded bands, previously attributed to SiO^+ and then to a $^1\Sigma^- - {}^1\Sigma^-$ transition of SiO, now assigned as above, occur in a high-frequency discharge. The following are origins of bands likely to be strong: λ_0 $3540\cdot 7$ (4, 5), $3501\cdot 0$ (3, 4), $3461\cdot 9$ (2, 3), $3423\cdot 3$ (1, 2), $3385\cdot 2$ (0, 1), $3315\cdot 0$ (1, 1), $3278\cdot 1$ (0, 0), $3248\cdot 6$ (2, 1), $3212\cdot 3$ (1, 0).

Further systems below 2000 Å are known.

SiO^+

Occurrence. Heavy-current discharge through a silica vacuum tube.
Appearance. A close double-headed band, degraded to the red.
Transition. $A^2\Sigma - {}^2\Sigma$.
References. R. C. Pankhurst, *Proc. Phys. Soc.*, **52**, 707 (1940).
L. H. Woods, *P.R.*, **63**, 426 (1943)†.
M. Singh *et al.*, *J. Phys.*, B, **6**, 2656 (1973).

λ	v', v''
3832·9	0, 0

Bands of SiN at 4116 Å and the $E^1\Sigma^+ - X^1\Sigma^+$ system of SiO were at one time attributed to SiO^+.

SiO_2 ?

Bands of complex structure 4500–3700 Å emitted by a silicon arc in O_2 and in heavy-current discharges in a silica tube were thought to be due to SiO_2, but most are now assigned to the $^3\Sigma - {}^3\Pi$ system of SiO; some, however, remain unassigned, including red-degraded heads at $\lambda\lambda 4468\cdot 5$, $4466\cdot 6$, $4417\cdot 3$, 4408, $3791\cdot 4$, $3776\cdot 9$ and $3713\cdot 5$.

References. W. H. B. Cameron, *Phil. Mag.*, **3**, 110 (1927)†.
R. C. Pankhurst, *Proc. Phys. Soc.*, **52**, 707 (1940)†.
L. H. Woods, *P.R.*, **63**, 426 (1943).

SiS

Three band-systems are ascribed to this molecule. Two of the systems are in the ultra-violet region. The third, a weak emission system in the visible region, may be due to impurity.

STRONG SYSTEM, 3959–2585 Å, $D^1\Pi - X^1\Sigma^+$

Occurrence. High-density discharge through a quartz tube containing silicon sulphide. Also in absorption.

References. R. F. Barrow and W. Jevons, *P.R.S.*, **169**, 45 (1938)†.
R. F. Barrow, *Nature*, **154**, 364 (1944).
R. F. Barrow, *Proc. Phys. Soc.*, **58**, 606 (1946).
A. Lagerqvist, G. Nilheden and R. F. Barrow, *Proc. Phys. Soc.*, A, **65**, 419 (1952).
G. Nilheden, *Ark. Fys.*, **10**, 19 (1956).

Appearance. Degraded to the red. Apparently single-headed bands. The following are the strong bands:

λ	I	v', v''	λ	I	v', v''	λ	I	v', v''
3506·3	4	3, 11	3244·2	5	2, 7	2883·2	8	1, 1
3471·8	4	5, 12	3221·8	8	1, 6	2863·7	8	0, 0
3447·4	4	4, 11	3149·0	7	1, 5	2822·7	9	1, 0
3423·6	5	3, 10	3127·7	8	0, 4	2783·2	9	2, 0
3399·7	5	2, 9	3078·8	7	1, 4	2764·7	6	4, 1
3343·4	5	3, 9	3057·9	9	0, 3	2745·3	8	3, 0
3320·1	6	2, 8	2990·8	10	0, 2	2708·8	7	4, 0
3297·6	7	1, 7	2926·1	10	0, 1	2673·8	6	5, 0

2500–1995 SYSTEM, $E^1\Sigma^+ - X^1\Sigma^+$

Occurrence. In absorption. In emission from an electrodeless discharge through a stream of silicon tetrachloride vapour containing a small amount of carbon disulphide.

Appearance. Degraded to the red. An extensive system which has been followed up to $v' = 28$. Heads of stronger bands are:

λ	v', v''	λ	v', v''
2394·5	0, 0	2286·6	5, 0
2371·7	1, 0	2266·6	6, 0
2349·6	2, 0	2247·4	7, 0
2328·0	3, 0	2228·6	8, 0
2306·9	4, 0	2210·4	9, 0

References. E. E. Vago and R. F. Barrow, *Proc. Phys. Soc.*, **58**, 538 (1946)†.
S. J. Q. Robinson and R. F. Barrow, *Proc. Phys. Soc.*, A, **67**, 95 (1954).
R. F. Barrow *et al.*, *Proc. Phys. Soc.*, **78**, 1306 (1961).

VISIBLE SYSTEM, λλ6169–3491

An extensive system of weak red-degraded bands 6169–3491 Å was reported by Barrow and Jevons (*P.R.S.*, **169**, 45 (1938)) but Barrow (see Rosen) later suggested that they were probably due to an impurity. See also H. Bredohl *et al.*, *J. Phys.*, B, **8**, L 259 (1975).

SiSe

Occurrence. Heavy-current discharge through quartz tube containing alumium selenide. Also in absorption.

Appearance. Degraded to the red. Single-headed bands.

Reference. R. F. Barrow, *Proc. Phys. Soc.*, **51**, 267 (1939)†.

Strong bands:

λ	I	v', v''	λ	I	v', v''
3406·0	7	1, 6	3145·3	9	0, 1
3342·2	9	1, 5	3106·4	9	1, 1
3323·7	8	0, 4	3051·8	8	1, 0
3262·4	10	0, 3	3015·8	7	2, 0
3203·0	10	0, 2	2981·1	6	3, 0

A weak system around 2770–2450 Å has been reported in absorption by Vago and Barrow (*Nature*, **157**, 77 (1946)).

SiTe

Occurrence. Heavy-current through quartz tube containing aluminium and tellurium. Also in absorption.

Appearance. Degraded to the red. Single-headed bands.

Reference. R. F. Barrow, *Proc. Phys. Soc.*, **51**, 45 (1939)†.

Strong bands:

λ	I	v', v''	λ	I	v', v''
3763·9	8	1, 5	3556·1	10	0, 1
3745·2	9	0, 4	3514·3	10	1, 1
3680·3	9	0, 3	3456·2	8	1, 0
3617·3	10	0, 2	3417·0	7	2, 0

A weak system around 3100–2800 Å has been reported in absorption by Vago and Barrow (*Nature*, **157**, 77 (1946)).

SmO

Occurrence. Samarium salts in oxy-hydrogen flame.

Reference. G. Piccardi, *Rend. Accad. Linc.*, **21**, 589 (1935).

The following are the strongest heads:

λ	I	λ	I
6570·1	6	6485·5	7
6557·2	8	6349·5	8
6533·5	9	6034·4	6
6510·9	10	5822·4	7

Herrmann and Alkemade (*Flame Photometry*, 1963) record over 80 heads, but the direction of degradation is not usually stated. The 6510·9 Å band is triple-headed with heads shaded both ways.

SnBr

Six systems have now been reported. $B - X$ is probably the strongest. $A - X$, $A' - X$ and $B - X$ occur in discharges through stannic bromide vapour, usually with argon as carrier. Systems $B - X$, $D - X$, $E - X$ and $F - X$ have been obtained by Oldershaw and Robinson in flash photolysis of $SnBr_2$, but the $A - X$ and $A' - X$ were not observed. Continua around 3600 and 2930–2850 Å have also been found in emission.

References. W. Jevons and L. A. Bashford, *Proc. Phys. Soc.*, **49**, 554 (1937)†.
P. R. K. Sarma and P. Venkateswarlu, *J. Mol. Spec.*, **17**, 203 (1965)†.
G. A. Oldershaw and K. Robinson, *Trans. Faraday Soc.*, **64**, 616 (1968); **67**, 2499 (1971)†.

VIOLET SYSTEM, 4255–3709 Å, $A^2\Delta - X^2\Pi$

Appearance. Degraded to the red.
Strong bands only:

λ	I	v', v''	λ	I	v', v''	λ	I	v', v''
4196·5	6	0, 3 i	3833·7	6	1, 3 ii	3785·8	7	0, 1 ii
4153·9	8	0, 2 i	3820·8	6	0, 2 ii	3750·8	8	0, 0 ii
4112·1	10	0, 1 i	3798·4	7	1, 2 ii	3729·4	6	1, 0 ii
4070·7	10	0, 0 i						

ULTRA-VIOLET SYSTEM, 3428–2915 Å, $B^2\Sigma - X^2\Pi$

Appearance. Degraded to shorter wavelengths. Two progressions which look rather like sequences.
Strong bands only:

λ	I	v', v''	λ	I	v', v''
3372·0	4	0, 4 i	3112·2	4	0, 4 ii
3344·6	5	0, 3 i	3089·2	5	0, 3 ii
3317·2	6	0, 2 i	3066·4	6	0, 2 ii
3290·4	7	0, 1 i	3043·6	7	0, 1 ii
3263·7	*	0, 0 i	3021·1	4	0, 0 ii

*Masked by Sn line 3262·33.

VISIBLE SYSTEM, 6400–5100 Å, $A' - X^2\Pi_{\frac{1}{2}}$

Degraded to the red. Sarma and Venkateswarlu record 35 bands; strongest ($I = 5$ or 4) $\lambda\lambda$6240·6 (0, 11), 6149·9 (0, 10), 5622·8 (2, 5), 5572·3 (0, 3), 5497·3 (0, 2), 5377·0 (1, 1), 5330·7 (2, 1).

FAR ULTRA-VIOLET SYSTEMS, 2500–2200 Å

Diffuse bands. The following are the stronger maxima.
$C - X^2\Pi$. Three weak bands 2483·4 (0, 2), 2468·4 (0, 1), 2453·5 (0, 0).
$D - X^2\Pi$. 2351·8 (0, 1), 2338·3 (0, 0) strong, 2323·7 (1, 0).
$E - X^2\Pi$. 2305·4 (0, 1), 2292·5 (0, 0) strong, 2275·8 (1, 0).

SnCl

Four or more ultra-violet systems and one in the visible are known. There has been some confusion in naming the states, both the visible and the first ultra-violet being referred to as $A - X$. We have retained $A - X$ for the longer known ultra-violet system and used $A' - X$ for the visible.

SnCl (*contd.*)

Occurrence. A′−X, A−X and B−X occur in discharges of various types through $SnCl_4$ vapour. A′−X has also been obtained by electron beam excitation. A−X, B−X and C−X occur in absorption by heated $SnCl_2$ while these, C′−X and two other ultra-violet bands have been found by Oldershaw and Robinson in flash photolysis of $SnCl_4$.

References. W. Jevons, *P.R.S.*, **110**, 365 (1926)†.

W. F. C. Ferguson, *P.R.*, **32**, 607 (1929).

C. A. Fowler, *P.R.*, **62**, 141 (1942)†.

P. R. K. Sarma and P. Venkateswarlu, *J. Mol. Spec.*, **17**, 252 (1965)†.

G. Pannetier and P. Deschamps, *J. Chim. Phys.*, **65**, 1164 (1968)†.

G. A. Oldershaw and K. Robinson, *J. Mol. Spec.*, **32**, 469 (1969)†.

A. Chatalic *et al.*, *J. Chim. Phys.*, **69**, 82 (1972); **70**, 481, 908 (1973).

VISIBLE SYSTEM, 6985–4880 Å, A′(?$^2\Sigma$)−X$^2\Pi_{\frac{1}{2}}$

About 50 bands, degraded to longer wavelengths. The following are mainly from Sarma and Venkateswarlu:

λ	I	v', v''	λ	I	v', v''	λ	I	v', v''
6013·1	4	0, 8	5561·1	5	0, 4	5389	6	1, 3
5893·9	4	0, 7	5490·4	10	1, 4	5323·2	4	2, 3
5779·3	4	0, 6	5457·1	7	0, 3	5227·5	4	2, 2
5668·3	4	0, 5						

3910–3486 Å SYSTEM, A($^2\Delta$ or $^4\Sigma^-$)−X$^2\Pi$

Appearance. Degraded to red. Two strong sequences.

λ	I	v', v''	λ	I	v', v''
3786·3	2	3, 3 i	3511·2	1	3, 3 ii
3776·6	5	2, 2 i	3502·5	2	2, 2 ii
3767·3	8	1, 1 i	3494·7	8	1, 1 ii
3758·5	10	0, 0 i	3487·8	10	0, 0 ii

3405–2830 Å SYSTEM, B$^2\Sigma$−X$^2\Pi$

Appearance. Degraded to shorter wavelengths. Strong SnCl^{35} heads with weaker SnCl^{37} heads visible in a few bands.

Heads of strong bands only:

λ	I	v', v''	λ	I	v', v''	λ	I	v', v''
3271·5	5	0, 2 i	3105·2	4	3, 1 i	2959·3	5	2, 2 ii
3234·4	7	0, 1 i	3036·4	4	0, 2 ii	2935·8	10	1, 0 ii
3197·8	7	0, 0 i	3004·5	6	0, 1 ii	2899·4	7	2, 0 ii
3154·5	7	1, 0 i	2973·4	8	0, 0 ii	2893·0	5	3, 1 ii
3112·4	5	2, 0 i	2966·3	3	1, 1 ii			

2450–2250 Å SYSTEM, C−X$^2\Pi_{\frac{1}{2}}$

Appearance. Degraded to shorter wavelengths.

The following are the strongest bands in absorption:

SnCl (*contd.*)

λ	I	v', v''
2307·4	5	0, 1
2288·9	9	0, 0
2268·6	10	1, 0
2248·9	8	2, 0

2465–2375 Å SYSTEM, $C' - X^2\Pi_{\frac{1}{2}}$

A rather weaker system, observed in flash photolysis. Degraded to shorter wavelengths. Clear sequences $\lambda\lambda$2464·7 (0, 2), 2443·6 (0, 1), 2422·6 (0, 0), 2398·7 (1, 0), 2375·2 (2, 0).

There are also diffuse absorption bands at 2185 and 2169 and continua 3500–3000 and 2950–1950 Å.

SnCl₂

There are reports of emission 4900–4200 Å in discharges and in chemiluminescence of $SnCl_4 + K$, and Naegeli and Palmer have given an analysis.

References. P. Deschamps and G. Pannetier, *J. Chim. Phys.*, **61**, 1547 (1964).

D. Naegeli and H. B. Palmer, *J. Mol. Spec.*, **21**, 325 (1966).

SnF

Occurrence. Systems A to F were first obtained by Jenkins and Rochester in absorption. The visible system was obtained by Yuasa in a discharge tube containing tin fluoride. Barrow *et al.* report that the spectrum is strongly excited in a hollow-cathode discharge with helium as carrier gas.

References. F. A. Jenkins and G. D. Rochester, *P.R.*, **52**, 1135 (1937)†.

T. Yuasa, *Proc. Phys.-Math. Soc. Japan*, **21**, 498 (1939)†.

R. F. Barrow, D. Butler, J. W. C. Johns and J. L. Powell, *Proc. Phys. Soc.*, **73**, 317 (1959).

A. N. Uzikov and Y. Y. Kuzyakov, *Vestnik. Moskov Univ., Ser. II, Khim.*, **23**, 33 (1968).

R. S. Ram, K. N. Upadhya and D. K. Rai, *J. Phys.*, B, **6**, L 372 (1973).

VISIBLE SYSTEM, 6301–4600 Å, $A^2\Sigma^+ - X^2\Pi$

Appearance. Double-headed bands, degraded to the red. Wavelengths of 45 bands are listed by Yuasa. The following are the strongest in the two sub-systems.

λ	I	v', v''	λ	I	v', v''	λ	I	v', v''
6301·0	3	1, 4 i	5506·8	9	1, 0 i	5132·6	10	0, 1 ii
6087·7	5	1, 3 i	5436·7	5	3, 1 i	4985·6	10	0, 0 ii
6032·6	7	0, 2 i	5384·8	7	2, 0 i	4883·3	10	1, 0 ii
5827·5	9	0, 1 i	5287·5	8	0, 2 ii	4786·4	8	2, 0 ii
5635·9	9	0, 0 i	5270·4	5	3, 0 i	4734·9	7	4, 1 ii
						4696·0	7	3, 0 ii

SYSTEM A, 3260–2660 Å, $B^2\Sigma^+ - X^2\Pi$

This is the strongest system. Degraded to shorter wavelengths. Close double-headed bands, separation 1 Å. P heads of strong bands:

SnF (*contd.*)

λ	v', v''	λ	v', v''	λ	v', v''
3199·7	0, 1 i	3020·7	0, 2 ii	2927·9	0, 0 ii
3141·2	0, 0 i	2978·2	0, 1 ii	2871·4	1, 0 ii
3076·2	1, 0 i	2969·4	1, 2 ii	2817·6	2, 0 ii

SYSTEM B, 2635–2450 Å, $C\,^2\Delta - X\,^2\Pi$

Degraded to shorter wavelengths. Close double-headed bands in two sub-systems. Prominent heads:

λ	v', v''	λ	v', v''
2635·4	0, 1 $^OP_{12}$	2556·3	1, 0 $^OP_{12}$
2595·5	0, 0 $^OP_{12}$	2452·3	0, 0 P_1

SYSTEM C, 2350–2100 Å, $F - X\,^2\Pi$

This is a weak system of single-headed bands, degraded to shorter wavelengths. P heads of strongest:

λ	v', v''
2348·2	1, 0 i
2222·8	0, 1 ii
2194·7	0, 0 ii
2162·5	1, 0 ii

SYSTEM D, 2300–2060 Å, $G\,^2\Delta - X\,^2\Pi$

A strong system of close double-headed bands, degraded to shorter wavelengths. Strongest P heads:

λ	v', v''	λ	v', v''
2296·7	0, 1	2184·3	0, 1 ii
2266·3	0, 0	2157·1	0, 0 ii
2236·1	1, 1	2129·3	1, 0 ii

There are probably additional systems $D - X$ and $E - X$ in the region 2700–2400 Å, two inter-combination systems $B\,^2\Sigma - A\,^2\Sigma$ 8300–6600 and $F - A\,^2\Sigma$ 4300–3950 Å and continua 2500–2370 and below 2100 Å.

SnH

BLUE SYSTEM, $^2\Delta - X\,^2\Pi$

Occurrence. In emission from a tin arc in H_2 at 5 atm. pressure and in absorption in a King furnace.

References. W. W. Watson and R. Simon, *P.R.*, **55**, 358 (1939); **57**, 708 (1940).
L. Klynning, B. Lindgren and N. Åslund, *Ark. Fys.*, **30**, 141 (1965)†.

Complex bands with open structure. Degraded to the red. The $^SR_{21}$ and R_2 heads are more prominent but the R_1 and Q_2 are stronger. The (0, 0) band has heads:

$$4466\cdot5\ Q_2,\ 4447\cdot3\ R_2,\ 4070\cdot7\ R_1,\ 4053\cdot9\ ^SR_{21}.$$

There is probably a (0, 1) band around 4370–4335.

RED SYSTEM, $^4\Sigma - X\,^2\Pi$

For occurrence and references see Blue system.

SnH (*contd.*)

The main heads of the (0, 0) band of the $^4\Sigma - {}^2\Pi_{\frac{1}{2}}$ and $^4\Sigma - {}^2\Pi_{\frac{3}{2}}$ components are at 6095 and 6892 Å. Most heads are degraded to longer wavelengths. Heads of (0, 0) from Klynning *et al.*:

6931·3	$^O P_{42}$ deg. V	6025·0	$^R P_4$ deg. V
6891·6	$^Q R_{22}$ deg. R	6021·9	$^R P_4$ deg. R
6095·0	$^Q R_1$ deg. R	5918·9	$^S Q_4$ deg. R
6063·3	Q_2 deg. R		

There is a (1, 1) band of $^4\Sigma - {}^2\Pi_{\frac{1}{2}}$ at 6214. Watson and Simon also note heads at 7030 and 6745 Å.

ULTRA-VIOLET BANDS

Reference. W. R. S. Garton, *Proc. Phys. Soc.*, A, **64**, 591 (1951).

In the furnace absorption spectrum of tin vapour, certain bands were obtained only in the presence of hydrogen. A double-headed, red-degraded, diffuse band with heads at λ2259·5 and λ2255·3 and a diffuse region of absorption around λ2750 were the most prominent of these features.

SnI

VISIBLE SYSTEM, 6100–4750 Å, A − X$^2\Pi$

Reference. A. A. Murthy and P. B. V. Haranath, *Curr. Sci. (India)*, **38**, 211 (1969)†.

The photograph shows an extensive system of red-degraded bands, none particularly outstanding. They occur in a radio-frequency discharge. Measurements are not available.

ULTRA-VIOLET SYSTEMS

Occurrence. In flash photolysis of SnI_4.

References. G. A. Oldershaw and K. Robinson, *Trans. Faraday Soc.*, **64**, 616 (1968)†; *J. Mol. Spec.*, **45**, 489 (1973).

All bands are degraded to shorter wavelengths except F, G and H − X which are diffuse.

$B^2\Sigma - X^2\Pi$. Own estimates of intensity:

λ	I	v', v''	λ	I	v', v''	λ	I	v', v''
3163·3	5	0, 3	3105·2	10	0, 0	3059·6	8	2, 0
3144·0	7	0, 2	3101·4	9	1, 1	3037·2	6	3, 0
3124·6	10	0, 1	3082·3	10	1, 0	3015·4	5	4, 0

$C - X^2\Pi$.	2449·6 (0, 0).
$D - X^2\Pi$.	2401·9 (0, 0), 2389·3 (1, 0).
$E - X^2\Pi$.	2409·6 (0, 1), 2398·2 (0, 0).
$F - X^2\Pi$.	2379·8 (0, 0).
$G - X^2\Pi$.	2343·8 (0, 1), 2333·0 (0, 0).
$H - X^2\Pi$.	2178·8 (0, 0).

SnO

Occurrence. In arcs and flames containing tin salts; Connelly used a high-tension discharge through a flame containing $SnCl_4$ vapour for the production of the main system, and Loomis and Watson used an arc at reduced pressure for their system. In absorption by Sharma.

SnO (*contd.*)

References. F. C. Connelly, *Proc. Phys. Soc.*, **45**, 780 (1933)†.
F. W. Loomis and T. F. Watson, *P.R.*, **45**, 805 (1934).
D. Sharma, *Proc. Nat. Acad. Sci. India*, A, **14**, 133 (1944).
B. Eisler and R. F. Barrow, *Proc. Phys. Soc.*, A, **62**, 740 (1949)†.
R. F. Barrow and H. C. Rowlinson, *P.R.S.*, **224**, 374 (1954)†.
A. Lagerqvist, N. E. L. Nilsson and K. Wigartz, *Ark. Fys.*, **15**, 521 (1959)†.
J. L. Deutsch and R. F. Barrow, *Nature*, **201**, 815 (1964).
M. M. Joshi and R. Yamdagni, *Indian J. Phys.*, **41**, 275 (1967)†.

There has been some confusion over naming the systems. The strongest lies in the violet and near ultra-violet. Two weaker systems (or sub-systems of a $^3\Pi - {}^1\Sigma$ transition) also lie in the blue to violet. Loomis and Watson's system is a little further to the ultra-violet and there are others in the orange, in the region 2132–2037 Å and below 2000 Å.

MAIN SYSTEM, 4488–3072 Å, $D^1\Pi - X^1\Sigma^+$

Appearance. Degraded to the red.

Strong bands only; intensities on a scale of 8:

λ	I	v', v''	λ	I	v', v''	λ	I	v', v''
3691·4	5	0, 3	3415·8	5	1, 1	3262·4	6	2, 0
3585·4	7	0, 2	3388·3	6	0, 0	3205·8	4	3, 0
3484·5	8	0, 1	3323·4	7	1, 0			

WEAK BLUE-VIOLET SYSTEMS

There is some doubt about the analysis. Deutsch and Barrow regard the bands as overlapping $A^3\Pi_{0^+} - X^1\Sigma^+$ and $B^3\Pi_1 - X^1\Sigma^+$ sub-systems. Joshi and Yamdagni, whose data are given here, treat them as separate systems.

5840–3920 Å. Degraded to the red. 33 bands; strongest:

λ	I	v', v''	λ	I	v', v''
5302·5	5	3, 9	4763·9	5	0, 4
5148·0	6	0, 6	4590·7	5	0, 3

5010–3700 Å. Degraded to the red. 41 bands:

λ	I	v', v''	λ	I	v', v''	λ	I	v', v''
4411·6	7	1, 4	4218·8	8	0, 2	3948·5	8	0, 0
4302·2	6	2, 4	4079·9	10	0, 1	3863·5	9	1, 0
4262·3	9	1, 3	3978·8	6	4, 3			

LOOMIS AND WATSON'S SYSTEM, $E(?{}^1\Sigma^+) - X^1\Sigma^+$

Appearance. Degraded to red.

Strong bands only are listed. Intensities I_e and I_a on a scale of 8 for emission and 10 for absorption, respectively.

λ	I_e	I_a	v', v''	λ	I_e	I_a	v', v''	λ	I_e	I_a	v', v''
3043·6	6		3, 6	2740·1	6	4	4, 2	2560·0	3	10	6, 0
2990·4	8		1, 4	2716·9	7	8	3, 1	2529·9	3	10	7, 0
2947·7	6		2, 4	2680·8	6	9	4, 1	2500·4	2	10	8, 0
2921·7	8	2	1, 3	2658·1	6	2	3, 0	2472·3	2	10	9, 0
2814·8	8	5	2, 2	2646·9	6	9	5, 1	2445·0	1	9	10, 0

SnO (*contd.*)

ORANGE SYSTEM, 6480–5250 Å

19 bands reported by Joshi and Yamdagni in a flame. Strongly degraded to the red. The analysis may be doubtful. Strongest: λλ6069·9 (0, 3), 5789·6 (0, 2), 5643·9 (1, 2).

Eisler and Barrow report red-degraded bands λλ2132, 2117, 2102, 2084, 2073, 2069, 2054, 2050, 2037.

SnS

Several systems have been obtained in absorption and two of these have been observed by Barrow in a discharge tube.
References. G. D. Rochester, *P.R.S.*, **150**, 668 (1935)†.
D. Sharma, *Proc. Nat. Acad. Sci. India*, A, **14**, 217 (1945)†.
A. E. Douglas, L. L. Howe and J. R. Morton, *J. Mol. Spec.*, **7**, 161 (1961).
R. F. Barrow, G. Drummond and H. C. Rowlinson, *Proc. Phys. Soc.*, **66**, 685 (1953).
R. Yamdagni and M. M. Joshi, *Indian J. Phys.*, **40**, 495 (1966)†.

BLUE SYSTEMS, 5000–4000 Å

Degraded to the red. Rochester's system A has been divided into two by Yamdagni and Joshi.
5050–4270 Å.   4745·0 (1, 4), 4605·5 (0, 2), 4505·3 (0, 1), 4408·9 (0, 0).
4660–4100 Å.   4430·5 (0, 2), 4337·8 (0, 1), 4248·6 (0, 0), 4183·8 (1, 0).

4033–3198 Å SYSTEM, $D^1\Pi - X^1\Sigma$

Degraded to the red. Strongest λλ3865·3, 3728·2, 3662·8, 3599·3, 3557·1, 3496·6, 3418·8, 3381·6.

3325–2500 Å SYSTEM, $E^1\Sigma - X^1\Sigma$

Degraded to the red. Strongest bands ($I = 10$) from Sharma: λλ3163·9 (2, 4), 3116·8 (2, 3), 3089·1 (3, 3), 3062·4 (4, 3), 2992·6 (5, 2), 2967·7 (6, 2), 2902·1 (7, 1), 2879·2 (8, 1), 2735·8 (13, 0), 2715·9 (14, 0). Barrow *et al.* have extended this system, in absorption, to 2506·1 (29, 0).

SnSe

Six systems are now known in absorption, and systems D and E have been obtained by Barrow and Vago in emission in a heavy-current discharge. Bands of all systems are degraded to the red, but many of the heads are diffuse because of isotope splitting.
References. J. W. Walker, J. W. Straley and A. W. Smith, *P.R.*, **53**, 140 (1938).
R. F. Barrow and E. E. Vago, *Proc. Phys. Soc.*, **55**, 326 (1943)†.
D. Sharma, *Proc. Nat. Acad. Sci. India*, A, **14**, 224 (1945).
E. E. Vago and R. F. Barrow, *Proc. Phys. Soc.*, **58**, 707 (1946)†.
R. Yamdagni, *J. Mol. Spec.*, **33**, 531 (1970).

$A - X^1\Sigma$ SYSTEM, 5625–4882 Å

26 bands known in absorption.

SnSe (*contd.*)

λ	I	v', v''	λ	I	v', v''
5394·2	4	1, 1	5238·5	3	2, 0
5331·0	5	2, 1	5180·1	4	3, 0
5269·5	3	3, 1	5122·9	3	4, 0

B – X¹Σ SYSTEM, 4807–4396 Å

Strongest absorption bands from Yamdagni.

λ	I	v', v''	λ	I	v', v''
4665·1	3	1, 4	4504·6	7	0, 1
4595·3	3	1, 3	4438·8	5	0, 0
4572·5	6	0, 2	4396·4	9	1, 0

C – X¹Σ SYSTEM, 4630–4191 Å

16 bands, in absorption.

λ	I	v', v''	λ	I	v', v''
4472·3	7	0, 2	4344·7	10	0, 0
4407·6	7	0, 1	4304·3	8	1, 0

D – X¹Σ SYSTEM

Strongest bands in emission, with our estimates of intensity from published spectrogram:

λ	I	v', v''	λ	I	v', v''	λ	I	v', v''
3963·7	7	0, 7	3770·9	9	0, 3	3620·2	8	2, 1
3914·1	8	0, 6	3739·2	10	1, 3	3591·6	10	3, 1
3865·1	8	0, 5	3694·1	9	1, 2	3563·5	9	4, 1
3817·7	9	0, 4	3664·0	9	2, 3	3536·3	8	5, 1

SYSTEM E – X¹Σ

The following bands have been observed in emission: λλ3434·7, 3397·3, 3374·9, 3359·8, 3339·0, 3317·8, 3297·0, 3282·2, 3267·8, 3247·4, 3242·2, 3227·8, 3214·0, 3203·4, 3189·7, 3157·6, 3119·3, 3102·2; no intensities are available.

F – X¹Σ SYSTEM, 2195–2077 Å

A progression of absorption bands λλ2194·8 (0, 7), 2179·3 (0, 6), 2163·9 (0, 5), 2148·9 (0, 4), etc.

SnTe

Seven systems have been obtained in absorption. Two of these, D and E, occur in emission in an uncondensed discharge through a mixture of Sn, Te and Al vapours. The notation is that used by Vago and Barrow. All systems are degraded to the red.

References. R. F. Barrow, *Proc. Phys. Soc.*, **52**, 380 (1940)†.

R. F. Barrow and E. E. Vago, *Proc. Phys. Soc.*, **56**, 78 (1944)†.

D. Sharma, *Proc. Nat. Acad. Sci. India*, A, **14**, 232 (1945).

E. E. Vago and R. F. Barrow, *Proc. Phys. Soc.*, **58**, 707 (1946)†.

No intensities are available, but the following are prominent in published photographs:

SnTe (*contd.*)

SYSTEM A, A − X$^1\Sigma$

λ	v', v''	λ	v', v''	λ	v', v''
6236·2	0, 3	5916·0	2, 1	5653·7	5, 0
6167·6	1, 3	5854·8	3, 1	5598·8	6, 0
6071·6	1, 2	5795·8	4, 1	5545·5	7, 0
6007·3	2, 2	5710·1	4, 0		

SYSTEM B, B − X$^1\Sigma$

λ	v', v''	λ	v', v''	λ	v', v''
5165·7	0, 4	4975·6	1, 2	4807·7	3, 1
5098·6	0, 3	4968·5	0, 1	4798·5	2, 0
5040·0	1, 3	4912·8	1, 1	4748·5	3, 0
5033·0	0, 2	4859·2	2, 1	4699·4	4, 0

SYSTEM C, C − X$^1\Sigma$

λ	v', v''	λ	v', v''	λ	v', v''
4738·8	1, 2	4580·0	2, 0	4493·1	4, 0
4672·0	0, 0	4546·1	4, 1	4451·1	5, 0
4634·7	2, 1	4535·8	3, 0	4410·5	6, 0
4589·6	3, 1				

SYSTEM D, D − X$^1\Sigma$

λ	v', v''	λ	v', v''	λ	v', v''
4189·5	0, 6	4058·8	0, 3	3893·3	3, 1
4145·3	0, 5	3988·5	1, 2	3854·0	3, 0
4101·3	0, 4	3920·3	2, 1	3827·9	4, 0

SYSTEM E, E − X$^1\Sigma$, 3800−3327 Å

λ	v', v''
3765·0	0, 4
3728·9	0, 3
3693·9	0, 2

SYSTEM F, F − X$^1\Sigma$, 2353−2184 Å

26 bands.

SYSTEM G, G − X$^1\Sigma$, 2188−2117 Å

7 bands.

Sharma refers to the last two systems as K − X and I − X and notes other systems 3666−3450, 3512−3298 and 3309−3165 Å.

SrBr

Four systems are now known, two in the red, one in the violet and one in the near ultra-violet.

SrBr (*contd.*)

MAIN RED SYSTEM, $A^2\Pi - X^2\Sigma$

Occurrence. In absorption and in emission from flames and high-frequency discharges. They do not appear strongly in an arc.

Appearance. Close marked sequences; the bands appear to be degraded to shorter wavelengths under low dispersion.

References. K. Hedfeld, *Z.P.*, **68**, 610 (1931).
O. H. Walters and S. Barratt, *P.R.S.*, **118**, 120 (1928).

The following list of measurements is compiled from the above sources. Intensities I_a and I_f are for absorption and emission in a flame respectively.

λ	I_a	I_f	Sequence	λ	I_a	I_f	Sequence
6924	0			6605·4		0	0, 1 ii
6800·2	10			6572·4		0	1, 0 i
6763·6		0	0, 1 i	6513·0	5	10	0, 0 ii
6666·7	10	10	0, 0 i	6422·8		0	1, 0 ii

WEAKER RED SYSTEM, $B^2\Sigma - X^2\Sigma$

According to Rosen bands 6695–6310 Å were observed by R. E. Harrington (*Ph.D. Thesis, Berkeley*) in absorption. Violet-degraded sequences 6605·3 (0, 1), 6512·9 (0, 0), 6420·5 (1, 0).

VIOLET SYSTEM, $C^2\Pi - X^2\Sigma$

Occurrence. In absorption and in a flame.

Appearance. Close sequences which appear to be degraded to longer wavelengths with small dispersion.

References. O. H. Walters and S. Barratt, *P.R.S.*, **118**, 120 (1928).
C. M. Olmsted, *Z. wiss. Photogr.*, **4**, 255 (1906).
S. N. Puri and H. Mohan, *Curr. Sci. (India)*, **44**, 152 (1975).

The following measurements are by Walters and Barratt. Intensities I_a and I_f are for absorption and emission in a flame, the latter being by Olmsted.

λ	I_a	I_f	λ	I_a	I_f	λ	I_a	I_f
4186	0	1	4090	3	3	3992	0	1
4146	4	3	4073	2	3	3945	5	
4129	1	2	4053	10	6	3909	5	
4108	9	6	4019	2	3			

ULTRA-VIOLET SYSTEM $D^2\Sigma - X^2\Sigma$, 3505–3320 Å

Observed in absorption by R. E. Harrington (see Rosen) and by Reddy and Rao (see below). Degraded to shorter wavelengths. Close double-headed. Sequence heads $\lambda\lambda$3502·1 (0, 2), 3476·3 (0, 1), 3450·4 (0, 0), 3421·3 (1, 0), 3392·7 (2, 0).

ULTRA-VIOLET SYSTEM $E - X^2\Sigma$, 3205–3040 Å

Reference. Y. P. Reddy and P. T. Rao, *Indian J. Pure Appl. Phys.*, **4**, 251 (1966)†.

Degraded to shorter wavelengths. Found in a high-frequency discharge. 21 bands recorded. Sequence heads:

SrBr (*contd.*)

λ	I	v', v''
3159·9	6	0, 2
3138·6	9	0, 1
3117·5	10	0, 0
3093·7	8	1, 0

SrCl

Occurrence. In an arc or flame. Also in absorption.

References. K. Hedfeld, *Z.P.*, **68**, 610 (1931)†.

A. E. Parker, *P.R.*, **47**, 349 (1935).

R. E. Harrington, *Ph.D. Thesis, Berkeley* (after Rosen).

J. Singh *et al.*, *Opt. Pura Apl.* (*Spain*), **3**, 76 (1970).

There are two strong systems in the red and another in the violet, with weaker systems in the ultra-violet.

FIRST RED SYSTEM, 6895–6465 Å, $A^2\Pi - X^2\Sigma^+$

Appearance. Degraded to the violet. Long sequences of bands, close together. Strong sequence heads:

λ	I	Sequence	λ	I	Sequence
6893·8	0	0, 1 P_1	6619·9	5	0, 0 P_2
6755·9	3	0, 0 P_1	6613·7	10	0, 0 Q_2
6744·9	5	0, 0 Q_1			

SECOND RED SYSTEM, 6600–6230 Å, $B^2\Sigma^+ - X^2\Sigma^+$

Degraded to the violet. Close double-headed bands in long sequences.

λ	I	Sequence
6485·2	4	0, 1
6362·4	10	0, 0
6239	2	1, 0

VIOLET SYSTEM, 4136–3852 Å, $C^2\Pi - X^2\Sigma^+$

Appearance. Degraded to red.

No intensities given. The following are the Q heads of sequences, the R heads which are presumably weaker, lie about 1 Å to the violet:

λ

4009·4 Q_1 head of (1, 2) band, the first observed member of the (0, 1) sequence.

3983·4 Q_2 head of (1, 2) band.

*3961·6 Q_1 head of (0, 0) band and sequence.

3937·1 Q_2 ” ” ” ” ”

3918·3 Q_1 ” (1, 0) ” ” ”

3894·0 Q_2 ” ” ” ” ”

* Strongest head.

SrCl (*contd.*)

ULTRA-VIOLET SYSTEMS

Harrington (see Rosen) notes absorption systems, all degraded to shorter wavelengths:

$$D^2\Sigma^+ - X^2\Sigma^+, 3570-2925 \text{ Å}; \ 3465 \cdot 9 \ (0, 0).$$
$$E^2\Sigma^+ - X^2\Sigma^+, 3160-2925 \text{ Å}; \ 3102 \cdot 4 \ (0, 0).$$
$$F^2\Sigma^+ - X^2\Sigma^+, 2945-2785 \text{ Å}; \ 2915 \cdot 6 \ (0, 0).$$

SrF

Seven systems have been observed in absorption by heated SrF_2 vapour by Fowler, and three of these, the A system in the red, the B system in the yellow, and the C system, are well known in emission when SrF_2 is introduced into a carbon arc or flame.

References. S. Datta, *P.R.S.*, **99**, 436 (1921)†.
R. C. Johnson, *P.R.S.*, **122**, 161 (1929).
A. Harvey, *P.R.S.*, **133**, 336 (1931)†.
C. A. Fowler, *P.R.*, **59**, 645 (1941)†.
M. M. Novikov and L. V. Gurvich, *Opt. Spectrosc.*, **22**, 395 (1967)†.

RED SYSTEM, 6870–6283 Å, $A^2\Pi - X^2\Sigma^+$

Marked sequences. Widely spaced double-headed bands, mostly degraded to the violet. Appearance best seen from Datta's photograph. Sequence heads:

λ	I	Sequence	λ	I	Sequence	λ	I	Sequence
6874·8	–	0, 1 P_{12}	6655·6	7	0, 0 P_{12}	6419·0	8	1, 0 deg. V
6858·7	–	0, 1 Q_{12}	6632·7	10	0, 0 Q_{12}	6394·7	8	1, 0 deg. R
6741·0	–	0, 1 P_2	6527·6	7	0, 0 P_2	6306·1	8	1, 0 deg. V
6729·4	–	0, 1 Q_2	6512·0	10	0, 0 Q_2	6283·1	8	1, 0 deg. R

YELLOW SYSTEM, 5852–5021 Å, $B^2\Sigma^+ - X^2\Sigma^+$

Degraded to the red.

(0, 0) sequence, Q head 5779·5.
        "   "   R "   5772·0.
(1, 0) sequence, evenly spaced bands 5622–5670 Å.

C SYSTEM, 3795–3646 Å, $C^2\Pi - X^2\Sigma^+$

Degraded to the red. Double double-headed. Sequence heads λλ3641·0 (0, 1 R_2), 3654·1 (0, 0 Q_1), 3650·8 (0, 0 R_1), 3646·0 (0, 0 R_2), 3641·0 (0, 0 $^sR_{21}$), 3591·9 (1, 0 R_2).

D SYSTEM, 3592–3345 Å, $D^2\Sigma^+ - X^2\Sigma^+$

Degraded to shorter wavelengths. Strongest bands in absorption (calculated from Fowler's formula):

λ	I	v', v''
3529·8	9	0, 0
3522·9	10	1, 1
3517·1	8	2, 2
3457·2	7	2, 1
3451·6	7	3, 2

SrF (*contd.*)

E SYSTEM, 3218–3052 Å, $E^2\Pi - X^2\Sigma^+$

Degraded to shorter wavelengths. Strongest bands in absorption:

λ	I	v', v''
3218·2	5	0, 1
3167·6	10	0, 0
3112·6	7	1, 0
3106·4	6	2, 1

F SYSTEM, 3069–2916 Å, $F^2\Sigma^+ - X^2\Sigma^+$

Degraded to shorter wavelengths. Strongest bands in absorption:

λ	I	v', v''
3088·3	6	0, 1
3041·5	8	0, 0
2987·8	10	1, 0
2978·1	7	2, 1
2929·6	7	3, 1

G SYSTEM, 2915–2775 Å, $G^2\Pi - X^2\Sigma^+$

Degraded to shorter wavelengths. Strongest bands in absorption:

λ	I	v', v''
2873·2	10	0, 0
2826·7	8	1, 0
2821·0	5	2, 1
2782·0	6	2, 0

SrH

References. W. W. Watson and W. R. Fredrickson, *P.R.*, **39**, 765 (1932).
W. R. Fredrickson, M. E. Hogan and W. W. Watson, *P.R.*, **48**, 602 (1935); **49**, 150 (1936).
D. R. More and S. D. Cornell, *P.R.*, **53**, 806 (1938).
G. Edvinsson *et al.*, *Ark. Fys.*, **25**, 95 (1963)†.
M. A. Khan, *Proc. Phys. Soc.*, **89**, 165 (1966).
M. A. Khan and M. R. Butt, *J. Phys.*, B, **1**, 745 (1968).

7508 Å, $A^2\Pi - X^2\Sigma$

Degraded to the violet. Obtained from strontium arc in H_2. (0, 0) heads, 7508 P_1, 7505 $^PQ_{12}$, 7348·0 $^QP_{21}$, 7346·7 Q_2.

7020 Å, $B^2\Sigma - X^2\Sigma$

Degraded to the violet. Obtained in Sr arc in H_2. Heads 7018·1 (0, 0 P_1), 7009·5 (1, 1 P_1), 6984·7 (0, 0 P_2), 6974·6 (1, 1 P_2).

3808 Å, $C^2\Sigma - X^2\Sigma$

Degraded to the violet. Found in an arc and in absorption. P heads, λλ3986·1 (0, 1), 3808·3 (0, 0) strong, 3789·7 (1, 1), 3773·2 (2, 2), 3628·7 (1, 0).

SrH (*contd.*)

4818 Å, $D^2\Sigma - X^2\Sigma$

Observed in an arc and in absorption. Strongly degraded to the red, of many-line type with weak R heads. Origins: λ_o 5766·3 (0, 3), 5421·0 (0, 2), 5105·5 (0, 1), 4862·0 (1, 1), 4817·6 (0, 0), 4406·7 (2, 0), 4234·1 (3, 0), 4080·4 (4, 0). The last three are strong in absorption. The (1, 0) is masked by an Sr line.

5323 Å, $E^2\Pi - X^2\Sigma$

Degraded to the violet. In an arc and in absorption. The strong (0, 0) sequence has a head at 5323.

2927 Å, $F^2\Sigma - X^2\Sigma$

In absorption. Degraded to shorter wavelengths. 2927·3 (0, 0 P).

3115 Å, $G^2\Sigma - X^2\Sigma$

Degraded to shorter wavelengths. 3115·3 (0, 0 P).

2420 Å, $H^2\Sigma - X^2\Sigma$

Headless. 2420 (0, 0).

Bands at 2836, 2680 and 3620 have not yet been analysed.

SrI

Occurrence. In high-frequency discharges, flames and arcs. Also in absorption.
References. B. R. K. Reddy, Y. P. Reddy and P. T. Rao, *J. Phys.*, B, **4**, 574 (1971)†.
S. G. Shah, M. M. Patel and A. B. Darji, *J. Phys.*, B, **6**, 1344 (1973)†.
Ashrafunnisa, K. V. K. Rao and P. T. Rao, *J. Phys.*, B, **6**, 1506 and 2653 (1973)†.

$A^2\Pi - X^2\Sigma$ SYSTEM, 7100–6580 Å

Appearance. Long sequences of violet-degraded bands. Sequence heads from Ashrafunnisa *et al.*:

λ	I	v', v''		λ	I	v', v''		λ	I	v', v''
7016·1	1	0, 1 $^oP_{12}$		6925·5	9	0, 0 P_1		6745·7	7	0, 0 Q_2
7013·6	5	0, 1 P_1		6845·5	1	1, 0 $^oP_{12}$		6665·6	3	1, 0 P_2
6931·6	8	0, 0 $^oP_{12}$		6747·2	8	0, 0 P_2				

$B^2\Sigma - X^2\Sigma$ SYSTEM, 6845–6615 Å

Appearance. Violet-degraded sequences.
Heads from Ashrafunnisa *et al.*:

λ	I	v', v''		λ	I	v', v''
6859·0	2	0, 1 P_2		6697·8	2	1, 0 P_2
6778·8	10	0, 0 P_2		6695·8	6	1, 0 P_1
6777·0	10	0, 0 P_1				

$C^2\Pi - X^2\Sigma$, 4500–4200 Å

Appearance. Degraded to the red. Long sequences, some 70 heads being known. Only the (0, 0) sequences are strong.

SrI (*contd.*)

Heads from Reddy *et al.*:

λ	I	v', v''	λ	I	v', v''
4410·9	10	0, 0 Q_1	4305·3	6	0, 0 R_2
4410·3	4	0, 0 R_1	4304·8	4	0, 0 $^SR_{21}$

$D^2\Sigma - X^2\Sigma$ SYSTEM, 3560–3350 Å

Appearance. Degraded to shorter wavelengths. About 50 heads, forming well marked sequences. Data by Shah *et al.* (Ashrafunnisa *et al.* give wavelengths about 0·6 Å different):

λ	I	v', v''	λ	I	v', v''
3515·5	2	0, 3	3452·4	10	0, 0
3494·2	7	0, 2	3428·8	9	1, 0
3473·2	9	0, 1	3405·7	8	2, 0

SrO

Strontium salts give bright red banded radiation in flames and arcs, but the flame bands are mostly due to SrOH and the arc bands may be due to a polyatomic oxide. The diatomic oxide possesses a strong system in the infra-red and further systems in the blue and ultra-violet; all these have the same final ($X\,^1\Sigma$) state but it is not certain that this is the ground state.

Occurrence. In arcs.

INFRA-RED SYSTEM, $A^1\Sigma - X^1\Sigma$

Appearance. Single-headed, degraded to longer wavelengths.

References. K. Mahla, *Z.P.*, **81**, 625 (1933)†.

G. Almkvist and A. Lagerqvist, *Nature*, **164**, 665 (1949); *Ark. Fys.*, **2**, 233 (1950).

A. Lagerqvist and L. E. Selin, *Ark. Fys.*, **11**, 323 (1956)†.

L. Brewer and R. Hauge, *J. Mol. Spec.*, **25**, 330 (1968).

λ	v', v''	λ	v', v''	λ	v', v''
7861·8	3, 0	8700·0	1, 0	9776·1	0, 1
7884·0	4, 1	8722·5	2, 1	10426·2	0, 2
8257·8	2, 0	9195·8	0, 0	10437·1	0, 3
8282·2	3, 1				

BLUE SYSTEM, $B^1\Pi - X^1\Sigma$

Appearance. Degraded to the red.

References. P. C. Mahanti, *P.R.*, **42**, 609 (1932)†.

I. Kovács and A. Budó, *Acta. Phys. Hungar.*, **1**, 469 (1952); *Ann. Phys. (Lpz.)*, **6**, 17 (1953)†.

The following measurements of strong bands are from Mahanti, with intensities amended to a scale of 10.

λ	I	v', v''	λ	I	v', v''	λ	I	v', v''
4692·7	9	2, 7	4463·3	7	3, 6	4189·1	7	1, 2
4672·6	7	1, 6	4420·9	7	1, 4	4167·2	9	0, 1
4652·4	5	0, 5	4399·6	10	0, 3	4058·0	5	0, 0
4564·8	9	2, 6	4302·7	7	1, 3	3975·4	5	1, 0
4544·1	9	1, 5	4281·0	9	0, 2	3897·1	5	2, 0
4523·3	7	0, 4						

SrO (*contd.*)

ULTRA-VIOLET SYSTEM, $C^1\Sigma - X^1\Sigma$

Appearance. Degraded to red.

References. P. C. Mahanti, *P.R.*, **42**, 609 (1932)†.

A. Lagerqvist and G. Almkvist, *Ark. Fys.*, **8**, 481 (1954).

The following measurements of strong bands are from Mahanti, with intensities amended to a scale of 10.

λ	I	v', v"	λ	I	v', v"
3586·9	5	0, 1	3445·2	7	1, 0
3525·4	5	1, 1	3389·8	7	2, 0
3503·8	10	0, 0	3337·5	5	3, 0

NEAR INFRA-RED SYSTEM, $A'\ ^1\Pi - X^1\Sigma$

G. A. Capelle, H. P. Broida and R. W. Field, *J. Chem. Phys.*, **62**, 3131 (1975) have observed a weak system in the reaction of Sr atoms with O_3 or N_2O. Strongest heads 7723·0 (8, 0), 7472·9 (9, 0), 7238·0 (10, 0).

Strontium Oxide

Occurrence. Strongly in strontium arc in air; weakly in flames. These bands may be due either to SrO (possibly a triplet system) or to a more complex oxide.

Reference. M. Charton and A. G. Gaydon, *Proc. Phys. Soc.*, A, **69**, 520 (1956)†.

There are strong bands near 5950 and 6050 Å and more strong complex structure between 6400 and 6850 Å.

The 5950 band shows apparent heads degraded to the *red* at 5938 and 5969 Å under small dispersion. With larger dispersion heads degraded to the *violet* are located at λλ5945·4, 5950·7, 5952·8, 5954·2, 5955·4, 5957·6, 5962·4, 5963·1, 5965·1, 5972·5, 5976·3, 5977·0, 5980·6 and 5986·2. There are maxima at 5967·4, 5966·3, 5965·6 and a red-degraded head at 5974·8.

The 6050 Å band shows two diffuse red-degraded heads.

The red bands show weak violet-degraded heads at λλ6672·4, 6637·4, 6634·3, 6600·6, 6585·8, 6521·4, 6515·8, 6511·6, 6507·8, 6466·1, 6454·4, 6451·1, 6447·3. There is a broad diffuse band at 6457 and a red-degraded head at 6620·7.

SrOH

Occurrence. These bands are responsible for the strong red colour of flames and fireworks containing strontium. They also occur in an arc in water vapour at reduced pressure.

References. C. G. James and T. M. Sugden, *Nature*, **175**, 333 (1953).

A. Lagerqvist and L. Huldt, *Naturwiss.*, **42**, 365 (1955).

L. Huldt and A. Lagerqvist, *Ark. Fys.*, **11**, 347 (1956).

M. Charton and A. G. Gaydon, *Proc. Phys. Soc.*, A, **69**, 520 (1956)†.

J. van der Hurk, T. Hollander and C. T. J. Alkemade, *J. Quant. Spectrosc. Rad. Transfer*, **13**, 273 (1973); **14**, 1167 (1974).

The strongest band in the orange is centred at 6050 Å. On the longer wavelength side it shows some structure, with close double heads, shaded to the violet.

SrOH (*contd.*)

λ	I	λ	I
6109·6	1	6089·9	6
6101·1	3	6084·7	10
6095·9	4	6076·6	8

Lagerqvist and Huldt give heads at λλ6114·2, 6111·8, 6109·8, 6107·9, 6107·5, and 6105·2.

In the far red the bands are diffuse, the maxima of the strongest being at λλ6820, 6675, 6590 and 6460.

Reference. M. Marcano and R. F. Barrow, *Trans. Faraday Soc.*, **66**, 1917 (1970)†.

Bands of a red-degraded $B^1\Sigma - X^1\Sigma$ system 3900–3600 Å in absorption. Few details are available; heads are not well developed, but those at 3881·1 (0, 2) and 3728·3 (1, 0) are clearest.

TaO

References. C. C. Kiess and E. Z. Stowell, *J. Res. Nat. Bur. Stand.*, **12**, 459 (1934).

I. Fernando and S. G. Krishnamurty, *Curr. Sci. (India)*, **18**, 371 (1949).

D. Premaswarup, *Indian J. Phys.*, **29**, 109 (1955)†.

C. J. Cheetham and R. F. Barrow, *Trans. Faraday Soc.*, **63**, 1835 (1967).

Kiess and Stowell reported an extensive band spectrum emitted by the tantalum arc in air. The bands are shaded towards the red and are without prominent heads. Most intense of heads given are:

λ	I	λ	I	λ	I
9919·9	8	5567·0	4	4154·4	7 ?
9902·7	5	5385·2	3 ?	4092·1	3
9868·6	25	4901·6	3 ?	4006·2	4
9849·3	15	4810·4	4	3896·4	4
9242·7	5	4679·5	3	3747·2	4
9197·4	8	4651·9	3	3625·7	4

In a short letter Fernando and Krishnamurty report a band spectrum in the region 5500–3900 Å obtained with the arc in air between tantalum rods and in the arc between carbon rods containing tantalum oxide. Heads are given at λλ4909·6, 4744·5, 4652·0, 4590·9, 4503·5, 4415·9, 4361·7, 4281·7 and 4154·3.

Premaswarup (1955) gives an extensive table of more than sixty heads for the region 4950–3429 Å and divides them into two systems A and C.

Cheetham and Barrow, using an arc and a high-frequency discharge, have analysed bands into systems involving 19 electronic states. The 3747·2 Å band is stated to be strongest. There are (0, 0) bands of systems or sub-systems at λλ9195, 9039, 8470, 7779, 7362, 6961, 6291, 5980, 5552, 5150, 4804·5, 4787, 4525, 4518, 4476·5, 4420, 4283, 4202, 4154, 3906, 3803·5, 3747, 3419, 3096, 3087, 3019, 3003 Å.

TaO⁺

Reference. C. J. Cheetham and R. F. Barrow, *Trans. Faraday Soc.*, **63**, 1835 (1967).

A red-degraded band at 5589 Å (0, 0) with single P, Q and R branches, obtained in an arc between tantalum electrodes, is provisionally attributed to the ionized molecule TaO^+, although the emitter is not certain.

TbO

Occurrence. Terbium salts in a flame or arc.

Appearance. Numerous bands, probably involving many systems, mostly degraded to the red.

References. G. Piccardi, *Spectrochim. Acta*, **1**, 533 (1941).

R. Herrmann and C. T. J. Alkemade, *Flame Photometry*, (1963).

A. Gatterer *et al.*, see B. Rosen.

This list is based on Herrmann and Alkemade's which includes over 80 bands; some wavelengths are corrected to Gatterer's values:

λ	I	λ	I	λ	I	λ	I
6076·6	6	5999	4	5925	4	5639	4
6068	5	5979·9	9	5920·8	10	5350·5	3
6056	4	5940	9	5857	5	5340	4

Te$_2$

MAIN SYSTEM, B $0_u^+ - $ X 0_g^+

Occurrence. Absorption by tellurium vapour, fluorescence, and emission in microwave discharge.

Appearance. Degraded to the red. An extensive system consisting of a large number of bands. Well over 100 bands.

References. B. Rosen, *Z.P.*, **43**, 69 (1927).

E. Olsson, *Z.P.*, **95**, 215 (1935).

B. L. Jha, K. V. Subbaram and D. R. Rao, *J. Mol. Spec.*, **32**, 383 (1969)†.

The following measurements of the strong bands are by Olsson, with his vibrational quantum numbers. The intensities are from Rosen and from Jha *et al.* for Te$_2^{130}$, and are for absorption. The band heads are complex because of the isotope effect. Olsson's measurements are for the strongest head, Te^{128}Te^{128} + Te^{130}Te^{126}. Rosen's measurements are systematically lower than Olsson's by an amount increasing from zero at the red end of the system to 5 Å at the violet end, this probably being largely due to the isotope effect.

λ	I	v', v''	λ	I	v', v''	λ	I	v', v''
4849·0	3	1, 7	4448·8	3	5, 2	4211·8	3	10, 0
4793·6	3	1, 6	4435·7	3	7, 3	4202·9	4	12, 1
4740·7	3	1, 5	4418·3	3	6, 2	4185·7	3	11, 0
4703·5	3	2, 5	4388·4	3	7, 2	4159·6	4	12, 0
4686·7	3	4, 6	4369·4	3	6, 1	4134·5	4	13, 0
4666·8	3	3, 5	4358·9	4	8, 2	4110·0	5	14, 0
4649·4	3	2, 4	4341·3	3	7, 1	4082·8	5	15, 0
4617·2	3	6, 6	4330·4	4	9, 2	4060·6	5	16, 0
4581·0	4	4, 4	4312·5	3	8, 1	4040·2	5	17, 0
4564·9	3	6, 5	4302·2	4	10, 2	4018·2	5	18, 0
4548·1	4	5, 4	4284·5	4	9, 1	3996·8	5	19, 0
4530·0	4	4, 3	4256·8	4	10, 1	3976·0	5	20, 0
4498·0	4	5, 3	4238·1	3	9, 0	3956·0	4	21, 0
4466·4	3	6, 3	4230·2	4	11, 1			

Te₂ (*contd.*)

WEAKER SYSTEM, 5200–4250 Å, A $0_u^+ - X 0_g^+$

Reference. R. F. Barrow and R. P. du Parcq, *P.R.S.*, **327**, 279 (1972).

Weak absorption bands of this system overlap the main system. Term values have been listed but wavelengths and intensities have not been published.

OTHER SYSTEMS

References. M. Désirant and A. Minne, *C. R. Acad. Sci.* (*Paris*), **202**, 1272 (1936). Choong Shin-Piaw, *C. R. Acad. Sci.* (*Paris*), **203**, 239 (1936); and *Ann. Phys.* (*Paris*), **10**, 173 (1938). R. Migeotte, *Mem. Soc. Roy. Sci. Liège*, **5**, 1 (1942). N. Durga Prasad and P. Tiruvenganna Rao, *Indian J. Phys.*, **28**, 549 (1954)†.

Désirant and Minne record bands in the visible in a high-frequency discharge, analysed into two systems with origins of the (0, 0) bands at 5290 and 6107 Å.

Choong Shin-Piaw has studied the spectrum in the ultra-violet and gives a formula for bands in the region λλ2495–1975. Migeotte has arranged absorption bands in this region into four systems.

Prasad and Rao have reinvestigated the band spectrum of tellurium in the region λλ6400–4900, using a discharge tube excited by a high power oscillator. They compare their emission bands with the absorption and fluorescence bands obtained by Rosen and with the emission bands obtained by Désirant and Minne.

TeBr

Reference. G. A. Oldershaw and K. Robinson, *Chem. Comm.*, **540**, (1970).

A $B - X^2\Pi$ system of bands, degraded to shorter wavelengths, is reported in flash photolysis. (0, 0) sub-bands 2413·9 and 2316·9 Å.

TeBr₂

Diffuse bands, degraded to the red, in the region λλ6500–5300 have been observed in absorption.

Reference. M. Wehrli, *Helv. Phys. Acta*, **9**, 208 (1936)†.

Strongest bands:

λ	I	λ	I
6037	6	5816	9
5957	7	5744	9
5888	8	5676	8

TeCl

Reference. G. A. Oldershaw and K. Robinson, *Chem. Comm.*, **540** (1970).

In flash photolysis, a $B - X^2\Pi$ system of bands, degraded to shorter wavelengths, is reported. (0, 0) sub-bands 2345·4 and 2256·8 Å.

TeCl₂

VISIBLE SYSTEM

Diffuse bands, degraded to the red, in the region λλ6400–4725 have been observed in absorption.

TeCl$_2$ (*contd.*)

Reference. M. Wehrli, *Helv. Phys. Acta*, **9**, 208 (1936)†.

Strongest bands:

λ	I	λ	I
5758·6	3	5443·4	7
5720·0	2	5353·6	6
5656·1	3	5267·5	5
5634·3	3	5183·5	4
5536·4	5	5103	4

ULTRA-VIOLET SYSTEM

Bands 2050–2300 Å in absorption. Degraded to shorter wavelengths.

Reference. P. Müller and M. Wehrli, *Helv. Phys. Acta*, **15**, 307 (1942).

Strongest bands: λλ2079·6, 2084·6, 2105·4, 2111·9, 2118·6, 2138·6.

TeI

Reference. G. A. Oldershaw and K. Robinson, *Trans. Faraday Soc.*, **67**, 907 (1971)†.

In flash photolysis there are two systems, both shaded to shorter wavelengths.

$B - X^2\Pi$. 20 bands. The following sequence heads are just visible in the photograph: λλ2446·0 (0, 1), 2433·5 (0, 0), 2418·4 (1, 0), 2403·8 (2, 0), 2389·3 (3, 0).

$C - X^2\Pi$. 22 bands. λλ2330·2 (0, 1), 2289·0 (0, 0), 2275·9 (1, 0), 2263·0 (2, 0), 2250·2 (3, 0).

TeO

Occurrence. In discharge through tellurium vapour and oxygen in heated silica tube, in heavy current arc and in electrodeless discharge with tellurium, oxygen and argon mixtures. Also in absorption.

Appearance. Degraded to red, with long progressions.

References. Choong Shin-Piaw, *Ann. Phys.* (*Paris*), **10**, 173 (1938)†.

P. V. B. Haranath, P. T. Rao and V. Swaramamurty, *Z.P.*, **155**, 507 (1959)†.

R. L. Purbrick, *J. Chem. Phys.*, **30**, 962 (1959).

G. G. Chandler, H. J. Hurst and R. F. Barrow, *Proc. Phys. Soc.*, **86**, 105 (1965)†.

This $^3\Sigma^- - {}^3\Sigma^-$ system is best treated as two sub-systems, A $0^+ - X_1 0^+$ and A $1 - X_2 0^+$. The full analysis is not readily available, but the 4135·8 band is the (0, 5) of A $0^+ - X_1 0^+$ and 4205·1 is (0, 5) of A $1 - X_2 1$.

In emission Haranath *et al.* find about 80 bands λλ6200–3380. The following are strongest:

λ	I	λ	I	λ	I
·5924·8	5	4639·9	7	4268·2	9
5746·2	4	4625·1	4	4205·1	10
5503·8	4	4555·5	7	4135·8	9
4940·8	4	4544·4	4	4125·1	7
4862·5	4	4487·4	9	4075·4	8
4801·7	4	4459·6	4	4069·6	5
4710·7	6	4407·6	8	4059·6	6
4698·5	5	4342·8	10	4009·0	7

TeO (*contd.*)

λ	I	λ	I	λ	I
4002·9	6	3879·0	5	3607·0	4
3951·9	6	3819·8	4	3560·8	3
3940·4	5	3710·8	4	3516·4	2
3890·6	4	3661·7	5	3508·0	2

TeO$_2$

References. Choong Shin-Piaw, *Ann. Phys.* (*Paris*), **10**, 173 (1938)†.

J. Duchesne and B. Rosen, *J. Chem. Phys.*, **15**, 631 (1947).

I. Dubois, *Bull. Soc. Roy. Sci. Liège*, **39**, 63 (1970).

A complex system of diffuse bands 4000–3200 Å has been observed in absorption. Strongest λλ3960·1, 3942·7, 3922·8, 3891·1, 3799·9, 3772·9, 3771·0, 3706·4, 3617·8, 3590·8.

Duchesne and Rosen have listed 88 absorption bands 4555–3026 Å. Their values for band maxima show practically no agreement with those above. They also list 12 bands λλ2650–2450.

TeS

VIOLET SYSTEMS, 4000–3700 Å

Reference. R. F. Barrow *et al.*, *J. Phys.*, B, **5**, L 172 (1972).

17 red-degraded absorption bands have been assigned to three systems. No wavelengths are available, but term values indicate the following as probable bands (origins, close to the heads), $A\,0^+ - X\,0^+$. 3916·2 (11, 0), 3888·8 (12, 0), 3862·2 (13, 0), 3836·0 (14, 0), 3810·6 (15, 0). $B\,0^+ - X\,0^+$. 3847·7 (7, 0), 3818·8 (8, 0). $C\,0^+ - X\,0^+$. 3718·2.

ULTRA-VIOLET SYSTEMS

Reference. H. Mohan and K. Majumdar, *Proc. Phys. Soc.*, **77**, 147 (1961)†.

Two systems of bands in the region 2450–2190 have been observed in absorption. Degraded to shorter wavelengths. Strong (0, 0) bands at 2367·5 and 2308·2 Å.

TeSe

Reference. M. Joshi and D. Sharma, *Proc. Phys. Soc.*, **90**, 1159 (1967)†.

42 absorption bands 2520–2140 Å, all shaded to shorter wavelengths, form two sub-systems and a possible third system.

Sub-system I			Sub-system II		
λ	I	v', v''	λ	I	v', v''
2475·6	8	0, 1	2387·8	8	0, 1
2456·4	10	0, 0	2369·9	10	0, 0
2435·4	8	1, 0	2329·4	9	3, 1
2415·2	6	2, 0	2312·3	6	3, 0

For II the (2, 0) band is perturbed and bands with $v' = 3$ are abnormally strong.

ThO

References. S. G. Krishnamurty, *Proc. Phys. Soc.*, A, **64**, 852 (1951).
G. Edvinsson, L. E. Selin and N. Åslund, *Ark. Fys.*, **30**, 283 (1965)†.
G. Edvinsson, A. von Bornstedt and P. Nylén, *Ark. Fys.*, **38**, 193 (1968)†.
A. von Bornstedt and G. Edvinsson, *Phys. Scripta*, **2**, 205 (1970)†.

Occurrence. Best obtained from a discharge through ThI_4 in $N_2 + O_2$. Also reported by Krishnamurty with thorium salts in a graphite arc.

Appearance. Ten red-degraded singlet systems are known, in all of which the (0, 0) band is strongest. The $^1\Pi - {}^1\Sigma$ transitions have well separated double heads. There are also two strong bands of more complex structure at 7300 (deg. V) and 6400 (deg. R).

(0, 0) bands of singlet systems:

λ	System	λ	System
9420·7	$A^1\Sigma - X^1\Sigma$	6121·4	$E^1\Sigma - X^1\Sigma$
8971·7	$B^1\Pi - X^1\Sigma$	5448·8	$F^1\Sigma - X^1\Sigma$
7868·2	$H^1\Phi - G^1\Delta$	5107·2	$I^1\Pi - X^1\Sigma$
6894·4	$C^1\Pi - X^1\Sigma$	4596·4	$M^1\Pi - X^1\Sigma$
6265·6	$D^1\Pi - X^1\Sigma$	4414·9	$K^1\Pi - X^1\Sigma$

TiBr

Reference. C. Sivaji and P. T. Rao, *J. Phys.*, B, **3**, 720 (1970).

Occurrence. In high-frequency discharge and in flash photolysis.

Appearance. Two systems of violet-degraded bands form long strong (0, 0) sequences.

$A^4\Pi - X^4\Sigma$. 53 heads are listed. (0, 0) sequence heads (all $I = 10$) at $\lambda\lambda 4266\cdot0, 4244\cdot8, 4228\cdot0$ and 4214·9.

$C^4\Pi - X^4\Sigma$. 18 heads are listed. (0, 0) sequence heads (and I), $\lambda\lambda 3837\cdot2$ (7), 3824·2 (4), 3808·4 (4), 3791·8 (3).

There are a few unassigned weaker band groups 4180–3940 Å.

TiCl

VIOLET SYSTEM, $A^4\Pi - X^4\Sigma$

References. K. R. More and A. H. Parker, *P.R.*, **52**, 1150 (1937).
V. R. Rao, *Indian J. Phys.*, **23**, 535 (1949)†.
E. A. Shenyavskaya, Y. Y. Kuzyakov and V. M. Tatevskii, *Opt. Spectrosc.*, **12**, 197 (1962).

Occurrence. Discharge tubes (including high-frequency discharge) containing flowing titanium chloride ($TiCl_4$) vapour.

Appearance. Degraded to the violet. A long (0, 0) sequence. Each band has 4 strong Q heads and weaker P heads. About 80 heads are known. The following are the main heads of the (0, 0) band:

λ	I	head	λ	I	head
4199·5	1	O_4	4184·5	4	P_4
4192·7	10	Q_1	4183·1	9	Q_3
4188·0	8	Q_2	4179·4	6	Q_4

The (1, 0) sequence has its Q_1 head at 4118·0.

TiCl (*contd.*)

OTHER SYSTEMS

Rao has found a second violet-degraded system, probably $^4\Sigma - X^4\Sigma$, with some 30 heads 4072–3787 Å, with main (0, 0) band at 3857 Å; (0, 1) at 3935, (1, 0) at 3800.

Weaker band groups 4050–4000, 3935–3840 and 3750–3720 Å may belong to these or other systems.

TiF

Reference.   R. L. Dibner and J. G. Kay, *J. Chem. Phys.*, **51**, 3547 (1969)†.

Observed in absorption following flash heating. The main system, 4088–3930 Å, probably $^4\Pi - {}^4\Sigma$, is degraded to shorter wavelengths, has a weak P_1 head at 4076·9 and first strong Q_1 head of the (0, 0) sequence at 4076·1. There is a weak (1, 0) band at 3967·5.

Weaker systems with line-like heads, 3906–3855 (strongest heads 3894·9, 3892·4) and 3775–3650 (3705·9, 3696·5) are also reported.

TiH

Occurrence.   In absorption by shock-heated titanium powder in H_2 + Ar. Also in M-type stars.

Appearance.   Complex structure 4800–4700, 5110–5030 and 5500–5210 Å. The last band shows more open structure, with weak violet-degraded heads at 5382, 5370 and 5361 Å, probably due to a $^4\Delta - {}^4\Phi$ transition.

References.   R. E. Smith and A. G. Gaydon, *J. Phys.*, B, **4**, 797 (1971)†.
A. G. Gaydon, *J. Phys.*, B, **7**, 2429 (1974).

TiN

RED SYSTEM, $^2\Pi - {}^2\Sigma$

Occurrence.   In microwave discharge through $TiCl_4$ + N_2 + H_2. The assignment has been confirmed by isotope studies.

Appearance.   Degraded to the red. A single sequence of double-headed bands.

Reference.   T. M. Dunn, L. K. Hanson and K. A. Rubinson, *Canad. J. Phys.*, **48**, 1657 (1970)†.

The strong (0, 0) band has heads at 6199·5 ($^2\Pi_{\frac{1}{2}} - {}^2\Sigma$) and 6138·6 ($^2\Pi_{\frac{3}{2}} - {}^2\Sigma$). The weaker (1, 1) band has heads 6234·6 and 6173·7 Å.

ULTRA-VIOLET SYSTEM

Reference.   W. H. Parkinson and E. M. Reeves, *Canad. J. Phys.*, **41**, 702 (1963)†.

Observed in absorption in a shock tube and also in emission in active nitrogen and in an arc. The bands have not been confirmed by other authors or found in a shock tube by Gaydon. Diffuse red-degraded bands. Strong heads 3043, 3014 Å. Less strong 3140, 3117; weak 3285, 3244, 3216 and 2958 Å.

TiO

Three strong triplet systems in the visible region and probably two in the ultra-violet are now known, and five weaker singlet systems in the visible and infra-red.

TiO (*contd.*)

Occurrence. In arcs, flames and furnaces containing titanium dioxide and in discharge tubes containing titanium chloride and oxygen. In exploded wires and shock tubes. With arcs the presence of water vapour frees the spectrum from many of the atomic lines which confuse the detection of weak heads. TiO bands are a prominent feature of the spectra of M-type stars.

TRIPLET SYSTEMS

RED SYSTEM, γ, $A^3\Phi - X^3\Delta$

Appearance. Degraded to longer wavelengths. The bands have rather widely spaced triple heads, but the appearance is confused by overlapping. See Plate 8.

References. F. Lowater, *Proc. Phys. Soc.*, **41**, 557 (1929)†.
K. Wurm and H. J. Meister, *Z. Astrophys.*, **13**, 199 (1937).
C. C. Kiess, *Publ. Astron. Soc. Pac.*, **60**, 252 (1948).
J. G. Phillips, *Astrophys. J.*, **144**, 152 (1951)†.

The following are the strong heads as listed by Lowater. Intensities have been reduced to a scale of 10.

λ	I	v', v''	λ	I	v', v''
7948·6	7	4, 5 Q_b	7197·7	7	1, 1 R_c, 2, 2 R_a
7907·3	5	4, 5 Q_a, 3, 4 R_c	7159·0	5	1, 1 R_b
7861·0	5	3, 4 R_b	7125·6	10	0, 0 R_c, 1, 1 R_a
7828·0	8	2, 3 R_c, 3, 4 Q_a	7087·9	9	0, 0 R_b
7820·1	7	3, 4 R_a	7054·5	7	0, 0 R_a
7705·2	7	1, 2 R_b	6852·3	5	4, 3 R_a
7672·1	8	0, 1 R_c, 1, 2 Q_a	6719·3	5	2, 1 R_a, 1, 0 Q_c
7666·4	5	1, 2 R_a	6714·4	5	1, 0 R_c
7628·1	7	0, 1 R_b	6681·1	5	1, 0 R_b
7589·6	7	0, 1 R_a	6651·5	4	1, 0 R_a
7269·0	5	2, 2 R_c, 3, 3 R_a	6215·2	5	6, 3 Q_c
7219·4	5	1, 1 Q_c			

Wurm and Meister (1937) and Kiess (1948) have extended observations in the infra-red. Phillips (1951) has made rotational analyses of the (1, 0), (0, 0) and (0, 1) bands, and later (see γ' system) has established the transition as $^3\Phi - {}^3\Delta$.

BLUE-GREEN SYSTEM, α, $C^3\Delta - X^3\Delta$

Appearance. Degraded to the red; close triple-headed bands with well marked sequences.

References. A. Fowler, *P.R.S.*, **79**, 509 (1907)†.
A. Christy, *P.R.*, **33**, 701 (1929).

First heads of strong bands:

λ	I	v', v''	λ	I	v', v''	λ	I	v', v''
6214·9	8	2, 5	5448·3	7	0, 1	4848·0	3	4, 2
6159·1	10	1, 4	5240·5	5	1, 1	4804·3	5	3, 1
5861·7	4	2, 4	5166·9	7	0, 0	4761·2	5	2, 0
5810·0	4	1, 3	5050·3	3	3, 2	4626·1	4	4, 1
5758·5	4	0, 2	5002·7	3	2, 1	4584·1	3	3, 0
5497·0	5	1, 2	4954·5	6	1, 0	4462·1	3	4, 0

TiO (*contd.*)

ORANGE-RED SYSTEM, γ', $B^3\Pi - X^3\Delta$

First observed by Coheur in an exploding wire source, and later in other sources. Now analysed in detail by Phillips. Complex multi-headed bands, degraded to the red.

    References. F. P. Coheur, *Bull. Soc. Roy. Sci. Liège*, 98 (1943).

J. G. Phillips, *Astrophys. J.*, **157**, 449 (1969); **169**, 185 (1971).

    The strong (0, 0) band has the following heads: 6221·9 (Q_3), 6214·9 (R_3), 6190·1 (Q_2), 6186·9 (R_2), 6160·9 (Q_1), 6158·5 (R_1). There are also satellite heads at 6174 and 6148·9 Å.

    The following are the first main (R_1) heads of other bands; no intensities are available.

λ	v', v''	λ	v', v''	λ	v', v''
6569	0, 1	6003	4, 3	5664	4, 2
6293	2, 2	5651	3, 2	5615	3, 1
6217	1, 1	5904	2, 1	5569	2, 0
6158·5	0, 0	5847	1, 0		

ULTRA-VIOLET SYSTEMS

    References. C. M. Pathak and H. B. Palmer, *J. Mol. Spec.*, **33**, 137 (1970)†.

H. B. Palmer and C. J. Hsu, *J. Mol. Spec.*, **43**, 320 (1972).

    In the flame of $TiCl_4 + O_2$ burning in potassium vapour, bands 3260–2900 Å are found and attributed to a $D - X^3\Delta$ transition. Double double-headed and degraded to the red. Strongest:

λ	I	v', v''
3214·5	7	0, 1
3113·7	10	0, 0
3014·7	9	1, 0

    The (0, 0) has heads at 3147·9, 3141·7, 3117·1 and 3113·7 Å. A further system in the same source has heads:

λ	I	v', v''
3634·9	1	0, 3
3511·9	3	0, 2
3395·6	4	0, 1
3286·2	5	0, 0

    The (0, 0) has its second head at 3291·2.

    Other bands in this source are now attributed to SnO.

SINGLET SYSTEMS

ORANGE SYSTEM, β, $c^1\Phi - a^1\Delta$

    Appearance. Degraded to the red. Double-headed bands.

    References. F. Lowater, *Proc. Phys. Soc.*, **41**, 557 (1929)†.

J. G. Phillips, *Astrophys. J.*, **111**, 314 (1950)†.

C. Linton and R. W. Nicholls, *J. Phys.*, B, **2**, 490 (1969)†.

C. Linton, *J. Mol. Spec.*, **50**, 235 (1974)†.

    This is the strongest of the singlet systems and occurs in a variety of sources. Strongest heads:

TiO (*contd.*)

λ	I	v', v''	λ	I	v', v''
5733·4	2	4, 4 Q	5661·5	7	2, 2 R
5727·4	3	4, 4 R	5635·3	2	1, 1 Q
5700·6	2	3, 3 Q	5629·3	7	1, 1 R
5694·4	4	3, 3 R	5603·6	3	0, 0 Q
5667·6	2	2, 2 Q	5597·6	8	0, 0 R
			5326·9	−	1, 0 R

INFRA-RED SYSTEM, 9094–8860 Å, δ, $b\,^1\Pi - a\,^1\Delta$

Appearance. Double-headed bands, degraded to the red.

References. J. G. Phillips, *Astrophys. J.*, **111**, 314 (1950)†.

G. W. Lockwood, *Astrophys. J.*, **157**, 278 (1969).

λ	v', v''	λ	I	v', v''	λ	I	v', v''
9986	3, 4	9094·5	2	3, 3 R	8937·4	15	1, 1 R
9899	2, 3	9014·6	2	2, 2 R	8868·5	10	0, 0 Q
9814	1, 2	8949·8	3	1, 1 Q	8859·6	50	0, 0 R
9725	0, 1						

FAR INFRA-RED SYSTEM, 11032–10025 Å, $b\,^1\Pi - d\,^1\Sigma$

There is a further singlet system with (0, 0) band at 11032·2 and (1, 0) at 10025 Å. Degraded to longer wavelengths.

References. A. V. Pettersson, *Ark. Fys.*, **16**, 185 (1959)†.

A. V. Pettersson and B. Lindgren, *Naturwiss.*, **48**, 128 (1961).

5240 Å BAND, $f\,^1\Delta - a\,^1\Delta$

Reference. C. Linton, *Canad. J. Phys.*, **50**, 312 (1972).

A strong (0, 0) band at 5240·5, shaded to the red, has been found in a microwave discharge.

4114 Å BAND, $e\,^1\Sigma - d\,^1\Sigma$

Reference. J. G. Phillips and S. P. Davis, *Astrophys. J.*, **167**, 209 (1971)†.

A weak red-degraded band at 4113·7.

TiS

INFRA-RED SYSTEM, $C\,^3\Delta - X\,^3\Delta$

Occurrence. In absorption in a King furnace containing Ti and ZnS.

Appearance. Degraded to the red. Triple-headed bands.

Reference. R. M. Clements and R. F. Barrow, *Trans. Faraday Soc.*, **65**, 1163 (1969).

No intensities are available. The (0, 0) band has heads at $\lambda\lambda$8629·6, 8627·3 and 8626·0. First heads of other bands $\lambda\lambda$9062·4 (0, 1), 8283·6 (1, 0), 7970·5 (2, 0).

Tl_2

References. H. Hamada, *Nature*, **127**, 535 (1931).

D. S. Ginter, M. L. Ginter and K. K. Innes, *J. Phys. Chem.*, **69**, 2480 (1965).

Tl$_2$ (*contd.*)

In thermal emission, thallium vapour shows red-shaded bands 6535–6250, strongest $\lambda\lambda$6373·8, 6318·6, 6283·6. In absorption, Hamada reports asymmetrical broadening of resonance lines, giving band systems around 5350, 3776 and 2768 Å, while Ginter *et al.* note heads 4353·0 (deg. V), 4321·0 (deg. V), 4315·7 (deg. V), 4297·9 (deg. V), 4293·7 (deg. R), 4292·7 (deg. V), 4289·6 (deg. R) and 4254·1 (deg. V).

TlBr

Occurrence. In absorption and in emission from uncondensed and high-frequency discharges.
References. K. Butkow, *Z.P.*, **58**, 232 (1929)†.
H. G. Howell and N. Coulson, *Proc. Phys. Soc.*, **53**, 706 (1941)†.
P. T. Rao, *Indian J. Phys.*, **23**, 265, 425 (1949); **24**, 434 (1950).

MAIN SYSTEM, A(? $^3\Pi_{0^+}$) − X$^1\Sigma^+$

Butkow (1929) observed fifteen bands, degraded to the red, in absorption in the region λ3600–3400 and obtained the isotope effect for bromine. Howell and Coulson (1941) extended the absorption to λ4000 and also obtained emission over the region λ4500–3400. Rao (1949) obtained additional bands in the region λ4050–3800 and observed a splitting of the bands attributed to the isotope Tl^{203}.

Strongest bands in emission from a list of about 70 bands given by Howell and Coulson extending to $v'' = 14$. Heads for TlBr^{81}.

λ	I	v', v''	λ	I	v', v''	λ	I	v', v''
3705·9	5	3, 13	3509·6	5	1, 4	3452·3	9	0, 1
3641·4	4	2, 10	3498·2	4	0, 3	3440·9	1	1, 1
3617·1	5	2, 9	3486·3	5	1, 3	3430·7	4	2, 1
3593·1	5	2, 8	3475·2	6	0, 2	3429·6	10	0, 0
3569·2	4	2, 7	3463·4	5	1, 2	3418·3	4	1, 0

3950 Å SYSTEM, D − A

Howell and Coulson (1941) also report a very intense group of bands, red degraded, in the region around λ3950, obtained in emission but not in absorption. This is attributed to another system for which the lower state is the upper state of the main system. Rao (1950) confirms occurrence only in emission.

λ	I	v', v''	λ	I	v', v''
3984·5	2	0, 3	3949·1	4	1, 1
3973·8	6	0, 2	3945·2	0	0, 0
3960·2	8	0, 1	3933·7	10	1, 0

Continua. Butkow also observed two regions of continuous absorption:
(1) λ3387–3272. Max. λ3332.
(2) λ2910–2630. Max. λ2690.

Howell and Coulson confirm these and observe two further regions (3) a narrow diffuse band λ3392–3370. Max. λ3380 and (4) a region beginning about λ1900 and stretching to longer wavelengths.

Howell and Coulson also report regions of continua in emission:
(1)   a region about λ3330 corresponding to that in absorption.
(2)   a region from λ2990–2850. Max. λ2960.
(3)   a continuum superposed on most of the band system.

TlCl

In absorption thallium chloride shows strong bands in the region λ3575–3176 and also a number of regions of continua. In emission from discharge tubes, including high-frequency discharges, further bands are obtained in the violet region.

References.   H. G. Howell and N. Coulson, *P.R.S.*, **166**, 238 (1938)†.

E. Miescher, *Helv. Phys. Acta*, **14**, 148 (1941)†.

P. T. Rao, *Indian J. Phys.*, **23**, 393 (1949)†.

P. T. Rao, *Indian J. Phys.*, **24**, 434 (1950)†.

R. F. Barrow, *Proc. Phys. Soc.*, **70**, 622 (1957).

MAIN SYSTEM, 3575–3176 Å, $A^3\Pi_{0^+} - X^1\Sigma^+$

Appearance.   Degraded to the red. Strong (0, 0) sequence. The bands show chlorine isotope effect. Howell and Coulson give extensive tables of heads.

The following are the R heads of the more abundant isotope $TlCl^{35}$ for the strong bands obtained in emission:

λ	I	v′, v″	λ	I	v′, v″	λ	I	v′, v″
3372·0	2	3, 7	3299·7	7	2, 4	3240·6	9*	2, 2
3360·4	1	2, 6	3289·6	8	1, 3	3230·4	10*	1, 1
3341·4	2	3, 6	3270·0	7	2, 3	3220·9	10*	0, 0
3329·9	2	2, 5	3263·6	2*	4, 4	3201·3	2	1, 0
3323·5	3	4, 6	3259·9	8	1, 2	3193·4	1*	3, 1
3319·6	3	1, 4	3250·8	6*	0, 1	3182·6	2*	2, 0

* Relatively strong in absorption.

VIOLET SYSTEM, 4300–3800 Å

Appearance.   A complex system of bands obtained in emission, strongest in the region λ4120–4050, with heads degraded in either direction.

Strongest heads from Miescher; letters R, V and M denote degraded to red, violet or maximum of headless structure.

λ	I	λ	I	λ	I	λ	I
4244·6 R	6	4108·3 R	10	4060 M	10	4021 M	6
4223·6 R	8	4096·8 R	8	4051 M	8	4015·2 R	5
4218·0 R	6	4088·7 V	6	4045·6 M	9	3991·9 R	4
4210·7 V	6	4086·6 R	8	4038·4 R	7	3972 M	4
4182 V	5	4085·9 R	10	4025 M	5	3947·5 R	4

Rao (1949) has extended the range of observation of these bands by exciting thallium chloride in a high-frequency discharge. He divided them into System A λ4300–4150, comprising some 18 bands mostly degraded to the red, and system B λ4150–3800 containing some 75 bands degraded in either direction with some headless and some diffuse. He found a common lower state for the two systems which is also the upper state of the main system. Rao (1950) examined the absorption over the range λ6500–2200.

CONTINUA

Four regions of continuum have been reported by Howell and Coulson (1938) and Rao (1950).

TlCl (contd.)

(1)   a narrow region of continuum around λ3110.

(2)   a continuum with a maximum at λ2520 which can be extended with increasing temperature and is associated with a number of diffuse bands.

(3)   a continuum beginning in the λ1900 region and extending to longer wavelengths.

(4)   a narrow continuum at λ2890 appearing at higher temperatures.

TlF

Three systems of bands have been obtained in absorption in thallium fluoride vapour. Two of these, $A - X$ and $B - X$, also appear in emission in a high-frequency discharge. There are also a number of regions of continuum.

References.   H. G. Howell, *P.R.S.*, **160**, 242 (1937)†.

J. V. R. Rao and P. T. Rao, *Indian J. Phys.*, **29**, 20 (1955)†.

R. F. Barrow, H. F. K. Cheall, P. M. Thomas and P. B. Zeeman, *Proc. Phys. Soc.*, A, **71**, 128 (1958).

SYSTEM A − X, 3105−2809 Å, $A^3\Pi_{0^+} - X^1\Sigma^+$

Appearance.   Bands degraded in either direction. The following are the strongest heads, with the direction of degradation indicated by R (red) and V (violet) as given by Howell (1937). Howell observes some 57 bands.

λ	I	v', v''	λ	I	v', v''	λ	I	v', v''
2924·7 V	5	1, 3	2882·0 V	6	0, 1	2843·5 V	10	0, 0
2921·3 V	5	0, 2	2852·2 V	9	2, 2	2818·4 R	5	3, 2
2886·0 V	6	1, 2	2848·0 V	10	1, 1	2810·6 V	7	1, 0

J. V. R. Rao and P. T. Rao (1955) report a further group of bands in the region λ3500−3400 similar to the above.

SYSTEM B − X, 2808−2668, $B^3\Pi_1 - X^1\Sigma^+$

Appearance.   Close double-headed bands, degraded to the red. Howell gives some 24 heads of which the strongest are:

λ	I	v', v''	λ	I	v', v''
2759·7	7	1, 2	2724·9	10	1, 1
2748·1	6	0, 1	2713·7	10	0, 0
2737·1	9	2, 2	2690·8	3	1, 0

SYSTEM C − X, 2347−2198 Å, $C(?^1\Pi) - X^1\Sigma^+$

Appearance.   Degraded to the red. Howell gives 11 heads. The strongest are as follows:

λ	I	v', v''	λ	I	v', v''
2250·4	4	1, 3	2221·0	9	0, 1
2244·3	5	0, 2	2203·9	8	1, 1
2227·0	8	1, 2	2198·0	10	0, 0

CONTINUA

(1)   At lower temperatures a continuum is observed around λ2200 which extends to the red as the temperature is increased.

TlF (*contd.*)

(2)   At higher temperatures further continua are observed with red edges about $\lambda 2680$, $\lambda 2516$ and $\lambda 2240$.

TlH

YELLOW-GREEN SYSTEM, $a\,^3\Pi_{0^+}-X\,^1\Sigma^+$

Occurrence. In an arc between thallium poles in H_2 at 500 torr, and in emission and absorption in a furnace.

Appearance. Weakly degraded. Some bands shade to the red.

References.   B. Grundström and P. Valberg, *Z.P.*, **108**, 326 (1938)†.
H. Neuhaus and V. Muld, *Z.P.*, **153**, 412 (1959).

Grundström and Valberg's measurements are given here; they analysed the bands into three systems but Neuhaus and Muld, who also studied TlD, have given the analysis below, as one system. The vibrational levels of the upper state are irregular. No intensities are available; the (0, 0) band is probably far the strongest, but the intensity distribution is likely to be abnormal. Origins and, where available, R heads:

λ_0	λ_H	v', v''	λ_0	λ_H	v', v''	λ_0	λ_H	v', v''
7327	7320	2, 4	6223·2	6220·2	5, 3	5903·8	5895·2	1, 1
6952·8	6935·9	1, 3	6180·9	–	0, 1	5772·3	–	5, 2
6730·6	6725·0	2, 3	6062·5	–	6, 3	5742·7	5739·2	2, 1
6557·4	6553·2	3, 3	6058·6	6055·2	3, 2	5706·2	–	0, 0
6394·7	6382·9	1, 2	5914·4	5911·7	4, 2			

VIOLET SYSTEMS, 4300–4200 Å

Occurrence. Absorption in a King furnace.

Appearance. Degraded to the red.

Reference.   T. Larsson and H. Neuhaus, *Ark. Fys.*, **23**, 461 (1963)†.

R heads:

$A\,^1\Pi-X\,^1\Sigma^+$.   4243·2 (0, 0), 4225·7 (1, 0), 4216·1 (2, 0).

$a\,^3\Pi_2-X\,^1\Sigma^+$.   4226·5 (0, 0). This is a weak forbidden transition, produced by the perturbations.

TlI

Occurrence. In absorption and in emission from high-frequency discharges.

References.   K. Butkow, *Z.P.*, **58**, 232 (1929)†.
P. T. Rao and K. R. Rao, *Indian J. Phys.*, **23**, 185 (1949)†.
P. T. Rao, *Indian J. Phys.*, **24**, 434 (1950).
J. V. R. Rao and P. T. Rao, *Indian J. Phys.*, **29**, 20 (1955)†.

SYSTEM A – X, $\lambda\lambda 5300$–3800

Butkow obtained a number of diffuse headless bands in absorption: $\lambda\lambda 4211$, 4181, 4152, 4120, 4092, 4062, 4030, 4000, 3967, 3933, 3896, 3836, 3808.

P. T. Rao and K. R. Rao obtained in emission, in a high-frequency discharge, an extensive band system over the range $\lambda 5300$–3800. At the violet end the bands are red-degraded, in the central portion headless and at the red end broad and diffuse; about 200 bands are listed. The following are given as the strongest bands:

Tll (*contd.*)

λ	I	λ	I	λ	I
4620·8	5	4422·6	5	4113·2	5
4591·3	5	4342·6	5	4088·8	5
4557·1	5	4315·9	5	3900·8	5
4511·6	5	4269·6	5	3877·9	5
4477·9	5	4213·3	5		

P. T. Rao has studied the absorption spectrum over the range λ6500–2200. He extends the range of Butkow's bands to λ4500–3700 giving 23 heads of which the strongest are: 3817·6 (9), 3813·2 (10), 3808·6 (10), 3804·2 (10), 3799·5 (10), 3795·5 (10).

SYSTEM B − X, λλ3690–3580

Also in emission from high-frequency discharge.

P. T. Rao and K. R. Rao obtained a second system consisting of 25 slightly violet-degraded bands. Four diffuse groups overlaid by continuum were measured:

(1)　λλ3688, 3684, 3681, 3677, 3674, 3670, 3667, 3663, 3660, 3657, 3653, 3650.
(2)　λλ3644, 3641, 3637, 3634, 3630, 3626.
(3)　λλ3619, 3616, 3612, 3609, 3605.
(4)　λλ3588, 3585.

REGION λ6340–5550

J. V. R. Rao and P. T. Rao found in the high-frequency discharge about 25 broad diffuse bands from λ6338–5577 which they have arranged into a tentative vibrational scheme. Strongest bands: λλ6013·1 (0, 4), 5985·6 (1, 4), 5962·0, 5940·2 (1, 3), 5922·3, 5871·8, 5853·3 (1, 1), 5827·1 (2, 1), 5785·6 (2, 0), 5663·7.

TmO

References. R. Herrmann and C. T. J. Alkemade, *Flame Photometry*, (1963).
R. Mavrodineanu and H. Boiteux, *Flame Spectroscopy*, (1965).

About 40 bands, mostly degraded to the red, occur in arcs and flames and are listed under thulium oxide in collected data. Strongest flame bands λλ7200, 5557·7, 5531, 5523·6, 5500·4, 5472·6, 5415·8, 5352, 5347, 5329·2, 4937·5, 4891·6, 4814.

VCl

Occurrence. In He atom bombardment of VCl_4, in high-frequency discharges and weakly in flash photolysis.

Appearance. A number of compact sequences, each of 10 to 20 heads.

Reference. D. Iacocca *et al.*, *C. R. Acad. Sci.* (*Paris*), C, **271**, 669 (1970).

System A. Triple-headed, with moderate P heads and stronger Q heads. Degraded to shorter wavelengths. The (0, 0) sequence shows P heads at 3796·9, 3790·4, 3783·9 and Q heads at 3795·5, 3788·9 and 3782·7 Å.

System B. Degraded to shorter wavelengths. 3776·5 (0, 0 P), 3774·6 (0, 0 Q).

System C. Degraded to shorter wavelengths. A close group of heads 3765·1–3753·8 Å.

System D. A strong system. Degraded to shorter wavelengths. Q heads of (0, 0) 3750·6, 3746·9, 3743·1.

VCl (*contd.*)

System E. Degraded to the red. 3711·1–3676·4 Å.
System F. Degraded to the red. 3565·7–3522·4 Å.
System G. Degraded to shorter wavelengths. 3576·5–3489·9 Å.

VH

Reference. R. E. Smith, *P.R.S.*, **332**, 113 (1973)†.
Occurrence. Absorption in shock-heated vanadium powder in H_2 + Ar.
Appearance. Complex headless structure 4400 to 4800 Å, with strong features as 4685, 4689 and 4709 Å.

VO

References. P. C. Mahanti, *Proc. Phys. Soc.*, **47**, 433 (1935)†.
P. C. Keenan and L. W. Schroeder, *Astrophys. J.*, **115**, 82 (1952)†.
A. Lagerqvist and L. E. Selin, *Ark. Fys.*, **11**, 429 (1957); **12**, 553 (1957).
B. B. Laud and D. R. Kalsulkar, *Indian J. Phys.*, **42**, 61 (1968)†.
D. Richards, *Thesis, Oxford* (1969).

RED SYSTEM, $C^4\Sigma^- - X^4\Sigma^-$

Occurrence. In arcs containing metallic vanadium or vanadium salts, and weakly in flames.

Mahanti has photographed a system of well-marked double-headed bands, degraded to the red, in the region $\lambda\lambda 8643$–4466. Some sixty heads are given of which the strongest are:

λ	I	v', v''	λ	I	v', v''	λ	I	v', v''
8643·4	2	0, 6	6477·8	6	0, 2	5517·3	5	2, 1
7986·0	2	0, 5	6086·4	8	0, 1	5469·3	9	1, 0
7418·2	3	0, 4	5837·3	4	2, 2	5275·8	6	3, 1
6976·2	4	1, 4	5786·4	1	1, 1	5228·2	6	2, 0
6919·0	4	0, 3	5736·7	10	0, 0	5010·5	4	3, 0
6532·8	6	1, 3						

Laud and Kalsulkar have found many more bands.

EXTREME RED SYSTEM, $B^4\Pi - X^4\Sigma^-$

Occurrence. Observed by Keenan and Schroeder with vanadium on lower pole of an iron arc, and in hot flames. Conspicuous in late M-type stars.

Appearance. Degraded to longer wavelengths. Multi-headed bands. About 60 heads have been measured; strongest:

λ	I	v', v''	λ	I	v', v''	λ	I	v', v''
8666·6	3	0, 1	7960·4	4n	1, 1	7472·1	5	2, 1
8624·0	5	0, 1	7938·9	5	0, 0	7405·2	4	2, 1
8604·0	3	1, 2	7918·4	4	0, 0	7393·2	5	1, 0
8575·3	6	–	7896·0	8	0, 0	7372·4	4	1, 0
8537·7	5	0, 1	7865·0	7	0, 0	7333·8	3	1, 0
7973·1	5n	0, 0	7850·9	4	0. 0			

VO (*contd.*)

Red-degraded bands in the far infra-red have been photographed by Lagerqvist and Selin (1957b). Strongest heads: $\lambda\lambda 10\,462\cdot6$ (3), $10\,482\cdot2$ (3), $10\,509\cdot2$ (2), $10\,530\cdot0$ (2) and further diffuse heads $\lambda9590-9530$.

WO

Occurrence. Arcs and high-tension arcs in air between tungsten electrodes.

Appearance. Complex spectrum of mostly single-headed bands, degraded to longer wave-lengths.

References. A. Gatterer and S. G. Krishnamurty, *Nature*, **169**, 543 (1952).

V. Vittalachar and S. G. Krishnamurty, *Curr. Sci. (India)*, **23**, 357 (1954).

Bands have been provisionally assigned to six or more overlapping systems. Measurements are from the above references for the more prominent bands; intensities are mostly from Foster and Gaydon:

λ	I	System	λ	I	System	λ	I	System
7743·2	–	VI (0, 0)	4806·3	9	I (0, 0)	4414·0	5	I (3, 1)
7059·8	–	V (0, 0)	4709·7	10	IV (0, 0)	4395·1	4	I (2, 0)
6260·2	4	III (1, 1)	4609·6	8	I (2, 1)	4338·3	5	IV (3, 1)
6219·8	6	III (0, 0)	4590·2	8	I (1, 0)	4313·3	4	IV (2, 0)
5865·2	4	III (1, 0)	4522·4	5	IV (2, 1)	4283·9	4	II (2, 1)
5210	5		4473·3	5	II (1, 1)	4271·2	4	II (1, 0)
4823·3	6	I (1, 1)	4459·5	5	II (0, 0)	4110·0	2	II (3, 1)

Xe_2 (or Xe_2^+)

References. Y. Tanaka, *J. Opt. Soc. Amer.*, **45**, 710 (1955).

C. Brocklehurst, *Trans. Faraday Soc.*, **63**, 274 (1967).

Brocklehurst studied excitation of xenon by soft X-rays. Above 30 torr pressure, diffuse bands, shaded to shorter wavelengths, were found at 2867 and 2721 Å. At higher pressure, complex diffuse structure extended from 4900 to the far ultra-violet. Tanaka, using a transformer discharge, found continua in the vacuum ultra-violet, extending weakly to 2250 Å and bands 1488–1483 Å.

XeF_2

Reference. S. L. N. G. Krishnamachari, N. A. Narasimham and M. Singh, *Curr. Sci. (India)*, **34**, 75 (1965)†.

In microwave discharges through xenon and a fluoride (BF_3, LiF or NaF), strong bands were found around 3510 and 3080 Å.

The 3510 band shows four groups, each with 3 or 4 heads. The clearest heads in the photograph are 3510·4 (deg. R) and 3502·3 (deg. V).

The 3080 structure consists of two groups of diffuse bands. There are weaker structures around 2635 and 2355 Å.

The emitter appears to be polyatomic and is likely to be XeF_2.

XeN

References.   L. Herman and R. Herman, *Nature*, **191**, 346 (1961)†; **193**, 156 (1962); *J. Phys. Radium*, **24**, 73 (1963).

In the positive column of a D.C. discharge through xenon with a little N_2 there is a band, apparently shaded to the violet from a head at λ4925; the rotational fine structure is, however, degraded to the red.

XeO

Reference.   C. D. Cooper, G. C. Cobb and E. L. Tolnas, *J. Mol. Spec.*, **7**, 223 (1961)†.

A system of red-degraded bands near the forbidden oxygen line λ5577 obtained from a discharge through xenon at 1 atm. with a trace of oxygen present. Strong heads with own estimates of intensity.

λ	I	v', v''	λ	I	v', v''
5143·7	2	0, 2	5304·1	7	0, 4
5191·6	2	1, 3	5376·0	10	0, 5
5228·8	4	0, 3	5441·7	8	0, 6
5267·0	3	1, 4			

YCl

Reference.   G. M. Janney, *J. Opt. Soc. Amer.*, **56**, 1706 (1966).

Occurrence.   Microwave discharge through yttrium chloride with krypton in a heated quartz tube.

Appearance.   Several sequences of single-headed bands ($^1\Sigma - X^1\Sigma$). Degraded to the red. 16 bands recorded.

λ	I	v', v''	λ	I	v', v''
6893·8	4	0, 1	6576·3	7	1, 0
6718·8	10	0, 0	6440·9	4	2, 0
6601·6	4	2, 1	6311·3	1	3, 0

YF

Occurrence.   In emission using a heated Pt hollow-cathode discharge in helium. Also in absorption.

References.   E. A. Shenyavskaya, A. A. Mal'tsev and L. V. Gurvich, *Opt. Spectrosc.*, **21**, 374 (1966).
R. F. Barrow *et al.*, *Nature*, **215**, 1072 (1967).

Eight systems, all degraded to the red, with marked sequences.

(0, 0) heads, λλ

6830·8, 6735·4, 6652·4	$^3\Phi - {}^3\Delta$
6291·9	$B^1\Pi - X^1\Sigma^+$
5208·6	$C^1\Sigma - X^1\Sigma^+$
3947	$^1\Pi - X^1\Sigma^+$
3925	$^1\Pi - X^1\Sigma^+$
3838	$^1\Pi - X^1\Sigma^+$
3572·2	$^1\Sigma - X^1\Sigma^+$
3203·0	$^1\Pi - X^1\Sigma^+$.

YO

Occurrence. In flames and arcs containing yttrium salts. Also in absorption using powder in shock-heated argon.

References. L. W. Johnson and R. C. Johnson, *P.R.S.*, **133**, 207 (1931).

W. F. Meggers and J. A. Wheeler, *J. Res. Nat. Bur. Stand.*, **6**, 239 (1931)†.

U. Uhler and L. Åkerlind, *Ark. Fys.*, **19**, 1 (1961).

U. Uhler and L. Åkerlind, *Naturwiss.*, **46**, 488 (1959).

ORANGE SYSTEM, $A^2\Pi - X^2\Sigma$

Appearance. Degraded to the red. Long sequences.

The following measurements of the strong heads at the beginnings of the main sequences are from Johnson and Johnson:

λ	I	v', v''	λ	I	v', v''
5697·8	5	1, 0 i Q	5939·1	8	0, 1 i R
5713·9	6	2, 1 i Q	5956·4	7	1, 1 i R
5730·2	7	3, 2 i Q	5972·2	10	0, 0 i Q
5747·0	8	4, 3 i Q	5987·7	10	1, 1 i Q
5764·3	7	5, 4 i Q	6003·6	10	2, 2 i Q
5842·0	4	1, 0 ii Q	6096·8	8	0, 0 ii R
5858·9	5	2, 1 ii Q	6114·8	7	1, 1 ii R
5876·2	4	3, 2 ii Q	6132·1	10	0, 1 ii Q
5912·3	6	5, 4 ii Q	6148·4	10	1, 1 ii Q
5931·1	5	6, 5 ii Q	6165·1	10	2, 2 ii Q

BLUE-GREEN SYSTEM, $B^2\Sigma - X^2\Sigma$

Appearance. Degraded to the red. Close double-headed bands, separation about 0·8 Å.

The following measurements of the first heads of the strong bands are by Johnson and Johnson:

λ	I	v', v''	λ	I	v', v''
5077·9	4	2, 3	4841·9	7	1, 1
5049·7	5	1, 2	4817·4	10	0, 0
5024·2	6	0, 1	4649·2	9	1, 0

YbF

Reference. R. F. Barrow and A. H. Chojnicki, *J. Chem. Soc., Faraday Trans. II*, **71**, 728 (1975)†.

Three violet-degraded systems have been observed in emission and absorption. P and Q heads of (0, 0) bands,

$A_1 - X$, 5532·7, 5521·5

$A_2 - X$, 5139·5, 5135·0

$B - X$, 4749·1, 4743·8.

YbH

Occurrence. Emission in a discharge and in absorption with ytterbium metal in a furnace at 1300°C.

YbH (*contd.*)

Appearance. Four systems, all degraded to the violet. No intensities are available but the (0, 0) bands at 6590, 5998 and 5640 are likely to be strongest.

References. L. Hagland, I Kopp and N. Åslund, *Ark. Fys.*, **32**, 321 (1966)†.
L. Hagland and I. Kopp, *Ark. Fys.*, **39**, 257 (1969).

$D^2\Pi_{\frac{1}{2}} - X^2\Sigma^+$. P heads: $\lambda\lambda 7756.2$ (0, 2), 7130.9 (0, 1), 6590.2 (0, 0), 6538.4 (1, 1).

$D^2\Pi_{\frac{3}{2}} - X^2\Sigma^+$. Triple-headed. All heads:

v', v''	P_2	P_{21}	Q_2
0, 1	6444.6	6404.0	6402.4
1, 2	6377.6	6339.8	6339.4
0, 0	5997.6	5947.7	5945.5
1, 1	5952.3	5907.6	5905.5
1, 0		5522.4	5518.8

$E\frac{1}{2} - X^2\Sigma^+$. Heads:

0, 1	6043.6	6041.0	6014.1
0, 0	5640.2	5637.0	5612.2

$F\frac{1}{2} - X^2\Sigma^+$. Only found for the deuteride, probably because the hydride is predissociated. (0, 0) around 4547 Å.

$a^4\Sigma^+ - X^2\Sigma^+$. Only known for the deuteride. (0, 0) around 7586 Å.

YbO

References. R. Herrmann and C. T. J. Alkemade, *Flame Photometry*, (1963).
R. Mavrodineanu and H. Boiteux, *Flame Spectroscopy*, (1965).

Bands, mostly degraded to the red, are listed in collected data. Occurring in flames and arcs. Strongest 5438, 4840*, 4824, 4778*, 4765, 4760, 4749, 4738 Å (two strongest indicated *).

YbOH

For references see YbO. Diffuse bands occurring in flames are attributed to the hydroxide by P. T. Gilbert. Strongest maxima $\lambda\lambda 6220$, 6020, 5870, 5725*, 5550, 5325*, 5174, 4981, 4850 (strongest *).

Zn₂

References. H. Hamada, *Phil. Mag.*, **12**, 50 (1931)†.
J. M. Walter and S. Barratt, *P.R.S.*, **122**, 201 (1929).
Y. Morimoto, *Proc. Phys. Soc. Japan*, **4**, 67 (1949).

Hamada has studied the emission by zinc in a hollow cathode. The zinc lines, especially $\lambda\lambda 2139$ and 3076 are broadened and show patches of continua and flutings attributed to incipient formation of Zn₂ molecules.

Walter and Barratt observe a diffuse band in absorption at about 3050 Å.

ZnBr

References. H. G. Howell, *P.R.S.*, **182**, 95 (1943).

C. Ramasastry and K. Sreeramurty, *Proc. Nat. Acad. Sci. India*, **16**, 305 (1950)†.

M. M. Patel and K. J. Rajan, *Indian J. Phys.*, **42**, 125 (1968)†.

K. J. Rajan and N. R. Singh, *Indian J. Pure Appl. Phys.*, **7**, 61 (1969).

VISIBLE SYSTEM, 8470–3300 Å, $B^2\Sigma - X^2\Sigma$

Occurrence. Discharge, including high-frequency, through $ZnBr_2$.

Appearance. Crowded bands, degraded to the red, on a continuum with intensity maximum at 8300 Å. At short wavelengths, around 3800 Å, the bands are very faint but a partial resolution into rotational fine structure is visible.

Patel and Rajan record about 100 heads; strongest:

λ	I	v', v''	λ	I	v', v''	λ	I	v', v''
3986·5	8	4, 2	3885·4	8	7, 1	3825·9	9	8, 0
3966·0	8	5, 2	3865·2	9	8, 1	3809·0	10	11, 1
3945·4	8	6, 2	3846·1	9	9, 1	3807·3	10	9, 0
3904·5	8	6, 1	3827·7	10	10, 1	3788·9	9	10, 0

ULTRA-VIOLET BANDS, 3113–2965 Å

Occurrence. In a high-frequency discharge and in absorption.

Appearance. Probably shaded to the red, although Rajan and Shah say no definite degradation. Analysed as two systems, $C - X^2\Sigma$ and $D - X^2\Sigma$; these may be components of a $^2\Pi$ but Rajan and Shah doubt this. They record 47 bands. Strongest:

$C - X^2\Sigma$

λ	I	v', v''	λ	I	v', v''	λ	I	v', v''
3108·0	7	0, 3	3081·3	6	1, 3	3030·7	5	0, 0
3105·4	5	3, 6	3080·2	5	3, 5	3030·5	5	1, 1
3102·4	5	6, 8	3055·7	7	1, 2	3030·2	5	2, 2

$D - X^2\Sigma$

λ	I	v', v''	λ	I	v', v''
3040·2	1	0, 1	3014·5	4	1, 1
3015·0	2	0, 0	2989·7	2	1, 0

$ZnBr_2$

No bands due to $ZnBr_2$ are known. Emission bands previously ascribed to this molecule by Wieland are attributed to ZnBr. In absorption only continuous spectra have been observed (E. Oeser, *Z.P.*, **95**, 699 (1935)).

ZnCl

References. K. Wieland, *Helv. Phys. Acta*, **2**, 46 (1929).

J. M. Walter and S. Barratt, *P.R.S.*, **122**, 201 (1929).

S. D. Cornell, *P.R.*, **54**, 341 (1938).

M. M. Patel and K. J. Rajan, *Indian J. Pure Appl. Phys.*, **5**, 330 (1967)†.

ZnCl (*contd.*)

VISIBLE SYSTEM, 8650–3200 Å, $B^2\Sigma - X^2\Sigma$

Occurrence. In discharges, especially high-frequency.

Appearance. Crowded bands on a continuum with intensity maximum at 8200 Å. Between 4700 and 3800 bands are most distinct and Patel and Shah have measured about 200 red-degraded heads. Bands are of even intensity, uniformly spaced, and it is not possible to select meaningful features for identification.

ULTRA-VIOLET SYSTEM, 2993–2903 Å

Occurrence. Observed by Cornell in a high-frequency discharge and by Walter and Barratt in absorption.

Appearance. Degraded to the red. Strong sequences at 2976·2 and 2942·6 Å in emission are probably (0, 0) heads of $C^2\Pi_{\frac{1}{2}} - X^2\Sigma$ and $D^2\Pi_{\frac{3}{2}} - X^2\Sigma$ sub-systems. There is a weaker sequence at 2910·0. The sub-systems overlap closely.

The following are the strongest absorption bands: $\lambda\lambda2956$ ($I = 8$), 2943 (5), 2934 (10), 2923 (5) and 2911 (2).

2074 Å SEQUENCE, $E - X^2\Sigma$

A single red-degraded sequence observed by Cornell in emission: $\lambda\lambda2075\cdot6$ (0, 0), 2077·9 (1, 1), etc.

ZnF

ULTRA-VIOLET SYSTEMS, 2700–2588 Å

Occurrence. In absorption.

Appearance. Degraded to the red.

Reference.   G. D. Rochester and E. Olsson, *Z.P.*, **114**, 495 (1939).

R heads:

$C^2\Pi_{\frac{1}{2}} - X^2\Sigma$.   2703·2 (0, 0), 2660·0 (1, 0).

$C^2\Pi_{\frac{3}{2}} - X^2\Sigma$.   2676·3 (0, 0), 2634·3 (1, 0).

$D - X^2\Sigma$.   2673·1 (0, 2), 2630·0 (0, 1), 2587·7 (0, 0).

ZnH

References.   G. Stenvinkel, *Dissertation, Stockholm* (1936).
M. A. Khan, *Proc. Phys. Soc.*, **80**, 599 (1962)†.

4300 Å SYSTEM, $A^2\Pi - X^2\Sigma$

Bands with P and Q heads degraded to the violet, obtained with zinc arc in hydrogen at reduced pressure and in quartz discharge tube containing zinc vapour and hydrogen. See Plate 4. Data from Stenvinkel:

	$^2\Pi_{\frac{1}{2}} - {}^2\Sigma$	Heads (*I*)			$^2\Pi_{\frac{3}{2}} - {}^2\Sigma$	Heads (*I*)	
v', *v''*	*Origins*	Q	P	*v'*, *v''*	*Origins*	Q	P
0, 3	5223·0			0, 3	5131·5		
0, 2	4905·3			0, 2	4824·5		
0, 1	4578·6			0, 1	4523·9		
0, 0	4299·1	4301 (10)	4326 (5)	0, 0	4237·0	4240 (10)	4260 (8)
1, 1	4238·3			1, 1	4178·2		
1, 0	3985·6	3989 (3)		1, 0	3932·2	3935 (3)	
2, 0	3726·0			2, 0	3679·4		

ZnH (*contd.*)

ULTRA-VIOLET SYSTEM, 4700–2800 Å, $B^2\Sigma - X^2\Sigma$

An extensive system of bands, strongly degraded to the red, with heads not well developed. Origins, from Stenvinkel:

v', v''	λ_0	v', v''	λ_0	v', v''	λ_0
0, 5	4703·1	2, 0	3418·3	7, 0	2981·8
0, 4	4520·0	3, 0	3314·3	8, 0	2916·1
0, 3	4310·9	4, 0	3219·3	9, 0	2855·7
1, 2	3934·0	5, 0	3132·7	10, 0	2800·7
1, 1	3731·7	6, 0	3054·1		

ULTRA-VIOLET SYSTEM, 2450–2300 Å, $C^2\Sigma - X^2\Sigma$

Observed by Khan in a capillary discharge. Degraded to shorter wavelengths.

Heads: $\lambda\lambda 2426\cdot5$ (0, 0), $2410\cdot0$ (1, 1), $2328\cdot8$ (1, 0), $2318\cdot8$ (2, 1).

ZnH⁺

Reference.   E. Bengtsson and B. Grundström, *Z.P.*, **57**, 1 (1929).

2152 Å SYSTEM, $A^1\Sigma - X^1\Sigma$

An extensive system of single-headed bands, degraded to the red. Obtained from a zinc arc in H_2 at low pressure. See Plate 5.

λ	v', v''	λ	v', v''	λ	v', v''
2366·5	2, 4	2261·5	1, 2	2115·1	2, 1
2350·7	1, 3	2240·2	0, 1	2091·7	1, 0
2332·0	0, 2	2151·9	0, 0		

ZnI

References.   A. Terenin, *Z.P.*, **44**, 713 (1927).

K. Wieland, *Helv. Phys. Acta*, **2**, 46 (1929).

E. Oeser, *Z.P.*, **95**, 699 (1935)†.

P. T. Rao and K. R. Rao, *Indian J. Phys.*, **20**, 49 (1946)†.

C. Ramasastry, *Indian J. Phys.*, **22**, 119 (1948)†; **23**, 35 (1949).

VISIBLE SYSTEM, 6140–3500 Å, $B(?^2\Sigma) - X^2\Sigma$

Occurrence.   In low-pressure discharge-tubes (Wieland) and in fluorescence of ZnI_2 (Terenin, Oeser).

Appearance.   This system, usually referred to as system B, consists of diffuse line-like bands on a continuum with pronounced intensity maximum at 6050 Å. If the spectrum is obtained from ZnI_2 vapour excited in the presence of a large excess of an inert gas, it shows a simplified vibrational structure of bands degraded to the red (Wieland).

SYSTEM C, 3393–3258 Å, $C(?^2\Pi_{\frac{1}{2}}) - X^2\Sigma$

Appearance.   Degraded to shorter wavelengths.

Heads of strongest sequences observed by Wieland using a discharge-tube containing ZnI_2:

ZnI (*contd.*)

λ	I	v', v''
3367·3	5	0, 2
3342·6	8	0, 1
3318·0	10	0, 0
3291·1	7	1, 0

SYSTEM D, 3277–3193 Å, $D(?\,^2\Pi_{\frac{3}{2}}) - X^2\Sigma$

Appearance. Diffuse bands.

These bands are considered by Rao and Rao to form with system C the two components of a $^2\Pi - {}^2\Sigma$ transition. Strongest bands:

λ	I	v', v''	λ	I	v', v''
3277·7	0	0, 0	3232·5	3	2, 0
3265·7	2		3215·0	3	4, 1
3262·5	3		3212·5	3	3, 0
3257·2	2	3, 2	3196·5	2	5, 1
3254·8	2	1, 0	3193·5	3	4, 0
3235·7	4	3, 1			

SYSTEM D′, 2990–2715 Å, $D' - X^2\Sigma$

A large number of bands in this region have been observed by Ramasastry using a high-frequency discharge. He designates them system D_1. The bands are degraded to the red. Strongest bands: $\lambda\lambda(I)$; 2989·0 (6), 2983·2 (5), 2971·4 (4), 2952·6 (4), 2946·4 (4), 2872·2 (4), 2823·9 (5), 2817·4 (4), 2815·7 (4), 2764·4 (4).

SYSTEM E, 2450–2250 Å, $E - X^2\Sigma$

A weak system, degraded to the red, observed in a discharge-tube by Wieland. The system has been studied more extensively by Ramasastry who refers to it as system E. Strongest bands: $\lambda\lambda(I)$: 2384·5 (4), 2376·8 (4), 2373·1 (4), 2369·4 (4), 2365·2 (4), 2357·8 (4), 2353·5 (5), 2341·9 (5), 2330·1 (3).

ZnIn

References. C. Santaram and J. G. Winans, *P.R.*, **136**, 57 (1964)†; *J. Mol. Spec.*, **16**, 309 (1965)†.

A high-frequency discharge through mixed zinc and indium vapours gives a strong violet-degraded $^2\Pi - {}^2\Sigma$ system with (0, 0) sequence at 5308, and a less strong system 5670–5470 with red-degraded sequence at 5615·7. There are also diffuse bands at 4511 and 4102 Å.

ZnO ?

References. A. C. Egerton and S. Rudrakanchana, *P.R.S.*, **225**, 431 (1954)†.
W. H. Parkinson, *Proc. Phys. Soc.*, **78**, 705 (1961)†.
D. Pešić, *Croatica Chem. Acta*, **38**, 313 (1966)†.

Zinc oxide does not readily give a good band spectrum. Some Zn lines are readily broadened asymmetrically to resemble diffuse bands. A red-degraded band at 3435·8 Å occurs in an arc in air, in a flame of zinc dimethyl and in emission from shock-heated zinc oxide in argon.

ZnO ? (*contd.*)

Pesić has also found bands in the yellow-green in a low-pressure Zn arc in O_2; these show an isotope effect with O^{18}. They are diffuse and mostly degraded to the red:

λ	I	λ	I	λ	I
5774·0	4	5523·5	4	5326·9	8
5678·9	10	5461·6	4	5312·3	3
5619·3	4	5437·3	7	5264·7	3*
5581·2	7	5369·4	8	5224·6	3*

*The last two bands are headless; maxima given.

It is not certain whether the emitter of the 3435 band or the yellow-green system is ZnO or a polyatomic emitter; the latter seems more probable.

ZnS

Reference. P. K. Sen Gupta, *P.R.S.*, **143**, 438 (1933–4).
Continuous absorption from 2800 Å to shorter wavelengths with maximum around 2300 Å.

ZnTl

Reference. C. Santaram, V. K. Vaidyan and J. G. Winans, *J. Phys.*, B, **4**, 133 (1971)†.
In emission in a high frequency discharge there is a diffuse violet-degraded sequence of bands at 4663·7 Å. There are also continua with maxima at 5560 and 3930 Å.

$ZnTl_2$

Reference. C. Santaram, V. K. Vaidyan and J. G. Winans, *J. Phys.*, B, **4**, 133 (1971)†.
In emission in a high-frequency discharge there is a complex system of bands at 6270–5910 Å. 40 heads, mostly degraded to the violet, have been included in an analysis. Strongest: λλ6221·7, 6210·7, 6208·6, 6207·6, 6196·0, 6187·5, 6186·2.

ZrBr

Reference. C. Sivaji and P. T. Rao, *Proc. Roy. Irish Acad.*, A, **70**, 1 (1970)†.
Occurrence. High-frequency discharge through Br_2 + Ar over metallic zirconium.

SYSTEM C, 4250–4050 Å

Four groups of bands with rotational structure degraded to the red, and vibrational structure to the violet. Attributed to a quartet transition:

Regions of band maxima	λ (0, 0) heads	I
4215	4216·7	3
4175	4175·8	4
4130	4135·0	9
4095	4098·1	10

ZrBr (*contd.*)

SYSTEM B, 3850–3775 Å

Four complex groups of red-degraded bands. Wavelengths and intensities of (0, 0) sequence heads: 3851·6 (6), 3825·4 (8), 3800·4 (6), 3775·2 (10).

There are also many unclassified heads around 3757 and 3708 Å.

ZrCl

Reference. P. K. Carroll and P. J. Daly, *Proc. Roy. Irish Acad.*, A, **61**, 101 (1961)†.
Occurrence. $ZrCl_4$ in high-frequency discharge.

SYSTEM C, 4150–4040 Å

Four groups of heads, maxima 4137, 4106, 4077 and 4046 Å; 75 heads, degraded to the red are listed; strongest 4138·7, 4105·4, 4077·6, 4046·7.

SYSTEM B, 3800–3620 Å

Very complex, with rotational structure mostly degraded to the red and vibrational structure to the violet. About 40 heads are listed including 3713·9 (10), 3713·8 (9), 3704·1 (5, deg. V), 3696·3 (4), 3691·3 (5, deg. V).

SYSTEM A, 2910–2840 Å

Three groups of sharp line-like heads. The (0, 0) sub-band heads are at 2910·0 ($I = 10$), 2871·4 (9), 2837·5 (7).

ZrF

Not known. Bands reported by Afaf are believed to be due to CuF.

ZrI

Reference. C. Sivaji and P. T. Rao, *Proc. Roy. Irish Acad.*, A, **70**, 7 (1970)†.
Occurrence. High-frequency discharge through I_2 + Ar over metallic Zr.

SYSTEM C, 4420–4225 Å

Four sub-systems with bands mostly degraded to the red but the vibrational structure the other way. 34 heads are listed:

Group maxima	λ (0, 0) heads	I
4400	4396·6	10
4345	4347·5	8
4290	4293·5	8
4240	4242·4	6

SYSTEM B, 4015–3930 Å

Degraded to the red. Complex but clear sequence heads of (0, 0) sub-systems at 4005·5 ($I = 7$), 3980·2 (6), 3955·1 (8), 3933·3 (10).

There are a number of unclassified bands, strongest 3910·7 ($I = 8$), 3861·7 (8), 3851·0 (10), 3851·8 (10), 3852·1 (8).

ZrN

Occurrence. In emission from $ZrCl_4$ in active nitrogen (by S. Rienzi and A. G. Gaydon, unpublished) and in absorption by shock-heated Zr powder in N_2 + Ar (R. E. Smith and A. G. Gaydon).

Complex structure with prominent red-degraded sequences at 5835 and 5648 Å.

ZrO

Occurrence. Zirconium oxide in carbon arc or furnace. Also in absorption in a furnace at $2650 \pm 50°$ C and behind shock waves. A prominent feature of S-type stars.

References. F. Lowater, *Proc. Phys. Soc.*, **44**, 51 (1932)†; *Phil. Trans. Roy. Soc.*, A, **234**, 355 (1935)†.

W. F. Meggers and C. C. Kiess, *J. Res. Nat. Bur. Stand.*, **9**, 325 (1932).

C. C. Kiess, *Pub. Astronom. Soc. Pacific*, **60**, 252 (1948).

G. H. Herbig, *Astrophys. J.*, **109**, 109 (1949)†.

M. Afaf, *Proc. Phys. Soc.*, A, **63**, 674 and 1156 (1950)†.

A. Lagerqvist, U. Uhler and R. F. Barrow, *Ark. Fys.*, **8**, 281 (1954)†.

U. Uhler, *Ark. Fys.*, **8**, 295 (1954)†.

U. Uhler and A. Åkerlind, *Ark. Fys.*, **10**, 431 (1956)†.

L. Akerlind, *Ark. Fys.*, **11**, 395 (1956).

J. B. Tatum, *J. Mol. Spec.*, **48**, 292 (1973)†.

B. Lindgren, *J. Mol. Spec.*, **48**, 322 (1973)†.

I. V. Veits *et al.*, *J. Quant. Spectrsoc. Rad. Transfer*, **14**, 221 (1974).

L. Schoonveld and S. Sundaram, *Astrophys. J.*, **192**, 207 (1974).

The ground state is believed to be $X^1\Sigma^+$ with the lowest triplet level $X'^3\Delta$ (at one time believed to be $X^3\Pi$), about $1700 \, cm.^{-1}$ higher. Triplet systems are, however, strong.

All systems are degraded to the red.

VISIBLE TRIPLET SYSTEMS

The following are the wavelengths of the strongest heads of three triplet systems, degraded to the red, obtained by Lowater. The intensities have been reduced to a scale of 10.

RED SYSTEM, γ, $A^3\Phi - X'^3\Delta$

λ	I	v', v''	λ	I	v', v''	λ	I	v', v''
6996·3	3	3, 4 R_3	6412·3	6	2, 2 R_2	6260·9	8	1, 1 R_1
6959·9	3	2, 3 R_3	6378·3	8	1, 1 R_2	6229·4	9	0, 0 R_1
6543·0	5	2, 2 R_3	6344·9	9	0, 0 R_2	6021·3	3	1, 0 R_2
6508·1	9	1, 1 R_3	6324·3	3	3, 3 R_1	5977·7	3	3, 2 R_1
6473·7	·10	0, 0 R_3	6292·8	7	2, 2 R_1	5439·4	4	5, 2 R_1

YELLOW SYSTEM, β, $B^3\Pi - X'^3\Delta$

λ	I	v', v''	λ	I	v', v''	λ	I	v', v''
6070·0	3	1, 2 R_3	5718·1	10	0, 0 R_3	5515·3	3	3, 2 R_3
5809·2	4	3, 3 R_3	5658·1	5	1, 1 R_2	5491·7	3	5, 4 R_2
5778·5	5	2, 2 R_3	5629·0	6	0, 0 R_2	5485·7	3	2, 1 R_3
5748·1	8	1, 1 R_3	5551·7	5	0, 0 R_1	5456·5	4	1, 0 R_3
5724·0	6	0, 0 Q_3	5545·2	3	4, 3 R_3			

ZrO (*contd.*)

BLUE SYSTEM, α, $C\,^3\Delta - X'\,^3\Delta$

λ	I	v', v''		λ	I	v', v''		λ	I	v', v''
4850·1	3	0, 1 R_1		4637·9	9	0, 0 R_2		4493·8	4	2, 1 R_2
4847·2	3	0, 1 R_2		4619·8	8	0, 0 R_3		4471·5	5	1, 0 R_1
4827·5	4	0, 1 R_3		4521·3	3	3, 2 R_1		4469·5	4	1, 0 R_2
4644·7	5	1, 1 R_3		4519·3	3	3, 2 R_2				
4640·6	10	0, 0 R_1		4496·2	4	2, 1 R_1				

ULTRA-VIOLET TRIPLET SYSTEMS

Data from Afaf.

SYSTEM δ, 3507–3472 Å, $D(?\,^3\Pi) - X'\,^3\Delta$

λ	I	v', v''		λ	I	v', v''
3507·6	1	0, 0 Q_3		3491·8	3	0, 0 R_2
3506·3	3	0, 0 R_3		3473·6	1	0, 0 Q_1
3493·1	1	0, 0 Q_2		3472·4	3	0, 0 R_1

SYSTEM ϕ, 3121–2940 Å, $E(?\,^3\Delta) - X'\,^3\Delta$

λ	I	v', v''		λ	I	v', v''
3121·0	1	0, 2 R_2		3023·8	1	0, 1 R_1
3110·9	1	0, 2 R_1		2967·8	2	0, 0 R_3
3052·4	1	0, 1 R_3		2950·0	2	0, 0 R_2
3033·5	1	0, 1 R_2		2940·9	2	0, 0 R_1

SINGLET SYSTEMS

SYSTEM B, 5480–4800 Å, $b\,^1\Sigma^+ - X\,^1\Sigma^+$

Data from Afaf.

λ	I	v', v''		λ	I	v', v''		λ	I	v', v''
5478·4	0	1, 2		5212·2	1	1, 1		4996·3	2	2, 1
5448·9	1	0, 1		5185·0	6	0, 0		4969·8	2	1, 0
5240·2	0	2, 2		5022·9	1	3, 2		4798·9	0	3, 1

SYSTEM A, 4000–3390 Å, $b\,^1\Sigma^+ - c\,^1\Sigma^+$

Data from Afaf.

λ	I	v', v''		λ	I	v', v''		λ	I	v', v''
3981·4	3	1, 3		3682·1	10	0, 0		3486·7	2	3, 1
3818·3	4	0, 1		3589·9	6	2, 1		3390·0	1	4, 1
3700·1	1	1, 1		3572·0	8	1, 0				

6495 Å BAND, $^1\Pi - X\,^1\Sigma^+$

Found by Balfour and Tatum among stronger β bands. Also prominent in S-type stars. Head $\lambda 6495\cdot3$.

ZrO (*contd.*)

INFRA-RED BANDS

Meggers and Kiess found a complex group of weak bands 9500–9200 Å and Afaf studied these and a similar group (probably the (1, 0) sequence) 8950–8500 Å. This system, I, for which Afaf made a tentative vibrational analysis, may have the $X'\,^1\Delta$ state as lower level. Strongest heads:
$\lambda\lambda 9370 \cdot 7$ ($I = 3$), $9360 \cdot 3$ (3), $9356 \cdot 1$ (3), $9343 \cdot 2$ (4), $9329 \cdot 9$ (5), $9315 \cdot 9$ (5), $9299 \cdot 6$ (5), $8744 \cdot 2$ (2), $8734 \cdot 0$ (2), $8721 \cdot 3$ (3), $8709 \cdot 3$ (3), $8695 \cdot 2$ (3).

Afaf also reported a strong band with a conspicuous though somewhat weak head at 8192 Å, accompanied by a weaker band at 8210 Å. A provisional rotational analysis has been made by Åkerlind. Afaf and Kiess also observed bands 7900–7600 without conspicuous heads.

Spectra of Deuterides

In general isotopic displacements of band heads are not very important for identification, because such displacements are usually small and positions of band heads for the most abundant isotopic components are sufficient. For deuterides, however, the displacements are very much greater and deuterium itself is frequently used as a tracer for studying molecular structure, flame processes, etc. The following section gives references and some data for spectra or deuterides, mostly diatomic. The details of occurrence, appearance and direction of degradation are in general the same as for the corresponding hydride systems and are not repeated here. The rotational fine structure is less open for deuterides than for hydrides and the bands are displaced towards the origin of the system. In a few cases predissociation is less strong for deuterides than for hydrides so that the fine structure of bands may be sharper.

AgD

P. G. Koontz, *P.R.*, **48**, 138 (1935).
U. Ringström, *Ark. Fys.*, **21**, 145 (1962)†.
R. C. M. Learner, *Thesis, London* (1961)†.
U. Ringström and N. Åslund, *Ark. Fys.*, **32**, 19 (1966)†.

$A^1\Sigma - X^1\Sigma$. Deg. R. 3532·2 (3, 4), 3505·6 (2, 3), 3485·2 (1, 2), 3468·4 (0, 1), 3371·6 (2, 2), 3350·3 (1, 1), 3333·9 (0, 0).
$B^1\Sigma - X^1\Sigma$. Deg. R. 2321·6 (0, 1), 2257·8 (0, 0).
$C^1\Pi - X^1\Sigma$. Headless. λ_0 2436·2 (1, 1), 2426·7 (0, 0).
$b^3\Delta_1 - X^1\Sigma$. Deg. R. 2241·4 (0, 0).
$D^1\Pi - X^1\Sigma$. Deg. R. 2233·3 (1, 2).
$c_2{}^3\Pi_2 - X^1\Sigma$. Deg. R. 2254·1 (0, 2).
$c_1{}^3\Pi_1 - X^1\Sigma$. Deg. R. 2278·8 (1, 3), 2251·8 (0, 2).

AlD

W. Holst and E. Hulthen, *Z.P.*, **90**, 712 (1934)†.
B. Grabe and E. Hulthen, *Z.P.*, **114**, 470 (1939).
M. A. Khan, *Proc. Phys. Soc.*, **79**, 745 (1962).

$A^1\Pi - X^1\Sigma$. Deg. R. 4247·5 (0, 0 R head), 4235·6 (0, 0 Q). Other origins, close to Q heads, 4472·1 (0, 1), 4309·6 (1, 1), 4100·7 (1, 0).
$C^1\Sigma - X^1\Sigma$. Deg. R. 2229 (0, 0), 2187·2 (1, 0).
$D^1\Sigma - X^1\Sigma$. Headless. λ_0 2033 (0, 0).

ArD

J. W. C. Johns, *J. Mol. Spec.*, **36**, 488 (1970)†.

$^2\Pi - A^2\Sigma$. Deg. R. Q heads 7666·6 (0, 0), 7706·1 (1, 1).

AsD

R. N. Dixon and H. M. Lamberton, *J. Mol. Spec.*, **25**, 12 (1968)†.

Red-degraded $A^3\Pi - X^3\Sigma^-$ system. Heads of (0, 0), 3259·5, 3262·7, 3270·2, 3380·8, 3389·8, 3399·1, 3400·9; (1, 0) 3163·8, 3226·6.

AsD$_2$

R. N. Dixon, G. Duxbury and H. M. Lamberton, *P.R.S.*, **305**, 271 (1968).

Degraded to the violet. Heads λλ4730·8, 4598·8, 4474·6, 4357·6, 4247·0, 4143·2, 4045·0, 3952·1.

AuD

T. Heimer, *Z.P.*, **104**, 303 (1937).
S. Imanishi, *Sci. Pap. Inst. Phys. Chem. Res. (Tokyo)*, **31**, 247 (1937)†.
U. Ringström, *Ark. Fys.*, **27**, 227 (1964)†.

$A^1\Sigma^+ - X^1\Sigma^+$. R heads λλ4384 (0, 3), 4183 (1, 3), 4117 (0, 2), 4005 (2, 3), 3935 (1, 2), 3872 (0, 1), 3780 (2, 2), 3646 (0, 0), 3585 (2, 1), 3502 (1, 0), 3362 (2, 0).

BD

S. F. Thunberg, *Z.P.*, **100**, 471 (1936).
S. H. Bauer, G. Herzberg and J. W. C. Johns, *J. Mol. Spec.*, **13**, 256 (1964)†.

$A^1\Pi - X^1\Sigma$. Deg. red. Q heads 4327·6 (0, 0), 4348·6 (1, 1).

BaD

G. Edvinsson *et al.*, *Ark. Fys.*, **25**, 95 (1963).
I. Kopp *et al.*, *Ark. Fys.*, **30**, 321 (1965).
I. Kopp, M. Kronekvist and A. Guntsch, *Ark. Fys.*, **32**, 371 (1966)†.
I. Kopp and R. Wirked, *Ark. Fys.*, **32**, 307 (1966)†; **38**, 277 (1968).

$A^2\Pi - X^2\Sigma$. 9754 (2, 1 $^2\Pi_{\frac{1}{2}} - {}^2\Sigma$), 9340, 9284 (1, 0 $^2\Pi_{\frac{3}{2}} - {}^2\Sigma$).
$B^2\Sigma - X^2\Sigma$. Deg. longer wavelengths. Heads, 9779·9 (1, 2 R$_1$), 9737·9 (0, 1 R$_1$), 9666·8 (0, 1 R$_2$), 9123·1 (2, 2 R$_1$), 9081·5 (2, 2 R$_2$), 9074·4 (1, 1 R$_1$), 9030·9 (1, 1 R$_2$), 9026·1 (0, 0 R$_1$), 8981·7 (0, 0 R$_2$), 8558·2 (3, 2 R$_1$), 8530·3 (3, 2 R$_2$), 8504·1 (2, 1 R$_1$), 8475·3 (2, 1 R$_2$), 8451·2 (1, 0 R$_1$), 8421·6 (1, 0 R$_2$).
$E^2\Pi - X^2\Sigma$. Deg. violet. $^0P_{12}$ heads 7277·1 (0, 1), 6881·9 (0, 0), 6856·7 (1, 1), 6515·5 (1, 0); P$_2$ heads 7038·6 (0, 1), 6664·7 (0, 0), 6646·9 (1, 1), 6318·6 (1, 0).
$F^2\Sigma - X^2\Sigma$. Deg. shorter wavelengths, 3256·0 (0, 0), 3166 (1, 0).

BeD

P. G. Koontz, *P.R.*, **48**, 707 (1935)†.

$A^2\Pi - X^2\Sigma$. Q heads, 4988·7 (0, 0), 4984·7 (1, 1).

BeD$^+$

P. G. Koontz, *P.R.*, **48**, 707 (1935).

$A^1\Sigma - X^1\Sigma$. Deg. red. Origins 2900·6 (0, 3), 2770·7 (0, 2), 2663·0 (0, 1), 2588·6 (1, 1), 2552·7 (1, 0), 2421·6 (2, 0), 2362·0 (3, 0).

BiD

A. Heimer, *Z.P.*, **103**, 621 (1936).

B $0^+ - X 0^+$. Weakly deg. violet. Origins 4699 (0, 0), 4696 (1, 1), 4697 (2, 2), 4451 (1, 0), 4458 (2, 1).

CD

T. Shidei, *Japan. J. Phys.*, **11**, 23 (1936)†.
L. Gerö, *Z.P.*, **117**, 709 (1941).

$A^2\Delta - X^2\Pi$.   Deg. violet, Q head 4308·2 (0, 0). Origins, 4308·9 (0, 0), 4306·1 (1, 1), 4319·7 (2, 2).
$B^2\Sigma - X^2\Pi$.   Deg. red. Origins 3970·5 (1, 1), 3874·1 (0, 0), 3674·2 (1, 0).
$C^2\Sigma - X^2\Pi$.   Origins (close to Q heads) 3142·8 (0, 0), 3150·0 (1, 1), 3162·3 (2, 2).

CD⁺

H. Cisak and M. Rytel, *Acta Phys. Polonica*, A, **39**, 627 (1971).

$A^1\Pi - X^1\Sigma$.   Deg. red. Heads 4594·6 (0, 1), 4203·5 (0, 0), 3995·3 (2, 0).

CaD

W. W. Watson, *P.R.*, **47**, 27 (1935)†.
B. Grundström, *Z.P.*, **97**, 171 (1935)†.
G. Edvinsson *et al.*, *Ark. Fys.*, **25**, 95 (1963).
M. A. Khan and M. K. Afridi, *J. Phys.*, B, **1**, 260 (1968).
M. A. Khan, *Proc. Phys. Soc.*, **87**, 569 (1966)†.
M. A. Khan and S. S. Hasnain, *Nuovo Cimento*, B, **18**, 384 (1973)†.

$B^2\Sigma - X^2\Sigma$.   6400–6200 Å. See Watson.
$C^2\Sigma - X^2\Sigma$.   P head of (0, 0) 3524 Å.
$F^2\Sigma - X^2\Sigma$.   P heads, 2720·4 (0, 0), 2711·1 (1, 1), 2646·1 (1, 0).
$G^2\Sigma - X^2\Sigma$.   P heads 2878·7 (0, 0).
$J^2\Sigma - X^2\Sigma$.   P head 2854·6 (0, 0).
$K^2\Sigma - X^2\Sigma$.   P head 2854·6 (0, 0), overlapping J − X.

CaOD

L. Brewer and R. Hauge, *J. Mol. Spec.*, **25**, 330 (1968).

Complex band, degraded to violet. Heads 5545·0, 5549·6, 5550·6, 5555·4, 5556·5.

CdD

O. Deile, *Z.P.*, **106**, 405 (1937).
M. A. Khan, *Proc. Phys. Soc.*, **80**, 1264 (1962)†.

$A^2\Pi_{\frac{1}{2}} - X^2\Sigma$.   (0, 0) P head 4508·0, Q head 4498·1.
$A^2\Pi_{\frac{3}{2}} - X^2\Sigma$.   (0, 0) P head 4312·2, Q head 4303·5.
$C^2\Sigma - X^2\Sigma$.   Deg. violet. Strong (0, 0) band 2483·1. Weaker bands 2414·0 (1, 0), 2407·0 (2, 1), 2400·1 (3, 2).

CdD⁺

R. V. Zumstein, J. W. Gabel and R. E. McKay, *P.R.*, **51**, 238 (1937).

$^1\Sigma - {}^1\Sigma$.   Deg. red. Origins, λλ2509·0 (2, 4), 2494·9 (1, 3), 2478·5 (0, 2), 2425·5 (1, 2), 2407·5 (0, 1), 2375·1 (2, 2), 2338·6 (0, 0), 2291·4 (1, 0).

CoD

L. Klynning and M. Kronekvist, *Phys. Scripta*, **6**, 61 (1973)†; **7**, 72 (1973)†.
R. E. Smith, *P.R.S.*, **332**, 113 (1973)†.

CoD (*contd.*)

$A^3\Phi_3 - X^3\Phi_3$. R heads 5551·2 (0, 0), 4333·5 (1, 0).
$A^3\Phi_4 - X^3\Phi_4$. R heads, 4810·4 (1, 2), 4766·5 (0, 1), 4527·5 (1, 0), 4481·8 (0, 0), 4269·4 (1, 0).

CsD

I. R. Bartky, *J. Mol. Spec.*, **21**, 25 (1966).
M. L. Csaszas and E. Koczkas, *Acta Phys. Hungar.*, **23**, 211 (1967).

$A^1\Sigma^+ - X^1\Sigma^+$. Strongly deg. red. Origins 5236·5 (10, 0), 5190·2 (11, 0).

CuD

U. Ringström, *Ark. Fys.*, **32**, 211 (1966).
M. A. Jeppesen, *P.R.*, **50**, 445 (1936).

$A^1\Sigma^+ - X^1\Sigma^+$.   Strongest R heads λλ4541·8 (0, 1), 4313·6 (1, 1), 4281·0 (0, 0), 4077·4 (1, 0),
    3899·0 (2, 0), 3742·1 (3, 0), 3603·7 (4, 0).
$B^3\Pi_{0^+} - X^1\Sigma^+$.   R head of (0, 0) 3799·1.
$a^3\Sigma^+ - X^1\Sigma^+$.   R head of (1, 0) 3642·4.
$C\ 1 - X^1\Sigma^+$.   R heads 3680·0 (0, 0), 3539 (1, 0).
$c\ 1 - X^1\Sigma^+$.   R heads 3750 (0, 1), 3569·8 (0, 0), 3305 (2, 0).

D₂

For data and references on the 'secondary' spectrum of deuterium see G. H. Dieke, *J. Mol. Spec.*, **2**, 494 (1958).

DCl⁺

F. Norling, *Z.P.*, **104**, 638 (1937)†.

$A^1\Sigma - X^2\Pi$.   Heads of (1, 0) λλ3363·5 $^sR_{21}$, 3366·5 R_1, 3368·2 Q_1; (0, 0) λλ3493·7 $^sR_{21}$,
    3497·4 R_1, 3499·2 Q_1, 3575·6 R_2, 3579·2 Q_2; (0, 1) λλ3830·6 R_2, 3835·0 Q_2.

GaD

H. Neuhaus, *Ark. Fys.*, **14**, 551 (1959).
M. Kronekvist, A. Lagerqvist and H. Neuhaus, *J. Mol. Spec.*, **39**, 516 (1971)†.

$A^1\Pi - X^1\Sigma^+$. Deg. red. (0, 1) R head 4392, Q 4395; (0, 2) diffuse Q head 4616 Å.
$a^3\Pi_1 - X^1\Sigma^+$. Deg. violet. (0, 0) P head 5682·1, Q head 5668·9 Å.

GeD

L. Klynning and B. Lindgren, *Ark. Fys.*, **32**, 575 (1966)†.

$^2\Delta - ^2\Pi$.   Deg. red. Heads λλ4095·4 (1, 1 R_{12}), 4023·9 (0, 0 R_{12}), 3950·6 (1, 1 R_1), 3885·0
    (0, 0 R_1), 3884·7 (1, 0 R_{12}), 3835·4 (2, 1 R_1), 3755·4 (1, 0 R_1).

HgD

Y. Fujioka and Y. Tanaka, *Sci. Pap. Inst. Phys. Chem. Res.* (*Tokyo*), **34**, 713 (1938).
D. M. Eakin and S. P. Davis, *J. Mol. Spec.*, **35**, 27 (1970)†.

$A_1\ ^2\Pi_{\frac{1}{2}} - X^2\Sigma^+$.   P heads 4602·2 (0, 5), 4547·6 (0, 4), 4447·5 (0, 3), 4321·2 (0, 2), 4179·3 (0, 1),
    4075·2 (1, 2), 4029·4 (0, 0), 3864·1 (2, 2), 3815·0 (1, 0); the (0, 0) is strongest.
$A_2\ ^2\Pi_{\frac{3}{2}} - X^2\Sigma^+$.   Heads λλ3821·2 (0, 3 P), 3808·1 (0, 3 Q), 3727·6 (0, 2 P), 3725·2 (0, 2 Q),
    3617·6 (0, 1 Q), 3503·9 (0, 0 Q).

HgD⁺

S. Mrozowski, *Acta Phys. Polonica*, **4**, 405 (1935).

T. Hori and J. Huriuti, *Z.P.*, **101**, 279 (1936).

$A^1\Sigma - X^1\Sigma$. R heads 2457·0 (3, 5), 2439·5 (2, 4), 2424·8 (1, 3), 2411·9 (0, 2), 2385·0 (3, 4), 2366·3 (2, 3), 2350·3 (1, 2), 2336·0 (0, 1), 2262·4 (0, 0).

KD

I. R. Bartky, *J. Mol. Spec.*, **20**, 299 (1966).

$A^1\Sigma - X^1\Sigma$. Deg. red. Origins (close to heads) λλ5421·5 (9, 3), 5122·7 (10, 2), 4835·4 (10, 0), 4787·3 (11, 0), 4740·5 (12, 0), 4694·2 (13, 0), 4648·4 (14, 0), 4602·7 (15, 0).

LiD

F. H. Crawford and T. Jorgensen, *P.R.*, **47**, 358 (1935).

A. R. Velasco, *Canad. J. Phys.*, **35**, 1204 (1957)†.

The $A^1\Sigma - X^1\Sigma$ system, 4300–3200 Å and the $B^1\Pi - X^1\Sigma$ system have been studied.

LuD

J. d'Incan, C. Effantin and R. Bacis, *J. Phys.*, B, **5**, L 187 (1972).

C. Effantin and J. d'Incan, *Canad. J. Phys.*, **52**, 523 (1974).

The C − X, E − X and H − X systems are strongest.

$A - X^1\Sigma$.   Origin 7666·6 (0, 0).

$B - X^1\Sigma$.   R heads 6541·5 (1, 1), 6509·4 (0, 0).

$C - X^1\Sigma$.   R heads 5991·6 (1, 1), 5958·5 (0, 0).

$E - X^1\Sigma$.   Origin 5636·4 (0, 0).

$F - X^1\Sigma$.   Head 5308·7 (0, 0).

$G - X^1\Sigma$.   R heads 5045·4 (1, 1), 5028·4 (0, 0).

$H - X^1\Sigma$.   Origins 4253·2 (1, 1), 4247·7 (0, 0).

MgD

Y. Fujioka and Y. Tanaka, *Sci. Pap. Inst. Phys. Chem. Res. (Tokyo)*, **30**, 121 (1936)†.

A. Guntsch, *Z.P.*, **93**, 534 (1934).

L. A. Turner and W. T. Harris, *P.R.*, **52**, 626 (1937).

M. A. Khan, *Proc. Phys. Soc.*, **77**, 1133 (1961)†; **80**, 209, 523 (1962)†.

W. J. Balfour, *J. Phys.*, B, **3**, 1749 (1970)†.

$A^2\Pi - X^2\Sigma$. P_1 heads λλ5497·1 (0, 1), 5465·0 (1, 2), 5201·9 (0, 0), 5181·3 (1, 1), 5161·5 (2, 2), 4921·6 (1, 0), 4911·9 (2, 1).

$C^2\Pi - X^2\Sigma$.   P heads λλ2428·3 (0, 0), 2422·7 (1, 1), 2363·9 (1, 0). The (0, 1) has a Q head at 2488·2 Å.

The systems at 2819, 2702 and 2172 have also been studied by Khan.

MgD⁺

H. Juraszynska and M. Szulc, *Acta Phys. Polonica*, **7**, 49 (1939).

A. Guntsch, *Ark. Mat. Astr. Fys.*, A, **31**, No. 22, 1, (1945).

$A^1\Sigma - X^1\Sigma$.   Origins (close to R heads) 3235·6 (1, 5), 3210·7 (0, 4), 3129·2 (1, ?), 3102·2 (0, 3), 2997·6 (0, 2), 2897·0 (0, 1), 2830·5 (1, 1), 2800·2 (0, 0), 2767·6 (2, 1), 2738·1 (1, 0), 2679·2 (2, 0), 2623·5 (3, 0), 2570·7 (4, 0), 2520·7 (5, 0), 2473·1 (6, 0).

MnD

T. E. Nevin, M. Conway and M. Crawley, *Proc. Phys. Soc.*, A, **65**, 115 (1952).

T. E. Nevin and D. V. Stephens, *Proc. Roy. Irish Acad.*, **55**, 109 (1953)†.

W. Hayes, P. D. McCarvill and T. E. Nevin, *Proc. Phys. Soc.*, A, **70**, 904 (1957).

$A^7\Pi - X^7\Sigma$. Deg. violet. Heads $\lambda\lambda$6854 (0, 3), 6475 (0, 2), 6073 (0, 1), 5704 (0, 0), 5352 (1, 0), 5065 (2, 0).

4800 Å System. Deg. red. Heads $\lambda\lambda$4734, 4719, 4698.

4500 Å System. Deg. red. Bands 4472 and 4430·6.

    Complex structure 9000–6900 Å has also been observed.

ND

G. Pannetier, H. Guenebaut and I. Hajal, *Bull. Soc. Chim. France*, **190**, 1159 (1959)†.

G. Pannetier, H. Guenebaut and A. G. Gaydon, *C. R. Acad. Sci. (Paris)*, **246**, 958 (1955)†.

H. Hanson *et al.*, *Ark. Fys.*, **30**, 1 (1965)†.

F. L. Whittaker, *Proc. Phys. Soc.*, **90**, 535 (1967)†; *J. Phys.*, B, **1**, 977 (1968)†.

$A^3\Pi - X^3\Sigma^-$. Narrow Q, maxima $\lambda\lambda$3357 (0, 0), 3364 (1, 1).

$c^1\Pi - b^1\Sigma^+$. R head 4484·5 (0, 0).

$c^1\Pi - A^1\Delta$. Deg. red. Heads 3509·3 (0, 1 Q), 3500·4 (0, 1 R), 3241·4 (0, 0 Q), 3234·8 (0, 0 R), 3079·8 (1, 0 Q), 3075·2 (1, 0 R).

$d^1\Sigma - c^1\Pi$. Heads, 2552·9 (0, 0 P), 2532·4 (0, 0 Q), 2526·5 (1, 1 P), 2515·5 (1, 1 Q).

ND$_2$

K. Dressler and D. A. Ramsay, *Phil. Trans. Roy. Soc.*, A, **251**, 553 (1959)†.

A progression of complex bands with ν_2' varying from 9 to 15. Origins, 6734·7, 6483·7, 6242·6, 6015·3, 5804·1, 5567·1, 5378·2.

NaD

E. Olsson, *Z.P.*, **93**, 206 (1955).

$A^1\Sigma - X^1\Sigma$. Origins (close to weak R heads) 4294·3 (7, 1), 4247·7 (8, 1), 4201·9 (9, 1), 4103·7 (8, 0), 4060·9 (9, 0), 4018·7 (10, 0), 3977·2 (11, 0), 3936·5 (12, 0), 3896·6 (13, 0), 3857·5 (14, 0), 3819·2 (15, 0), 3781·8 (16, 0), 3745·3 (17, 0).

NiD

N. Åslund *et al.*, *Ark. Fys.*, **28**, 271 (1965).

R. E. Smith, *P.R.S.*, **332**, 113 (1973)†.

6257 Å System. R heads $\lambda\lambda$6824·5 (0, 1), 6233·1 (0, 0), 5839·3 (1, 0), 5506·6 (2, 0); this is a $^2\Delta_{\frac{5}{2}} - {}^2\Delta_{\frac{5}{2}}$ transition.

4207 Å System. R heads $^2\Delta_{\frac{5}{2}} - {}^2\Delta_{\frac{5}{2}}$ 4450·9 (0, 1), 4191·5 (0, 0), 4036·5 (1, 0); $^2\Delta_{\frac{3}{2}} - {}^2\Delta_{\frac{3}{2}}$ 4310·8 (0, 0), 4178·7 (1, 0).

OD

M. Ishaq, *P.R.S.*, **159**, 110 (1937)†; *Proc. Phys. Soc.*, **53**, 355 (1941)†; *Indian J. Phys.*, **18**, 52 (1944).

M. G. Sastry and K. R. Rao, *Indian J. Phys.*, **15**, 27 (1941)†.

M. G. Sastry, *Indian J. Phys.*, **15**, 95 and 455 (1941)†.

P. Felenbok, *Ann. Astrophys.*, **26**, 393 (1963)†.

L. Herman, P. Felenbok and R. Herman, *J. Phys. Radium*, **22**, 83 (1961)†.

OD (*contd.*)

$A^2\Sigma - X^2\Pi$. R_1 heads $\lambda\lambda 3329\cdot9$ (0, 1), $3194\cdot3$ (3, 3), $3148\cdot6$ (2, 2), $3105\cdot5$ (1, 1), $3064\cdot8$ (0, 0), $2962\cdot7$ (3, 2), $2915\cdot9$ (2, 1), $2871\cdot8$ (1, 0), $2755\cdot2$ (3, 1), $2708\cdot4$ (2, 0), $2569\cdot1$ (3, 0).

$B^2\Sigma - A^2\Sigma$.   R heads $\lambda\lambda 5472\cdot7$ (0, 11), $5313\cdot6$ (1, 11), $5209\cdot4$ (0, 10), $4922\cdot9$ (0, 9), 4793 (1, 9), $4636\cdot1$ (0, 8).

Felenbok describes the region 2600–2300 Å as very rich in lines and mentions an unassigned band at $2673\cdot8$ Å.

OD⁺

S. O'Connor, *Proc. Roy. Irish Acad.*, **A, 62**, 73 (1963).

$A^3\Pi - X^3\Sigma^-$. R_3 heads 3845 (0, 1), 3547 (0, 0), 3465 (1, 0).

PD

M. Ishaq and R. W. B. Pearse, *P.R.S.*, **173**, 265 (1939)†.

$A^3\Pi - X^3\Sigma$.   Prominent heads of (0, 0) band, with intensities, R_3 3388 (10), Q_1 3413 (8).

PD⁺

N. A. Narasimham and M. N. Dixit, *Curr. Sci. (India)*, **36**, 1 (1967)†.

$A^2\Delta - X^2\Pi$.   First and third heads, $\lambda\lambda 4136\cdot9$ (0, 1 R_2), $4086\cdot8$ (0, 1 $^SR_{21}$), $3870\cdot4$ (0, 0 R_2), $3826\cdot7$ (0, 0 $^SR_{21}$), $3723\cdot7$ (1, 0 R_2), $3683\cdot5$ (1, 0 $^SR_{21}$).

PD₂

D. A. Ramsay, *Nature*, **178**, 374 (1956)†.

A violet-degraded progression with v_2' varying from 0 to 12. Heads $\lambda\lambda 5471\cdot4$, $5272\cdot2$, $5088\cdot6$, $4919\cdot3$, $4761\cdot5$, $4615\cdot2$, $4479\cdot3$, $4353\cdot0$, $4235\cdot6$, $4125\cdot9$, $4021\cdot9$, $3921\cdot9$, $3823\cdot6$.

PdH

A. Lagerqvist, H. Neuhaus and R. Scullman, *Proc. Phys. Soc.*, **83**, 498 (1964).

Red-degraded $^2\Sigma - {}^2\Sigma$ band with head $4048\cdot6$ Å.

PtD

B. Kaving and R. Scullman, *Phys. Scripta*, **9**, 33 (1974).

$A^2\Delta_{\frac{5}{2}} - X^2\Delta_{\frac{5}{2}}$.   Deg. red. Heads $\lambda\lambda 4897\cdot7$ (0, 1), $4532\cdot9$ (0, 0), $4310\cdot4$ (1, 0), $4232\cdot9$ (3, 1), $4120\cdot1$ (2, 0), $3957\cdot5$ (3, 0).

$B^2\Delta_{\frac{5}{2}} - X^2\Delta_{\frac{5}{2}}$.   Deg. red. Heads $\lambda\lambda 3986\cdot4$ (0, 1), $3741\cdot3$ (0, 0), $3589\cdot7$ (1, 0), $3459\cdot5$ (2, 0).

RbD

I. R. Bartky, *J. Mol. Spec.*, **21**, 1 (1966).

$A^1\Sigma - X^1\Sigma$.   Extensive system. Origins (close to weak R heads) $\lambda\lambda 5185\cdot8$ (11, 1), $5238\cdot2$ (10, 1).

SD

D. A. Ramsay, *J. Chem. Phys.*, **20**, 1920 (1952)†.

J. W. C. Johns and D. A. Ramsay, *Canad. J. Phys.*, **39**, 210 (1961)†.

$A^2\Sigma - X^2\Pi$. R_1 heads $\lambda\lambda 3227\cdot5$ (0, 0), $3095\cdot8$ (1, 0), $2983\cdot1$ (2, 0); R_2 heads, $3266\cdot9$ (0, 0), $3128\cdot8$ (1, 0), $3014\cdot7$ (2, 0).

SeD

B. Lindgren, *J. Mol. Spec.*, **28**, 536 (1968).

$A^2\Sigma - X^2\Pi$.   Diffuse red-degraded bands $\lambda\lambda 3206\cdot 5$ (0, 0), $3110\cdot 8$ (1, 0), 3037 (2, 0).

SiD

G. D. Rochester, *Z.P.*, **101**, 769 (1936)†.
R. D. Verma, *Canad. J. Phys.*, **43**, 2136 (1965)†.
L. Klynning and B. Lindgren, *Ark. Fys.*, **33**, 73 (1967).
P. Bollmark, L. Klynning and P. Pagès, *Phys. Scripta*, **3**, 219 (1971).

$A^2\Delta - X^2\Pi$.   Heads $\lambda\lambda 4098\cdot 1$ (0, 0 R_1), $4134\cdot 4$ (0, 0 Q_2), $4150\cdot 2$ (1, 1 R_2), $4170\cdot 6$ (1, 1 Q_2).
$B^2\Sigma - X^2\Pi$.   (0, 0) heads $3223\cdot 3$ $^S R_{21}$, $3236\cdot 7$ R_2, $3240\cdot 3$ Q_2.
$D^2\Delta - X^2\Pi$.   (0, 0) has Q_{11} head (deg. violet but almost line-like) at $2057\cdot 7$ Å.

SnD

L. Klynning, B. Lindgren and N. Åslund, *Ark. Fys.*, **30**, 141 (1965).

$^2\Delta - X^2\Pi$. 4447 (0, 0 Q_2), 4129 (1, 1 R_1), 4055 (0, 0 R_1), 3936 (1, 0 R_1).
$^4\Sigma^- - X^2\Pi$. 6153 (1, 1), 6080 (0, 0).

SrD

W. W. Watson, W. R. Fredrickson and M. E. Hogan, *P.R.*, **49**, 150 (1936).
G. Edvinsson *et al.*, *Ark. Fys.*, **25**, 95 (1963)†.
M. A. Khan, *Proc. Phys. Soc.*, **89**, 165 (1966).
M. A. Khan and M. R. Butt, *J. Phys.*, B, **1**, 745 (1968).

$B^2\Sigma - X^2\Sigma$.   Heads $\lambda\lambda 7003\cdot 9$ (0, 0 P_1), $6981\cdot 4$ (0, 0 P_2), $6996\cdot 4$ (1, 1 P_1), $6973\cdot 7$ (1, 1 P_2).
$C^2\Sigma - X^2\Sigma$. $3807\cdot 7$ (0, 0 P).
$D^2\Sigma - X^2\Sigma$. $5419\cdot 7$ (v', 2 R), $5764\cdot 7$ (v', 3 R).
$F^2\Sigma - X^2\Sigma$.   P heads $3001\cdot 5$ (0, 1), $2927\cdot 8$ (0, 0), $2917\cdot 3$ (1, 1), $2847\cdot 7$ (1, 0).
$G^2\Sigma - X^2\Sigma$. P_1 heads $3114\cdot 7$ (0, 0), $3022\cdot 5$ (1, 0).

TiD

R. E. Smith and A. G. Gaydon, *J. Phys.*, B, **4**, 797 (1971)†.

Three systems with complex structure. Maxima 5300–5340, 5065, 4750 and 4777 Å.

TlD

T. Larsson and H. Neuhaus, *Ark. Fys.*, **31**, 299 (1966).
H. Neuhaus and V. Muld, *Z.P.*, **153**, 412 (1959)†.

$^3\Pi_{0^+} - X^1\Sigma^+$.   R heads $\lambda\lambda 5989\cdot 7$ (5, 3), $5972\cdot 6$ (2, 2), $5863\cdot 7$ (3, 2), $5793\cdot 8$ (1, 1), $5769\cdot 4$ (4, 2),
      $5389\cdot 5$ (5, 1), $5362\cdot 7$ (2, 0), $5307\cdot 1$ (6, 1), $5035\cdot 5$ (3, 0).
$^1\Pi - X^1\Sigma^+$.   R heads, $\lambda\lambda 4196\cdot 1$ (1, 0), $4185\cdot 9$ (2, 0).

VD

R. E. Smith, *P.R.S.*, **332**, 113 (1973)†.

Complex headless structure 4800–4400 with strong features at 4703, 4688 and 4686 Å.

YbD

L. Hagland, I. Kopp and N. Åslund, *Ark. Fys.*, **32**, 321 (1966)†.

YbD (*contd.*)

$D^2\Pi_{\frac{1}{2}} - X^2\Sigma$. Heads $\lambda\lambda 6936\cdot1$ (0, 1 P_{12}), $6896\cdot5$ (0, 1 P_1), $6891\cdot6$ (1, 2 P_{12}), $6551\cdot6$ (0, 0 P_{12}), $6522\cdot9$ (1, 1 P_{12}), $6508\cdot2$ (0, 0 P_1), $6477\cdot1$ (1, 1 P_1).

$D^2\Pi_{\frac{3}{2}} - X^2\Sigma$. Heads $\lambda\lambda 6290\cdot0$ (0, 1 P_2), $6268\cdot6$ (0, 1 P_{21}), $5972\cdot6$ (0, 0 P_2), $5947\cdot1$ (0, 0 P_{21}), $5943\cdot4$ (1, 1 P_2), $5920\cdot2$ (1, 1 P_{21}), $5635\cdot1$ (1, 0 P_{21}).

$E\frac{1}{2} - X^2\Sigma$. Heads $\lambda\lambda 5914\cdot5$ (0, 1 P_1), $5905\cdot2$ (0, 1 P_2), $5887\cdot3$ (1, 2 P_1), $5879\cdot0$ (1, 2 P_2), $5628\cdot9$ (0, 0 P_1), $5618\cdot7$ (0, 0 P_2), $5601\cdot6$ (1, 1 P_2), $5343\cdot9$ (1, 0 P_2).

$F\frac{1}{2} - X^2\Sigma$. Heads $\lambda\lambda 4561\cdot1$ (0, 0 P_2), $4548\cdot2$ (0, 0 Q_2).

ZnD

Y. Fujioka and Y. Tanaka, *Sci. Pap. Inst. Phys. Chem. Res.* (*Tokyo*), **32**, 143 (1937).

M. A. Khan, *Proc. Phys. Soc.*, **80**, 599 (1962)†.

$A^2\Pi - X^2\Sigma$. Heads of (0, 0), $4306\cdot8$ P_1, $4303\cdot8$ Q_1, $4255\cdot8$ P_2, $4245\cdot3$ Q_2.

$C^2\Sigma - X^2\Sigma$. P heads $\lambda\lambda 2426\cdot2$ (0, 0), $2414\cdot9$ (1, 1), $2353\cdot9$ (1, 0), $2345\cdot9$ (2, 1), $2337\cdot6$ (3, 2), $2329\cdot2$ (4, 3).

ZnD⁺

J. W. Gabel and R. V. Zumstein, *P.R.*, **52**, 726 (1937).

$A^1\Sigma - X^1\Sigma$. Origins (very close to R heads) $\lambda\lambda 2396\cdot9$ (5, 8), $2389\cdot1$ (4, 7), $2379\cdot9$ (3, 6), $2369\cdot3$ (2, 5), $2357\cdot1$ (1, 4), $2343\cdot9$ (0, 3), $2305\cdot7$ (2, 4), $2292\cdot0$ (1, 3), $2277\cdot6$, $2212\cdot8$, $2183\cdot0$, $2149\cdot8$, $2106\cdot3$.

Practical Procedure and Precautions

The following section contains a few brief notes on various minor points which arise in the identification of molecular spectra, and which have been found to trouble the inexperienced, but which are not usually dealt with in the general textbooks.

On the Identification of Bands. It should be borne in mind that the most satisfactory comparison of two spectra is made by bringing plates or prints together, side by side. It is preferable that the spectra should be taken with the same instrument in the same state of adjustment, but if this is not possible, enlargements from the plates may be made to the same scale by means of the iron arc comparison spectra. That such direct comparison is not always necessary is of course true; in fact, the object of constructing these tables is largely to make this unnecessary; nevertheless, there will remain cases where one must resort to this method. This is especially true in dealing with sources of very low intensity, such as phosphorescent glows, fluorescence, the night sky, comet-tails, etc., where, in order to get a record in a reasonable time, instruments of small dispersion are used with wide slits. In such cases, while the wavelengths recorded by the observer are not infrequently useless for identification, much may be done with his published photograph. Direct comparison is also useful in dealing with spectra which consist of small regions of continuum, headless bands or other structures lacking outstanding features capable of accurate measurement, and in dealing with spectra which contain several band systems superimposed. Small differences in complicated spectra, otherwise the same, and points of resemblance in spectra mainly different are certainly most readily detected by direct comparison.

Where the dispersion is sufficient to allow of reasonably accurate measurements on well-defined heads, identification by means of wavelengths becomes practicable. In using these tables the following procedure is suggested as a guide:

(1) Select two or three of the strongest bands of the spectrum to be identified and compare their wavelengths with the list of persistent bands. If entries are found in close agreement with these wavelengths, and if the bands are degraded in the appropriate direction, refer to the detailed list for the corresponding system.

(2) If all the bands given in the detailed list are found to be present in the spectrum and the details of appearance and occurrence are applicable, the identity may be considered to be established. The approximate agreement of a few of the bands should not be accepted as identification unless the selection can be reasonably explained, e.g., in absorption it may happen that only those bands with $v'' = 0$ are obtained or in fluorescence only those arising from certain values of v'; a random selection should be rejected. If bands remain unaccounted for in the spectrum they may be an extension of the system, if they are the same type of band, or they may belong to another system of the same molecule. If bands are still outstanding after these possibilities have been examined, select the strongest of them and refer again to the list of persistent bands.

(3) Having identified as many systems as possible by this method, it is usually worth while to refer to the detailed lists for systems of other molecules which may be formed from the elements now known to be present. In this way weak bands which have escaped notice in a crowded spectrum are often detected and accounted for.

(4) Determine the origin of strong atomic lines if any are present. This may provide a clue to an identification, support one already made, or supply the clearest evidence of an unsuspected impurity.

(5) Consider whether the systems obtained are likely to occur in the given source. Such consideration often helps to eliminate erroneous identifications due to chance coincidences.

Sources. It is desirable to have some acquaintance with the properties of various sources commonly employed for the production of spectra, both in regard to choosing a source suitable for the production of a given spectrum, and also in regard to assessing the probability of a suspected system appearing in the given source.

The evaluation of the absolute intensities of the band systems of a molecule in different sources requires a knowledge of such quantities as the concentrations of the various atoms and molecules present, the proportion of each in their possible states of excitation and ionization, with their velocity distributions, as well as the concentration and velocity distribution of electrons, and, in addition, a knowledge of the collision processes which may occur. Such knowledge is not in general available, but fortunately it is possible to make a few generalizations of some value as a result of direct observation, without going into so much detail.

Flames. Many band systems are observed in flames; some by the direct combustion of inflammable substances; others by the introduction of additional substances into a flame already established. The general characteristic of the band systems obtained in this way is that they arise from transitions between a few of the lowest levels of the molecule concerned. The energy of the upper level involved rarely exceeds 5 e-volts, while the lower level is in most cases the ground state. Without exception, flame bands have been found to belong to molecules which are electrically neutral, but very frequently the molecules are not stable in the chemical sense, thus such combinations as CH, NH and OH are of very common occurrence. A few examples will serve to illustrate these points. The 4300 Å and 3900 Å bands of CH, the 3064 band of OH and the Swan bands of C_2 appear readily enough in the flames of hydrocarbons, but the Third Positive and the Ångström bands of CO, which require more than 10 electron volts for their excitation, are absent. Different systems occur most strongly in different parts of the flame; the OH bands are spread through the blue outer cone of a Bunsen flame using coal-gas, but the Swan bands are restricted to the greenish inner cone of the roaring flame, which in fact owes its colour mainly to the presence of these bands. The 3360 Å band of NH, the 3064 Å band of OH and the red and violet systems of CN are given by a flame of moist cyanogen. The 3360 Å band of NH is also obtained strongly from the oxy-ammonia flame, but the systems arising from more excited levels, which are known from other sources, do not appear as well. This is also true of the cyanogen flame. To obtain other systems by the introduction of additional substances, it is necessary that these should be brought to the gaseous state within the flame. Gases and vapours may be mixed directly with the gas being burnt; volatile liquids may be sprayed into the flame and volatile solids introduced on suitable supports. The number of spectra which may be obtained in this way is thus restricted by the necessity of finding suitable volatile substances to add to the flame. This restriction is not so far reaching as it appears at first sight, since the substance whose spectrum is required does not have to be introduced directly, but may be formed as a result of chemical reaction within the flame. Thus in the example mentioned above, although carbon is among the least volatile of all substances, yet the bands of C_2 are readily observed during the combustion of hydrocarbons, even being observed in a candle flame. Again, in cases where the metallic oxide is refractory, the spectrum of the oxide may be obtained by introducing the metal itself or, more generally, by introducing a volatile halide. Chemical action in the flame also allows the spectra of many metal hydrides to be obtained from flames in cases where the metal does not form a stable compound with hydrogen. Thus the spectra of MgH and CuH may be obtained by putting the finely divided metal into a hydrogen flame and that of NiH by allowing the vapour of nickel carbonyl to mix with the hydrogen. In the examples quoted so far the band systems obtained from flames are readily obtained in other ways, from the electric arc or the discharge tube for the most part, but there are a few systems known which appear readily in the flame yet are not obtained or are only obtained with difficulty in other sources. Such are the CO-flame bands, the hydrocarbon-flame bands and the α-bands of ammonia. It is unlikely that such systems arise from highly-excited states of the molecules concerned; it appears more probable that the equilibrium configurations for the excited states differ considerably from those for the normal state. According to the Franck-Condon theory excitation by the absorption of light or by electron impact takes place in such a way that the instantaneous kinetic energy and configuration of the nuclei are unchanged during the change of electronic state; therefore in absorption or in sources where excitation is mainly by electron impact, band systems arising from states in which the configuration of the nuclei differs markedly from that of the normal state may be expected to be weak. In the flame, where excitation occurs mainly as a result of collisions between atoms and molecules, these systems may be relatively strong.

The Arc. In general, arcs develop higher temperatures than flames and therefore are able to volatilise many substances which resist flames. The arc spectra of atoms contain many more lines than the flame spectra, for there is usually sufficient energy available in collision processes in the arc to excite all states up to ionization, and, in the case of readily ionized elements like calcium, even to excite a few states of the ion. With molecules the behaviour is similar; in the arc spectrum more band systems appear than in the flame spectrum as higher levels are excited. The number of additional systems is not usually great, however, for just as in atomic series the number of lines discernible is limited by the pressure, so other factors, including pressure, limit the number of band systems. The arc in air has been widely used for the production of the spectra of oxides and halides of the metals; in some cases, such as CuCl and TiO, the bands are more clearly shown in the flame of the arc than in the core. By enclosing the arc it may be run in various gases and at various pressures ranging from a few millimetres of Hg to several atmospheres. The spectra of many of the metallic hydrides have been obtained using arcs in hydrogen at a pressure of a few centimetres of mercury. Reduction of the pressure favours ionization; thus the spectra of Mg^+ and MgH^+ can be obtained easily from an arc between poles of magnesium in an atmosphere of hydrogen by reducing the pressure to a few millimetres of Hg. Increase of pressure up to several atmospheres is sometimes successful in producing band systems not otherwise obtained, such as those of SnH and PbH. This occurs where states of the molecular are subject to predissociation. Band spectra emitted by arcs do not necessarily arise from molecules containing the material of the poles, sometimes only the atmosphere is involved; many arcs produce the OH bands if water vapour is present, and several, notably the Cu arc, produce the NO γ-bands in air. Under reduced pressure such bands as those of PN, NH 3360 Å and the Second Positive system of nitrogen are produced when the appropriate elements are present in the atmosphere. (For high-tension arc see 'The Spark.')

Discharge Tubes. Although, as sources of illumination, flames and open arcs have the advantage of simplicity, discharge tubes offer greater scope for the variation of conditions. The discharge tubes formerly used were of low intensity but many of the types now in use compare favourably with the arc in this respect. Moreover, discharge tubes have the additional advantage of steadiness, so that continual readjustment of the image of the source on to the slit is avoided. In what is called the normal discharge seven different regions have been distinguished, viz.: (1) the anode glow, (2) the positive column, (3) the Faraday dark space, (4) the negative glow, (5) the cathode dark space, (6) the cathode glow and (7) the primary dark space. The most luminous parts and therefore those most used in spectroscopy are the positive column and the negative glow.

The Positive Column. With an uncondensed discharge the positive column presents a source which resembles the arc in many ways. The spectra obtained from it are usually those of uncharged atoms and molecules but the number of excited states reached is greater than in the flame or the arc in air. Thus with CO or CO_2 present the Fourth Positive and Ångström bands of CO appear readily in the positive column although they are not observed in the CO flame and only with difficulty in the arc. Excitation appears to be due mainly to electron impacts, the electrons having a velocity distribution of the Maxwell-Boltzmann type but for a temperature much higher than that of the gas molecules in the tube. The actual distribution depends very much on the nature and pressure of the gas in the tube and on the intensity of the electric field along the column. Both lowering the pressure and increasing the intensity of the field tend to favour higher stages of excitation; the variation takes place in such a way that the state of excitation appears to depend mainly on the ratio of the field to the pressure, X/p, or perhaps rather more accurately on $X.\lambda$, the product of the field and the mean free path of the electron.

The Negative Glow. In the region of the negative glow there accumulates a considerable positive space-charge. The ions are excited to emission by electrons from the direction of the cathode and as a result the negative glow gives largely the spectra of positively charged ions. A modification of the form of the cathode, known as the *hollow cathode*, allows fuller advantage to be taken of this peculiarity of the negative glow. The cathode takes the form of a hollow cylinder or a massive block through which a slot has been cut. For a certain range of pressure the negative glow passes into the recess, becoming at the same time more brilliant. For the production of molecular spectra the linear dimensions of the recess are usually greater than for the types of hollow cathode used for the production of fine lines for the study of hyperfine structure.

The Addition of Other Gases. Several molecules are known to emit somewhat different spectra in the presence of different gases. Thus, in the presence of excess of one of the rare gases, CO is found to give the Cameron bands and the Triplet bands. Again, whereas the positive column in pure nitrogen appears of an orange colour, the addition of oxygen causes the colour to change to pink, due to a weakening of the red and yellow bands of the First Positive system of N_2 relative to the blue and violet bands of the Second Positive system. The mechanism in most cases is still somewhat obscure. The following, however, are processes which may be expected to occur. If the excess is an inactive gas, such as one of the rare gases, then excited molecules which ordinarily lose their energy in a collision with their fellows may collide with rare gas molecules without loss of energy. There is then greater probability of the molecule radiating band systems arising from these particular excited states. This is especially likely to be true for metastable states of the molecule. On the other hand, if there are metastable states of the molecules of the added gas which are excited, these may in collision hand over their energy to the other molecules, thereby exciting states of these molecules which are not readily excited by electron impact. Also there is the possibility that the excess of other gas may so modify the velocity distribution of the electrons as to cause a marked change in the relative numbers of molecules excited to different levels.

High Frequency Discharges. Several methods of producing electrodeless discharges through low-pressure gases using high-frequency electromagnetic oscillations are now available. Early work was done with a 'ring discharge' using a spark and capacitor circuit, but this method has been superceded by either a tuned valve oscillator circuit giving radio frequencies or a microwave oscillator giving higher frequencies. The spectra excited in this way are in general rather similar to those of the positive column of an ordinary transformer-type discharge, although at low pressure with a strong discharge the excitation may bring up some spectra from higher electronic levels and even a few spectra of ionized molecules. High-frequency discharges provide a useful means of exciting afterglows and have the advantage that contamination with material from internal electrodes is avoided.

Tesla Discharge. The electrodeless discharge from a Tesla coil or 'leak tester' generally gives a rather similar spectrum to that of the positive column from a discharge from an induction coil, but it seems to produce less chemical dissociation. It is thus particularly suitable for exciting the electronic spectra of polyatomic molecules such as benzene, glyoxal and formaldehyde; for this purpose a continuous flow of gas through the discharge tube should be maintained.

Afterglows and 'Atomic Flames'. The afterglow of gases flowing through discharge tubes (or observed stroboscopically) is usually due to recombination of free atoms or radicals, and since the energy of the recombination process is a fairly well-defined quantity, varying only slightly with temperature due to the kinetic energy of the free atoms, there is a tendency for emission to arise from selected vibrational levels near the available excitation energy. This may result in an unusual vibrational intensity distribution. Similarly the glow or so-called 'atomic flame' due to reaction of free atoms produced in discharge tubes reaction with a second introduced gas, may also show an unusual intensity distribution, although in this case the selection of vibrational levels is usually less precise. Some afterglows of mixed gases may involve excitation to meta-stable energy levels, and the energy available for the afterglow is then again closely defined; some of the rare gases, such as argon, possess long-lived metastable states and can produce these afterglows.

Active Nitrogen. The ring discharge or condensed discharge through carefully purified nitrogen are both capable of giving rise to a strong orange-coloured afterglow. The spectrum of this after-glow consists of some of the bands of the First Positive system of N_2. If a small amount of oxygen is mixed with the nitrogen some of the bands of the β-system of NO also appear; in fact, this system is best obtained in this way. In the positive column of a discharge tube the γ-system is much stronger than the β-system, whereas in the afterglow the reverse is the case. The equilibrium constants for the upper state of the γ-system are much closer to those of the normal state than are those for the upper state of the β-system. Many other band systems can be excited by introducing appropriate gases or vapours into nitrogen thus activated, the excitation often being accompanied by chemical reaction. Thus organic compounds such as CCl_4 and C_2H_2 yield systems of CN, CH, C_2, and sometimes NH; $SiCl_4$ yields SiN systems; and, with a trace of oxygen, BCl_3 yields systems of BO. It is usually difficult to remove all trace of oxygen, so that band systems of the oxides often occur quite strongly when other compounds, such as the halides of metals, are added to active nitrogen. A band system produced in active nitrogen often differs considerably in appearance from the same system as observed in the arc or discharge tube; the violet CN bands form a good example. The bands produced in active nitrogen have much shorter branches than in the arc but many more bands of the system are observed; fewer states of rotation are excited, but more of vibration. The energy available for excitation is about $9\frac{1}{2}$ electron volts.

Controlled Electron Sources. The variation of intensity of the various band systems of a molecule with the velocity of impacting electrons can be studied more accurately than in discharge tubes if the velocities of the electrons are controlled. A heated wire (usually coated with suitable electron emitter) is used as source of thermionic electrons, and a grid, separated from it by a distance less than the mean free path, is used to accelerate the electrons by applying a variable voltage between the filament and grid. As the voltage is gradually increased the gas beyond the grid begins to glow. This indicates that molecules are being raised to excited states; as the voltage is further increased, other excited states are stimulated with the consequent emission of other band systems. Ionized molecules are readily excited by electron impact, and the method has the advantage that the effective vibrational and rotational temperatures are often fairly low so that the bands are well defined.

Fluorescence. Excitation by absorption of light by molecules in the lower or lowest vibrational level of the ground state leads to selective population of a few vibrational levels of excited electronic states. At low gas pressure, when collision processes are unimportant, the subsequent emission from the selected vibrational levels may give an emission spectrum of very abnormal vibrational intensity distribution. If excitation of the fluorescence is by a continuous source such as a high-pressure xenon arc lamp, then a number of rotational energy levels and a few vibrational levels may be involved. For monochromatic excitation, as by a laser or an atomic spectrum lamp, only a single rotational energy level of a single vibrational state may be excited and the subsequent emission may show just a few highly selected lines. At higher gas pressure, fluorescence is usually strongly quenched by collisions with other molecules, although there may be some redistribution of energy so that additional vibrational and rotational levels become populated and produce emission; rare-gas atoms are effective in producing this redistribution of population.

The Spark. The condensed spark discharge is not much used for the production of molecular spectra, since the violence of the discharge is such that many lines of atoms in various stages of ionization are produced, but few band systems. Sometimes, however, band systems are emitted in an afterglow following the passage of the spark and may be photographed if a synchronised shutter is adjusted to cut off the light of the spark itself from the spectrograph while exposing it to the afterglow. The uncondensed discharge is used in a variety of ways. The discharge from a high tension transformer between metal rods in air is useful for the production of the spectra of some metallic oxides, the bands being obtained with fewer atomic lines and with shorter branches than in the arc; this facilitates vibrational analysis. By enclosing the discharge it may be used to excite the spectra of various gases and vapours, e.g., the Schumann-Runge bands in O_2. The uncondensed discharge has also been used in conjunction with flames, in some cases to increase the intensity of bands emitted by the flame itself, and in other cases to introduce the vapour of metals of high melting point into the flame for the production of the spectra of their oxides or hydrides. The spectra of the hydrides of nickel, manganese and chromium have been produced in this way by passing the discharge between poles of the appropriate metal in a flame of hydrogen.

Absorption. It is frequently convenient to observe band systems in absorption and this is particularly true where polyatomic molecules are concerned, for these are usually decomposed in emission sources. Observed in absorption a molecular spectrum differs in some respects from one taken in emission. Unless the temperature is unusually high, absorption only occurs for those systems which have the normal state of the molecule as lower level, and only for those bands of these systems which start from the first two or three vibrational levels of the normal state. Thus the absorption spectrum is in general much more simple than the corresponding emission spectrum. If the nuclear configuration for the equilibrium position of the upper state of a system is very different from that for the lower state it may happen that the absorption spectrum shows few if any bands in common with the emission spectrum. In absorption, transitions take place from a few of the lowest vibrational levels of the lower electronic state to high vibrational levels of the upper electronic state, while in emission, transitions take place from a few of the lowest vibrational levels of the upper state to high vibrational levels of the lower state. Since the spacing of the vibrational levels is different for the two electronic states, it is sometimes difficult to recognise that the bands belong to the same system. This applies particularly to polyatomic molecules.

It is sometimes desired to establish proof of the presence of a molecule, especially if this is a radical not stable chemically, by attempting to observe its spectrum in absorption. Sufficient consideration is not, however, always given to the conditions which much be fulfilled for this observation to be possible. The individual lines of band structure are usually very sharp and will only be observed in absorption if the power of resolution of the spectrograph is comparable with the width of the lines themselves. This usually means that an instrument of high dispersion must be used and the slit kept as narrow as possible. This point is well illustrated by observing how the number of Fraunhofer lines which can be distinguished in the solar spectrum depends on the resolving power of the spectrograph used and on the width of the slit. Since the width of an absorption line depends on the number of absorbing molecules in the line of sight, an increase of the length of absorbing column improves the chance of observing the line, but often, as in dealing with flames and explosions, such increase is limited. Lines crowded together to form heads resemble wide lines and as such may be observed with smaller resolution than is necessary to show the individual lines. It often happens therefore that the head of a band may be observed in absorption but not the open branches which accompany it when it is observed in emission with the same spectrograph.

The effective path length in absorption may be increased by use of a multiple-reflection system of concave mirrors. For absorption through hot gases, or transitory sources such as shock-tubes, a flash-tube may be used as background. The background must, of course, always have a higher brightness temperature than the absorbing gas for absorption to be observed.

The King furnace. The emission spectra obtained by heating substances up to about $3000°$ K in carbon-tube furnaces involve only low-lying electronic levels. Emission bands of C_2, C_3, OH, and various metals (e.g. Bi_2) may be obtained. Furnaces are also used extensively for the study of absorption spectra.

Shock Tubes. Since 1955, an increasing use of the bursting-diaphragm shock tube for producing high temperatures has lead to interest in the spectra produced under such conditions. Generally it seems that thermal equilibrium is well maintained, and spectra produced by shock heating to temperatures $2000-3500°$ K resemble those of the King furnace. Bands of CN, C_2, OH and O_2 Schumann-Runge and of metal oxides appear readily. Systems involving highly excited levels, such as the Ångström and Third Positive bands of CO do not occur. In higher temperature shocks ($>3500°$ K) bands of N_2^+ and of the 2nd positive system of N_2 occur.

Flash Photolysis. When a gas which absorbs ultra-violet light is exposed to a very bright flash it may be dissociated to free atoms and radicals, and these may in turn react to form other species. By using a second flash tube, timed to discharge at a suitable delay after the first, it is possible to photograph the absorption spectrum of the radicals, etc. formed by the first flash. The absorption spectra of many free radicals, such as CS, ClO, NH_2, PH_2, OH, CH_2, CH_3 have been observed in this way. Normally the gas is heated by the photolysis flash to high temperature, but in the presence of an inert diluent the temperature may be kept down so that the absorption bands show low rotational temperature. In some cases flash photolysis leads to the formation of molecules or radicals with an abnormal vibrational energy distribution and absorption bands may then be obtained from high vibrational energy levels, e.g. of the Schumann-Runge system of O_2 in flash photolysis of O_3. Flash photolysis may also be used to initiate combustion or other chemical reactions, and species such as C_2, C_3 and CH may then be observed in absorption.

Collimation. Beginners are sometimes troubled by unduly long exposures, lack of definition, doubling and shading of the lines, these defects arising from poor collimation. Whenever a spectrograph is used, care should be taken to see that it is collimated so that it is used to the best advantage; and this includes the giving of due consideration to the selection and adjustment of optical parts, such as condensing lenses, placed between the source and the slit. Assuming that the optical parts of the spectrograph are without fault, it is essential, to obtain speed and good definition, that the dispersing system, prism or grating, should be uniformly filled with light. At the same time, it is undesirable that additional light should be admitted to the spectrgraph through the slit as this extra light, which does not pass through the optical system, is merely scattered within the instrument causing a background of fog on the plate. The ideal is therefore that the light entering the slit should diverge from it in the form of a cone with its axis along the optical axis of the collimator and its base just filling the optical system. If the source is sufficiently extended, it may be brought near enough to the slit for this condition to be fulfilled; if it is not, a condensing lens must be used. In either case, the first adjustment is to arrange the source so that it is on the axis of the collimator. A simple procedure for making this adjustment is as follows: The slit of the spectrograph is opened to about 1 mm. and the source moved (the standard iron arc is convenient for this purpose) both laterally and vertically until the narrow pencil of light entering the slit falls on the centre of the prism or grating of the spectrograph. If a condensing lens is to be used it is next put in place so as to focus an image of the source on the slit. It is an advantage to use the enlarged rather than the diminished image on the slit, provided care is taken to ensure that the full aperture of the spectrograph is used. Use of the diminished image does not give an increase of speed proportional to the brightness of the image, since the light is spread over a cone of larger solid angle, thus more than filling the optical system and flooding the spectrograph with light. The diminished image also has the additional disadvantage of giving a very narrow and uneven spectrum. When the source and lens have been set in position the adjustment should be checked by placing the eye in the plane of the spectrum and observing whether the optical system is completely and uniformly filled with light. When collimating it is often useful to remember that light travelling in the opposite direction takes the same path through an optical system. Thus with large concave gratings, where the grating and source rooms are separate, it is convenient to place a small strip of white paper in front of the grating, illuminate it so that it may be seen through the slit from the source room, and then to place the arc (with current off) in line with the paper and the slit. The lens may then be arranged to focus the image of the arc on the slit. Again, if a source is difficult to move when running it may be set in position, once the lens has been fixed, by illuminating the slit, finding the real image of the slit, and then adjusting the source to coincide with this image. Small gas lasers are very useful to assist collimation in this way. In the majority of experiments a condensing lens is used. Once the lens has been adjusted to be on the axis of the instrument it may be kept there and the source changed as required, the source automatically coming on to the axis when its image is focused on the slit. If the work is sufficiently routine, it may be worth while to arrange an optical bench in conjunction with the spectrograph. Concave mirrors may also be used to focus the source on to the slit and have the advantage that the image is achromatic. Mirrors, however, are otherwise inconvenient and lenses are generally preferred. It must be remembered, when using a lens, that the different wavelengths come to a focus at different distances from the lens. With large instruments, when only a small region of the spectrum is being photographed at a time, this is not serious if care is taken to focus for the wavelength region required, but with small instruments covering a large range, such as the usual quartz spectrographs, it may lead to great variations of intensity. If a particular region is required the lens may be adjusted to bring this to a focus, but if the whole range is required, as in exploratory work, it is usually of advantage to focus the farthest ultra-violet image on the slit. This may be done by using a fluorescent screen in front of the slit and a source, such as the copper arc, which is rich in lines about 2100 Å to adjust the lens. In this way a very uniform intensity may be obtained from the visible to the far ultra-violet.

Comparison Spectra. To obtain the wavelengths of features of a spectrum, a comparison spectrum is photographed alongside. The comparison most generally used is the iron arc. The spectrum of the iron arc contains a very large number of sharp strong lines distributed fairly evenly from about 2330 Å to the infra-red region; there are one or two gaps and the orange region is somewhat confused by bands of FeO, but on the whole it is good throughout the visible and near ultra-violet regions. The lines have been investigated by the International Astronomical Union and accurate standard values of their wavelengths set up. To obtain the highest accuracy the form of arc lamp used to produce the spectrum has been standardized following the recommendations of Pfund. The electrodes are vertical; the anode below, consisting of a bead of iron oxide supported on a massive rod of iron, and the cathode above, consisting of a rod of iron 6–7 mm. in diameter, having a massive cooling cylinder of copper or iron close to the end of the rod. The arc is operated on a 110–250 volts supply with a current of 5 amperes or less. For accurate measurements the arc should be 12–15 mm. long and light should be taken only from the central zone at right angles to the axis of the arc not exceeding 1·5 mm. in width. Recently developed hollow cathode lamps give very narrow Fe lines, but these sources are barely bright enough for use with large dispersion, when it is most advantageous to have narrow lines.

For the region 2000–2300 Å the copper arc is employed. Since the spectrum of the copper arc is relatively simple it is sometimes used as a general comparison spectrum for work with small dispersion. The practice is not, however, much to be recommended as the appearance of the spectrum is rather variable, there are few lines at the red end of the spectrum, and bands of NO often confuse the ultra-violet region.

Mercury shows far too few lines to be of much use as a comparison spectrum, but the frequent employment of the mercury arc as a source for fluorescence and photochemistry had led to its adoption for this purpose. The principal features of the spectrum are of course very easily recognised.

For the yellow and red regions a neon spectrum has been recommended. A neon discharge tube is a very convenient source but unfortunately the lines are rather far apart for use with high dispersion and the useful range is very limited. For the photographic infra-red, a barium arc (Ba salt on carbon poles) is a useful wavelength guide for work with small dispersion.

Reproductions of the spectra of iron, copper, mercury and neon are shown in Plates 11 and 12.

Measurement. If a spectrum is to be measured the comparison spectrum should be photographed alongside, so that there is a slight overlap. In doing this the adjustments of the spectrograph, the plate holder and the dark slide should not be disturbed between the two exposures; the spectra are brought into the desired positions by use of a Hartmann diaphragm in front of the slit for a prism instrument, and by use of a Rowland shutter in front of the plate for a grating instrument.

To obtain the wavelengths, a travelling microscope with a screw accurate to 0·001 mm. is used to measure the positions of the bands or lines under investigation and a sufficient number of lines of the comparison spectrum. Care should be taken always to approach the line or band from the same direction, and after completing the measurements in one range the plate should be reversed, end to end, and a second run made. Reversal of the plate is very important when the lines measured differ much in definition or intensity, as individual observers show, as a rule, a tendency to set regularly off centre by an amount depending on the character of the line. Reduction to the mean setting is greatly facilitated if the scales of the microscope are graduated in both directions, as subtraction is thereby eliminated and there is not difficulty in identifying corresponding readings.

The microscope scale readings are converted to wavelengths by use of a suitable interpolation formula. In the measurement of grating plates a linear formula,

$$\lambda = \lambda_0 + S.D.$$

is used, where λ_0 is a constant depending on the zero of the scale, S is the scale reading and D is the dispersion of the spectrograph. Two carefully chosen lines of the comparison spectrum, one at either end of the range, are used to determine the constants of the formula; the remaining comparison lines are used to construct a correction curve for deviations from this approximate formula. For prism spectrograms a three-constant formula,

$$\lambda = \lambda_0 + C/(S + S_0)$$

is generally used, where λ_0 is a constant depending on the material of the prism, C is a constant depending on the dimensions of the instrument, S is the scale reading and S_0 is a constant depending on the position of the scale zero. As before, the constants of the formula are obtained from carefully chosen standard lines, this time using one in the middle of the range as well as one at either end, and a correction curve is constructed from the remainder. In the measurement of short ranges the calculation is simplified by assuming an approximate round number for λ_0 and using two standard lines to give the other constants. With this procedure a correction curve is essential, and the corrections are larger than would be the case with the three constant formula, but there is the advantage that the curve does not contain a point of inflection. If the standard lines are converted to wave-numbers a linear formula,

$$\nu = \nu_0 + S.K.$$

for wave-numbers may be used. This is much easier for use with a calculating machine, but the formula is still less accurate than the last and places a correspondingly greater burden on the correction curve.

In measuring the band heads the crosswires of the microscope should not be set on the extreme edge of the band, but an attempt should be made to allow for the finite width of a line by setting half a line width within the head. Unless this is done the value obtained for the wavelength of the head may vary considerably from spectrograph to spectrograph, and if a wide slit is used may lead to discrepancies of several Ångströms.

In measuring very faint lines it is generally found to be of some slight advantage to use blue light to illuminate the plate; faint lines appear a little more distinctly with the shorter wavelength illumination, possibly on account of greater scattering by the grains of the plate.

Spurious Bands. When using a quartz spectrograph care should be taken to avoid the light from the source becoming polarized before reaching the spectrograph. With polarised light, interference between the ordinary and extraordinary rays may introduce into a continuum a banded structure not unlike a diffuse band spectrum. The bands are usually too regular to be mistaken for a real molecular spectrum when strong and isolated, but if superimposed on a band system may cause errors in measurements of wave-length and intensity. The light may be polarized if reflectors are used to bring the image on the slit. Light from a discharge tube is often slightly polarized, probably by reflections within the tube.

If plates are not rocked or brushed during development broad lines or bands may become more strongly developed at the edges than in the centre, giving a spurious resolution into two.

With long exposures in a well-lit room it is possible for sufficient diffuse daylight to enter the slit to record the solar spectrum with the stronger Fraunhofer lines, particularly the H and K lines of Ca^+. If their origin is not recognised, these may be attributed to absorption bands from the source in use.

Literature. For the basic theory of diatomic spectra the reader may be referred to the authorative book by G. Herzberg, *Spectra of Diatomic Molecules*, Van Nostrand, New York 1950, and for polyatomics to his *Electronic Spectra of Polyatomic Molecules*, Van Nostrand, 1966. The *Tables de Constants numeriques IV, Donneés spectroscopiques concernant les Molécules diatomiques*, 1970, edited by B. Rosen and colleagues, gives constants and wavelengths, in some cases even more fully than this book, for diatomic molecules to about 1969, but does not cover polyatomic ones or give reproductions. The series of special reports 'Identification Atlas of Molecular Spectra' by R. W. Nicholls and colleagues, first at Western Ontario and later at York University (Canada) gives full data for a few individual band systems of important molecules (C_2 Swan, CN Violet, O_2 Schumann-Runge, etc.). For detailed information about flame spectra see *The Spectroscopy of Flames*, by A. G. Gaydon (Chapman and Hall, 1974). For organic molecules there is a book *Organic Electronic Spectra*, Interscience, 1960, edited by M. J. Kamlet and H. B. Ungnade.

Appendix

Persistent Atomic Lines

In the following table the most persistent atomic lines are given for each element. In addition to the 'raies ultimes' usually quoted, we have given in many cases additional lines to cover regions in which there are no 'raies ultimes.'

For elements possessing a simple readily excited spectrum, *e.g.*, metals like Na, Ca, Al, intensities are quoted on a scale of 10 for the strongest line. For elements whose spectra are less distinctive and less easily excited, *e.g.*, Fe, W, the intensities are given on a scale of 8. For elements whose spectra appear with difficulty, *e.g.*, the non-metals, O, Cl, the intensities are given on a scale of only 5 for the strongest line. Lines which are readily observed in absorption are indicated by the letter *a* following the intensity. The intensities given are in most cases those for arc sources, or other mild conditions of excitation; for gases they refer to Geissler tube excitation.

Ag
5465·48   7
5209·06   8
3382·89   10a
3280·67   10a

Al
3961·54   10a
3944·02   9a
3092·72   8a
3082·16   8a

Ar
8115·31   5
7503·87   4
7067·22   3
6965·43   4
4848·0   5

As
2860·46   6
2780·20   6
2349·84   8
2288·14   6

Au
4792·62   6
2675·95   8a
2427·96   8a

B
2497·73   5

2496·78   4

Ba
5777·7   9
5535·53   10a
5519·11   6
3421·0   8

Ba⁺
4934·09   8
4554·04   10

Be
4573·69   5
3321·35   6
3321·09   6
2650·78   5
2348·62   10a

Be⁺
3131·06   8
3130·42   6

Bi
4722·55   8
3067·73   10
2989·04   7
2938·30   8
2897·98   9

Br
4816·72   4

4785·72   5
4704·83   5

C
2478·57   5

C⁺
4267·27   5

Ca
5602·84   5
to
5581·96
5270·27   4
to
5261·70
4454·78   8
4434·96   7
4425·44   6
4318·65   6
to
4238·10
4226·73   10a

Ca⁺
3968·47   10
3933·67   10

Cb (Nb)
4079·73   5
4058·94   8

Cd
6438·47   10
5085·82   6
4799·91   6
3610·51   9
3466·20   9
3403·65   8
3261·05   10
3288·03   9a

Ce
4628·15
to
4460·21
4186·60   8
4165·61   4
4040·76   7
4012·40   6

Cl
4819·46   4
4810·06   5
4794·54   5

Co
3465·80   6
3453·51   8
3405·12   6
2407·26   4

Cr
5208·43   6

5206·04	6	3719·94	7a	5460·66	6	**Li**	
5204·54	6	3581·20	7	4358·34	6	6707·86	10a
4289·72	9a	3570·10	7	4046·56	6	6103·64	5
4274·80	9a	3475·46	7a	3650·15	6	4602·99	4
4254·34	10a	3465·86	7a	2536·52	9a	3232·61	3a
3605·35	9	3020·65	7a	1849·6	10a		
3593·48	9	2522·86	7a			**Lu**	
3578·60	10	2488·15	7a	**Ho**		4518·54	5
		2483·28	8a	3891·02	5	2911·39	2
Cs				3748·19	1	2894·84	1
8943·50	10a	**Ga**					
8521·15	10a	4172·05	10a	**I**		**Mg**	
4593·18	7a	4033·01	10a	5464·61	3	5183·62	8
4555·36	8a			5161·19	5	5172·68	7
		Gd		3288·3	–	5167·33	6
Cu		3768·40	2	2062·38	–	3838·26	7
5220·06	4	3646·19	10			3832·31	6
5218·20	5			**In**		3096·92	8
5153·26	4	**Ge**		4511·31	10a	2852·13	10a
5105·55	4	4226·61	7	4101·76	8		
3273·96	9a	3269·49	7	3256·08	8	**Mg⁺**	
3247·55	10a	3039·08	8	3039·36	7	2802·71	8
		2754·59	8			2795·54	8
Dy		2709·61	8	**Ir**			
4211·72	4	2651·18	8	3513·67	5	**Mn**	
4167·97	1			3220·79	5	6021·79	3
4077·98	3			3133·34	5	6016·64	3
4046·00	3	**H**				6013·50	3
4000·50	8	6562·79	8	**K**		4034·49	10a
		4861·33	6	7698·98	9a	4033·07	10a
Er		4340·47	4	7664·91	10a	4030·76	10a
3906·34	5			4047·22	4a	2798·27	6
3692·65	4	**He**		4044·16	5a		
3499·12	3	5875·62	5			**Mo**	
		4685·75	2	**Kr**		3902·96	7a
Eu		3888·65	5	5870·92	5	3864·12	8a
4205·03	5			5570·29	5	3798·26	8a
4129·73	3	**Hf**				3193·98	6a
		4093·17	2	**La**		3170·34	6a
F		3134·72	8	6249·93	4	3132·60	6a
6902·46	4	3072·88	8	5930·65	3		
6856·01	5	2940·77	6	4429·90	5	**N**	
		2916·48	5	4333·80	5	4935·03	–
Fe		2904·41	3	4123·23	6	4447·0	–
4957·61	5	2898·25	5	4086·71	6	4109·94	5
4920·52	5			3949·10	8	4099·94	2
4891·50	4	**Hg**		3337·49	7		
3734·87	7a	5790·66	5				
		5769·60	5				

N⁺

5679·5	5
5666·6	5

Na

5895·93	9a
5889·96	10a
3302·94	4a
3302·34	4a

Nd

4303·61	5
4177·34	1
3951·15	2

Ne

6402·25	8
5852·49	8
5400·56	8

Ni

3524·54	7a
3515·06	4a
3446·26	6a
3414·77	8a

O

7771·95	5
6158·21	3
5330·65	2
4368·30	2
3947·29	2

Os

4420·46	3
4260·85	3
3267·94	6
3058·66	6
2909·08	8
2488·55	6

P

2554·93	5
2553·28	8
2535·65	8
2534·01	6

Pb

4057·83	10
3683·47	8
3639·58	6
2833·07	5a
2170·00	−

Pd

3634·68	2
3609·55	2
3516·95	4
3421·23	6
3404·59	8

Pr

4225·34	1
4189·52	2
4179·43	5
4062·83	3

Pt

3064·71	8
2659·44	8

Ra

4825·94	

Rb

7947·63	10a
7800·23	10a
6298·6	5
4215·58	8a
4201·81	8a

Rh

4374·82	4
3692·35	4
3434·90	8

Ru

3728·02	4
3726·93	4
3498·95	8
3436·74	4

S

4696·25	3
4695·45	4

Sb

4694·13	5

3267·48	4
3232·52	2
2877·92	8
2598·08	8
2528·53	8
2311·50	6a
2175·89	8a
2068·38	8a

Sc

4023·72	8
4020·42	5
3911·81	8
3907·48	5

Se

4742·25	3
4739·03	4
4730·78	5
2062·79	4a
2039·85	5a

Si

3905·52	1
2881·59	8
2528·52	4
2524·12	4
2516·12	5
2506·90	4

Sm

4424·35	5
4390·87	2

Sn

5731·70	1
3262·33	9
3175·05	10a
3034·12	8a
2863·32	10a
2839·99	10a
2706·50	8a

Sr

4962·26	3

4872·49	2
4832·08	5
4607·33	10a

Sr⁺

4215·52	6
4077·71	10

Ta

6485·36	1
3318·85	5
3311·14	5
2714·68	8

Tb

3874·19	5
3848·76	2
3561·75	5
3509·18	5

Te

2769·65	4
2385·78	5
2383·27	5
2142·75	4a

Th

4019·14	
3601·05	
3538·75	

Ti

4999·51	2
4981·73	3
4536·05	4
to	
4533·25	
3998·64	6
3371·46	6
3341·87	8

Ti⁺

3372·80	6
3361·22	6
3349·41	6
3349·04	5

Tl				Y			
5350·47	10	4408·52	5	4674·85	4	4810·53	10
3775·73	9a	to		4643·69	2	4722·16	10
3519·24	7	4379·24	8	4374·95	4	4680·14	9
2918·32	2	4128·07	8	4142·87	6	3344·91	9
2767·87	2a	3185·41	8	4102·38	8	2138·61	5a
		3183·99	8	3774·33	6		
				3710·30	6	**Zr**	
Tu		**W**		3242·28	6	4710·07	1
3761·91	4	4302·11	6			4687·80	2
3761·34	5	4294·62	5	**Yb**		3601·19	5
3462·21	4	4008·76	8	3988·01	5	3547·68	3
		3617·52	3	3694·20	3	3519·61	2
U				3289·73	3	3481·15	1
5527·84		**Xe**					
4241·68		4671·22	5	**Zn**		**Zr⁺**	
3672·58		4624·27	2	6362·35	6	3572·47	1
		4500·98	1			3496·21	2
V						3438·23	3
4460·31	5					3391·98	5

For the most complete work on line spectra, the reader is referred to *Massachusetts Institute of Technology Wavelength Tables*, edited by G. R. Harrison (John Wiley and Sons, New York; Chapman and Hall, London; (1969)).

Conversion between wavelengths in air, wavelengths in vacuum and wave-numbers

Since measurements, especially with grating instruments or precision work with interferometers, usually lead to determination of wavelengths in air, while theoretical work usually requires wave-numbers in vacuum, it is often necessary to convert between wavelengths in air and in vacuum, or from wavelengths in air to vacuum wave-numbers; the wave-number is the number of waves per centimetre = frequency/velocity of light (in cm./sec.).

Below are listed corrections (corr.) from wavelength in air to wavelength in vacuum (the correction in angstroms is to be added), and the refractive index of air, n.

$$\lambda_{air} = \lambda_{vac}/n.$$

$$\lambda_{air}(\text{Å}) = 10^8/n\nu(\text{cm.}^{-1}).$$

λ_{air}	corr.	n	λ_{air}	corr.	n	λ_{air}	corr.	n
10,000	2·74	1·000274	6400	1·77	1·000276	4000	1·13	1·000282
9500	2·60	1·000274	6200	1·71	1·000276	3800	1·07	1·000283
9000	2·46	1·000274	6000	1·66	1·000276	3600	1·02	1·000284
8500	2·33	1·000274	5800	1·60	1·000276	3400	0·97	1·000286
8000	2·20	1·000275	5600	1·55	1·000277	3200	0·92	1·000288
7800	2·14	1·000275	5400	1·50	1·000277	3000	0·87	1·000291
7600	2·09	1·000275	5200	1·44	1·000278	2800	0·82	1·000294
7400	2·04	1·000275	5000	1·39	1·000278	2600	0·78	1·000299
7200	1·98	1·000275	4800	1·34	1·000279	2400	0·73	1·000305
7000	1·93	1·000275	4600	1·28	1·000279	2200	0·69	1·000313
6800	1·87	1·000276	4400	1·23	1·000280	2000	0·65	1·000326
6600	1·66	1·000276	4200	1·18	1·000281			

Description of Plates

Below are given brief indications of the source used and of the type of spectrograph employed, vis., concave *grating, glass* prismatic instrument, or *E. 1* (large Littrow type) *E. 2* (medium) or *E. 3* (small) quartz spectrograph. Where the plate has been taken by other than one of the present authors this is indicated by the name in brackets.

Plates 1 to 10 all show positive enlargements, while the comparison spectra shown in Plates 11 and 12 are reproductions of negatives.

Plate 1.   CaF. Calcium fluoride in carbon arc; glass.
CaO. Calcium carbonate in carbon arc; glass.
AlO. High tension arc between Al electrodes; grating. (W. Jevons).
AlCl. Discharge tube; E. 1. (B. N. Bhaduri.)
SiO. Heated silica discharge tube; grating. (R. F. Barrow.)

Plate 2.   Angström, Herzberg and Triplet Systems of CO. Positive column of discharge through CO_2; glass.
 (A. Fowler.)
Third positive and 5B bands of CO. Positive column of discharge through CO_2; E. 2. (A. Fowler.)
Fourth positive bands of CO. Positive column of discharge through CO_2; E. 2. (A. Fowler.)
CO^+, first negative. Negative glow of discharge through CO_2; E. 2. (A. Fowler.)
CO_2^+. Negative glow of discharge through flowing CO_2; E. 2. (A. Fowler.)

Plate 3. N_2, first positive. Positive column of discharge through N_2; glass.
N_2, second positive. Positive column of discharge through N_2; E. 1. (R. C. Pankhurst.)
N_2^+. Negative glow of discharge through N_2; E. 2.
NO β. Active nitrogen; E. 2. (A. Fowler.)
NO γ. Positive column of discharge through air; E. 2.

Plate 4. H_2, blue region. Discharge through H_2; glass.
H_2, red region. Discharge through H_2; glass.
OH. Bunsen flame; E. 2.
CH. Discharge (? acetylene).
NH. Discharge through flowing NH_3; E. 2. (R. W. Lunt.)
CuH. High-tension arc in hydrogen flame; E. 2.
ZnH. Zn in discharge through H_2; E. 2.
CdH. Hollow cathode; E. 2. (E. W. Foster and A. G. G.)

Plate 5.   NaH. Discharge through H_2 and Na vapour; E. 2. (R. C. Pankhurst.)
MgH. Arc with Mg electrodes in H_2; glass. (A. Fowler.)
MgH^+. Arc with Mg electrodes in H_2; E. 1.
ZnH^+. Arc with Zn electrodes in H_2; E. 1.
PH. Discharge through H_2 and P_2 vapour; E. 1.
MnH violet and blue systems. High tension arc between Mn electrodes in H_2 flame; glass.
MnH green system. As above.
NiH. High-tension arc between Ni electrodes in H_2 flame; glass.

Plate 6.   Condensed discharge through N_2; E. 2.
NS. Uncondensed discharge through N_2 and sulphur vapour; E. 1. (C. J. Bakker.)
NS. As above continued.
PO. P_2O_5 in arc in air between Cu poles; E. 1.
P_2. Uncondensed discharge through $P_2 + H_2$; E. 2.

Plate 7. O_2. Schumann-Runge. High-tension arc in O_2 at atmospheric pressure; E. 1. (M. W. Feast.)
O_2. Schumann-Runge continued.
O_2^+. Second negative. High-frequency discharge through O_2; E. 1. (M. W. Feast.)
O_2^+. Second negative continued.
O_2^+. Second negative continued, and first negative.

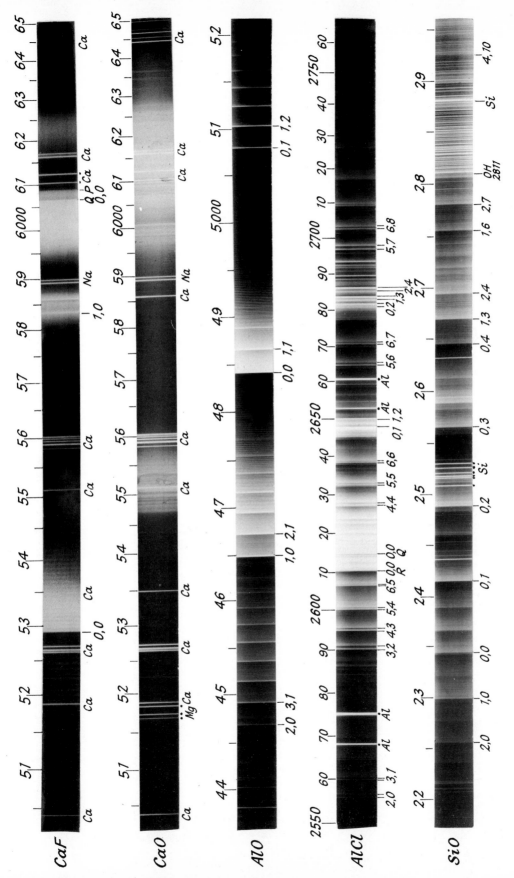

PLATE 1

PLATE 2

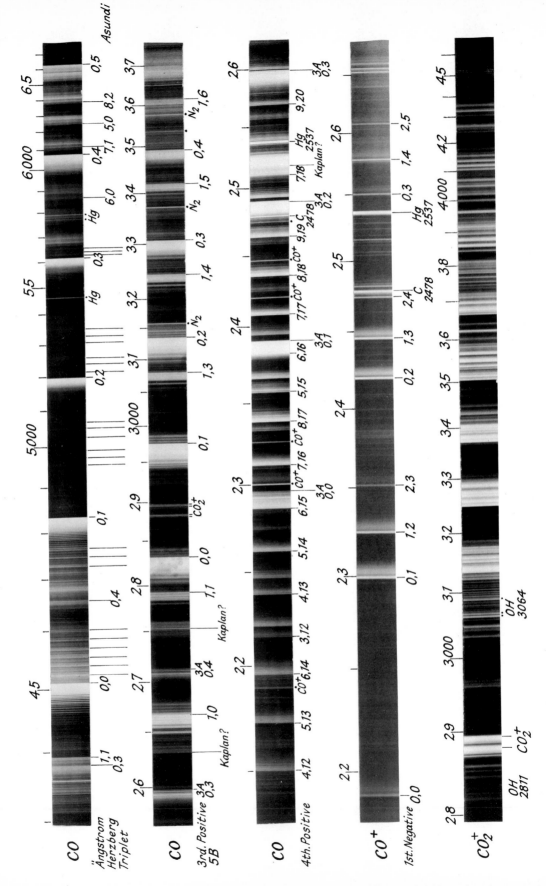

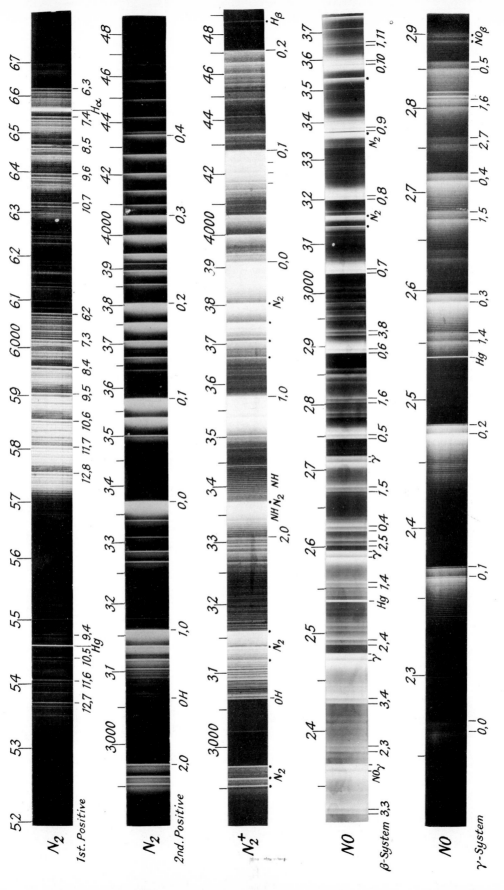

PLATE 3

PLATE 4

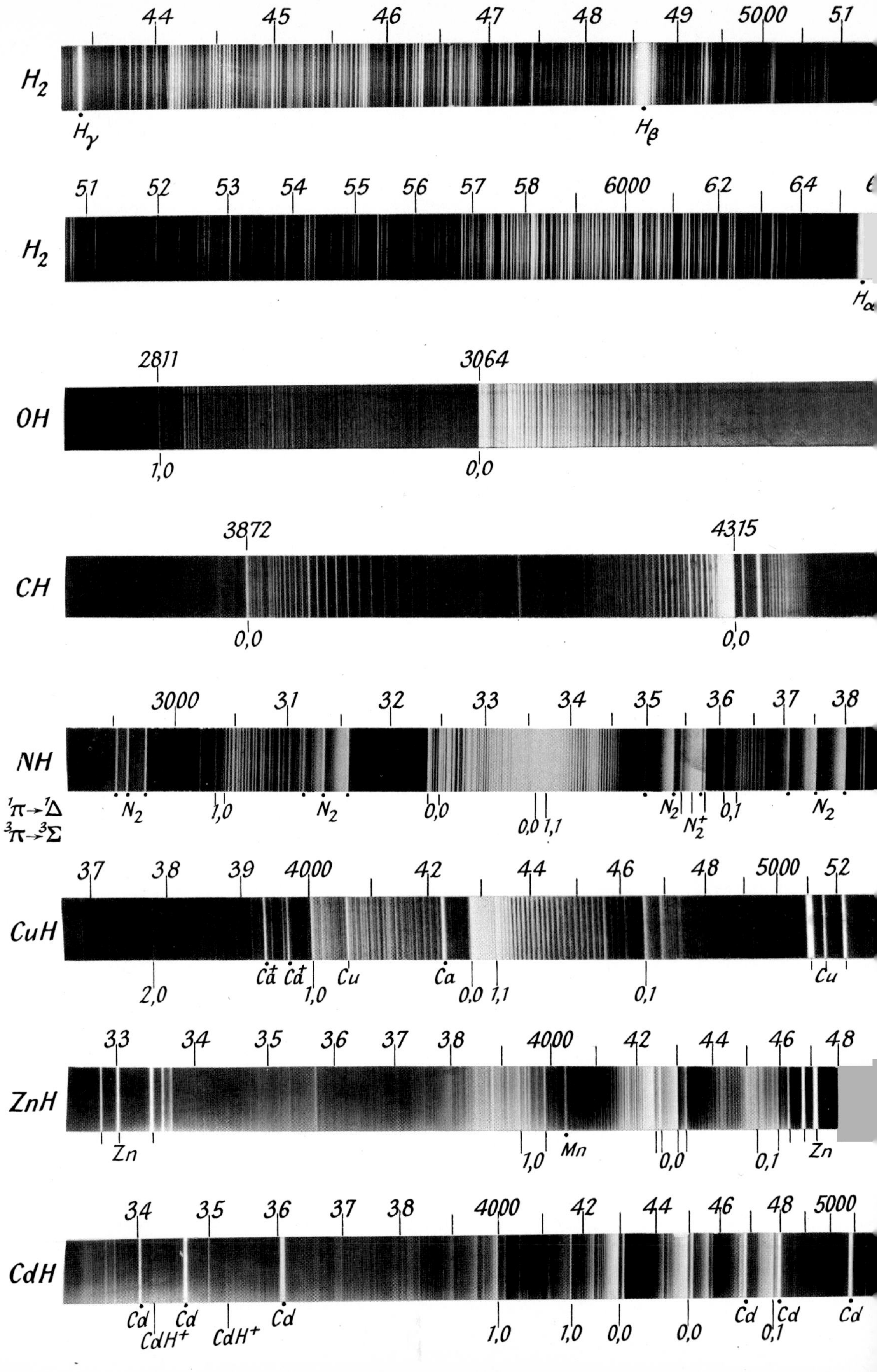
44 45 46 47 48 49 5000 51
H2
Hγ
Hβ
51 52 53 54 55 56 57 58 6000 62 64
H2
Hα
2811 3064
OH
1,0 0,0
3872 4315
CH
0,0 0,0
3000 31 32 33 34 35 36 37 38
NH
1π→1Δ
3π→3Σ
N2 1,0 N2 0,0 0,0 1,1 N2 N2+ 0,1 N2
37 38 39 4000 42 44 46 48 5000 52
CuH
2,0 Ca+ Ca+ 1,0 Cu Ca 0,0 1,1 0,1 Cu
33 34 35 36 37 38 4000 42 44 46 48
ZnH
Zn 1,0 Mn 0,0 0,1 Zn
34 35 36 37 38 4000 42 44 46 48 5000
CdH
Cd CdH+ Cd CdH+ Cd 1,0 1,0 0,0 0,0 Cd 0,1 Cd Cd

PLATE 5

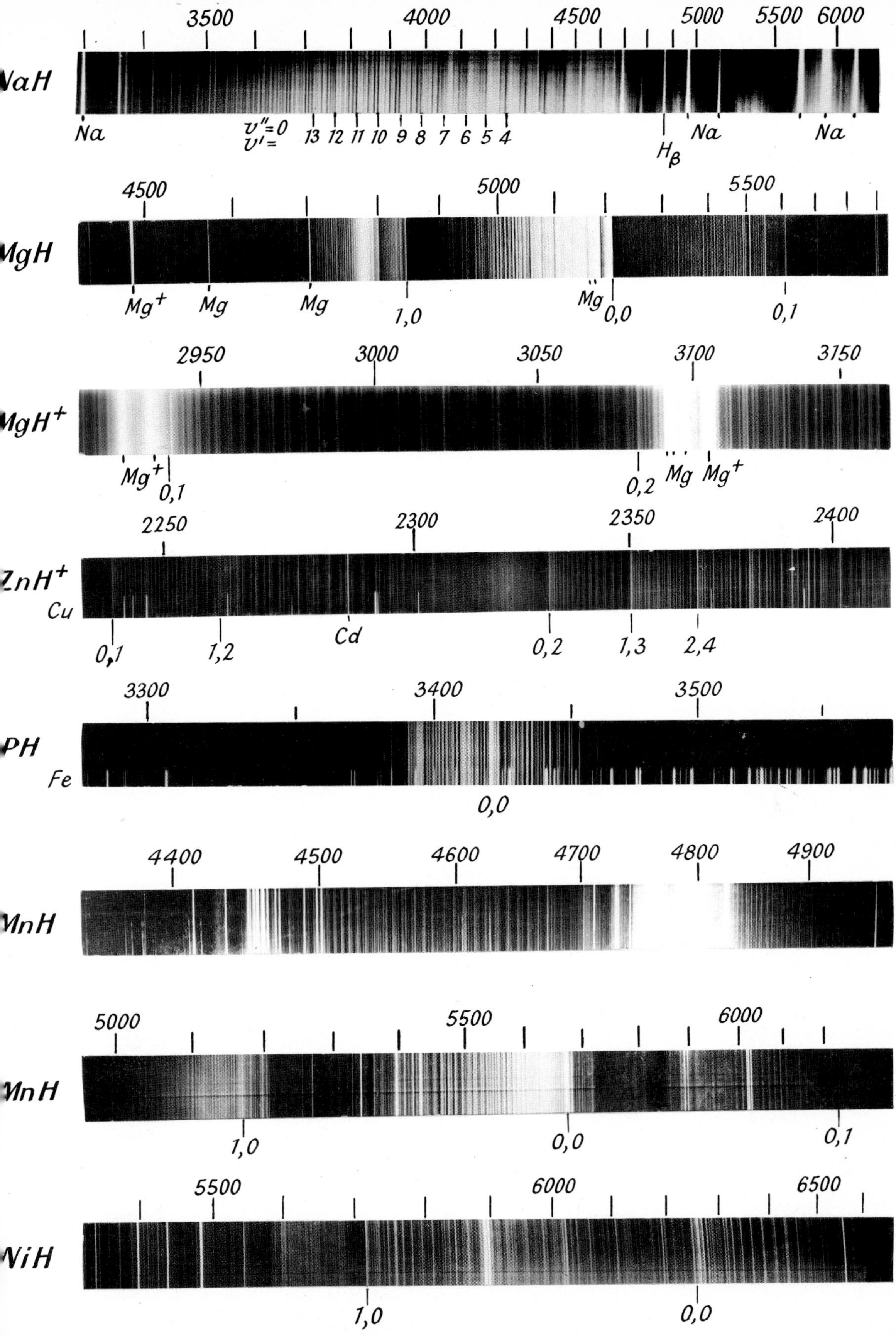
NaH
3500
4000
4500
5000
5500
6000
Na
v″=0
v′=
13 12 11 10 9 8 7 6 5 4
Hβ
Na
Na
MgH
4500
5000
5500
Mg+
Mg
Mg
1,0
Mg
0,0
0,1
MgH+
2950
3000
3050
3100
3150
Mg+
0,1
0,2
Mg
Mg+
ZnH+
Cu
2250
2300
2350
2400
0,1
1,2
Cd
0,2
1,3
2,4
PH
Fe
3300
3400
3500
0,0
MnH
4400
4500
4600
4700
4800
4900
MnH
5000
5500
6000
1,0
0,0
0,1
NiH
5500
6000
6500
1,0
0,0

PLATE 6

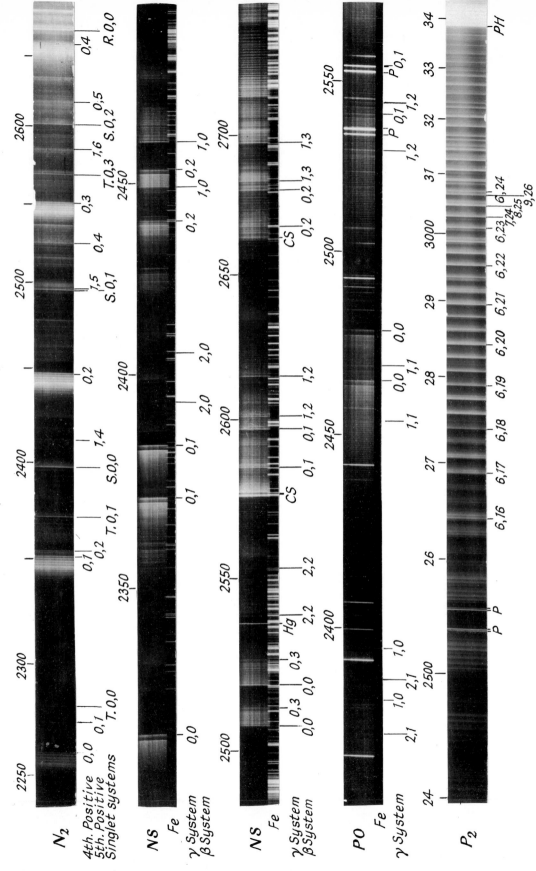

PLATE 7

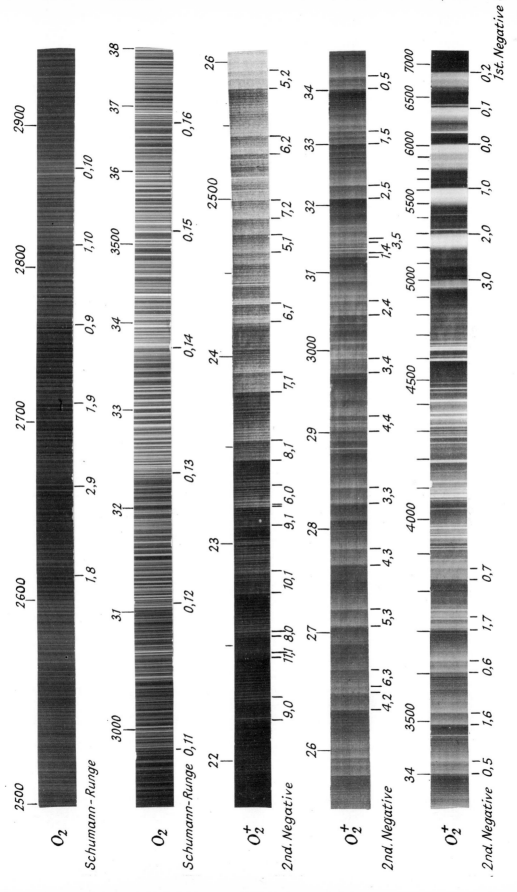

PLATE 8

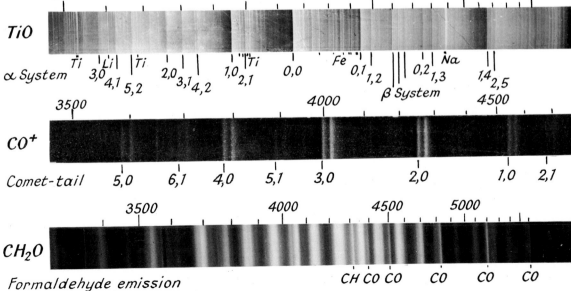

O_2

1775   1800   1825   1850

14,0 13,0 12,0 11,0 10,0 9,0 8,0 7,0

Schumann-Runge absorption

O_2

1850 1900 1950 2000 2100

7,0 6,0 5,0 4,0 3,0

Schumann-Runge atmospheric absorption

O_4

2600 2650 2700

Absorption by oxygen at high pressure

NO_2

42   43   44   4500   46   47   48

Absorption

SiF
Fe

43   44   4500   46   47

F   F   F Si^{++}

TiO

4500   5000   5500   6000

α System   Ti 3,0 4,1 Li 5,2 Ti 2,0 3,1 4,2 Ti 1,0 2,1   0,0   Fe 0,1 1,2   0,2 1,3 Na   1,4 2,5
β System

CO^+

3500   4000   4500

Comet-tail   5,0   6,1   4,0   5,1   3,0   2,0   1,0   2,1

CH_2O

3500   4000   4500   5000

CH CO CO   CO   CO   CO

Formaldehyde emission

PLATE 9

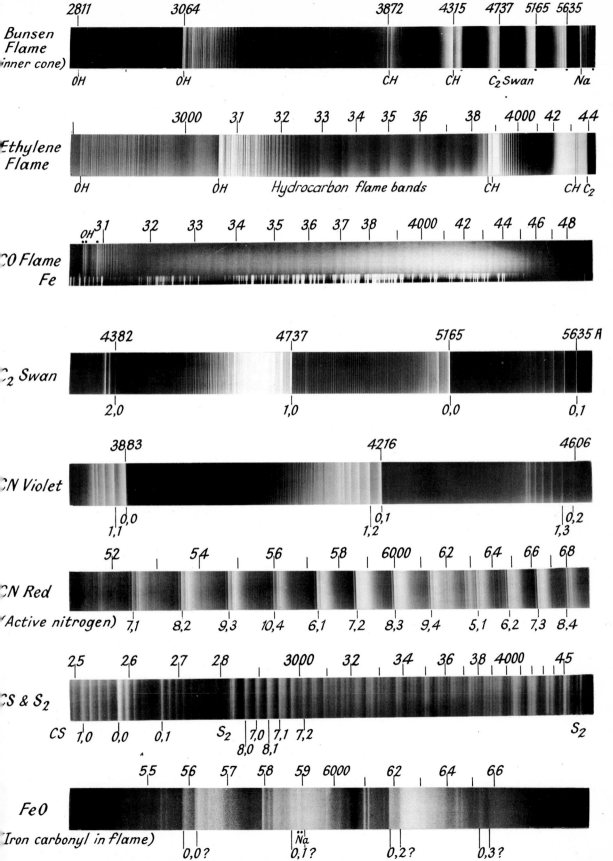

Bunsen Flame (inner cone) — 2811, 3064, 3872, 4315, 4737, 5165, 5635 — OH, OH, CH, CH, C₂ Swan, Na

Ethylene Flame — 3000, 3.1, 3.2, 3.3, 3.4, 3.5, 3.6, 3.8, 4000, 4.2, 4.4 — OH, OH, Hydrocarbon flame bands, CH, CH C₂

CO Flame Fe — OH, 3.1, 3.2, 3.3, 3.4, 3.5, 3.6, 3.7, 3.8, 4000, 4.2, 4.4, 4.6, 4.8

C₂ Swan — 4382, 4737, 5165, 5635 Å — 2,0 1,0 0,0 0,1

CN Violet — 3883, 4216, 4606 — 1,1 0,0 1,2 0,1 1,3 0,2

CN Red (Active nitrogen) — 5.2, 5.4, 5.6, 5.8, 6000, 6.2, 6.4, 6.6, 6.8 — 7,1 8,2 9,3 10,4 6,1 7,2 8,3 9,4 5,1 6,2 7,3 8,4

CS & S₂ — 2.5, 2.6, 2.7, 2.8, 3000, 3.2, 3.4, 3.6, 3.8, 4000, 4.5 — CS 1,0 0,0 0,1 S₂ 7,0 7,1 7,2 8,0 8,1 S₂

FeO (Iron carbonyl in flame) — 5.5, 5.6, 5.7, 5.8, 5.9, 6000, 6.2, 6.4, 6.6 — 0,0? Na 0,1? 0,2? 0,3?

PLATE 10

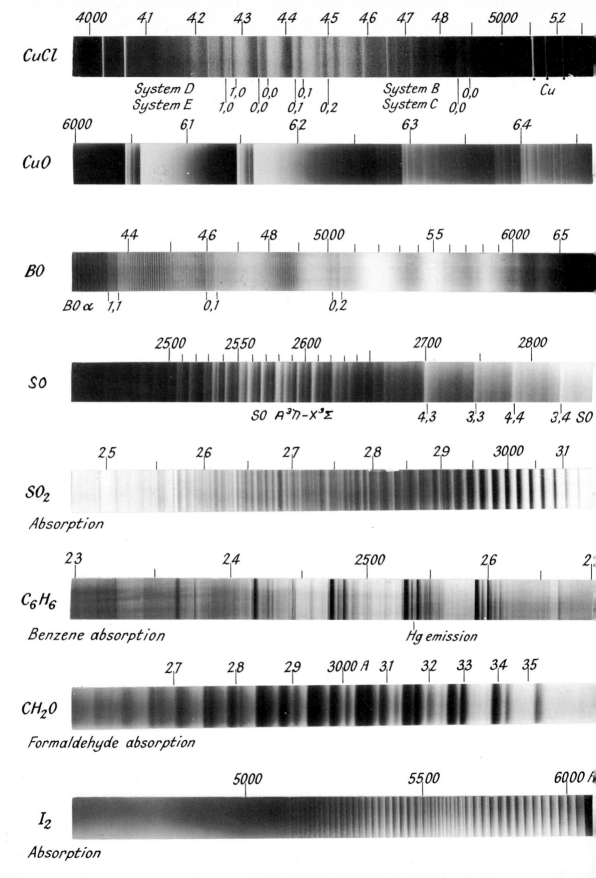

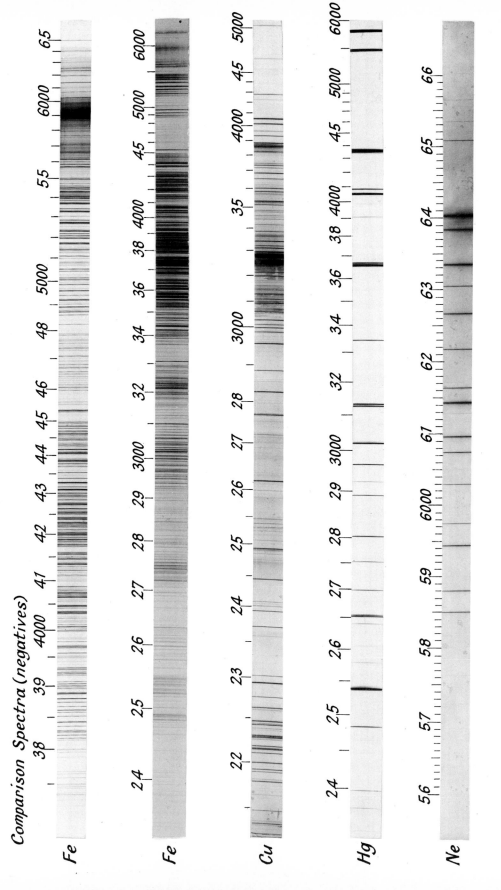

PLATE 11

Comparison Spectra (negatives)

Fe   Fe   Cu   Hg   Ne

PLATE 12

Iron Arc Spectrum (negatives)

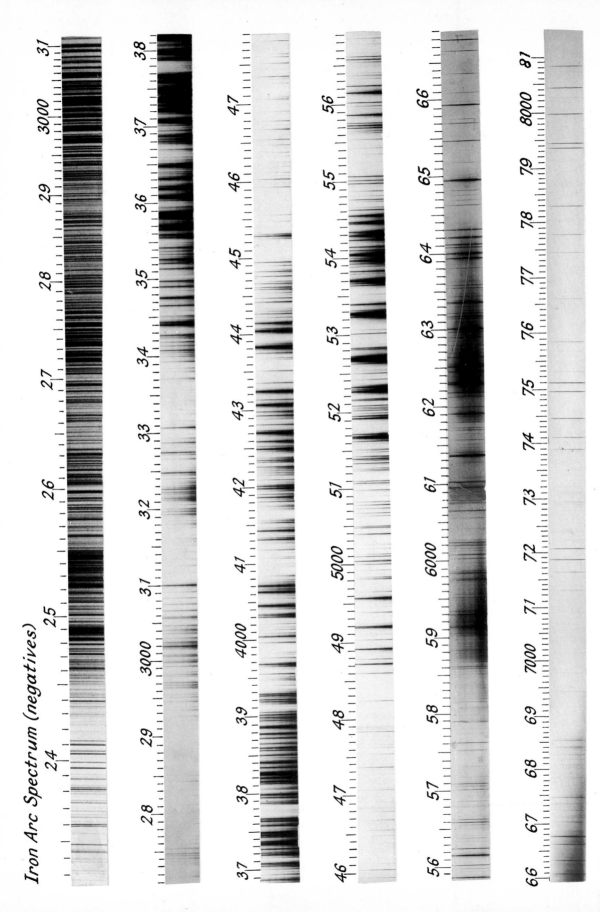

Plate 8. O_2 Schumann-Runge absorption. 3 m. vacuum grating. (H. P. Knauss and S. S. Ballard, *Phys. Rev.*, **48**, 796 (1935).)

O_2 Schumann-Runge. Atmospheric absorption; E. 3.

O_4. Absorption by oxygen at 50 atm. pressure; E. 2. (Enlarged from a print kindly supplied by K. Wieland.)

NO_2. Absorption by heated NO_2; glass. (L. Harris and R. W. B. Pearse.)

SiF. Discharge through SiF_4; glass.

TiO. TiO_2 in arc with Fe electrodes; glass. (A. Fowler.)

CO^+. Discharge at very low pressure; glass. (A. Fowler.)

CH_2O. Tesla discharge through flowing formaldehyde vapour; E. 2. (J. C. D. Brand.)

Plate 9. Bunsen flame. Inner cone; E. 2.

Ethylene flame. Inner cone; E. 2.

CO flame. E. 2; on process plate.

C_2, Swan. Discharge (? acetylene); glass.

CN, violet system; glass.

CN, red system. CCl_4 in active nitrogen; glass. (A. Fowler.)

CS and S_2. Sulphur in carbon arc; E. 3. (L. C. Martin.)

FeO. Iron carbonyl in flame; glass.

Plate 10. CuCl. Cuprous chloride in carbon arc; glass.

CuO. Arc in air between Cu electrodes; grating.

BO. Boric acid in carbon arc; glass.

SO. Discharge through flowing SO_2; E. 3. (B. N. Bhaduri.)

SO_2 absorption. Hydrogen continuum; E. 2.

C_6H_6. Absorption by vapour; hydrogen continuum; E. 2.

CH_2O. Absorption by formaldehyde vapour; hydrogen continuum; E. 2. (G. H. Young.)

I_2. Absorption by iodine vapour; incandescent filament; glass.

Plate 11. Comparison spectra. Iron, copper and quartz, mercury arcs, and neon discharge tube. Negatives.

Plate 12. Comparison spectra. Iron arc. Negatives.

Subject Index

with key to abbreviations

Chemical symbols of molecules are given in the tables of 'Individual Band Systems' in alphabetical order and are therefore not included in this index. Some organic molecules and radicals are, however, listed here by name to facilitate finding them.

Author Index